# Handbook
# of Real-Time
# Fast Fourier Transforms

# Handbook
# of Real-Time
# Fast Fourier Transforms

## Algorithms to Product Testing

**Winthrop W. Smith**
**Joanne M. Smith**

IEEE
PRESS

The Institute of Electrical and Electronics Engineers, Inc., New York

This book may be purchased at a discount from the publisher
when ordered in bulk quantities. For more information, contact

IEEE PRESS Marketing
Attn: Special Sales
P.O. Box 1331
445 Hoes Lane
Piscataway, NJ 08855-1331 USA
Fax: (908) 981-9334

Printed in the United States of America
10   9   8   7   6   5   4   3   2   1

**ISBN 0-7803-1091-8**
**IEEE Order Number: PC4440**


**Library of Congress Cataloging-in-Publication Data**
Smith, Winthrop W., (date)
    Handbook of real-time fast Fourier transforms / Winthrop W. Smith
Joanne M. Smith
        p.   cm.
    Includes bibliographical references and index.
    ISBN 0-7803-1091-8
    1.  Signal processing—Digital techniques.    2.  Fourier
transformations.    3.  Integrated circuits    I.  Smith, Joanne M.,
(date)  .  II.  Title.
TK5102.9.S58   1995
621.382′2′0285416—dc20                                      94-12936
                                                            CIP

**To our family and friends,
who encouraged us**

# Contents

## 10    Arithmetic Building Blocks for Architectures    245

## 11    Multiprocessor Architectures    255

# Preface

This book gives engineers and other technical innovators the foundation and facts they need to construct and implement fast Fourier transforms (FFTs) that synthesize, recognize, enhance, compress, modify, or analyze signals. Because of special integrated circuits, known as digital signal processing (DSP) chips, a wide array of applications is affordably done, from magnetic resonance imaging (MRI) to Doppler weather radar. Increased demand for wireless communication, multimedia, and consumer products has created the need for high-volume, low-cost, multifunction, DSP-based products that use FFTs for their signal processing or data manipulation.

In 1974, E. Oran Brigham lived and worked in the small East Texas town of Greenville. He was employed by a little-known aerospace company named E-Systems, Inc. when his 230-page book, *The Fast Fourier Transform* [1], was published. Over the years it has helped thousands of engineers learn the fundamentals of that analytical tool. After moving to Greenville in 1991 for Win to join E-Systems, we decided to write a book that continued the efforts begun here two decades before—putting practical information about FFTs into the hands of practicing professionals and engineering students.

The explosion of digital products, ignited by the proliferation of integrated circuits in the 21 years since Brigham's book came out, marks the coming of age for computing FFTs. Because of personal computers, with chips or plug-in boards for doing DSP functions, including FFTs, thousands of engineers, scientists, and students now work with and develop new FFT techniques and products. The National Information Infrastructure, popularly called "The Information Superhighway," and other digital-based goods and services now provide the impetus for sophisticated new products, once driven by the Department of Defense.

The book addresses the following areas of real-time FFT implementation:

- How to compute an FFT of any length with a wide variety of algorithms
- How to convert algorithms to assembly or high-level language code
- How to map algorithms onto several architectures

- How to select DSP chips and commercial off-the-shelf (COTS) boards for FFT applications
- How to detect and isolate errors in every phase of development

The goal of the book is to provide a single-source reference for the elements used in programming real-time FFT algorithms on DSP and special-purpose chips. It uses a building-block approach to constructing several FFT algorithms. Extensive use is made of examples and spreadsheet-style comparison charts. With hundreds of figures, tables, and Algorithm Steps, its practical features are geared to assist design engineers, scientists, researchers, and students. The book may even open the design of FFT-based products to innovators with no prior FFT experience, if they have microprocessor programming, engineering, or mathematics backgrounds. Though useful as a handy reference book by topic, it is laid out in a logical sequence that can be a textbook for a course on applied FFTs.

Sid Burrus's and Tom Park's book *DFT/FFT and Convolution Algorithms* [2], written a decade ago, met the mushrooming hunger of engineers for TMS32010 code, which would make it easier to use the new Texas Instruments chip for computing FFT algorithms. Mainstream applications for consumer products incorporating FFTs, precipitated by recent advances in integrated circuits, especially ASICs, have fostered a need to:

- Create versatile FFT algorithms of any length, to overcome the power-of-two constraints
- Understand how to map algorithms efficiently onto single and multiprocessor architectures
- Program in assembly language to optimize [3] code, in order to reduce power consumption and lower the cost of high-volume consumer products
- Shorten the design cycle and lower development costs to compete in global markets

Unique features include:

- Performance measure Comparison Matrices for selection of weighting functions, algorithm building blocks, algorithms, algorithm mappings, arithmetic formats, and DSP chips
- Extensive algorithm examples, with step-by-step instructions for memory mapping and conversion to high-level or assembly language code
- A "generic" programmable DSP chip block diagram, to which 24 chip vendor block diagrams are standardized and compared, to illustrate differences that affect FFT performance
- Unbiased description of the FFT-related features of 51 fixed-point DSP chips, including ASIC and multiple-processor chips, 13 floating-point DSP chips, and 6 dedicated FFT chips
- Test signals with instructions and examples for detecting and isolating errors during FFT algorithm development, code development and debugging, and product operation
- A list of questions and answers for selecting COTS boards
- Four design examples that do frequency analysis, power spectrum estimation, linear filtering, and two-dimensional processing

Win's 28-year DSP career in both military and commercial companies, teaching courses and seminars nationwide, has repeatedly shown him that engineers need to be able to work easily with any length of FFTs to do real-time signal conversion and analysis. Joanne's 12 years experience as founder and president of two DSP companies has given her exposure to the rapidly changing technology, market, and economic realities of this industry. Coauthoring a book seemed the logical way to combine our diverse talents and complementary perspectives to comprehensively address the topic of real-time fast Fourier transform algorithms.

This book is only one of several tools for expanding the knowledge base of the DSP community. A service called DSP Net provides access to the latest vendor information in this field through InterNet. *DSP and Multimedia Technology* magazine addresses this growing market, as do two annual applications-oriented conferences—DSPx and the International Conference on Signal Processing Applications & Technology. The IEEE International Conference on Acoustics, Speech and Signal Processing holds its 20th annual gathering in 1995. The chip vendors have free bulletin boards for algorithms, code, and other pertinent information. Additional information on resources available to design engineers should be sent to the authors, in care of the publisher, for possible inclusion in follow-up publications.

## ACKNOWLEDGMENTS

We are pleased to thank Frank J. Thomas, Rosalie Sinnett, Thomas L. Loposer, Randy Davis, and Wayne Yuhasz, who convinced us we could accomplish this effort; Ross A. McClain, Jr., Jeffrey W. Marquis, Vito J. Sisto, V. Rex Tanakit, and Joel Morris, Ph.D., for their contributions during the editing process; Harold W. Cates, Ph.D., and Robert H. Whalen, for their mentoring of Win's career; the many friends and colleagues who have encouraged us throughout our careers; and our daughters Patricia and Paula for not letting us give up. Most of all we thank God for His inspiration, guidance, and strength throughout this seemingly impossible task.

## REFERENCES

[1] E. Oran Brigham, *The Fast Fourier Transform*, Prentice-Hall, Englewood Cliffs, NJ, 1974.

[2] C. S. Burrus and T. W. Parks, *DFT/FFT and Convolution Algorithms*, Wiley, New York, 1985.

[3] John P. Sweeney, "Mainstream Applications Require Optimized Assembly Language for Fast DSPs," EDN, April 28, 1994.

# Handbook
# of Real-Time
# Fast Fourier Transforms

# 1

# Overview

## 1.0 INTRODUCTION

The increased demand for communication, multimedia, and other consumer products has created the need for high-volume, low-cost, multifunction DSP-based products that can use fast Fourier transforms (FFTs) for their signal processing or data manipulation. This book is the first to cover FFTs from algorithms to product testing, with the information needed to create and convert to code FFT algorithms of any length on 10 different architectures. It uses a building-block approach for constructing the algorithms. Included are recommended Memory Maps to streamline assembly and high-level language coding of 17 small-point FFTs, four general algorithms, and seven FFT algorithm examples. To ensure that the algorithms work properly, a test approach for the detection and isolation of errors, refined over many years of time consuming searches for mistakes in FFT algorithms, is detailed.

Spreadsheet-style comparison matrices provide easy to use inventories of the comprehensive array of key FFT elements and performance measures. Dozens of digital signal processing (DSP) chips and criteria for selecting DSP boards are covered. Four design examples at the end of the book show how to apply most of what has been explained.

## 1.1 LAYING THE FOUNDATION

Chapters 2 and 3 provide the technical foundation and mathematical equations for the algorithms in Chapters 8 and 9. The discrete Fourier transform (DFT) is an equation for converting time domain data into its frequency components. The DFT equation is implemented with FFT algorithms because they are computationally efficient ways of calculating it. All the properties and strengths of the DFT are shared by the wide variety of FFTs that have been developed over the years. However, only three of the five weaknesses of the DFT are also weaknesses of FFT algorithms.

In the beginning of the design process, comparison of the uses and properties of the DFT with the technical specifications of the application will determine if the DFT is a good match. If so, then it makes sense to examine the FFT algorithms, hardware architectures, arithmetic formats, and mappings in this book to decide which combination is best for a specific design.

## 1.2 DESIGN DECISIONS

The decisions listed are the ones related to real-time FFT selection and implementation. They are listed in an order which differs from the sequence of the chapters, because learning the facts happens more easily in an order that is different from applying them.

- Choosing the number of dimensions (Chapters 5–7)
- Picking a type of processing (Chapters 5–7)
- Selecting the arithmetic format (Chapter 13)
- Deciding on a weighting function (Chapter 4)
- Determining the transform length (Chapter 5)
- Selecting algorithm building blocks (Chapter 8)
- Constructing the algorithm (Chapter 9)
- Choosing a chip (Chapter 14)
- Selecting the architecture (Chapters 10 and 11)
- Mapping the algorithm onto the architecture (Chapter 12)
- Selecting an off-the-shelf board (Chapter 15)
- Creating the test signal and procedures (Chapter 16)

### 1.2.1 Number of Dimensions

All multidimensional FFTs are done as a sequence of one-dimensional FFTs. The importance of knowing how many dimensions (one, two, or three, usually) there are determines how many FFTs will be needed and how the data must be organized to do the multiple dimensions. This will affect chip processing load and the choice of architecture.

### 1.2.2 Type of Processing

The type of processing (frequency analysis, convolution, or correlation) will also affect the chip processing load. Frequency analysis requires one FFT for every group of samples, while the other two types require an FFT and an inverse FFT for every group of samples.

### 1.2.3 Arithmetic Format

The choice of fixed-point, floating-point, or block-floating-point arithmetic format will affect the numerical accuracy of the results. Fixed-point DSP chips were the first available and are generally less expensive than floating-point, because this arithmetic takes less silicon area. Floating-point has grown in popularity as semiconductor manufacturers advanced to smaller micron wafers and high-level language compilers became available. Block-floating-point is a compromise approach that provides better accuracy than fixed-point and takes less silicon area than floating-point. It is only available in chips designed specifically for computing FFTs.

### 1.2.4 Weighting Functions

The selection of one of more than a dozen weighting functions will affect frequency location accuracy while controlling sidelobe effects. They also modify coherent gain, bandwidth, and frequency straddle loss. The selection depends on what combination of these effects matters most in an application.

### 1.2.5 Transform Length

Choosing a transform length closest to the number of data points to be analyzed will improve the accuracy of the computation, thereby improving frequency accuracy. The size of the transform will directly affect frequency resolution, memory requirements, and the speed at which the computation can be done. A unique feature of this book is the choice of more than one algorithm to compute an FFT of any length.

### 1.2.6 Algorithm Building Blocks

The algorithm building blocks used will affect the computational load the algorithm requires and the complexity of code to implement that algorithm. This chapter provides 17 small-point transform algorithms for constructing larger algorithms. The choice depends on whether computational load or code complexity is the deciding factor in a specific design.

### 1.2.7 Algorithm Construction

The way in which the algorithm building blocks are connected to create a larger algorithm will affect the complexity and amount of the code needed to implement it. This chapter details the Bluestein, Winograd, prime factor, and mixed-radix methods for assembling small-point transforms into larger algorithms.

### 1.2.8 DSP Chips

The selection of which Harvard architecture DSP chip to actually compute the algorithm is determined by the cost and speed considerations of the application, the number of chips needed, a suitable architecture (for multiple-processor designs), and available peripheral hardware to handle some of the functions. This chapter covers the FFT-related features of 51 fixed-point DSP chips, including ASIC and multiple-processor chips, 13 floating-point DSP chips, and 6 dedicated FFT chips.

### 1.2.9 Architectures

Bit-slice, arithmetic chips were used to construct FFT applications prior to the introduction of DSP chips. However, advances in silicon technology have replaced bit-slice building blocks with DSP chips that include a complete fixed- or floating-point multiplier and adder, as well as memory and program control logic.

All of the DSP chips in this book use a Harvard architecture for interconnecting these elements. FFT-specific chips interconnect several arithmetic building blocks into a small-point FFT to increase performance. Multiprocessor interconnections (pipeline, linear bus, ring bus, crossbar, two- and three-dimensional massively parallel, star, hypercube, and hybrid architectures) of DSP chips are used when a single chip is not adequate. In fact, up to four Harvard processors are now available on a single chip (SPROC 1000 and TMS320C80 families). Chapter 10 describes bit slice, integrated arithmetic and FFT-specific hardware building blocks. Then Chapter 11 shows how to use them in single and multiprocessor architectures. These two chapters prepare the reader for mapping the algorithms in Chapter 9 onto these architectures.

### 1.2.10 Mapping Algorithms onto Architectures

How an algorithm is mapped onto the chosen architecture will affect the throughput (how many FFTs per second) and the latency (the delay between input and output) of that algorithm. This chapter explains how to map FFT algorithms onto single and multiprocessor architectures to attain either maximum throughput or minimum latency performance.

### 1.2.11 Board Decisions and Selection

A commercial, off-the-shelf (COTS) board can reduce the time and cost of getting to market with a board-level FFT product. With several dozen manufacturers selling a wide variety of DSP boards suitable for doing FFTs, board selection is a complex decision. Whether the chip selection process has narrowed the choice to a chip or to multiple acceptable chips, the following five areas cover the main issues of choosing or developing a board:

1. Algorithm performance
2. I/O Performance
3. Architecture
4. Software support
5. Expansion capability

### 1.2.12 Test Signals and Procedures

The design process can bog down in algorithm development and conversion to code if there are no easy ways to detect and isolate errors. Having an efficient set of test signals to use as inputs to an FFT algorithm or its code allows quick detection and precise isolation of errors. In combination with these signals, flow graphs of the algorithm and code are needed to trace an error back to its source. The same signals can be used to do end-product and built-in testing.

## 1.3 TYPES OF EXAMPLES

The extensive use of examples is one of the unique features of the book. In addition to the four design examples in Chapter 17, six kinds of algorithm examples are used to demonstrate the wide array of concepts and facts the book contains. The particular lengths were chosen because they are large enough to show the pattern of an algorithm yet small enough to easily follow.

### 1.3.1  Eight-Point DFT to FFT Example

Section 3.3 explains that all of the FFT algorithms presented in this book are based on ways to remove redundant computations from the DFT equations without changing the final result of the equations. While deriving an FFT algorithm from its DFT origins is a theoretical process, using an example is a practical way of seeing the principle.

### 1.3.2  Algorithm Steps and Memory Maps

Sections 8.3 through 8.10 contain 17 examples of building-block algorithms that are most likely to be used to construct larger algorithms. These are the most efficient small-point transforms to implement. For each example every arithmetic operation (Algorithm Step) is given, with a memory address (Memory Map) beside it, for the results. Instructions are given for converting these small-point transforms into code. This coding can be in any of the chip vendors' assembly languages or in a high-level language. To convert to assembly language, both the Algorithm Steps and their companion Memory Map will be needed. Conversion to high-level languages, such as versions of C or FORTRAN, only require use of the Algorithms Steps.

### 1.3.3  Fifteen-Point or 16-Point FFT Algorithm Examples

In Chapter 9 seven 15-point or 16-point FFT algorithm examples, using the building blocks from Chapter 8, show how to implement the general types of FFT algorithms. A technique for relabeling Memory Maps from Chapter 8 is given and illustrated in these examples. Power-of-two and non-power-of-two examples are used to illustrate the range of algorithms that cover computing any transform length.

### 1.3.4  Sixteen-Point Radix-4 FFT Algorithm Examples

In Chapter 12 a 16-point, radix-4 FFT algorithm is used in one single-processor and nine multiprocessor examples. Maximum throughput and minimum latency examples are done for mapping the algorithm and its data, for a total of 20 examples. A 16-point example is used because it is a typical power-of-two length and familiar from Chapter 9. The reader is given all the input, intermediate, and output steps needed to code the algorithm.

### 1.3.5  Four-Point FFT and 16-Point Radix-4 FFT Algorithm Examples

In Chapter 16 the 4-point FFT (a small-point building-block algorithm in Chapter 8) and 16-point, radix-4 FFT examples are used again to explain how to detect and isolate errors in FFT algorithm development, code development and debugging, and end-product operation. Flow graphs are used to show how to track an error through an algorithm. Equations show how to verify Algorithm Step accuracy. Algorithm Steps and Memory Maps are used with test signals to show how the results are altered by an error in an algorithm. The altered results illustrate how to isolate a detected error.

## 1.4 DESIGN EXAMPLES

In Chapter 17, frequency analysis, power spectrum estimation, linear filtering, and two-dimensional processing examples were chosen to illustrate:

- Three common uses of the DFT from Chapter 2
- Single and multiprocessor architectures from Chapter 11
- Three algorithms from Chapter 9
- Three classes of chips (fixed-point, floating-point, and FFT-specific) from Chapter 14

Whether the design will be single or multiple chip on single or multiple boards may not be determined until far into the design process. In this chapter both multiple-chip and multiple-board applications are developed to illustrate making those decisions. These are not intended to be full-scale product designs. They are taken far enough into a design to show how to use the wide array of information in the book.

### 1.4.1 Doppler Radar

Example 1 is the Doppler processing portion of a ground-based air surveillance radar. This can be used for commercial airport air traffic control or for Doppler weather radar, as well as defense applications. Doppler weather radar has become a household word in the 1990s, through its use in daily weather forecasting and broadcasts. Doppler processing is a classical use of frequency analysis, the first common use of the DFT.

### 1.4.2 Power Spectrum Estimator

Example 2 is a power spectrum estimator personal computer (PC) plug-in board. Commonly used to modify PCs for use as sophisticated instrumentation, plug-in boards generate hundreds of millions of dollars of business. Earthquake prediction, satellite communication, and magnetic fields are areas of intense public interest, where the signals a board like this can analyze are found. There are countless other applications where recognizing signals and the patterns in them can have a life-saving effect. This is the third common use of DFTs—frequency domain conversion.

### 1.4.3 Speech Recognition

Example 3 is the signal processing portion of a voice-activated number recognition system. Voice dialing of car phones, one of many products for the burgeoning consumer electronics market, is a use for this. This technique can also be applied to other numerical data entry situations, where hands are not free to use a keypad; speaker verification for security systems; and credit card fraud protection. This speech application taps DFT's ability to provide a numerical shorthand of a signal, its second common use, and its use for frequency analysis.

### 1.4.4 Image Deblurring

Example 4 is another PC plug-in board, this one for doing image deblurring. The PC housing this board could be found at a police station, crime lab, or as instrumentation for an engineer or researcher. Though deblurring images does not have the widespread uses of the first three examples, the image processing principles it employs do. Some of them are CAT scans and MRIs, seismic exploration, and multimedia applications. Like Example 2, this product does frequency domain conversion, the third common use of the DFT.

## 1.5  CONCLUSIONS

This chapter provides an overview of the contents of the book. From a foundation in the DFT through design examples, the authors have tried to present a logical, easy to follow explanation of how to implement real-time FFTs on commercially available processors. Digital signal processing is a mushrooming field of technology. The FFT is a valuable technique for synthesizing, recognizing, enhancing, compressing, modifying, or analyzing digital signals from many sources.

The next chapter, on the DFT, lays the foundation for all that is said about the FFT in subsequent chapters.

# 2

# The Discrete Fourier Transform

## 2.0 INTRODUCTION

The discrete Fourier transform (DFT) is an equation for converting time domain data into frequency domain data. Discrete means that the signal is sampled in time rather than being continuous. Therefore, the DFT is an approximation for the continuous Fourier transform [1]. This approximation works well when the frequencies in the signal are all less than half the sampling rate (Section 2.3.1) and do not vary more than the filter spacing (Section 2.3.2).

Because of heat-transfer work done by the French mathematician J. B. Fourier in the early 1800s, many fields of science and engineering have benefited from the use of his mathematical link between time and frequency domains, called the Fourier transform. This link is valuable because many natural or man-made signals (waveforms) are periodic and thus can be expressed in terms of a sum of sine waves. Mathematicians realized that rather than compute continuous spectra, they could take discrete data points in the time domain and translate that information into the frequency domain, and so the discrete Fourier transform came into being.

The DFT equation, unlike the continuous Fourier transform, covers a finite time and frequency span. These data points may be collected from the output of an analog-to-digital (A/D) converter, generated by a digital computer, or output from another signal processing algorithm. They can be the plotted points of the performance of any numerical data, such as stock prices. The DFT equation is implemented with FFT algorithms because they are computationally efficient ways of calculating it. The properties (Section 2.3) and strengths (Section 2.5) of the DFT also belong to the FFT. However, only three of the weaknesses (Section 2.6) of the DFT are also weaknesses of FFT algorithms.

Comparison of the uses and properties of the DFT, with the technical specifications of the application, determines if the DFT will be useful. If so, it makes sense to examine the FFT algorithms, hardware architectures, arithmetic formats, and mappings in this book to decide which combination of them will provide the specified performance. This chapter lays the technical foundation for the FFT algorithms in Chapters 8 and 9.

## 2.1 COMMON USES OF THE DFT

The three common uses of the DFT are:

1. Frequency analysis, which is determining the size and location of frequencies in a signal. See Chapter 5 for details.

2. Reduction of adds and multiplies in linear filtering (convolution) and pattern matching (correlation). See Chapter 6 for details.

3. Numerical shorthand as a way of describing a signal. For example, the power coming out of an electrical outlet is described as 120 volts at 60 cycles. This is Fourier transform shorthand using only two numbers to describe a continuously changing waveform. The same shorthand is used in signal processing to describe any time domain signal as a sum of sine waves. The speech analyzer example in Chapter 17 takes advantage of this use of the DFT.

## 2.2 EQUATION AND BLOCK DIAGRAM

Equation 2-1 is the standard description of the DFT of $N$ complex data points, $a(n)$.

$$A(k) = \sum_{n=0}^{N-1} a(n) * W_N^{k*n} \qquad \text{where } W_N = \cos(2\pi/N) - j\sin(2\pi/N) \qquad (2\text{-}1)$$

Before the DFT properties are described, it is useful to have a simple picture of the function that Equation 2-1 is performing.

Since Equation 2-1 takes the same set of $N$ input data points, $a(n)$, and produces $N$ output signals, $A(k)$, each representing a different frequency, the $N$-point DFT can be modeled as an array of $N$ narrowband filters, each providing an output if the input signal has frequency components in its passband. Since a narrowband filter can be implemented with a multiplier and a low-pass filter (LPF), Figure 2-1, on page 11, can be used to represent the DFT. The only difference between the DFT and this array of narrowband filters is that the DFT only produces an output from each filter every $N$ input samples. A narrowband filter produces an output for every new input data point.

## 2.3 PROPERTIES

All FFT algorithms are just faster ways of computing the DFT equations; they are not approximations for the DFT equations. Thus the DFT properties described in this section apply to all FFT algorithms. These properties have been derived in detail in many textbooks [1–4].

### 2.3.1 Frequency Limits

The first property to be understood about the DFT is the frequencies that it can unambiguously determine. That range is defined by the sampling theorem [5], also called

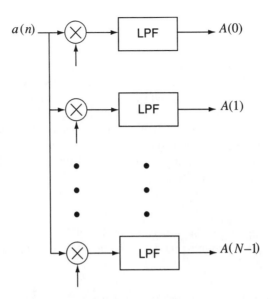

**Figure 2-1**    Block diagram of the DFT as an array of narrowband filters.

the Nyquist rate [6]. The DFT determines the presence of zero-frequency signals in the input data points by calculating $A(0)$. The $A(1)$ term in Equation 2-1 determines the presence of a sine wave that goes through exactly one 360° cycle during the $N$ data points. Similarly, the $A(k)$ term determines the presence of sine waves that go through exactly $k$ 360° cycles during the $N$ data samples.

The frequencies $A(k)$ in Equation 2-1 are the only ones that the DFT computes. When the frequency of a signal is higher than the sampling rate, the sampled version of the signal appears to be at the signal's frequency minus the sampling rate. To illustrate this, consider a sine-wave signal that goes through exactly $N$ 360° cycles during the $N$ input data points. That means it goes through exactly one 360° cycle between each data point. Therefore, every time it is sampled it has the same data value. However, a zero-frequency signal also has the same value each time it is sampled. Therefore, the DFT cannot distinguish between zero-frequency sine waves and sine waves that go through $N$ 360° cycles during the $N$ samples.

The Nyquist rate is a formal mathematical description of this phenomena. For a DFT to accurately represent frequencies up to $F$ samples per second, a sample rate of at least $2 * F$ samples per second is required. Further, frequencies that are higher will appear to be lower-frequency signals (ambiguous), just as the sine waves in the previous paragraph that had $N$ 360° cycles in $N$ samples looked the same as the zero-frequency sine wave. A sine wave with $2 * N$ 360° cycles in $N$ samples also looks the same as a zero-frequency sine wave.

For real signals, the sampling theorem, as stated above and by Shannon, holds directly. If the samples are complex, real and imaginary samples are taken at the sampling rate. The result is two samples at the sampling rate or samples taken at twice the sampling rate. This implies that, for complex sampling, frequencies are unambiguously analyzed by the DFT up to the complex sampling rate $F$.

### 2.3.2 DFT Filter Spacing/Nulls

Since there are $N$ equally spaced DFT filters between zero and the sampling rate, the spacing between the filters is $1/N$ times the sampling rate. It is important to note that $1/N$ times the sampling rate is also the total time period over which the $N$ samples were taken. Therefore, the filter spacing is equal to 1/(total time for data collected for the DFT input). Further, the DFT filters are designed so that, if a signal has an input frequency in the center of one of the filters, the other filters do not respond. Therefore, the spacing between the center of a DFT filter and its first null response is equal to the 1/(total time for data collected for the DFT input). In filtering terms, each DFT filter has a null in its response at the input frequencies of the other filters.

### 2.3.3 Linearity

Linearity means that the output of the DFT for the sum of two input signals is exactly the same as summing the DFT outputs of two individual input signals, as shown in Equation 2-2.

$$C(k) = \sum_{n=0}^{N-1} [a(n) + b(n)] W_N^{kn} = \sum_{n=0}^{N-1} a(n) W_N^{kn} + \sum_{n=0}^{N-1} b(n) W_N^{kn} = A(k) + B(k) \quad (2\text{-}2)$$

### 2.3.4 Symmetry

The symmetry property is helpful in understanding the response of a DFT to a particular waveform. It states that if $A(k) = $ DFT of $a(n)$, then an input waveform with the shape of $A(n)$ will have a DFT equal to $a(N - k)$.

### 2.3.5 Inverse DFT

The inverse discrete Fourier transform (IDFT), shown in Equation 2-3, is used to convert frequency information into time domain data points. This property allows the DFT to be used to perform linear filtering and pattern matching in the frequency domain. These frequency domain algorithms are described in Chapter 6 and often require fewer adds and multiplies than doing linear filtering and pattern matching directly in the time domain.

$$a(n) = [1/N] \sum_{k=0}^{N-1} A(k) W_N^{-kn} \qquad \text{where } W_N^{-1} = \cos(2\pi/N) + j \sin(2\pi/N) \quad (2\text{-}3)$$

### 2.3.6 Ease of IDFT Computation

Notice that the IDFT, Equation 2-3, is similar to Equation 2-1, which describes the DFT. This similarity makes it possible to use almost the same algorithm to compute the IDFT as is used for the DFT. This is most simply illustrated by Equations 2-4 and 2-5. Except for the factor of $1/N$, the difference between the IDFT equation and the DFT equation is the sign of the sine terms of $W^{kn}$.

$$W_N^{kn} = \cos(2\pi kn/N) - j \sin(2\pi kn/N) \tag{2-4}$$

$$W_N^{-kn} = \cos(2\pi kn/N) + j \sin(2\pi kn/N) \tag{2-5}$$

Therefore, any DFT or FFT algorithm can be converted to its comparable IDFT algorithm by changing the sign of the coefficient multipliers formed by the sine terms and dividing the results by $N$. This becomes important when using the frequency domain algorithms in Chapter 6 to perform linear filtering and pattern matching. In those algorithms, FFTs and IFFTs are required. This property allows the same FFT algorithm to be used for both the FFT and IFFT portions of the computations.

### 2.3.7 Time and Frequency Scaling

The DFT performs frequency analysis on sequences of digital data points, independent of the source of these data points or how fast the A/D was that took the samples. Therefore, it determines only the presence of frequency components that repeat 0, 1, ... up to $N$-1 times during the $N$ data points. This means that, if the same sequence of numbers is collected from A/D converters with different sampling rates, the DFT outputs, $A(k)$, will be identical. However, the output $A(1)$ represents the presence of a higher frequency from the A/D output that was sampled at the higher rate.

Summarizing, if the time between A/D samples is scaled (i.e., the sampling rate is changed), then the frequency represented by each DFT output is also scaled (i.e., the frequency it represents is changed). For example, if the A/D rate is doubled, each DFT output $A(k)$ represents the presence of a frequency that is also doubled.

### 2.3.8 Time and Frequency Shifting

This property of the DFT is most easily illustrated by using a sine wave at frequency $k$ as the input signal. Then DFT filter $k$ will output the amplitude and phase $A(k)$ of that sine wave in the input signal. The phase of the sine wave at sample 5 is different than at sample 0. Therefore, if the DFT is performed on samples 5, 6, ... up to $N + 4$ (i.e., a time shift of five samples) of the same input signal, the phase in the output of DFT filter $k$ will be changed by the difference in phase between samples 0 and 5. Since the DFT is linear, this phenomena is true regardless of the number of sine waves that comprise the input signal.

Figure 2-2 shows this phenomena for a signal that is a single sine wave that repeats once during 16 samples. This signal has one DFT output response, in filter $A(1)$. Since the

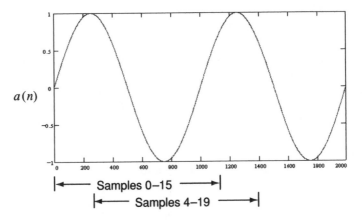

**Figure 2-2**    Time shift example.

sine-wave phase for samples 0–15 is zero, the $A(1)$ FFT output has zero phase. Since the sine-wave phase for samples 4–19 is 90°, the $A(1)$ FFT output has 90° phase.

Similarly, if a frequency component $A(k)$ is shifted to a new frequency $A(k-i)$, then the IDFT of the shifted frequency is a sine wave at frequency $k - i$. This sine wave can also be obtained by multiplying a sine wave at frequency $k$ by a sine wave at frequency $i$. This is mathematically described by multiplying the original input signal by a complex sine wave. Again, since the IDFT is linear, this phenomena is true regardless of the number of sine waves that comprise the sampled signal.

Time and frequency shifting are represented mathematically by Equations 2-6 and 2-7.

$$a(n + i) \Leftrightarrow A(k)e^{-j2\pi ki/N} \tag{2-6}$$

$$A(k - i) \Leftrightarrow a(n)e^{+j2\pi ni/N} \tag{2-7}$$

### 2.3.9 Parseval's Theorem

The power of a sequence of input data points is defined as the sum of squares of all the values of the data points. Parseval's theorem is a way of computing the signal's power after it has been converted by an FFT to its frequency components $A(k)$ as shown in Equation 2-8.

$$\sum_{n=0}^{N-1} a^2(n) = 1/N \sum_{k=0}^{N-1} |A(k)|^2 \tag{2-8}$$

Therefore, except for a factor of $1/N$, the sum of the magnitudes of the FFT outputs is the same as the sum of the magnitudes of the input samples. Therefore, the forms of the outputs of an FFT allow the power in a signal to be calculated as easily in the frequency domain as in the time domain.

### 2.3.10 Zero Padding

Zero padding is a technique used when a signal does not have as many samples as the FFT to be used for analyzing the signal. For example, if the application requires analyzing 12 input samples, but the engineer wanted to use a 16-point FFT, four zeros are added to the 12 samples to produce the 16 samples needed by the FFT. The advantage of zero padding is that it allows variable data collection lengths to be input to a single FFT algorithm designed to calculate the FFT of a longer sample length. The disadvantage is that the center frequencies of the 16-point FFT filters are not at the same frequencies as those of a 12-point FFT that was matched to the data collection needs of the application.

There is a subtle effect of using zeros, or any other numbers, to fill in uncollected data samples. From the sampling theorem, the unambiguous frequency range of the 12- or 16-point FFTs can only be from zero to the sampling rate, or half that rate if the input signal is real rather than complex. However, from Section 2.3.2, the spacing from the center of each filter to its first null response is equal to 1/(total time for data collected for the FFT input). Since the total collection time for the data in the 12- and 16-point FFTs is the same, the spacing to each filter's first null response must be the same. For the 12-point FFT this occurs at the location of the center of the adjacent filter. For the 16-point FFT this is not true because 16 filters are equally spaced in the same frequency range as the 12 filters. The result is that each of the 16-point FFT filters will have responses to signals that are at the centers of the other 16-point FFT filters.

Figures 2-3 and 2-4 illustrate the effects zero padding has on the real and imaginary parts of the responses of 12- and 16-point FFTs, for a 1-kHz sine wave that has been sampled at 12 kHz. In Figure 2-3 the real part has an amplitude of zero and the imaginary part has a nonzero amplitude at filters 1 and 11. This is because the sine wave has a 270° phase. This particular phase was used so that the real parts would be obviously different between the 12- and 16-point transforms. In Figure 2-4 the real and imaginary parts have nonzero responses in most of the filters because four zeros are appended to the 12 actual samples, and a 16-point FFT is performed.

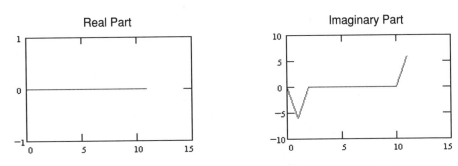

**Figure 2-3**    Twelve-point FFT response to 1-kHz input samples.

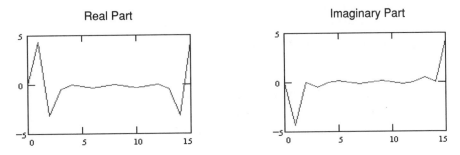

**Figure 2-4**    Sixteen-point FFT response to 12 samples and four zeros
of 1-kHz input samples.

The 16 FFT filter outputs in Figure 2-4 only span a 12-kHz frequency range because 12 kHz is the sample rate. With 16 filters to span the 12 kHz, the frequency spacing between them is smaller. This example shows that appending zeros to the end of the periodic sine wave, to make it a power-of-two length, alters the real and imaginary responses of the FFT filters. The weighting functions in Chapter 4 are used to minimize zero-padding effects.

### 2.3.11 Resolution

The resolution of two sine waves is defined as how close they can be in frequency before they can no longer be distinguished. If two frequencies are positioned at adjacent DFT filter outputs, namely $A(k)$ and $A(k+1)$, then they are distinguishable. If the frequency at $k + 1$ moves closer to frequency $k$, then it will start to appear as part of the passband of

$A(k)$, as well as $A(k+1)$, and it is no longer clear whether there is one signal at a frequency between $k$ and $k+1$ or two separate signals near $k$ and $k+1$.

Therefore, the frequency resolution of the DFT is the separation between adjacent filters. Since there are $N$ filters that cover the region from zero to the sampling frequency, the DFT resolution is the sampling frequency divided by $N$. This implies that, for a given sampling rate, the longer the transform length the better the frequency resolution of the analysis.

### 2.3.12 Periodicity

Section 2.3.1 showed that the DFT correctly analyzes frequencies from zero to half the sampling frequency. All other frequencies appear to be frequencies between zero and half the sampling rate. For complex inputs the real sampling rate is actually twice the sampling rate for the real or imaginary parts because both are being sampled at the same time. This leads to the two rules for the way frequencies below zero and above the sampling rate are analyzed by the DFT, one for complex signals and the other for real signals.

For complex input signals, periodicity means that frequencies that are higher than the sampling frequency appear at frequencies that are less than the sampling frequency $(A(N+k) \Rightarrow A(k))$. Similarly, negative frequencies appear as if they are at the sampling frequency minus their frequency $(A(-k) \Rightarrow A(N-k))$.

For real input signals with frequencies, $k$, below half the sampling rate, DFT filters $k$ and $N-k$ respond. Note that these two responding filters are symmetric about half the sampling rate. If the frequency is less than zero, add twice the sampling rate to the frequency and then apply the rule in the first sentence of this paragraph.

### 2.3.13 Summary of Properties

These 12 DFT properties:

- Apply to all of the FFT algorithms in Chapters 8 and 9
- Provide the framework for the capabilities of FFTs described in Chapters 5, 6, and 7
- Allow multiple mapping options for FFTs onto the multiprocessor architectures in Chapter 12
- Underlie the capabilities of the test signals in Chapter 16
- Provide the basis for using the FFT in the examples in Chapter 17

## 2.4 REAL INPUT SIGNALS

The DFT (Equation 2-1) produces complex frequency response outputs based on an input data sequence that is complex. However, many applications that can take advantage of the DFT have only real input data. The speech analyzer (Example 3) in Chapter 17 is one such application.

The DFT of a real data sequence can be computed directly by setting the imaginary part of the input sequence to zero. However, since the DFT is a linear algorithm, and a complex signal is the sum of a real signal and an imaginary one, it is possible to process a second real signal by entering it as the imaginary part of the input signal. The DFT output for this combined input is the sum of the output for the real input plus $j$ times the DFT output for the second real signal [1,2].

Equations 2-9 to 2-11 define the process of combining real signals $a(n)$ and $b(n)$ to form a complex input to the DFT. Since both $A(k)$ and $B(k)$ are complex sets of numbers, an additional step must be performed on the output of the DFT algorithm to separate these two real input signals. The algorithms in this section show two ways of utilizing the DFT for frequency analysis of real signals. The first is for the case of two independent real signals. The second is to more rapidly compute the frequency content in a single real signal.

$$A(k) = \sum_{n=0}^{N-1} a(n) * W^{kn} \tag{2-9}$$

$$B(k) = \sum_{n=0}^{N-1} b(n) * W^{kn} \tag{2-10}$$

$$C(k) = A(k) + jB(k) = \sum_{n=0}^{N-1} [a(n) + jb(n)] * W^{kn} \tag{2-11}$$

### 2.4.1 Two-Signal Algorithm

If an application has more than one real signal for which the frequency components need to be computed, an algorithm has been constructed to combine pairs of these signals into one FFT computation. A vital constraint of this algorithm is that the transform lengths must be the same for both real input signals. If there are an even number of real signals to be transformed, the signals can be paired off into FFTs that all operate on artificially created complex input signals.

The stages of the two-signal algorithm are presented using real input signals $a(n)$ and $b(n)$ as examples and assuming both $a(n)$ and $b(n)$ have the same number of samples to be converted. Stage 3 is different for $N$ an odd integer than for $N$ an even integer. The odd and even versions of the two-signal algorithm are presented as Cases 1 and 2 in Stage 3 of the algorithm.

### Stage 1: Form the Complex Input Signal

For each $n = 0, 1, 2, \ldots, N - 1$, combine $a(n)$ and $b(n)$ into the complex input function $c(n)$:

$$c(n) = a(n) + j * b(n)$$

### Stage 2: Compute an N-Point FFT

Compute the $N$-point FFT of $c(n)$ to obtain the $N$ frequency components $C(k), k = 0, 1, 2, \ldots, N - 1$, and identify the real and imaginary parts of $C(k)$ as $R(k)$ and $I(k)$, respectively, where $R(k)$ and $I(k)$ are real:

$$C(k) = \sum_{n=0}^{N-1} c(n) * e^{-j2\pi kn/N} = R(k) + j * I(k)$$

In Equation 2-11, $C(k) = A(k) + j * B(k)$, but both $A(k)$ and $B(k)$ are complex numbers. This is why Stages 3 and 4 are needed to compute $A(k)$ and $B(k)$ from the outputs of this stage. The variables $RP(k)$, $RP(N-k)$, $RM(k)$, $RM(N-k)$, $IP(k)$, $IP(N-k)$, $IM(k)$, and $IM(N-k)$ are used to compute the intermediate results necessary to convert $R(k)$ and $I(k)$ to $A(k)$ and $B(k)$.

### Stage 3: Separate Outputs into Real and Imaginary Parts

*Case 1: N Is an Odd Integer*

If $N$ is odd, then for each $k = 1, 2, \ldots, (N-1)/2$, compute

$$RP(k) = RP(N-k) = 0.5 * [R(k) + R(N-k)]$$
$$RM(k) = -RM(N-k) = 0.5 * [R(k) - R(N-k)]$$
$$IP(k) = IP(N-k) = 0.5 * [I(k) + I(N-k)]$$
$$IM(k) = -IM(N-k) = 0.5 * [I(k) - I(N-k)]$$
$$RP(0) = R(0)$$
$$IP(0) = I(0)$$
$$RM(0) = IM(0) = 0$$

This requires $2(N-1)$ adds and no multiplies because multiplying by 0.5 is just shifting the binary point to the left 1 bit. Note that this algorithm does require each computed answer to be stored in two places. This puts an additional burden on the memory address generators of the DSP chips (Chapter 14) used to compute the answers.

*Case 2: N Is an Even Integer*

If $N$ is even, then for each $k = 1, 2, \ldots, (N-2)/2$, compute

$$RP(k) = RP(N-k) = 0.5 * [R(k) + R(N-k)]$$
$$RM(k) = -RM(N-k) = 0.5 * [R(k) - R(N-k)]$$
$$IP(k) = IP(N-k) = 0.5 * [I(k) + I(N-k)]$$
$$IM(k) = -IM(N-k) = 0.5 * [I(k) - I(N-k)]$$
$$RP(0) = R(0)$$
$$IP(0) = I(0)$$
$$RM(0) = IM(0) = RM(N/2) = IM(N/2) = 0$$
$$RP(N/2) = R(N/2)$$
$$IP(N/2) = I(N/2)$$

This requires $2(N-2)$ adds and no multiplies because multiplying by 0.5 is just shifting the binary point to the left 1 bit. Note that this algorithm also requires each computed answer to be stored in two places. This puts an additional burden on the memory address generators of the DSP chips (Chapter 14) used to compute the answers.

### Stage 4: Compute the FFT Outputs for Each Real Input Signal

For each $k = 0, 1, 2, \ldots, N-1$, identify the FFT output $A(k)$ and $B(k)$ for each of the real input signals $a(n)$ and $b(n)$, respectively, as

$$A(k) = RP(k) + j * IM(k)$$
$$B(k) = IP(k) + j * RM(k)$$

The total number of computations for the two-signal algorithm is the number of adds and multiplies required by the FFT algorithm plus the $2 * (N-1)$ or $2 * (N-2)$ adds in Stage 3, depending on whether $N$ is odd or even.

### 2.4.2 Double-Length Algorithm

If an application requires computing the $M$ frequency components of only one real signal, then an algorithm has been developed to compute that $M$-point transform using an

$M/2 = N$-point FFT. This algorithm significantly reduces the computational requirements over simply assuming that the imaginary portion of the signal is zero in Equation 2-1.

The stages of this algorithm are presented for the input data sequence $a(n)$. A vital constraint of this algorithm is that it is restricted to transform lengths, $M$, that have a factor of 2 so that $M/2 = N$ is an integer. Stage 3 is different for $N$ an odd integer than for $N$ an even integer. The odd and even versions of the double-length algorithm are presented as Cases 1 and 2 in Stage 3 of the algorithm.

### Stage 1: Form Complex Input Signal

For $n = 0, 1, 2, \ldots, N - 1$, divide the input sequence $a(n)$ into sequences $b(n)$ and $c(n)$, and form the complex FFT input $d(n)$ by using $b(n)$ for the real part and $c(n)$ for the imaginary part:

$$b(n) = a(2 * n)$$
$$c(n) = a(2 * n + 1)$$
$$d(n) = b(n) + j * c(n)$$

### Stage 2: Compute an *N*-Point FFT

Compute the $N$-point FFT of $d(n)$ to obtain the complex frequency components $D(k)$, and identify the real part of these components as $R(k)$ and the imaginary part as $I(k)$.

$$D(k) = \sum_{n=0}^{N-1} d(n) * e^{-j2\pi kn/N}$$

$$D(k) = R(k) + j * I(k)$$

Note that $R(k)$ and $I(k)$ are real numbers equal to the real and imaginary parts of $D(k)$ respectively. This is why Stages 3 and 4 are needed to compute $A(k)$ from the outputs of this stage. The variables $RP(k)$, $RP(N - k)$, $RM(k)$, $RM(N - k)$, $IP(k)$, $IP(N - k)$, $IM(k)$, $IM(N - k)$, $AR(k)$, $AR(M - k)$, $AI(k)$, and $AI(M - k)$ are used to compute the intermediate results necessary to convert $R(k)$ and $I(k)$ to $A(k)$.

### Stage 3: Separate Outputs into Real and Imaginary Parts

#### Case 1: *N Is an Odd Integer*

If $N$ is odd, then for each $k = 1, 2, \ldots, (N - 1)/2$, compute

$$RP(k) = RP(N - k) = 0.5 * [R(k) + R(N - k)]$$
$$RM(k) = -RM(N - k) = 0.5 * [R(k) - R(N - k)]$$
$$IP(k) = IP(N - k) = 0.5 * [I(k) + I(N - k)]$$
$$IM(k) = -IM(N - k) = 0.5 * [I(k) - I(N - k)]$$
$$RP(0) = R(0)$$
$$IP(0) = I(0)$$
$$RM(0) = IM(0) = 0$$

This requires $2(N - 1)$ adds and no multiplies because multiplying by 0.5 is just shifting the binary point to the left 1 bit. Note that this algorithm does require each computed answer to be stored in two places. This puts an additional burden on the memory address generators of the DSP chips (Chapter 14) used to compute the answers.

*Case* 2: *N Is an Even Integer*

If $N$ is even, then for each $k = 1, 2, \ldots, (N-2)/2$, compute

$$RP(k) = RP(N-k) = 0.5 * [R(k) + R(N-k)]$$
$$RM(k) = -RM(N-k) = 0.5 * [R(k) - R(N-k)]$$
$$IP(k) = IP(N-k) = 0.5 * [I(k) + I(N-k)]$$
$$IM(k) = -IM(N-k) = 0.5 * [I(k) - I(N-k)]$$
$$RP(0) = R(0)$$
$$IP(0) = I(0)$$
$$RM(0) = IM(0) = RM(N/2) = IM(N/2) = 0$$
$$RP(N/2) = R(N/2)$$
$$IP(N/2) = I(N/2)$$

This requires $2(N-2)$ adds and no multiplies because multiplying by 0.5 is just shifting the binary point to the left 1 bit. Note that this algorithm also requires each computed answer to be stored in two places. This also puts an additional burden on the memory address generators of the DSP chips (Chapter 14) used to compute the answers.

**Stage 4: Compute the FFT Outputs for Each Real Input Signal**

For each $k = 1, 2, \ldots, N-1$, identify the FFT output $A(k)$ as

$$AR(k) = AR(M-k) = RP(k) + \cos(k\pi/N) * IP(k) - \sin(k\pi/N) * RM(k)$$
$$AI(k) = -AI(M-k) = IM(k) - \cos(k\pi/N) * RM(k) - \sin(k\pi/N) * IP(k)$$
$$AR(0) = RP(0) + IP(0)$$
$$AI(0) = IM(0) - RM(0)$$
$$AR(N) = R(0) - I(0)$$
$$AI(N) = 0$$
$$A(k) = A(M-k) = AR(k) + j * AI(k)$$

This requires $4 * N - 1$ adds and $4 * (N-1)$ multiplies. Note that this algorithm requires each computed answer to be stored in two places. This puts an additional burden on the memory address generators of the DSP chips (Chapter 14) used to compute the answers.

The total number of computations for the double-length algorithm is the adds and multiplies required by the FFT algorithm, $N_F$, plus $5 * M - 7$ or $5 * M - 9$, depending on whether $N$ is odd or even.

## 2.5 STRENGTHS

The DFT has four types of strengths. The first two are associated with the types of data the DFT analyzes. The third is associated with the way data (complex samples) must be collected and processed by a DFT. The fourth is associated with the signal-to-noise improvement offered by the DFT.

### 2.5.1 Periodic Signals

The DFT is an equation for converting time domain data into its frequency components. However, it only converts the signal to the specific frequency components $A(k)$ in Equation 2-1. Since the signals associated with these frequency components go through $0, 1, \ldots, N-1$ 360° cycles during the $N$ input data points, any sum of them must also repeat itself a whole number of times during the $N$ input data points. Therefore, the DFT is ideal for analyzing the sine waves in a signal when the signal repeats an integer number of times (i.e., is periodic) during the $N$ input data samples.

Even if the data is not periodic during the $N$ samples, the DFT output is still the amplitude and phase of a set of frequencies that can be used to reconstruct the time domain signal. However, the DFT's output frequencies are not the actual ones in the signal. The frequency-shift-keyed (FSK) modem example in Section 2.6.5 is a good illustration of this phenomena. Therefore, the DFT is not particularly well suited for signals that are either never periodic (random or transient) or are periodic at a rate different from the number of samples in the transform. Example 2 in Chapter 17 shows how to use the DFT to analyze random signals. The ability to choose any DFT length allows the DFT to match the period of the transient input signals.

### 2.5.2  Real or Complex Input Data

Equation 2-1 shows $W_N$ as a complex number. Therefore, even if the input data $a(n)$ is real, the output frequency data $A(k)$ is complex. In fact, this is how the DFT provides both amplitude and phase information for the $k$th frequency component in a signal. This fact permits the DFT to be used in the analysis of real and complex input signals. Example 3 in Chapter 17 uses real input signals, and Example 1 in Chapter 17 uses complex inputs.

### 2.5.3  Sets of Data

Equation 2-1 shows that the frequency components, $A(k)$, are computed on the last $N$ data points. In many applications the DFT is computed for multiple sets of $N$ data samples. These sets of data may be contiguous (i.e., samples 0 through $N - 1$ followed by samples $N$ through $2 * N - 1$), or they may be overlapped by any number of points (i.e., samples 0 through $N - 1$ followed by samples $N/2$ through $3 * N/2 - 1$ are overlapped by half of the samples). Since the DFT equation can be computed for any length $N$ and for any overlap of the sets, it provides a versatile method for performing and comparing the frequency analysis of data sequences. Figure 2-5 shows this overlapping of data sets by $(N - P)$-samples.

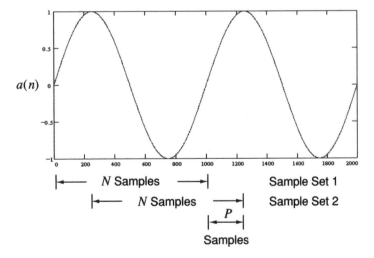

**Figure 2-5**    Overlapping data sets by $(N - P)$-samples.

### 2.5.4 Coherent Integration Gain

Equation 2-1 shows that $N$ input samples are summed to obtain each frequency component value. If the input samples contain a frequency that is in the center of one of the DFT's narrowband filters (Figure 2-1), then the frequency component at the output of the appropriate filter will have an amplitude that is $N$ times the amplitude of that input sine wave. For example, the zero-frequency component $A(k)$ sums the $N$ samples with $k = 0$. If those samples are all the same, the output of $A(0)$ is $N$ times larger than the amplitude of the input samples. This is one aspect of coherent integration.

The second aspect of coherent integration exhibited by each DFT output is a reduction in noise bandwidth by a factor of $N$ over the input signal. This is most easily understood by using the sampling theorem (Nyquist rate) in Section 2.3.1. Namely, a signal that is properly prepared for the DFT will have frequency components that go no higher than the sampling rate. Therefore, the noise bandwidth into the DFT will be limited to the sampling rate. Since this allowable bandwidth is divided into $N$ pieces by the $N$ DFT bandpass filters, the output of any one of the filters can only have $1/N$ of the input noise power. Since white noise is equally distributed across the available bandwidth by definition, the noise bandwidth of each DFT filter is $1/N$ of the input bandwidth. The result is an improvement of a factor of $N$ in the signal-to-noise ratio of a single sine wave plus noise at the output of the DFT.

## 2.6 WEAKNESSES

The DFT has five weaknesses. The first two are improved through the use of FFT algorithms. The second two are improved by applying a weighting function to data before computing an FFT of it. The fifth, inaccurate identification of frequencies in a transient signal, is not improved by FFT algorithms. Transforms that do identify transient signals are not addressed in this book.

### 2.6.1 Computational Load

Computational load is the number of adds and multiplies that must be performed. Equation 2-1 shows that $N$ complex multiplies and $N - 1$ complex adds are required to compute each of the $N$ DFT outputs. Since a complex multiply requires four real multiplies and two real adds and a complex add requires two real adds, the total computational load for an $N$-point DFT is

$$\text{\# Adds} = N(2N + 2(N - 1)) = 4N^2 - 2N$$
$$\text{\# Multiplies} = N(4N) = 4N^2 \tag{2-12}$$

For a 1024-point DFT this is roughly 4 million adds and 4 million multiplies. Even for audio rates on the order of 20,000 samples per second, twenty 1024-point DFTs per second corresponds to 80 million adds and 80 million multiplies, a significant computational load. All of the FFT algorithms presented in this book require computations on the order of $N * \log_2 N$ computations rather than $N^2$. For a 1024-point DFT this is a reduction by a factor of $N^2/(N * \log_2 N)$ or roughly 100:1. This is the fundamental motivation behind developing and using FFT algorithms.

### 2.6.2 Quantization Noise Error

In a digital computer all numbers are represented by some number of bits either as fixed- or floating-point numbers. When these numbers are used in multiplication, the resulting number has more bits than either of the input numbers. Because the number of bits used to represent a number must be controlled, to avoid running out of memory to store the numbers, the outputs from arithmetic computations must be rounded off at some point.

The round-off process introduces an error that changes the results of all of the rest of the computations that use the rounded-off results. This is called quantization noise error. The numerous computations required by the DFT result in a lot of quantization noise error. One of the advantages of FFT algorithms is that the reduced number of computations reduces quantization noise error. This will be discussed quantitatively in Chapter 13.

### 2.6.3 High Sidelobes

Sidelobes are a way of describing how a filter responds to signals at frequencies that are not in its main lobe, commonly called its passband. Specific details on the DFT's sidelobes are discussed in Section 4.1.1, because weighting functions are used to control the sidelobe behavior of DFT filters. Each DFT filter's first sidelobe is only 13 dB below the main lobe (therefore considered high), and subsequent sidelobes fall off very slowly. The result is that a signal with strong frequency, far away from the center frequency of a DFT filter, will not be completely removed by that filter and can look like a significant signal at the output of that filter.

### 2.6.4 Frequency Straddle Loss

Frequency straddle loss is the reduced output of a DFT filter caused by the input signal not being at the filter's center frequency. The coherent gain of the DFT is $N$ when the input frequency is located at the center of one of the narrowband filters whose output is $A(k)$. If the input frequency is halfway between two of the narrowband filters, the coherent gain is reduced, because half of the signal will appear in one filter and half in the other. The difference between the full coherent gain of $N$ and this lower gain is called frequency straddle loss. This subject is explained in more detail in Section 4.1.3.

### 2.6.5 Transient Signals

In Section 2.5.1 the DFT was shown to be ideal for analyzing signals that are periodic within the number of samples being analyzed. Transient signals are not well analyzed by the DFT. This is true regardless of whether the signal is a true transient or a transient sine wave. An example of a transient sine wave is an FSK modem signal, which changes frequency during the set of data points being analyzed. An FSK modem signal is a sum of two sine waves, each of which lasts for a portion of the sequence of input samples. Figures 2-6 and 2-7 show an FSK modem signal and its DFT.

While the time waveform in Figure 2-6 shows just two frequencies, the DFT of the time waveform in Figure 2-7 suggests there are five prominent frequencies and some smaller ones. This is a result of the DFT analyzing transient signals as if they were periodic signals.

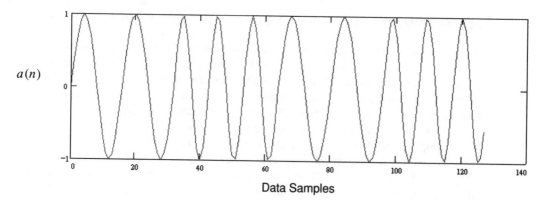

$a(n)$

Data Samples

**Figure 2-6** One hundred twenty-eight samples of an FSK modem signal.

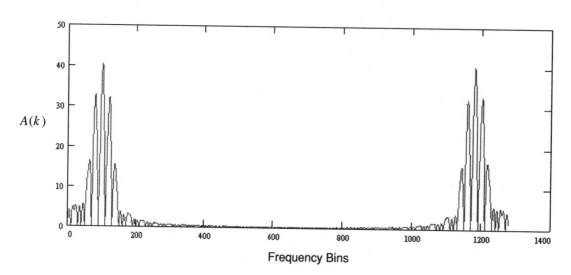

$A(k)$

Frequency Bins

**Figure 2-7** One hundred twenty-eight-point DFT of the FSK modem signal.

## 2.7 CONCLUSIONS

The DFT is a sound computational method, whose characteristics make it useful in manipulating periodic signals and poor at dealing with transient signals, though it is used on the latter when applied carefully with a thorough understanding of its limitations. Even though the DFT equation assumes complex input signals, it is frequently used to analyze real signals by doing input data reorganization and performing additional computations on the output data.

Because the FFT inherits all the properties and strengths of the DFT, a firm foundation about the DFT must be laid in order to see why FFTs are so useful and versatile. Its property of linearity appears throughout the book in the implementation of many FFT algorithms.

The next two chapters deal with the ways that four of the five weaknesses of the DFT are minimized. The fifth drawback—being poor at analyzing transient signals—requires transforms not covered in this book, such as wavelet and joint time frequency.

## REFERENCES

[1] E. Oran Brigham, *The Fast Fourier Transform*, Prentice-Hall, Englewood Cliffs, NJ, 1974.

[2] A. V. Oppenheim and R. W. Schafer, *Digital Signal Processing*, Prentice-Hall, Englewood Cliffs, NJ, 1975.

[3] L. R. Rabiner and B. Gold, *Theory and Application of Digital Signal Processing*, Prentice-Hall, Englewood Cliffs, NJ, 1975.

[4] E. Oran Brigham, *The Fast Fourier Transform and Its Applications*, Prentice-Hall, Englewood Cliffs, NJ, 1988.

[5] C. E. Shannon, "A Mathematical Theory of Communication," *The Bell System Technical Journal*, Vol. 27, pp. 379–423 (1948).

[6] H. Nyquist, "Certain Topics in Telegraph Transmission Theory," *AIEE Transactions*, Vol. 47, pp. 617–644 (1928).

# 3

# The Fast Fourier Transform

## 3.0 INTRODUCTION

Fast Fourier transforms (FFT) are a group of algorithms for significantly speeding up the computation of the DFT. The most widely known of these algorithms is attributed to Cooley and Tukey [1] and is used for a number of points $N$ equal to a power-of-two. A unique feature of this book is that it provides multiple FFT algorithms for fast computation of any length DFT. These are found in Chapters 8 and 9. In fact, the article by Cooley and Tukey presented a non-power-of-two algorithm which has mostly been ignored. Several of the algorithms in Chapter 9 are spin-offs of that work.

The most important fact about all FFT algorithms is that they are mathematically equivalent to the DFT, not an approximation of it. This means that all of the properties, strengths, and most of the weaknesses of the DFT apply to the FFT algorithms in this book. The FFT improves two weaknesses of the DFT: high number of adds and multiplies; and quantization noise.

An example of an 8-point DFT to FFT is used in this chapter to illustrate how FFTs actually speed up the DFT. The chapter concludes with a detailed explanation of how to use the building-block approach to construct FFTs.

## 3.1 IMPROVEMENTS TO THE DFT

The FFT improves the DFT by reducing the computational load and quantization noise of the DFT.

### 3.1.1 Computational Load

Section 2.6.4 established that the total computational load for an $N$-point DFT is

$$\# \text{Adds} = N(2N + 2(N - 1)) = 4N^2 - 2N$$
$$\# \text{Multiplies} = N(4N) = 4N^2$$

$$(3\text{-}1)$$

Chapter 9 establishes that the number of computations required for FFT algorithms, regardless of the transform length, can be expressed as a constant times $N*\log_2(N)$. Therefore, the computation reduction factor when using an FFT algorithm is a constant times $N/\log_2(N)$. The constant is different, but near 5, for each algorithm and nearly always provides a significant advantage for using the FFT.

### 3.1.2 Quantization Noise

The other improvement offered by FFT algorithms is a direct result of the reduction in the number of computations. Namely, the quantization noise generated by the FFT computations is smaller than if the DFT had been used. The reason is that there are fewer multiplications performed to compute each of the FFT output frequencies. This means there are fewer places where the multiplier output must be rounded off.

For example, a 1024-point DFT requires 1024 multiplies and 1023 adds to compute each frequency output. This presents 1024 places in the computations where results must be rounded off. The radix-4 1024-point FFT described in Chapter 9, has only four places in the computations where multiplications are performed. The DFT and FFT quantization noise generated by these round-off procedures is described in more detail in Chapter 13.

## 3.2 FFT-SPECIFIC WEAKNESS

In addition to the weakness associated with transient input signals, the reorganization of data and reduction of computations required by FFT algorithms leads to the need to compute all of the output frequencies, even if only a few are required. In contrast, DFT outputs can be computed one at a time. However, this is not generally a practical weakness for two reasons. First, FFTs are usually used for frequency analysis where all of the outputs are needed. The second reason is that the dramatic reduction in computational load makes the FFT algorithms more efficient, even when only a few output frequencies of the DFT need to be computed. For example, consider the radix-2 Cooley-Tukey algorithm. In Chapter 9 this algorithm is shown to require roughly $3 * N * \log_2 N$ adds and $2 * N * \log_2 N$ multiplies. For a 1024-point FFT this amounts to 30,720 adds and 20,480 multiplies for a total of 51,200 arithmetic operations. In contrast, each DFT output frequency requires $4 * N = 4096$ multiplies and $4 * N - 1 = 4095$ adds for a total of 8191 arithmetic operations. Therefore, if more than $51,200/8191 = 6.25$ of the 1024 potential DFT outputs are needed, it is more efficient to use an FFT algorithm to compute all 1024 outputs and throw away the unneeded ones.

## 3.3 EIGHT-POINT DFT TO FFT EXAMPLE

All of the FFT algorithms in this book are based on ways to remove redundant computations from the DFT equations without changing their final result. The simplest way to illustrate these techniques is to show the process for the 8-point DFT. This is the only place in this book where an FFT algorithm is actually derived from its DFT origins. The rest of the book focuses on choosing and applying the algorithms, not deriving them. The building-block algorithms described in Chapter 8 are the result of using techniques, such as those in this section, to remove redundant computations from small DFTs.

### 3.3.1 Eight-Point DFT Equations in Matrix Form

Equation 3-2 is a simplified matrix representation of the 8-point DFT, based on Equation 2-1. The simplification over the standard DFT equation is easily visualized by drawing the $W_8^{kn}$ terms as vectors on a unit circle (Figure 3-1). From Figure 3-1 it is clear that the $W_8^{kn}$ rotates around the unit circle as $k * n$ increases and the vector returns to the same location when $k * n$ is increased by multiples of 8.

$$
\begin{bmatrix} A_0 \\ A_1 \\ A_2 \\ A_3 \\ A_4 \\ A_5 \\ A_6 \\ A_7 \end{bmatrix}
=
\begin{bmatrix}
W^0 & W^0 & W^0 & W^0 & W^0 & W^0 & W^0 & W^0 \\
W^0 & W^1 & W^2 & W^3 & W^4 & W^5 & W^6 & W^7 \\
W^0 & W^2 & W^4 & W^6 & W^0 & W^2 & W^4 & W^6 \\
W^0 & W^3 & W^6 & W^1 & W^4 & W^7 & W^2 & W^5 \\
W^0 & W^4 & W^0 & W^4 & W^0 & W^4 & W^0 & W^4 \\
W^0 & W^5 & W^2 & W^7 & W^4 & W^1 & W^6 & W^3 \\
W^0 & W^6 & W^4 & W^2 & W^0 & W^6 & W^4 & W^2 \\
W^0 & W^7 & W^6 & W^5 & W^4 & W^3 & W^2 & W^1
\end{bmatrix}
\begin{bmatrix} a_0 \\ a_1 \\ a_2 \\ a_3 \\ a_4 \\ a_5 \\ a_6 \\ a_7 \end{bmatrix}
\tag{3-2}
$$

Simplified 8-point DFT matrix

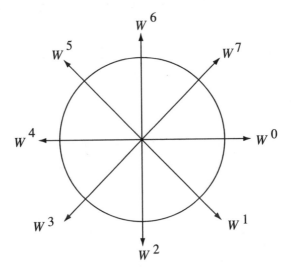

**Figure 3-1**   Vector representation of $W^{kn}$.

For example,

$$W_8^4 = W_8^{12} = W_8^{20} = W_8^{28} = W_8^{36} \tag{3-3}$$

This cyclic feature of $W_8^{kn}$ plays a primary role in the development of all of the FFT algorithms in this book. In Equation 3-1 all of the exponents $(k*n)$ of $W$ larger than 8 have been reduced to the equivalent power that is less than 8 by repeatedly subtracting 8 until the exponent is less than 8. Using the example in Equation 3-3, the powers of $k*n = 36, 28, 20,$ and 12 have all been replaced by $W_8^4$.

### 3.3.2 180° Redundant Computations

The first observation from Figure 3-1 is that $W_8^0 = -W_8^4$, $W_8^1 = -W_8^5$, $W_8^2 = -W_8^6$, and $W_8^3 = -W_8^7$. If these equalities are substituted into Equation 3-2, then it is clear that $a_0 + a_4, a_0 - a_4, a_1 + a_5, a_1 - a_5, a_2 + a_6, a_2 - a_6, a_3 + a_7,$ and $a_3 - a_7$ are each used four times in the DFT equations. Therefore, computations can be removed if there is an efficient way to compute these eight terms once and use the results in each of the other places they are required rather than recompute them. This can be done, and the result is matrix Equation 3-4.

$$
\begin{bmatrix}
A_0 \\ A_1 \\ A_2 \\ A_3 \\ A_4 \\ A_5 \\ A_6 \\ A_7
\end{bmatrix}
=
\begin{bmatrix}
1 & 0 & 1 & 0 & 1 & 0 & 1 & 0 \\
0 & 1 & 0 & W & 0 & -j & 0 & -jW \\
1 & 0 & -j & 0 & -1 & 0 & j & 0 \\
0 & 1 & 0 & -jW & 0 & j & 0 & W \\
1 & 0 & -1 & 0 & 1 & 0 & -1 & 0 \\
0 & 1 & 0 & -W & 0 & -j & 0 & jW \\
1 & 0 & j & 0 & -1 & 0 & -j & 0 \\
0 & 1 & 0 & jW & 0 & j & 0 & -W
\end{bmatrix}
\begin{bmatrix}
a_0 + a_4 \\ a_0 - a_4 \\ a_1 + a_5 \\ a_1 - a_5 \\ a_2 + a_6 \\ a_2 - a_6 \\ a_3 + a_7 \\ a_3 - a_7
\end{bmatrix}
\tag{3-4}
$$

Eight-point DFT with 180° redundancies removed

### 3.3.3 90° Redundant Computations

The next observation from Figure 3-1 is that $W_8^1$, $W_8^3$, $W_8^5$, and $W_8^7$ exhibit 90° symmetry, namely,

$$W_8^3 = W_8^2 * W_8^1 = -j * W_8^1$$
$$W_8^5 = W_8^2 * W_8^3 = (-j) * (-j) * W_8^1 = -W_8^1 \tag{3-5}$$
$$W_8^7 = W_8^2 * W_8^5 = -j * (-W_8^1) = j * W_8^1$$

The simplest example of using the property in Equation 3-5 to reduce computations is in columns 0 and 4 of the matrix in Equation 3-4. Notice that rows 0 and 4 have 1 to multiply the $a_0 + a_4$ and $a_2 + a_6$ terms in the right-hand column vector. Similarly, rows 2 and 6 both subtract the $a_1 + a_5$ and $a_3 + a_7$ terms. In both cases, redundant computations can be removed by performing the required computations once and using the results twice. Other symmetries similar to this illustration also exist in Equation 3-4. When all these are exploited, matrix Equation 3-4 is converted to matrix Equation 3-6.

$$
\begin{bmatrix} A_0 \\ A_1 \\ A_2 \\ A_3 \\ A_4 \\ A_5 \\ A_6 \\ A_7 \end{bmatrix} =
\begin{bmatrix}
1 & 0 & 0 & 0 & 1 & 0 & 0 & 0 \\
0 & 0 & 1 & 0 & 0 & 0 & W & 0 \\
0 & 1 & 0 & 0 & 0 & -j & 0 & 0 \\
0 & 0 & 0 & 1 & 0 & 0 & 0 & -jW \\
1 & 0 & 0 & 0 & -1 & 0 & 0 & 0 \\
0 & 0 & 1 & 0 & 0 & 0 & -W & 0 \\
0 & 1 & 0 & 0 & 0 & j & 0 & 0 \\
0 & 0 & 0 & 1 & 0 & 0 & 0 & jW
\end{bmatrix}
\begin{bmatrix}
(a_0 + a_4) + (a_2 + a_6) \\
(a_0 + a_4) - (a_2 + a_6) \\
(a_0 - a_4) - j(a_2 - a_6) \\
(a_0 - a_4) + j(a_2 - a_6) \\
(a_1 + a_5) + (a_3 + a_7) \\
(a_1 + a_5) - (a_3 + a_7) \\
(a_1 - a_5) - j(a_3 - a_7) \\
(a_1 - a_5) + j(a_3 - a_7)
\end{bmatrix}
\qquad (3\text{-}6)
$$

Eight-point DFT with 90° and 180° redundancies removed

In addition to removing redundant computations, the other important feature of this approach is that the required computations are performed in a way that allows them to be efficiently used later in the algorithm. Specifically, the first step in this version of the 8-point FFT algorithm is to compute the terms found in the right-hand vector in Equation 3-4. The second step is to combine these results as shown in the right-hand vector in Equation 3-6.

### 3.3.4  45° Redundant Computations

The final observation in this example is based on noticing columns 0 and 4 of rows 0 and 4 in Equation 3-6. Notice that these terms in the matrix require the sum and difference of terms in the right-hand vector. This does not reduce the overall computations. However, it does complete the computational symmetry of the algorithm. The advantage of this is that this algorithm needs only one computational building block, the sum and difference calculation of a pair of numbers, which is called a butterfly. Therefore, not only have this set of observations resulted in butterfly computations at each stage, but the number of computations has been reduced.

Figure 3-2 is a flowchart of the 8-point FFT. This algorithm's detailed equations are in Section 8.8.2. Each node in the flowchart represents a complex add, which is two real adds. There are 24 of these nodes, which corresponds to 48 adds. Similarly, there are two complex multiplies in the algorithm. Since these multipliers are applied to a complex number, the algorithm requires eight real multiplies and four additional real adds. Based on Equation 3-1, the 8-point DFT requires $4 * N^2 = 256$ multiplies and $4 * N^2 - 2 * N = 240$ adds. Therefore, this algorithm reduces the total number of arithmetic operations from $256 + 240 = 496$ to $48 + 8 + 4 = 60$, more than a factor of 8.

To be absolutely fair, the $W_8^0 = 1$ and $W_8^4 = -1$ terms in Equation 3-1 do not require complex multiplications. This reduces the DFT computational load by 16 com-

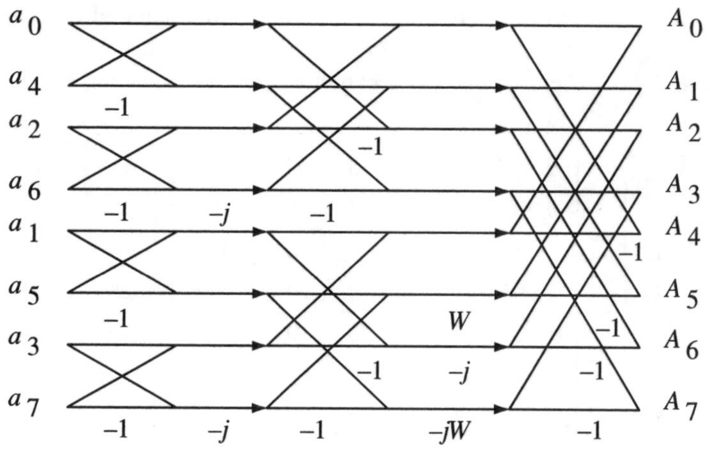

**Figure 3-2**    Eight-point FFT flow graph.

plex multiplies, which is 64 multiplies and 32 adds. Removing these 96 computations reduces the DFT arithmetic operations count to 400, which reduces the gain associated with using the FFT algorithm to a factor of 6.67 (400/60). This is still a significant savings. Other approaches to making the 8-point DFT fast are in Section 8.8. Most will have the same number of total computations, even though they are derived by different approaches.

## 3.4 BUILDING-BLOCK CONSTRUCTION OF FFT ALGORITHMS

The previous section described how the FFT speeds up the DFT by removing redundant computations. This section describes the building-block approach to constructing FFT algorithms. It is useful to have a simple picture of how the DFT is decomposed into sets of building blocks. One way to understand this is to return to the concept of the DFT being an array of narrowband filters.

The narrowband filters implemented by the $N$-point DFT equations are equally spaced in frequency and divide the frequency spectrum into $N$ equal increments. Suppose that $N$ is not a prime number so that it can be written as the product of at least two numbers, $N = P * Q$. Then, with a $P$-point DFT, it is possible to decompose the frequency spectrum into $P$ equal increments (i.e., create $P$ narrowband filters). On the left-hand side of Figure 3-3 is a block diagram of the $P$-point DFT based on the array of narrowband filters concept in Chapter 2. On the right-hand side of the figure the rectangle drawn with dotted lines is all of the narrowband filters inside the dotted lines on the left side of the figure. The block on the right is used again in Figure 3-4 to show how the $P$- and $Q$-point DFTs are combined to form the $N$-point DFT.

The outputs of each of these $P$ narrowband filters are also a signal. It just has a narrower bandwidth than the original input signal. Furthermore, since each narrowband filter has one output for each $P$ inputs, it has $N/P = Q$ outputs for each $N$ inputs. Therefore, each of these $P$ output signals can be further analyzed by decomposing its

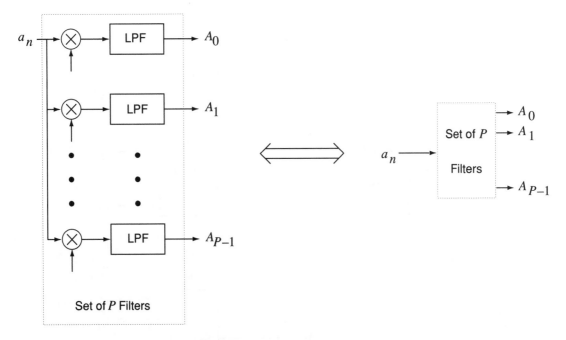

**Figure 3-3**    Block diagram of the $P$-point DFT as an array of narrowband filters.

frequency spectrum into $Q$ equally spaced increments by using a $Q$-point DFT to implement $Q$ narrowband filters. The result is $Q$ narrowband filters for each of the $P$ filters as shown in Figure 3-4. If Figure 3-4 were expanded by using the block diagram in Figure 3-3, there would be $Q$ narrowband filters for each of the $P$ narrowband filters. Since a narrowband filter connected to the output of a narrowband filter is also a narrowband filter, Figure 3-4 can be redrawn as $P * Q$ narrowband filters.

Since these $N = P * Q$ narrowband filter outputs are also equally spaced and cover the same frequency spectrum as an array of $N$ narrowband filters, they must be the same as the ones implemented by a $P * Q$-point DFT. This is the strategy used by each of the FFT algorithms in Chapter 9 to decompose the FFT into the smaller building blocks described in Chapter 8.

If Figure 3-4 is compared with the prime factor algorithm block diagrams (Figures 9-17 and 9-18) or the mixed-radix algorithm block diagrams (Figures 9-23, 9-24, and 9-25), two differences are noticed. First, the frequency component outputs are in different order in each of the figures. The details of the FFT algorithms result in these different output frequency orders. Second, while Figure 3-4 and all of the FFT algorithms have $PQ$-point FFTs, all of the FFT algorithms have $QP$-point FFTs on the input and Figure 3-4 only has one. This makes it look like Figure 3-4 requires fewer computations than the FFT algorithms in Chapter 9. The catch is that each of the $P$ narrowband input filters on the left-hand side of Figure 3-4 must process all $N$ of the input data samples. However, each of the $P$-point FFTs on the inputs to the FFT algorithms in Figures 9-17, 9-18, 9-23, 9-24, and 9-25 only processes $Q$ points. In all cases each of the $Q$-point output filters and

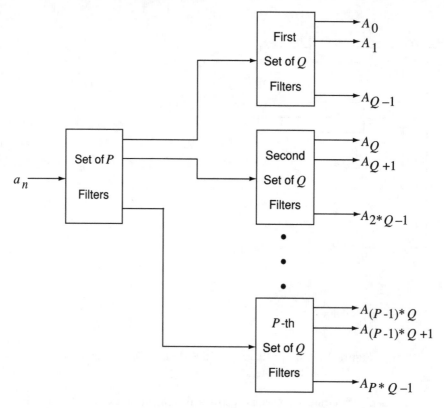

**Figure 3-4** Block diagram of the $N$-point DFT as an array of narrow-band filters.

FFTs only processes $P$ intermediate results. Section 3.3 shows how the FFT approach is used to reduce the total computational load over using the narrowband filter approach.

## 3.5 CONCLUSIONS

The fast versions of the DFT overcome two of its weaknesses. The FFT reduces computational load (adds and multiplies) by significantly reducing the redundancy that is inherent in the structure of the DFT equation. Quantization noise is also reduced by using FFTs because the number of computations is less than with the DFT.

While improving the DFT so dramatically that it is now used in hundreds of applications, the FFT does not add any drawbacks of its own, which cannot be said for the element covered in the next chapter. Weighting functions get teamed with FFTs to reduce two more weaknesses of the DFT.

## REFERENCES

[1] J. W. Cooley and J. W. Tukey, "An Algorithm for the Machine Calculation of Complex Fourier Series," *Mathematics of Computation*, Vol. 19, p. 297 (1965).

# 4

# Weighting Functions

## 4.0 INTRODUCTION

A weighting function, $w(n)$, is a sequence of numbers that is multiplied times input data prior to performing a DFT on that data. Weighting (also called window) functions reduce sidelobes of DFT filters and widen main lobes while, fortunately, not altering the locations of the centers of the filters. The weighting functions in this chapter provide options to reduce sidelobes from the $-13$-dB peak sidelobe of the DFT to as low as $-94$ dB.

Weighting function selection can be made early in the design process because the choices of FFT algorithm and weighting function are independent of each other. Choice of a weighting function to provide the specified sidelobe level is done without concern for the FFT algorithm that will be used because:

- They work for any length FFT.
- They work the same for any FFT algorithm.
- They do not alter the FFT's ability to distinguish two frequencies (resolution).

Weighting functions are applied three ways:

- As a rectangular function, which does not modify the input data
- By having all the weighting function coefficients stored in memory
- By computing each coefficient when it is needed

## 4.1 SIX PERFORMANCE MEASURES

The choice of weighting function depends on which of the features of the narrowband DFT filters are most important to the application. Those features are performance measures of the narrowband filters in order to analytically compare weighting functions. All these measures, except frequency straddle loss, refer to individual filters. Frequency straddle loss is associated with how filters work together.

### 4.1.1 Highest Sidelobe Level

Sidelobes are a way of describing how a filter responds to signals at frequencies that are not in its main lobe, commonly called its passband. Each FFT filter has several sidelobes. With rare exception, the highest one is closest in frequency to the main lobe and is the one that is most likely to cause the passband filter to respond when it should not. The higher a sidelobe level is, the lower is the amplitude of a signal outside the passband of the filter that produces a significant filter response. This response erroneously indicates the presence of a signal in the passband.

### 4.1.2 Sidelobe Fall-off Ratio

Sidelobes have peaks in response as a function of frequency. The peak (amplitude) of sidelobes decreases or remains level as they get further away in frequency from the passband. This performance measure describes how fast the sidelobe amplitude is reduced as a function of frequency. If the sidelobes reduce rapidly, then only signals that are close in frequency can cause the DFT filters to have erroneous responses. This performance measure is important for applications with multiple signals that are close in frequency.

### 4.1.3 Frequency Straddle Loss

Frequency straddle loss is the reduced output of a DFT filter caused by the input signal not being at the filter's center frequency. Frequencies seldom fall at the center of any of the filter passbands. When a frequency is halfway between two filters, the response of the FFT has its lowest amplitude. For a rectangular weighting function the frequency response halfway between two filters is 4 dB lower than if the frequency were in the center of a filter. Each of the other weighting functions in this chapter has less frequency straddle loss than the rectangular one. This performance measure is important in applications where maximum filter response is needed to detect the frequencies of interest.

### 4.1.4 Coherent Integration Gain

Coherent integration gain is the ratio of amplitude of the DFT filter output to the amplitude of the input frequency. $N$-point FFTs have a coherent gain of $N$ for frequencies at the centers of the filter passbands. Since most weighting function coefficients are less than 1, the coherent gain of a weighted FFT is less than $N$. While weighting functions reduce the coherent integration gain, the combination of this reduction and the improved straddle loss results in an overall signal response improvement halfway between two filters. Like frequency straddle loss, this performance measure is important in applications where maximum filter response is needed to detect the frequencies of interest.

### 4.1.5 Equivalent Noise Bandwidth

Equivalent noise bandwidth is the ratio of the input noise power to the noise power in the output of an FFT filter times the input data sampling rate. Every signal contains some noise. That noise is generally spread over the frequency spectrum of interest, and each narrowband filter passes a certain amount of that noise through its main lobe and sidelobes. White noise is used as the input signal and the noise power out of each filter is compared to the noise power into the filter to determine the equivalent noise bandwidth of each passband filter. In other words, equivalent noise bandwidth represents how much noise would come through the filter if it had an absolutely flat passband gain and no sidelobes.

### 4.1.6 Three-dB Main-Lobe Bandwidth

The standard definition of a filter's bandwidth is the frequency range over which sine waves can pass through the filter without being attenuated more than a factor of 2 (3 dB) relative to the gain of the filter at its center frequency. The narrower the main lobe, the smaller the range of frequencies that can contribute to the output of any FFT filter. This means that the accuracy of the FFT filter, in defining the frequencies in a waveform, is improved by having a narrower main lobe.

## 4.2  WEIGHTING FUNCTION EQUATIONS AND THEIR FFTS

This section gives the equations for 15 weighting functions and shows the plots of the frequency responses of their corresponding FFT narrowband filters. It also gives the best use of each weighting function. More details can be found in References 1 and 2.

### 4.2.1 Rectangular

$$\text{For } n = 0 \text{ to } N - 1, \, w(n) = 1$$

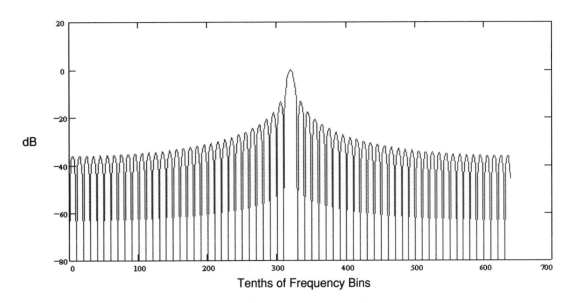

**Figure 4-1**    FFT of rectangular weighting function.

The rectangular weighting function is just the plain FFT without modifying the input data samples. The peak of the highest sidelobe is only 13 dB (a factor of roughly 5) below the main-lobe response, and the sidelobe peaks do not drop off rapidly. This makes it poor for signals with multiple frequency components that have amplitudes that are more than 6 dB different from each other.

In contrast to the poor sidelobe performance, the main lobe is narrower and the coherent gain higher than for any of the other weighting functions. This gives these FFT filters the highest amplitude response to a frequency in the main lobe (coherent gain) and the smallest output noise power (3-dB noise bandwidth). The narrow main lobe also causes these FFT filters to have the poorest response when the frequency is halfway between two adjacent filters (straddle loss). For these reasons, the rectangular weighting function is used when maximum signal-to-noise ratios are critical.

### 4.2.2 Triangular

For $n = 0$ to $N/2$, $w(n) = 2 * n/N$

For $n = N/2 + 1$ to $N - 1$, $w(n) = 2 * (N - n)/N$

The triangular weighting function is used to provide sidelobes and straddle loss lower than the rectangular weighting function and can be easily constructed as a sequence of two straight-line segments. Notice that the sidelobes start off lower than the rectangular weighting function by 14 dB and fall off faster than the rectangular weighting function. The outstanding characteristic of this weighting function is the smaller number of sidelobes

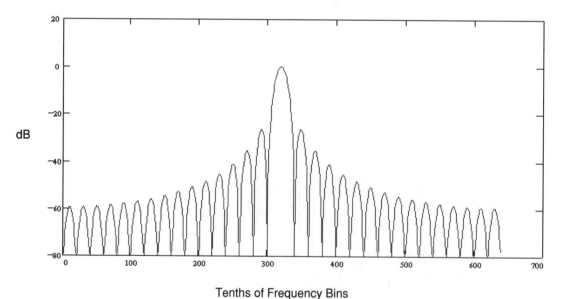

Tenths of Frequency Bins

**Figure 4-2** FFT of triangle weighting function.

than the others in this chapter. It is best used when additional sidelobe reduction, more than the rectangular weighting function, is required and when the weighting function must be computed by the processor because there is no room in its memory to store the values of the weighting function.

### 4.2.3 Sine Lobe

$$\text{For } n = 0 \text{ to } N - 1, w(n) = \sin(\pi n/N)$$

The sine-lobe weighting function can be stored in processor memory or determined by the processor using any of several algorithms for computing the sine function. It is popular because it provides improved sidelobe performance, more than the rectangular weighting function, while using multiplier constants already required for the complex multiplications between power-of-two FFT building blocks. The peak sidelobe level and fall-off rate are roughly the same as the triangular weighting function. Like that one, the sine lobe is most useful when some additional sidelobe reduction, more than the rectangular weighting function, is required and the weighting function must be computed because there is not room in the processor's memory to store the values. For power-of-two FFTs, this weighting function has a computational advantage over the triangular weighting function, because the coefficients are the same ones used to compute the FFT. Therefore, they do not require additional memory locations or computations.

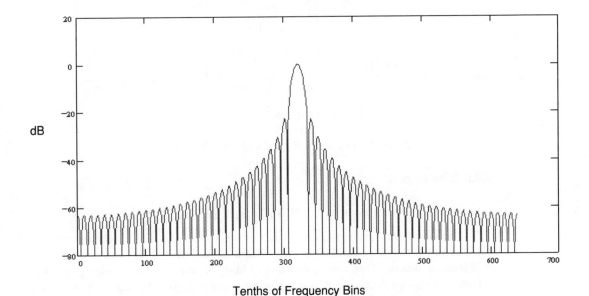

Tenths of Frequency Bins

**Figure 4-3**    FFT of sine-lobe weighting function.

### 4.2.4 Hanning

$$\text{For } n = 0 \text{ to } N - 1, \ w(n) = 0.5 * [1 - \cos(n\pi/N)]$$

The Hanning weighting function is slightly more complicated to compute than the sine lobe. However, it provides 9 dB of additional sidelobe attenuation and can be computed with constants that are already in memory for the complex multiplications between power-of-two FFT building blocks. The peaks of its sidelobes fall off 50% faster than the triangular and sine lobe weighting functions. This weighting function has better 3-dB bandwidth and equivalent noise bandwidth than 16 of the 22 weighting functions in this chapter. These features make it most useful when better than 32-dB sidelobe attenuation is needed, along with 3-dB bandwidth that is less than 1.5 filter widths.

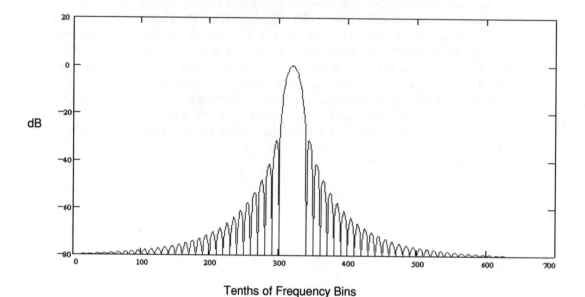

Tenths of Frequency Bins

**Figure 4-4** FFT of Hanning weighting function.

### 4.2.5 Sine Cubed

$$\text{For } n = 0 \text{ to } N - 1, \ w(n) = \sin^3(n\pi/N)$$

The sine-cubed function is a natural extension to the sine-lobe weighting function, but with values that are not used for the complex multiplications between power-of-two FFT building blocks. Therefore, if constant memory is available, the weighting function constants are stored there. If not, two multiplies are needed to cube values $(\sin(n\pi/N) * \sin(n\pi/N) * \sin(n\pi/N))$ from the FFT multiplier constants. Notice that the peak sidelobe is 39 dB below the main lobe, and the peaks of the other sidelobes drop off twice as fast as the triangular and sine-lobe weighting functions. This weighting function is most useful when better than 39 dB of sidelobe attenuation is needed, and the weighting function must be utilized without adding to memory allocated for constants but can afford adding to the computational load for the arithmetic processor.

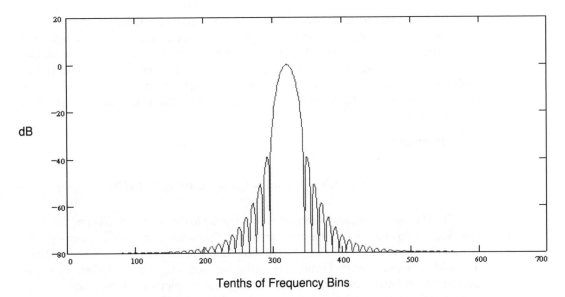

**Figure 4-5**    FFT of sine-cubed weighting function.

## 4.2.6  Sine to the Fourth

For $n = 0$ to $N - 1$, $w(n) = \sin^4(n\pi/N)$

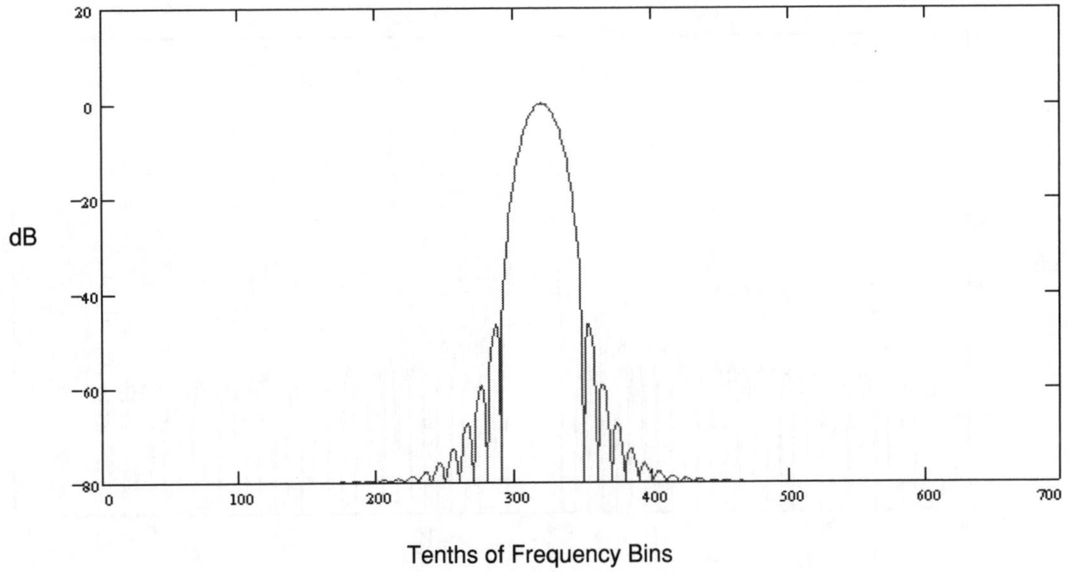

**Figure 4-6**    FFT of sine to the fourth weighting function.

The sine to the fourth, like the sine-cubed weighting function, is one whose values are not used as multiplier constants between power-of-two FFT building blocks. Therefore, if constant memory is available, the weighting function constants are stored there. If not, two multiplies are needed to square values from the multiplier constant values and then square those results $(\sin^2(n\pi/N) * \sin^2(n\pi/N))$. Notice that the peak sidelobe is 47 dB below the main lobe, and the peaks of the other sidelobes drop off 2.5 times as fast as the triangular and sine-lobe weighting functions. This weighting function is most useful when better than 47 dB of sidelobe attenuation is needed, and the weighting function must be utilized without adding to memory allocated for constants but can afford adding to the computational load for the arithmetic processor.

### 4.2.7 Hamming

$$\text{For } n = 0 \text{ to } N - 1, \ w(n) = 0.54 - 0.46 * \cos(2\pi n/N)$$

The Hamming weighting function is very similar to the Hanning weighting function. It provides 11 dB of more sidelobe attenuation than the Hanning and can be computed with constants that are already in memory for the complex multiplications between power-of-two FFT building blocks. Like the Hanning weighting function, it has better 3-dB bandwidth and equivalent noise bandwidth than 17 of the 22 weighting functions in this chapter. However, the peaks of the other sidelobes do not fall off as fast as the Hanning weighting function. These features make this weighting function most useful when better than 43-dB sidelobe attenuation is needed along with 3-dB bandwidth that is less than 1.5 filter widths.

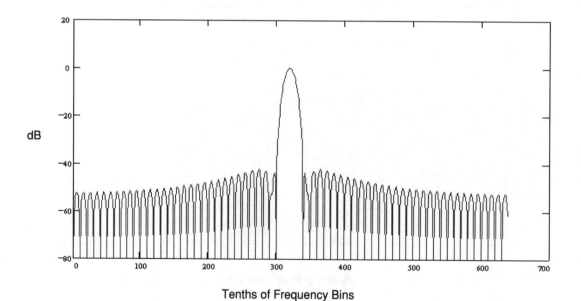

Tenths of Frequency Bins

**Figure 4-7** FFT of Hamming weighting function.

### 4.2.8 Blackman

For $n = 0$ to $N - 1$,

$$w(n) = 0.42 - 0.50 * \cos(2\pi n/N) + 0.08 * \cos(4\pi n/N)$$

The Blackman weighting function is an extension of the Hamming and Hanning approaches of using multiplier constants that are already in memory for complex multiplications between FFT stages. This weighting function also provides the best fall-off ratio of any of the weighting functions with peak sidelobes below $-50$ dB. If the FFT multiplier constants are used, two multiplies and two adds are required to compute each value to be multiplied times the complex FFT input data. This increases the weighting function computational load from two to six arithmetic operations per complex input data point, if it is computed rather than stored in memory. This weighting function is most useful when over 50 dB of sidelobe attenuation is needed close to the main lobe and rapid sidelobe fall-off is required to attenuate frequency components, with large amplitudes, that are separated from each other by more than three to four FFT filters.

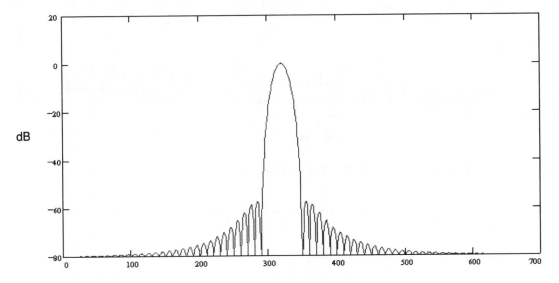

Tenths of Frequency Bins

**Figure 4-8**    FFT of Blackman weighting function.

### 4.2.9 Three-Sample Blackman-Harris

(a) For $n = 0$ to $N - 1$,

$$w(n) = 0.44959 - 0.49364 * \cos(2\pi n/N) + 0.05677 * \cos(4\pi n/N)$$

(b) For $n = 0$ to $N - 1$,

$$w(n) = 0.42323 - 0.49755 * \cos(2\pi n/N) + 0.07922 * \cos(4\pi n/N)$$

The three-sample Blackman-Harris weighting functions can also be computed by using constants that are already in memory for complex multiplications between FFT stages. The computation requires two multiplies and two adds for each input data sample.

Figures 4-9 and 4-10 show two of these weighting functions. Both provide over 60 dB of peak sidelobe attenuation. Note one peculiarity of both: there is a dip in the peaks of the sidelobes near the main lobe and then the sidelobes drop off monotonically. The difference between (a) and (b) is that (b) provides additional sidelobe attenuation but requires a wider 3-dB main-lobe bandwidth. These weighting functions are most useful when over 60 dB of attenuation is required and the width of the main lobe (frequency accuracy) is not critical.

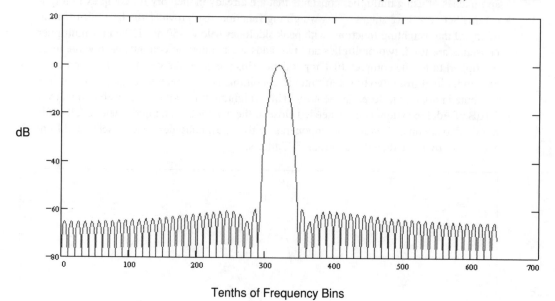

Tenths of Frequency Bins

**Figure 4-9** FFT of three-sample Blackman-Harris (a) weighting function.

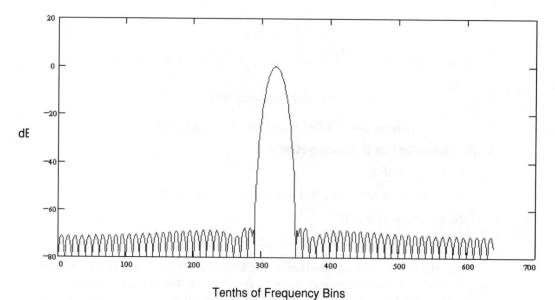

Tenths of Frequency Bins

**Figure 4-10** FFT of three-sample Blackman-Harris (b) weighting function.

### 4.2.10  Four-Sample Blackman-Harris

(a) For $n = 0$ to $N - 1$,

$$w(n) = 0.40217 - 0.49703*\cos(2\pi n/N) + 0.09892*\cos(4\pi n/N) - 0.00188*\cos(6\pi n/N)$$

(b) For $n = 0$ to $N - 1$,

$$w(n) = 0.35875 - 0.48829*\cos(2\pi n/N) + 0.14128*\cos(4\pi n/N) - 0.01168*\cos(6\pi n/N)$$

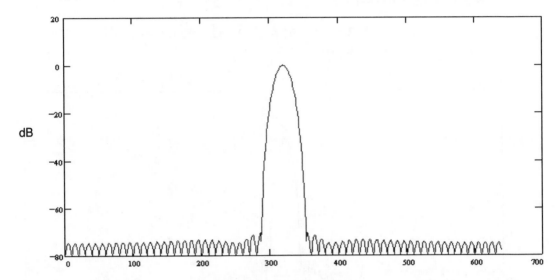

Tenths of Frequency Bins

**Figure 4-11**    FFT of four-sample Blackman-Harris (a) weighting function.

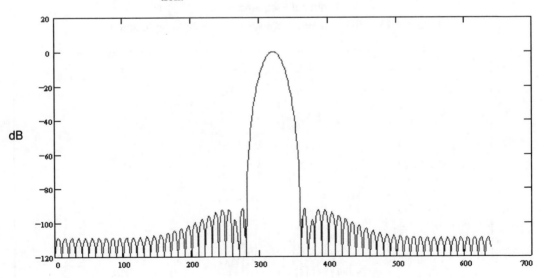

Tenths of Frequency Bins

**Figure 4-12**    FFT of four-sample Blackman-Harris (b) weighting function.

Four-sample Blackman-Harris weighting functions can be computed by using constants already in memory for complex multiplications between FFT stages. Figures 4-11 and 4-12 show these weighting functions. One peculiarity of both is a dip in the peaks of the sidelobes near the main lobe. These weighting functions are most useful when over 70 dB of attenuation is required and the width of the main lobe (frequency accuracy) is not critical.

### 4.2.11 Kaiser-Bessel

For $/n/ = 0$ to $N/2$, $w(n) = I_0[\pi\alpha\sqrt{1.0 - (2n/N)^2}]/I_0[\pi\alpha]$

where $I_0(x) = \sum\limits_{k=0}^{\infty}[(x/2)^k/k!]^2)$

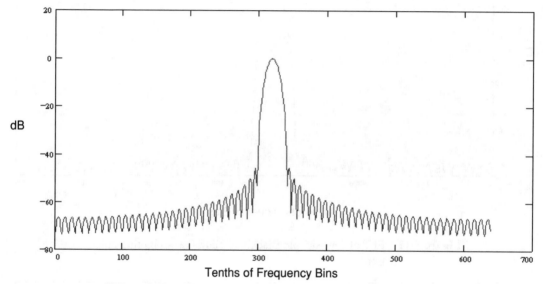

**Figure 4-13**    FFT of $\alpha = 2.0$ Kaiser-Bessel weighting function.

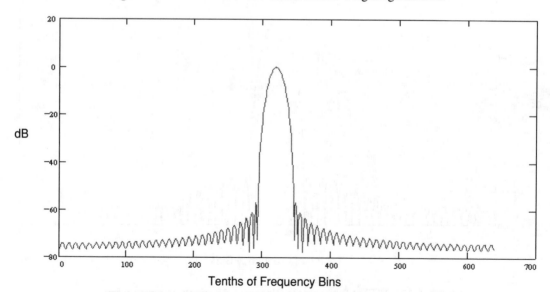

**Figure 4-14**    FFT of $\alpha = 2.5$ Kaiser-Bessel weighting function.

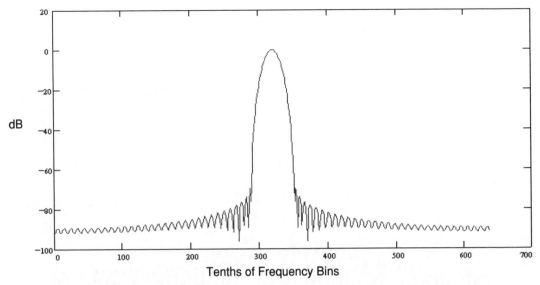

**Figure 4-15**    FFT of $\alpha = 3.0$ Kaiser-Bessel weighting function.

The Kaiser-Bessel weighting function is the ratio of two zero-order Bessel functions of the first kind $(I_0(x))$. Even though the summation that defines these Bessel functions has an infinite number of terms, the functions have finite values [3]. In particular, these Bessel functions have a value of 1 when $x = 0$ and they increase as $x$ gets larger. Figures 4-13 to 4-16 show Kaiser-Bessel weighting functions for different values of $\alpha$. These weighting functions have the most energy in the main lobe for a given peak sidelobe level. The peaks of the sidelobes only fall-off at 6 dB per octave. Therefore, this set of weighting functions is most useful when the filters are being used to distinguish multiple frequencies that have amplitudes that must be attenuated by the filter sidelobes by 46 to 82 dB, depending on which $\alpha$ is chosen.

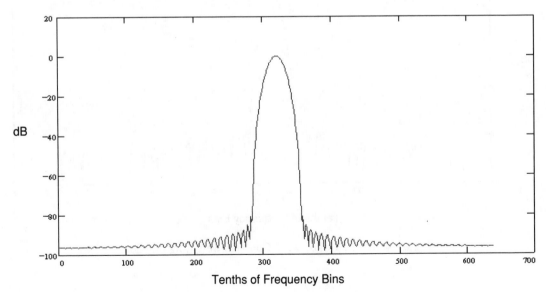

**Figure 4-16**    FFT of $\alpha = 3.5$ Kaiser-Bessel weighting function.

### 4.2.12 Gaussian

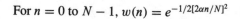

For $n = 0$ to $N - 1$, $w(n) = e^{-1/2[2\alpha n/N]^2}$

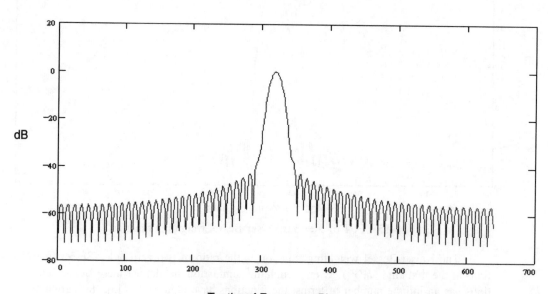

**Figure 4-17** FFT of $\alpha = 2.5$ Gaussian weighting function.

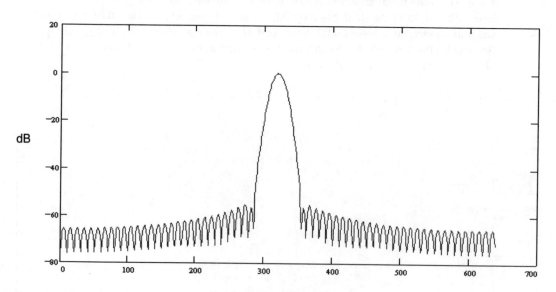

**Figure 4-18** FFT of $\alpha = 3.0$ Gaussian weighting function.

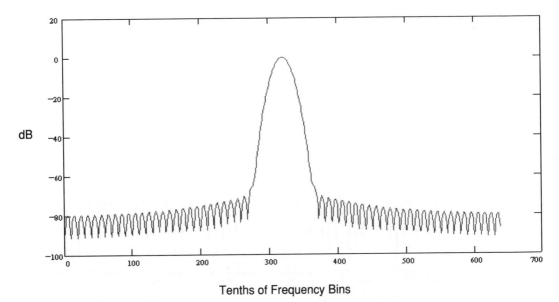

Tenths of Frequency Bins

**Figure 4-19**   FFT of $\alpha = 3.5$ Gaussian weighting function.

The next three weighting functions are derived by optimizing the weighting function for the minimum time-bandwidth product for a given sidelobe level. The narrower a signal in the time domain, the wider it appears in the frequency domain. Likewise, signals that are represented with a narrow set of frequency components do not vary rapidly in the time domain. For a given narrow signal (i.e., a sine wave that lasts less than the number of samples in the FFT) in the time domain, the Gaussian windows provide the tightest concentration of energy in the frequency domain. This means that the Gaussian weighting function is most useful in converting transient signals to the frequency domain. Figures 4-17 to 4-19 show Gaussian weighting functions for different values of $\alpha$.

### 4.2.13 Dolph-Chebyshev

For $k = 0$ to $N - 1$,

$$W(k) = (-1)^k \cos\{N \cos^{-1}[\beta \cos(k\pi/N)]\} / \cosh[N \cosh^{-1}(\beta)]$$

$$\text{where } \beta = \cosh[(1/N) \cosh^{-1}(10^{\alpha})]$$

This equation is the FFT of the Dolph-Chebyshev polynomials that form this weighting function. Figures 4-20 to 4-23 show Dolph-Chebyshev weighting functions for different values of $\alpha$. These weighting functions are a result of minimizing the main-lobe width for a given sidelobe level. The sidelobes do not decrease as they get further away from the main lobe. This makes these weighting functions most useful when multiple frequencies are present and the sidelobes must attenuate each equally while minimizing the chance for one frequency contributing to the output of more than one FFT filter.

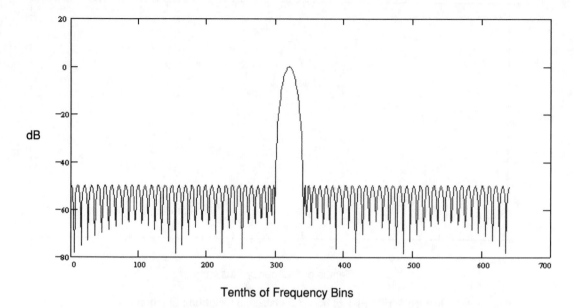

Tenths of Frequency Bins

**Figure 4-20** FFT of $\alpha = 2.5$ Dolph-Chebyshev weighting function.

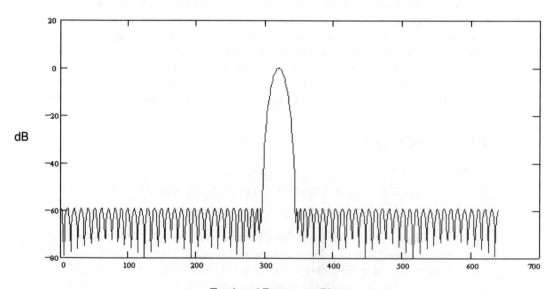

Tenths of Frequency Bins

**Figure 4-21** FFT of $\alpha = 3.0$ Dolph-Chebyshev weighting function.

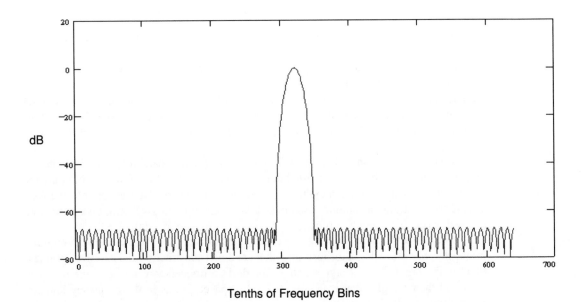

**Figure 4-22**    FFT of $\alpha = 3.5$ Dolph-Chebyshev weighting function.

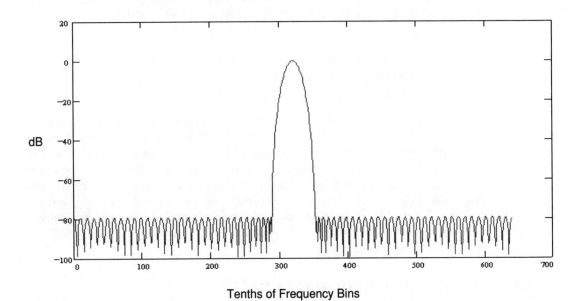

**Figure 4-23**    FFT of $\alpha = 4.0$ Dolph-Chebyshev weighting function.

### 4.2.14 Finite Impulse Response Filter Design Techniques

Linear finite impulse response (FIR) filter-based weighting functions are not popular for two reasons. First, the memory required to store all of the coefficients has only recently become inexpensive as part of DSP chips (Chapter 14). Second, for a given set of frequency response characteristics, the optimal FIR filter is rarely $2^N$ points long. This means that commercially available filter design software is not well suited for computing weighting functions for standard power-of-two FFTs. The numerous nonpower-of-two FFT algorithms in this book remove this barrier and make it practical to use off-the-shelf filter design software.

Chapter 2 established that the DFT is an array of narrowband filters implemented as multipliers followed by low-pass FIR filters. The coefficients of the low-pass filter are the same as the weighting function multiplier coefficients used in the DFT to control sidelobes. Therefore, the design techniques used to develop optimal low-pass FIR filters can be used to develop optimal weighting functions for specific applications of the DFT.

The two approaches to designing coefficients of an FIR filter, based on the required frequency characteristics of the resulting filter are direct construction and iterative optimization. In the first [4], the designer defines the desired frequency response of the low-pass filter and then samples that response at equally spaced points in the frequency domain. By applying the IFFT to the sequence of frequency samples, one computes the unit pulse response of the equivalent filter. If the computed unit pulse response decays rapidly, it can be accurately represented by an FIR filter by truncating the unit pulse response and using the nonzero terms as the weighting function sequence.

In the second approach the designer also starts by defining the desired frequency response of the equivalent low-pass filter. The definition consists of the

- Passband width
- Width of the transition between the passband and sidelobes
- Stopband maximum sidelobe level
- Ripple in the filter's gain across the passband

Algorithms have been developed to construct an FIR filter with a frequency response with the least-mean-squared error relative to these desired frequency response requirements. The problem with this optimization approach is that it produces filters with gain that peaks up at the edges of the filter passbands. This is called the Gibbs effect. The Gibbs effect is reduced by designing the filter with an optimization criterion that minimizes the maximum, rather than mean-squared, error. Chebychev polynomial-based filter design uses this approach. Filters that exhibit this property also have equiripple behavior in the sidelobes. The most popular of these optimization algorithms was published by Parks and McClellan and has been named for them [5].

## 4.3 WEIGHTING FUNCTION COMPARISON MATRIX

Coherent integration gain is normalized to the gain of the rectangular weighting function. Equivalent noise and 3-dB bandwidths are expressed as number of frequency bins. Table 4-1 compares characteristics of various weighting functions.

**Table 4-1**    Weighting Function Comparison Matrix

| Weighting function | Highest sidelobe level (dB) | Sidelobe fall-off ratio | Frequency straddle loss (dB) | Coherent integration gain | Equivalent noise bandwidth | 3-dB bandwidth |
|---|---|---|---|---|---|---|
| Rectangular | −13 | −6 | 3.92 | 1.00 | 1.00 | 0.89 |
| Triangle | −27 | −12 | 1.82 | 0.50 | 1.33 | 1.28 |
| Sine lobe | −23 | −12 | 2.10 | 0.64 | 1.23 | 1.20 |
| Hanning | −32 | −18 | 1.42 | 0.50 | 1.50 | 1.44 |
| Sine cubed | −39 | −24 | 1.08 | 0.42 | 1.73 | 1.66 |
| Sine to the fourth | −47 | −30 | 0.86 | 0.38 | 1.94 | 1.86 |
| Hamming | −43 | −6 | 1.78 | 0.54 | 1.36 | 1.30 |
| Blackman | −58 | −18 | 1.10 | 0.42 | 1.73 | 1.68 |
| Three-sample Blackman-Harris (a) | −61 | −6 | 1.27 | 0.45 | 1.61 | 1.56 |
| Three-sample Blackman-Harris (b) | −67 | −6 | 1.13 | 0.42 | 1.71 | 1.66 |
| Four-sample Blackman-Harris (a) | −74 | −6 | 1.03 | 0.40 | 1.79 | 1.74 |
| Four-sample Blackman-Harris (b) | −92 | −6 | 0.83 | 0.36 | 2.00 | 1.90 |
| Kaiser-Bessel (a) $\alpha = 2.0$ | −46 | −6 | 1.46 | 0.49 | 1.50 | 1.43 |
| (b) $\alpha = 2.5$ | −57 | −6 | 1.20 | 0.44 | 1.65 | 1.57 |
| (c) $\alpha = 3.0$ | −69 | −6 | 1.02 | 0.40 | 1.80 | 1.71 |
| (d) $\alpha = 3.5$ | −82 | −6 | 0.89 | 0.37 | 1.93 | 1.83 |
| Gaussian (a) $\alpha = 2.5$ | −42 | −6 | 1.69 | 0.51 | 1.39 | 1.33 |
| (b) $\alpha = 3.0$ | −55 | −6 | 1.25 | 0.43 | 1.64 | 1.55 |
| (c) $\alpha = 3.5$ | −69 | −6 | 0.94 | 0.37 | 1.90 | 1.79 |
| Dolph-Cheb. (a) $\alpha = 2.5$ | −50 | 0 | 1.70 | 0.53 | 1.39 | 1.33 |
| (b) $\alpha = 3.0$ | −60 | 0 | 1.44 | 0.48 | 1.51 | 1.44 |
| (c) $\alpha = 3.5$ | −70 | 0 | 1.55 | 0.45 | 1.62 | 1.55 |
| (d) $\alpha = 4.0$ | −80 | 0 | 1.65 | 0.42 | 1.73 | 1.65 |

## 4.4 CONCLUSIONS

Because of the third and fourth weaknesses of the DFT, weight functions are applied before data is processed with FFTs to lower high sidelobes and reduce frequency straddle loss. The trade-off for those improvements to the DFT is the introduction of coherent gain reduction and increasing the 3-dB bandwidth of each FFT filter. Fortunately, a wide selection of weighting functions allows users to choose one that offers the balance between benefits and drawbacks needed in a specific application. Chapters 2–4 cover fundamentals of FFTs. The next three chapters address what can be done well with them.

# REFERENCES

[1] F. J. Harris, "On the Use of Windows for Harmonic Analysis with the Discrete Fourier Transform," *Proceedings of the IEEE*, Vol. 66, No. 1 (1978).

[2] A. H. Nuttal, "Some Windows with Very Good Sidelobe Behavior," *IEEE Transactions on Acoustics, Speech, and Signal Processing*, Vol. ASSP-29, No. 1 (1981).

[3] A. N. Lowan, *Table of Bessel Functions for Complex Arguments*, Columbia University Press, New York, pp. 362–381, 1943.

[4] T. W. Parks and C. S. Burrus, *Digital Filter Design*, Wiley, New York, 1987.

[5] T. W. Parks and J. H. McClellan, "Chebyshev Approximation for Nonrecursive Digital Filters with Linear Phase," *IEEE Transactions on Circuit Theory*, Vol. CT-20, pp. 697–701 (1973).

# 5

# Frequency Analysis

## 5.0 INTRODUCTION

Frequency analysis is the process of determining the amplitude and phase of the frequencies that comprise a real or complex sequence of data samples in one and more dimensions. Based on the Nyquist (also called Shannon) sampling theorem (Chapter 2), those frequencies span from zero to half the sampling rate for real signals and from zero to the sampling rate for complex signals. The span of frequencies detected by an FFT is called the frequency spectrum of the data samples. If the output of the FFT is used to catalogue the frequencies in a signal, it is performing the first of the common uses of the DFTs listed in Section 2.1. If the output is used as a shorthand way of describing the signal, because of its small number of frequencies, the FFT is performing the second common use of the DFT. This chapter presents the steps required for one-dimensional frequency analysis. Chapter 7 presents the additional steps required for multidimensional frequency analysis.

## 5.1 FIVE PERFORMANCE MEASURES

Frequency analysis can be done with overlapped or nonoverlapped data sets. In either case the computations can be performed with or without a weighting function. For each of the four possible cases, five measures can be used to describe the performance of the FFT algorithm.

### 5.1.1 Input Sample Overlap

When frequency analysis is performed on data sequences larger than the chosen transform length, the sequence gets divided into smaller segments and transforms are computed on each segment. If the FFT is being used to detect the presence of a frequency that is not always present, the FFT length is chosen to match the expected duration of the frequency of interest. If the frequency of interest is present and aligned with a segment of data samples, the maximum improvement in signal-to-noise ratio is provided by the FFT because the frequency is amplified by a factor of the transform length and the noise by the square root of the transform length. The maximum signal-to-noise ratio provides the highest probability of signal detection.

If the frequency appears in two segments, the signal-to-noise improvement is not as great in either of the two segments, hence a lower probability of detection. The worst case is when the frequency appears half the time in each of the two segments. Segments are overlapped to increase the probability of detecting a frequency of interest. For example, if the segments are overlapped 50%, the frequency of interest lines up with the straddling segment when it is half in each of the two contiguous segments. When segments are overlapped, some of the data points in the sequence are the input to more than one transform. In the example, if the data segments overlap 50%, each data sample is used twice, except for the first and last segments. The larger the overlap, the larger the number of computations; the more complex data addressing; and the larger the data memory required.

### 5.1.2 Sidelobe Level

Sidelobe level is the ratio of the amplitude response of a filter to a frequency in one of its sidelobes to the response it would have if the frequency were in the center of the filter. A filter has a sidelobe level for every frequency outside its main lobe. It is important to ensure that the sidelobe response is attenuated far enough by the filter sidelobes that the filter only gets a significant output when a frequency in its passband is present. These requirements change radically from application to application.

### 5.1.3 Frequency Straddle Loss

Frequency straddle loss is the reduced output of a filter caused by the input signal not being at the filter's center frequency but still in its main lobe. Frequencies to be detected in an application seldom fall at the center of any of the filter passbands. When a frequency is halfway between two filters, the response of the FFT has its lowest amplitude. For a rectangular weighting function the frequency response halfway between two filters is 4 dB lower than if the frequency were in the center of a filter. Each of the other weighting functions in this chapter has less frequency straddle loss than the rectangular one. This performance measure is important in applications where maximum filter response is needed to detect the frequency of interest.

### 5.1.4 Frequency Resolution

Frequency resolution is the measure of how close two frequencies can be before they can no longer be distinguished by the FFT. Frequencies closer than the separation between filter center frequencies are generally considered unresolvable. Weighting functions do not change the separation between the centers of the FFT filters.

### 5.1.5 Coherent Integration Gain

Coherent integration gain is the ratio of amplitude of the filter output to the amplitude of the input frequency. $N$-point FFTs have a coherent gain of $N$ for frequencies at the centers of the filter passbands. Since most of weighting function coefficients are less than 1, the coherent gain of a weighted FFT is less than $N$. Like frequency straddle loss, this performance measure is important in applications where maximum filter response is needed to detect the frequency of interest.

## 5.2 COMPUTATIONAL TECHNIQUES

There are four basic ways that the $N$-point DFT, in any of its fast implementation forms (FFTs), is used. The first two are associated with the spacing between the starting samples in the computation of $N$-point FFTs on data sequences that are longer than $N$ samples. The third and fourth are modifications that can be made to the input data prior to using either of the first two techniques. Each of these is described in this section.

### 5.2.1 Nonoverlapped

Nonoverlapped frequency analysis is generally performed for two types of input sequences. The first is where there are only $N$ points in the data sequence to be analyzed by the $N$-point FFT. In this case one $N$-point FFT is performed. The second case is the analysis of a data sequence that is longer than $N$ points where the frequency components are assumed not to change or to change very slowly over the entire data sequence. In this case, the starting sample for the $N$-point FFT computations can be separated by $N$ or more samples without losing frequency content information.

A practical reason for nonoverlapped processing is the inability of the processor to compute the FFT of the most recent set of $N$ points before the next set of $N$ points is collected. If the data cannot be recorded and processed later, then one common approach is to ignore some number of samples while the present set of $N$ points is being processed. Figure 5-1 is an example of the nonoverlapped method where there are M samples ignored between $N$-point transforms.

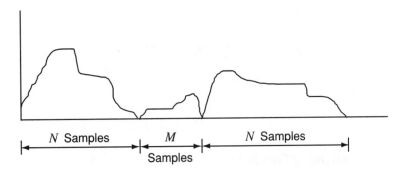

**Figure 5-1**   Nonoverlapped frequency analysis.

### 5.2.2 Overlapped

Chapter 2 discusses the weakness of using the DFT to analyze transient signals. However, there are applications where the frequency content of the data sequence is known to be constant, but only for a specific number of samples. If the goal of the application is to detect when this signal is present in a long data sequence, then the best DFT approach is to use an FFT that matches the expected number of signal samples at the frequency of interest.

However, choosing the correct transform length is not sufficient. If the $N$-point FFT does not start when the transient sequence starts, then two effects occur. First, the coherent gain will not be $N$ because some of the samples integrated by the FFT are noise not signal. Second, the transient that is caused when the signal appears will distort the FFT's ability to recognize the signal. When the $N$-point FFT matches up with the signal, all $N$ samples are integrated and the FFT does not see the transient of the signal turning on and off and therefore performs the analysis without artifacts. An example is a Doppler radar where the antenna beam is scanning at a constant rate to find a target. Since the antenna beamwidth is fixed, the radar receives returns from the target for a fixed period of time as the beam passes by. Until the target is detected, there is no way to know when this time period starts. The theoretically best, but computationally most costly, solution is the start a new $N$-point FFT every time a new sample arrives.

If the FFT is not overlapped, the worst-case situation is to have half of the returns in one set of samples and half in the other. The loss of coherent gain associated with this case is reduced by starting a new $N$-point FFT every $N/2$ samples. Figure 5-2 illustrates this process with an overlap of $P$ samples. With a 2:1 overlap each input data point is used in two FFT computations. This increases the required computational load by a factor of 2. For an overlap of $P$ out of $N$ samples, the increase in computational load is $N/(N - P)$.

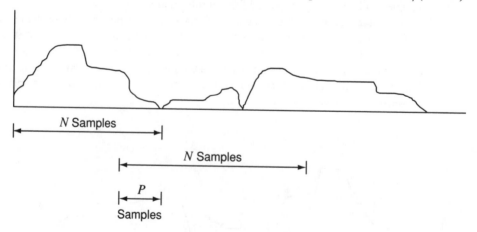

$N$ Samples

$N$ Samples

$P$
Samples

**Figure 5-2**    Overlapped frequency analysis.

### 5.2.3 Weighting Functions

Weighting functions were presented in Chapter 4. The primary value of weighting functions in frequency analysis is to reduce the effects of sidelobes and frequency straddle losses described in Section 5.1. Weighting functions can be used in either the overlapped or nonoverlapped processing approaches. For a slowly varying signal the FFT provides the sidelobe and straddle loss improvements described in Chapter 4.

However, for transient signals the weighting function only improves the performance of the FFT if the FFT is aligned with the signal. In that case the FFT calculates as if the signal is always present and processes it just like slowly varying signals. When input samples to an FFT do not align with the time when the transient signal is present, the transient occurs somewhere in the middle of the set of samples. Then the FFT thinks there is a transient at that point and also one at the end of the data set. The effect of the transient at the end of the data set is minimized by the weighting function, but the effects of the transient in the middle of the data set are virtually unaffected because the transients are not attenuated (see Chapter 4 for more details).

Figure 5-3 shows an example of a transient signal. The first and third sets of $N$ data samples match the transient signals exactly. In the first set there is a transient at the beginning of the data set because the first sample is not zero. For this set of samples a weighting function will reduce the sidelobe effects associated with this transient.

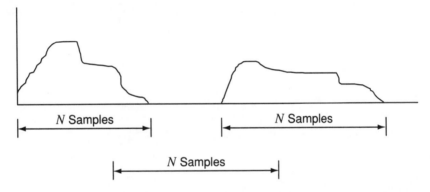

**Figure 5-3**    Effect of weighting functions on frequency analysis of transient signals.

In the third set of samples the first and last samples are zero. Therefore, adding a weighting function to the FFT computations provides no improvement because there are no transient conditions to reduce at the ends of the data set. In fact, the weighting function has a detrimental effect in this case because the coherent gain of the FFT is reduced by the weighting function, and the main lobe of each FFT filter is widened.

The second set of samples has transient effects at both ends of the data set and straddles the two transient signals. Therefore, a weighting function will reduce the transient effects at the ends of the data set. However, the FFT will provide little useful data about either of the transients because it straddles them.

## 5.3 CONCLUSIONS

This chapter covers one of the two functions where FFTs are primarily used. As can be seen in the Doppler radar and speech processing design examples in Chapter 17, frequency analysis and the use of FFTs to create a shorthand version of a signal have wide application in aviation and consumer products. Frequency analysis and the functions explained in the next chapter get used separately or together in almost every place an FFT is used. This chapter contains no algorithms because frequency analysis is performed with the algorithms in Chapters 8 and 9.

# 6

# Linear Filtering
# and Pattern Matching

## 6.0 INTRODUCTION

Linear filtering and pattern matching are techniques for determining the presence of specific waveforms in a signal of one or more dimensions. Generally, linear filtering is used to pass certain bands of frequencies and block others. Pattern matching is the process of finding a pattern in a signal, whether it is a sine-wave frequency or an arbitrary sequence of data samples that do not resemble any easily defined function.

While neither a linear filter nor a pattern matcher is the same as an FFT, FFT algorithms are often able to speed up their computation. The purpose of this chapter is to present algorithms for using an FFT to perform one-dimensional linear filtering and pattern matching. It also shows how to determine when using an FFT requires fewer adds and multiplies than performing those functions in the time domain. The additional steps required to perform multidimensional versions of this processing are in Chapter 7.

## 6.1 EQUATIONS

Linear filtering and pattern matching, also known as convolution and correlation, respectively, are defined by Equations 6-1 and 6-2. For linear filtering applications, $x(k - i)$ is the input sequence to the filter and $h(i)$ is the unit pulse response of the filter. For pattern matching applications, $x(k + i)$ is still the input signal and $h(i)$ is the pattern to be found in the signal. This chapter presents two FFT-based approaches for computing these two equations because there are many instances when the FFT approach is more efficient than computing the equations directly. Both approaches can be implemented with any of the FFT algorithms in Chapters 8 and 9.

$$y(k) = \sum_{i=0}^{M-1} x(k-i) * h(i) \qquad (6\text{-}1)$$

$$y(k) = \sum_{i=0}^{M-1} x(k+i) * h(i) \qquad (6\text{-}2)$$

Figure 6-1 shows the steps needed to implement Equations 6-1 and 6-2 in the frequency domain. Derivations of this approach can be found in several DSP textbooks [1–4].

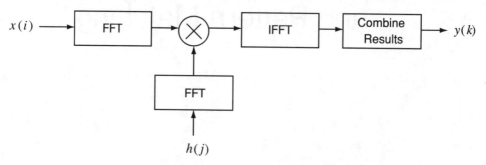

**Figure 6-1**    General frequency domain processing block diagram.

## 6.2 THREE PERFORMANCE MEASURES

These three performance measures provide a way to compare the one direct and two frequency domain methods for computing Equations 6-1 and 6-2 (linear filtering and pattern matching).

### 6.2.1 Number of Computations per Data Point

The direct method for computing Equations 6-1 and 6-2 requires $M$ complex multiplies and $M - 1$ complex adds for each complex input data point. For each frequency domain method the number of computations per input data point is shown in Sections 6.4 and 6.5 to be

$$\# \text{Comp.} = 2 * N_F + 6 * N - 2 * (N - L)$$

### 6.2.2 Number of Data Memory Locations

For the direct method the required data memory is the total of twice the number of stages, $M$, plus two for the next complex input sample and two for the most recent complex output. For the frequency domain methods the memory required is larger than twice the FFT length, $N$, depending on the algorithm. The Comparison Matrices in Chapters 8 and 9 give the amount of data memory locations for every algorithm in the matrix.

### 6.2.3 Computational Latency

Computational latency is the time between the start of computations and when output of results begins. Computational latency is considerably different for frequency domain methods of computing Equations 6-1 and 6-2 than for the direct method. For the direct method a new output is computed for each new input by performing $M$ multiplies and $M - 1$ adds. This is a latency of one input sample. In the frequency domain methods, $M$ new pieces of data are collected and less than $M$ new output values are produced because of the required input data overlapping. Therefore, the latency is at least $M$ data samples.

## 6.3 DIRECT METHOD

To determine when the approach in Figure 6-1 is computationally advantageous, equations must be developed for the number of computations required by Equations 6-1 and 6-2 for the direct and frequency domain approaches. When $x(i)$ is an input sequence that is much larger than $M$, Equations 6-1 and 6-2 require $M$ multiplies and $M - 1$ adds for each new output $y(k)$. Since a new output occurs every time a new input $x(i)$ is processed, $M$ multiplies and $M - 1$ adds are needed to process each new input sample.

### 6.3.1 Complex Input Signal

For a finite input sequence of length $L$, Equation 6-1 does not require $M$ complex multiplies and $M - 1$ adds for each value of $k$. In particular:

- For $k = 0$, the only term in the summation is $i = 0$, so there are one multiply and no adds.
- For $k = 1$, there are two terms to compute and add in Equation 6-1. Namely, a multiply is required for $i = 0$ and 1, and an add is required to combine these two multiplications.
- Each time $k$ increases by 1, the number of adds ($k$ adds) and multiplies ($k + 1$ multiplies) increases by 1 until $k = M - 1$.

This totals $M * (M - 1)/2$ complex adds and $M * (M + 1)/2$ complex multiplies.

For $k = M$ to $L - 1$, Equation 6-1 requires $M$ multiplies and $M - 1$ adds for each value of $k$. This is a total of $(L - M) * M$ multiplies and $(L - M) * (M - 1)$ adds.

From $k = L$ through $k = L + M - 1$, a similar phenomenon occurs for $k = 0, 1, \ldots, M-1$. These terms also require $M*(M-1)/2$ adds and $M*(M+1)/2$ multiplies. Adding all of these computational requirements shows that Equation 6-1 requires $L*(M-1)$ half-complex adds and $(L+1)*M$ half-complex multiplies to compute all $N = L+M-1$ outputs $y(k)$ if the input data is complex and the filter is real. Since each half-complex add requires one real add and each complex multiply requires two real multiplies, this case requires $2 * (L + 1) * M$ real multiplies and $L * (M - 1)$ real adds.

### 6.3.2 Real Input Signal

If the input data is real and the unit pulse response remains real, the basic logic for determining the number of computations remains unchanged. The only difference is that the half-complex adds and multiplies are replaced with real adds and real multiplies. Adding all of these computational requirements, Equation 6-1 requires $L * (M - 1)$ real adds and $(L + 1) * M$ real multiplies to compute all $N = L + M - 1$ outputs $y(k)$ if the input data is complex and the filter is real.

## 6.4 SINGLE-STEP FREQUENCY DOMAIN METHOD

If the input sequence, $x(i)$, and the unit pulse response, $h(j)$, are finite, then one frequency domain solution is to compute an FFT whose length is large enough to encompass the entire response of the linear filter. If the input sequence is length $L$, (i.e., $x(l)$ exists for $l = 0, 1, 2, \ldots, L - 1$) and the unit pulse response is length $M$ (i.e., $h(m)$ exists for $m = 0, 1, 2, \ldots, M - 1$), then Equation 6-1 shows that $y(k)$ will have an output starting at $k = 0$ and ending after $k = L + M - 1$. Therefore, if the FFT in Figure 6-1 is at least $L + M - 1$ points, only one is required to convert all of the needed data to the frequency domain. The number of computations for the FFT is determined from the Comparison Matrices in Chapters 8 and 9.

### 6.4.1 Complex Input Signal

The complex multiplications (four multiplies and two adds per complex multiply) required after each $N$-point FFT total $6 * N$ computations. Finally, the IFFT takes the same number of computations as the FFT. If the number of computations for an $N$-point FFT is $N_F$, the frequency domain approach requires fewer computations for complex input sequences when

$$2 * \{L * (M - 1) + (L + 1) * M\} > 2 * N_F + 6 * N$$

### 6.4.2 Real Input Signal

If the input signal is real, then all of the FFT computations are reduced by using the double-length algorithm from Section 2.4. If $N/2$ is odd, this reduces the input FFT computations to

$$\# \text{Comp.} = N_F + 5 * N - 7$$

Likewise, if $N/2$ is even, Chapter 2 shows the total input FFT computations are:

$$\# \text{Comp.} = N_F + 5 * N - 9$$

Then the outputs of the input FFT are multiplied by complex numbers to provide the filter shaping. Since the FFT input and the unit pulse response are real, the FFT outputs of both are symmetric around the center filter. This means the only complex multiplies to be performed are those below the center filter.

*Case* 1: *Real Input Signal with N/2 an Even Number*

If $N/2$ is even, this is $N/2$ complex multiplies, which is $2 * N$ real multiplies and $N$ real adds. If $N/2$ is odd, the total number of filters to be multiplied is the $(N - 1)/2$ below the center filter and the center filter. This is $(N - 1)/2$ complex multiplies plus one real multiply for the center filter (see the symmetry properties of DFTs in Chapter 2). This is a total of $2 * N - 1$ real multiplies and $N - 1$ real adds.

The output of the complex multiplication step is then fed into an $N$-point IFFT that requires $2 * N_F$ computations. Therefore, the equation to determine when the total computations for $N/2$ even is less in the frequency domain for real input signals is

$$3 * N_F + 8 * N - 9 < L * (M - 1) + (L + 1) * M$$

*Case* 2: *Real Input Signals with N/2 an Odd Integer*

For $N/2$ odd,

$$3 * N_F + 8 * N - 5 < L * (M - 1) + (L + 1) * M$$

## 6.5  MULTIPLE-STEP FREQUENCY DOMAIN METHOD

If the length of the input sequence $L$ is too long to practically compute as a single transform length, a means must be found to segment the input sequence into manageable lengths and perform the functions in Figure 6-1 several times. Once these several sets of operations are performed, the results must be recombined to form the complete output sequence. There are two algorithms for performing the frequency domain method on long sequences of input data. These algorithms are described, and the total number of computations determined and compared with the time domain approach for real and complex input sequences.

## 6.6  OVERLAP-AND-ADD FREQUENCY DOMAIN ALGORITHM

### 6.6.1  Introduction

The overlap-and-add approach to filtering in the frequency domain requires additions to combine the results from consecutive data sequence computations to reconstruct the output sequence $y(k)$ in Equation 6-1. In this approach, perfect finite convolutions as described in Section 6.4 are obtained by choosing $L$ samples of the input sequence and appending $N - L$ zeros so that the $M$ nonzero values of $h(i)$ do not overlap using an $N$-point FFT. Then the $N$-point FFT frequency domain processing provides all valid outputs. The next step is to move over and use the next $L$ samples and append $N - L$ zeros. When the frequency domain processing of this second set of data is complete, all of its results are also correct (Figure 6-2). Since the two input sample sequences add to form the actual input sequence, the linearity property of FFTs guarantees that adding the $N$ overlapped outputs provides the actual $y(k)$ results. If this process is continued, the correct outputs continue to be obtained for $y(k)$.

### 6.6.2  Complex Input Signals

For complex input signals, the specific overlap-and-add algorithm stages are as follows.

### Stage 1: Choose a Transform Length *N*

### Stage 2: Compute *N*-Point FFT of the Unit Pulse Response *h*(*i*) One Time

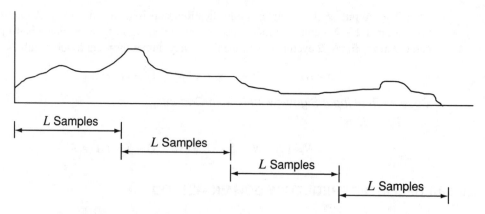

**Figure 6-2** Sample sequence for the overlap-and-add algorithm.

Compute the $N$-point FFT of the $M$ members of the sequence for $h(i)$, after $N - M$ zeros are appended to the end and label the results $H(k)$.

$$H(k) = \sum_{i=0}^{N-1} h(i) * W_N^{ki}$$

This computation only happens once, and the results are stored in memory for use in multiplying all of the transformed data sets as shown in Figure 6-1.

**Stage 3: Set *t* = 0**

**Stage 4: Load and Augment the Next Set of Input Data Points for Processing**

Collect $L$ data points, $x[i + t * L]$, and store in the input data memory along with $N - L$ zeros to occupy the last $N - L$ samples in the sequence of $N$ data points, $x_t(i)$.

$$x_t(i) = x[i + t * L] \qquad \text{for } i = 0, 1, 2, \ldots, (L - 1)$$

$$x_t(i) = 0 \qquad \text{for } i = L, L + 1, \ldots, (N - 1)$$

**Stage 5: Transform the Next Set of Data Points to the Frequency Domain**

Compute the $N$-point FFT of $x_t(i)$, using one of the appropriate algorithms from Chapters 8 and 9.

$$X_t(k) = \sum_{i=0}^{N-1} x_t(i) * W_N^{ki}$$

This stage requires $N_F$ arithmetic computations. However, the first stage in all of the algorithms in Chapters 8 and 9 is the sums and differences of the input samples. Therefore, $2 * (N - L)$ of the input complex adds can be removed from the FFT algorithm because $N - L$ of the input data points are known to be zero. Therefore, the first time these samples need to be added to other samples the addition can be omitted. This reduces the total to $N_F - 4 * (N - L)$ computations.

## Stage 6: Perform Frequency Domain Filtering

For each $k = 0, 1, 2, \ldots, (N-1)$, compute the product $P(k)$. This requires $4 * N$ multiplies and $2 * N$ adds since both numbers are complex.

$$P(k) = H(k) * X_t(k)$$

## Stage 7: Transform the Results Back to the Time Domain

Compute the IFFT of $P(k)$ and divide each result by $N$ to obtain $y_t(n)$ for $n = 0, 1, 2, \ldots, (N-1)$ and store the results in $N$ complex memory locations. Use the appropriate algorithms from Chapters 8 and 9 with the sign of the imaginary multiplier terms reversed as described in Chapter 2.

$$y_t(n) = 1/N * \sum_{k=0}^{N-1} P(k) * W_N^{-kn}$$

This stage requires $N_F$ arithmetic computations because the IFFT takes the same number of computations as the FFT.

## Stage 8: Perform Output Adds

1. If $t = 0$, then for $i = 0, 1, 2, \ldots, (L-1)$, set $y(i) = y_t(i)$.
2. If $t > 0$, then for $i = 0, 1, 2, \ldots, (N-L-1)$, set $y[i+t*L] = y_{t-1}[i+L] + y_t(i)$, and for $i = (N-L), (N-L+1), \ldots, (N-1)$ set $y[i+t*L] = y_t(i)$.

This requires $2 * (N-L)$ adds if the input data sequence is complex.

## Stage 9: Set *t = t + 1* and Repeat Stages 4 through 8

If the computations from Stages 5–8 are added, the total number of arithmetic computations for a complex input signal is:

$$\# \, \text{Comp.} = 2 * N_F + 4 * N + 2 * L$$

Since these computations are performed every time $L$ new data samples are used, the number of computations per complex input data sample is

$$\# \, \text{Comp.} = \{2 * N_F + 4 * N + 2 * L\}/L$$

### 6.6.3 Real Input Signals

If the input signal to the overlap-and-add algorithm is real, then all of the FFT computations are reduced by using the double-length algorithm from Chapter 2. The exact answer depends on whether $N/2$ is odd or even. If $N/2$ is odd, the input FFT computations per data point are

$$\# \, \text{Comp.} = \{N_{F2} + 5 * N - 7\}/L$$

where $N_{F2}$ is the number of computations for the $N/2$-point FFT algorithm chosen from Chapters 8 and 9. If $N/2$ is even, the input FFT computations per data point are

$$\# \, \text{Comp.} = \{N_{F2} + 5 * N - 9\}/L$$

Then the outputs of the input FFT are multiplied by complex numbers to provide the filter shaping. Since the FFT input and the unit pulse response are real, the FFT outputs of both are symmetric around the center filter. This means the only complex multiplies to be performed are those below the center filter. If $N/2$ is even, this is $N/2$ complex multiplies, which is $2 * N$ real multiplies and $N$ real adds. If $N/2$ is odd, the total number of filters to be multiplied is the $(N-1)/2$ below the center filter and the center filter. This is $(N-1)/2$ complex multiplies plus one real multiply for the center filter (see the symmetry properties of DFTs in Chapter 2). This is a total of $2 * N - 1$ real multiplies and $N - 1$ real adds. The output of the complex multiplication stage is then fed into an $N$-point IFFT that requires $N_{F2}$ computations.

The total number of computations per data point is:

$$\# \text{Comp.} = 2 * N_{F2} + 13 * N - 18$$

## 6.7 OVERLAP-AND-SAVE FREQUENCY DOMAIN ALGORITHM

### 6.7.1 Introduction

The overlap-and-save algorithm overlaps the data sequences into the FFT rather than artificially creating the overlap by adding zeros (Figure 6-3). The process starts by taking the first $N$ samples in the sequence $xt(i)$ and computing its FFT. These results are multiplied by the $N$-point FFT of $h(j)$, and the result is transformed back to the time domain by an IFFT. The result is only accurate starting at the first sample in the sequence until the unit pulse response $h(j)$ of $M$ samples no longer completely overlaps the data sequence $x_t(i)$. Therefore, each set of computations generates $(N - M + 1)$ new valid outputs. To cover the last $M - 1$ outputs, the next input sequence overlaps the previous one by $M - 1$ samples. If this process is continued, the correct outputs are always obtained for $y(k)$.

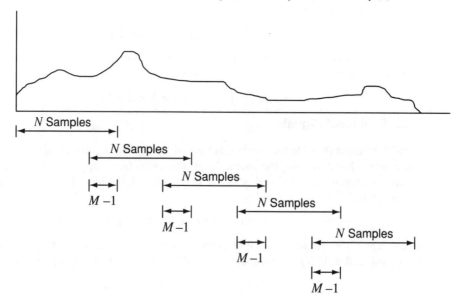

**Figure 6-3** Sample sequence for the overlap-and-save algorithm.

## 6.7.2  Complex Input Signals

For complex input signals, the specific overlap-and-add algorithm stages are:

### Stage 1:  Choose a Transform Length *N*

### Stage 2:  Compute *N*-Point FFT of the Unit Pulse Response *h(i)* One Time

Compute the $N$-point FFT of the $M$ members of the sequence for $h(i)$ after $N - M$ zeros are appended to the end, and label the results $H(k)$.

$$H(k) = \sum_{i=0}^{N-1} h(i) * W_N^{ki}$$

This computation only happens once, and the results are stored in memory for use in multiplying all of the transformed data sets.

### Stage 3:  Set *t* = 0

### Stage 4:  Load and Augment the Next Set of Input Data Points for Processing

Collect $N$ data points, $x[i + t * (N - M + 1)]$, and store in the input data memory, $x_t(i)$. Note that this means this algorithm will use $M - 1$ of every $N$ input data points twice. This makes the input data addressing nonsequential.

$$x_t(i) = x[i + t * (N - M + 1)] \qquad \text{for } i = 0, 1, 2, \ldots, (N - 1)$$

### Stage 5:  Transform the Next Set of Data Points to the Frequency Domain

Compute the $N$-point FFT of $x_t(i)$, using one of the appropriate algorithms from Chapters 8 and 9.

$$X_t(k) = \sum_{i=0}^{N-1} x_t(i) * W_N^{ki}$$

This stage requires $N_F$ arithmetic computations, where $N_F$ is computed based on the algorithm chosen from Chapters 8 and 9.

### Stage 6:  Perform Frequency Domain Filtering

For each $k = 0, 1, 2, \ldots, (N - 1)$, compute the product $P(k)$:

$$P(k) = H(k) * X_t(k)$$

This requires $4 * N$ multiplies and $2 * N$ adds since both numbers are complex.

### Stage 7:  Transform the Results Back to the Time Domain

Compute the IFFT of $P(k)$ to obtain $y_t(n)$ for $n = 0, 1, 2, \ldots, (N - 1)$ and store the results in $N$ complex memory locations.

$$y_t(n) = [1/N] * \sum_{k=0}^{N-1} P(k) * W_N^{-kn}$$

This stage requires $N_F$ arithmetic computations.

### Stage 8: Append the First $N - M + 1$ Outputs to the Output Sequence

Keep the first $N - M + 1$ of these outputs and append them to the previous valid outputs. Namely, for $i = 0, 1, 2, \ldots, (N - M + 1)$:

$$y[i + t * (N - M + 1)] = y_t(i)$$

This means that $M - 1$ of the final adds in the Stage 7 IFFT need not be computed. This is a total of $2 * (M - 1)$ adds. If $N$ is chosen such that $N = L + M - 1$, then $M - 1 = N - L$.

### Stage 9: Set $t = t + 1$ and Repeat Stages 4 through 8

Totaling the arithmetic computations from Stages 5 to 7 and dividing by the $N - M + 1$ new output samples yield the same number of arithmetic computations per complex input data point as the overlap-and-add algorithm:

$$\# \text{ Comp.} = \{2 * N_F + 6 * N - 2 * (N - L)\}/(N - M + 1) = \{2 * N_F + 4 * N + 2 * L\}/(N - M + 1)$$

If $N = L + M - 1$, then $N - M + 1 = L$, and this is the same number of computations required for the overlap-and-add algorithm in Section 6.6.

### 6.7.3 Real Input Signals

If the input signal is real, then all of the FFT computations are reduced by using the double-length algorithm from Chapter 2. The exact answer depends on whether $N/2$ is even or odd. If $N/2$ is odd, this reduces the input FFT computations per data point to

$$\# \text{ Comp.} = \{N_{F2} + 5 * N - 7\}/(N - M + 1)$$

where $N_{F2}$ is the number of computations required for the $N/2$-point algorithm chosen from Chapters 8 and 9. If $N/2$ is even, the input FFT computations per data point are

$$\# \text{ Comp.} = \{N_{F2} + 5 * N - 9\}/(N - M + 1)$$

Then the outputs of the input FFT are multiplied by complex numbers to provide the filter shaping. Since the FFT input and the unit pulse response are real, the FFT outputs of both are symmetric around the center filter. This means the only complex multiplies to be performed are those below the center filter.

If $N/2$ is even, this is $N/2$ complex multiplies, which is $2 * N$ real multiplies and $N$ real adds. If $N/2$ is odd, the total number of filters to be multiplied is the $(N - 1)/2$ below the center filter and the center filter. This is $(N - 1)/2$ complex multiplies plus one real multiply for the center filter (see the symmetry properties of DFTs in Chapter 2). This is a total of $2 * N - 1$ real multiplies and $N - 1$ real adds.

The output of the complex multiplication stage is then fed into an $M$-point IFFT that requires $N_{F2}$ computations. The total number of computations for real input data is

$$\# \text{ Comp.} = 2 * N_{F2} + 13 * N - 18$$

## 6.8 LINEAR FILTERING AND PATTERN MATCHING COMPARISON MATRIX

The Comparison Matrix of Table 6-1 summarizes the key performance measures that can be used to determine the best way to implement Equations 6-1 and 6-2. The important point to note is that the performance measures for both frequency domain methods are the same. Therefore, this matrix is only useful in determining if Equations 6-1 and 6-2 should be implemented directly in the time domain or in the frequency domain.

**Table 6-1**  Linear Filtering and Pattern Matching Comparison Matrix

| Algorithm | # of computations per data point | # of data locations | Comp. latency |
|---|---|---|---|
| **Real Input Data** | | | |
| Direct | $2 * M - 1$ | $M + 2$ | 1 |
| Overlap-and-add ($N/2$ odd) | $(2 * N_{F2} + 13 * N - 16)/L$ | $N$ | $3 * N$ |
| Overlap-and-save ($N/2$ odd) | $(2 * N_{F2} + 13 * N - 16)/(N - M + 1)$ | $N$ | $3 * N$ |
| Overlap-and-add ($N/2$ even) | $(2 * N_{F2} + 13 * N - 18)/L$ | $N$ | $3 * N$ |
| Overlap-and-save ($N/2$ even) | $(2 * N_{F2} + 13 * N - 18)/(N - M + 1)$ | $N$ | $3 * N$ |
| **Complex Input Data** | | | |
| Direct | $4 * M - 2$ | $2 * M + 4$ | 2 |
| Overlap-and-add | $(2 * N_F + 4 * N + 2 * L)/L$ | $2 * N$ | $6 * N$ |
| Overlap-and-save | $(2 * N_F + 4 * N + 2 * L)/(N - M + 1)$ | $2 * N$ | $6 * N$ |

*Key to Variables*

$N$ = FFT length

$M$ = number of stages in direct implementation

$L$ = number of new outputs per set of computational stages

$N_{F2}$ = number of computations in the $N/2$-point FFT chosen from Chapters 8 and 9

$N_F$ = number of computations in the $N$-point FFT chosen from Chapters 8 and 9

## 6.9 CONCLUSIONS

While linear filtering and pattern matching can be done in the time domain, and often are, frequency domain implementation using FFTs often requires fewer adds and multiplies. The algorithms in this chapter, in combination with the FFT algorithms in Chapters 8 and 9, provide all the steps necessary to implement linear filtering and pattern matching in the frequency domain.

The next chapter describes how to perform these functions and those from Chapter 5 in more than one dimension by simply converting the multidimensional processing to a sequence of one-dimensional processes.

## REFERENCES

[1] L. R. Rabiner and B. Gold, *Theory and Application of Digital Signal Processing*, Prentice-Hall, Englewood Cliffs, NJ, 1975.

[2] A. V. Oppenheim and R. W. Schafer, *Digital Signal Processing*, Prentice-Hall, Englewood Cliffs, NJ, 1975.

[3] E. Oran Brigham, *The Fast Fourier Transform*, Prentice-Hall, Englewood Cliffs, NJ, 1974.

[4] E. Oran Brigham, *The Fast Fourier Transform and Its Applications*, Prentice-Hall, Englewood Cliffs, NJ, 1988.

# 7

# Multidimensional Processing

## 7.0 INTRODUCTION

To this point the book has only addressed the use of the DFT and its fast versions (FFTs) to convert one-dimensional signals to their frequency components. Signals such as music, speech, radar, and sonar are waveforms that change as a function of one variable, time. They are usually analyzed with one-dimensional FFTs. However, some signals have more than one dimension or can be turned into waveforms with more than one dimension. The most obvious example is an image, a two-dimensional waveform, which is analyzed with two-dimensional FFTs. Video is described in three-dimensional terms, some number of two-dimensional pictures per second, with time as the third dimension.

The most important fact about multidimensional DFTs is that they can be decomposed into a sequence of one-dimensional DFTs. The results of this fact are twofold:

- Understanding how to choose and implement one-dimensional FFTs is most of the work in implementing an $N$-dimensional FFT.
- Any of the one-dimensional FFTs can be used to compute multidimensional FFTs.

Mathematically, the multidimensional DFT is called a separable function because its implementation can be separated into multiple, one-dimensional DFTs. There are three properties of multidimensional DFT processing:

- Each dimension of a multidimensional DFT has all the properties of a one-dimensional DFT.
- Any of the one-dimensional FFTs can be used to compute multidimensional FFTs.
- Each dimension of a multidimensional DFT can be a transform of any length.

These three separable function properties significantly reduce the number of computations required for multidimensional DFTs. This, combined with FFT algorithms that provide fast computation of one-dimensional DFTs, has led to uses of two- and three-dimensional FFTs for applications such as image formation (synthetic aperture radar and magnetic resonance imaging) and image analysis (deblurring).

## 7.1 FREQUENCY ANALYSIS

This section starts by giving the algorithm for using one-dimensional DFTs to compute two-dimensional DFTs [1, 2, 3]. It then expands the algorithm to more than two dimensions so that any dimension of a DFT can be computed by just using the algorithms in this chapter.

### 7.1.1 Two Dimensions

At first glance, frequency analysis in more than one dimension seems a bit strange because the common definition of frequency is associated with a signal, like electric power, that changes over time. However, if the concept of dimension is expanded to include space, then images certainly change as a function of the $x$ and $y$ positions in the image. The result is the concept of spatial frequency. Therefore, two-dimensional frequency analysis measures the spatial frequency content of an image. The equation for frequency analysis in two dimensions is

$$A(k_1, k_2) = \sum_{n_1=0}^{N_1-1} \sum_{n_2=0}^{N_2-1} a(n_1, n_2) e^{-j2\pi[n_1 k_1/N_1 + n_2 k_2/N_2]} \tag{7-1}$$

The conversion of this equation to a sequence of two one-dimensional DFTs is accomplished by noting that the exponential term can be factored into two terms, each with its own subscripted set of $n$, $k$, and $N$ variables that are independent of each other:

$$e^{-j2\pi[n_1 k_1/N_1 + n_2 k_2/N_2]} = e^{-j2\pi(n_1 k_1/N_1)} * e^{-j2\pi(n_2 k_2/N_2)} \tag{7-2}$$

Once the exponential is factored, it can be separated between the two summation signs to produce

$$\sum_{n_1=0}^{N_1-1} \sum_{n_2=0}^{N_2-1} a(n_1, n_2) * e^{-j2\pi[n_1 k_1/N_1 + n_2 k_2/N_2]}$$

$$= \sum_{n_1=0}^{N_1-1} \left\{ \sum_{n_2=0}^{N_2-1} a(n_1, n_2) * e^{-j2\pi(n_2 k_2/N_2)} \right\} * e^{-j2\pi(n_1 k_1/N_1)} \tag{7-3}$$

The inner summation is the $N_2$-point one-dimensional DFT of $a(n_1, n_2)$. Since $a(n_1, n_2)$ is different for each value of $n_1$, this DFT must be computed for each $n_1 = 0, 1, 2, \ldots,$ $(N_1 - 1)$. Those results become the terms used to compute the second set of one-dimensional DFTs described by the outer summation to the right of the equals sign in Equation 7-3. To summarize, if this two-dimensional image described by $a(n_1, n_2)$ is to be transformed, then:

1. For each row: $n_1 = 0, 1, 2, \ldots, (N_1 - 1)$, compute its $N_2$-point DFT and place the results back in the same row.

2. For each column of the results from 1): $n_2 = 0, 1, 2, \ldots, (N_2 - 1)$, in this interim two-dimensional set of numbers, compute its $N_1$-point DFT and place the results back in the same column.

Each of these $N_1 * N_2$ one-dimensional DFTs can be computed using any of the FFT algorithms in Chapters 8 and 9 to improve the computation time. If the input data is complex, the complex version of the algorithms is most efficient. If the input is real, then the overlap-and-add or overlap-and-save approaches from Chapter 6 can also be applied to the chosen FFT algorithm to further reduce the computational load.

### 7.1.2 Three or More Dimensions

The technique in Section 7.1.1 can be extended to any number of dimensions by using the same strategy. For three dimensions, factor the exponential and then separate one of the dimensions as shown in Equation 7-4. Then the three-dimensional DFT is a sequence of two-dimensional DFTs on the results of the one-dimensional transform that has been separated. Then the two-dimensional DFT can be decomposed as described in Section 7.1.1. The same logic follows to convert an $N$-dimensional DFT into a sequence of one-dimensional DFTs and $(N - 1)$-dimensional DFTs:

$$
\sum_{n_1=0}^{N_1-1} \sum_{n_2=0}^{N_2-1} \sum_{n_3=0}^{N_3-1} a(n_1, n_2, n_3) * e^{-j2\pi[n_1k_1/N_1+n_2k_2/N_2+n_3k_3/N_3]}
$$

$$
= \sum_{n_1=0}^{N_1-1} \sum_{n_2=0}^{N_2-1} \left\{ \sum_{n_3=0}^{N_3-1} a(n_1, n_2, n_3) * e^{-j2\pi[n_3k_3/N_3]} \right\} * e^{-j2\pi[n_1k_1/N_1+n_2k_2/N_2]}
$$

(7-4)

## 7.2 LINEAR FILTERING

One-dimensional linear filtering is defined in Chapter 6 by using Equation 6-1. Just as one-dimensional filtering, two-dimensional filtering (spatial filtering) can be performed in the spatial frequency domain as well as the spatial domain [1, 2, 3]. For example, the sharp edges in an image can be softened by passing the image through a two-dimensional low-pass filter, just as the sharp edges of a square wave are smoothed by passing it through a low-pass filter. Further, a two-dimensional low-pass filter can be implemented in the frequency domain, just as for one-dimensional filters by using a generalized version of one of the two techniques in Chapter 6.

If $h(j, i)$ is the two-dimensional equivalent of the unit pulse response of the linear filter and $x(j, i)$ is the two-dimensional array of data points in the image, the equation for two-dimensional linear filtering is

$$
y(k_1, k_2) = \sum_{j=0}^{N_1-1} \sum_{i=0}^{N_2-1} x(k_1 - j, k_2 - i) * h(j, i)
$$

(7-5)

For a general unit pulse response this equation requires an enormous number of computations. Suppose the image has $P$ rows and $Q$ columns of pixels, and the two-dimensional unit pulse response has $N_1$ rows and $N_2$ columns. Generally, $N_1$ and $N_2$ are much smaller than $P$ and $Q$.

Equation 7-5 is computed for each value of $k_1 = 0, 1, 2, \ldots, (P - 1)$ and $k_2 = 0, 1, 2, \ldots, (Q - 1)$. Since $P \gg N_1$ and $Q \gg N_2$, almost all of the $P * Q$ computations of Equation 7-5 require the full $(N_1 * N_2)$ multiplies and $(N_1 * N_2 - 1)$ adds. Therefore, $P * Q * \{2 * N_1 * N_2 - 1\}$ computations is a good estimate for real input sequences and unit pulse responses. If the input sequence is complex and the unit pulse response remains real, these numbers double.

### 7.2.1 Separable Two-Dimensional Filter

One of the most popular techniques to reduce the computational requirements of the two-dimensional linear filter is to require the two-dimensional unit pulse response to be the product of two one-dimensional unit pulse responses. This dramatically reduces the computational load because it allows Equation 7-5 to be rewritten as

$$y(k_1, k_2) = \sum_{j=0}^{N_1-1} \left\{ \sum_{i=0}^{N_2-1} x(k_1 - j, k_2 - i) * h(i) \right\} * h(j) \qquad (7\text{-}6)$$

The inner summation is a one-dimensional linear filter that is computed for each value of $j = 0, 1, 2, \ldots, (N_1 - 1)$ in each row $k_1 = 0, 1, 2, \ldots, (P - 1)$. Since each one-dimensional linear filter requires $N_2$ multiplies and $(N_2 - 1)$ adds, the inner summation requires $N_1 * P * [2 * N_2 - 1]$ arithmetic computations and produces the signal used by the outer summation which is now also only a one-dimensional linear filter. Similarly, the outer summation requires $N_2 * Q * [2 * N_1 - 1]$ arithmetic computations. The total computations for Equation 7-6 are then reduced to $N_1 * P * [2 * N_2 - 1] + N_2 * Q * [2 * N_1 - 1]$. This total can be roughly approximated as $2 * N_1 * N_2 * (P + Q)$. The ratio of the number of computations required for the two-dimensional approach to the separable one-dimensional approach is roughly

$$(P + Q)/(P * Q) \qquad (7\text{-}7)$$

For a $512 \times 512$ image this ratio is $(512 + 512)/(512 * 512) = 1/256$, which is why this approach to the unit pulse response is commonly found in image processing. Note that Equation 7-7 is not dependent on the size of the unit pulse response. There actually is a weak dependence that has been lost in the equation because of the approximations made on the number of computations near the edge of the image.

### 7.2.2 Frequency Domain Approach

The frequency domain linear filtering algorithms in Chapter 6 can be used on Equation 7-6 to further reduce the computational requirements. Namely, each linear filter can be replaced by the three-step process in Chapter 6 for computing linear filters in the frequency domain. The frequency domain algorithm stages for computing the two-dimensional linear filter are as follows.

### Stage 1: Choose Inner Filter Transform Length

Choose a transform length $M_2$ for the inner summation in Equation 7-6 based on the criteria in Chapter 6. Using a larger number than $M_2 = N_2 + Q - 1$ requires adding zeros (zero padding), which is equivalent to adding a border of zeros at the ends of the rows of the image.

## Stage 2: Perform Inner Filter Frequency Domain Processing

For each row $k_1 = 0, 1, 2, \ldots, (P - 1)$, compute either the overlap-and-add or overlap-and-save algorithm from Chapter 6 and replace the $x(j, i)$ with the results $X(j, k_2)$. This approach requires

$$\# \text{ Comp.} = P * \{2 * N_{M2} + 13 * M_2 - 16\}$$

for real input sequences $x(j, i)$ and $M_2/2$ odd. If $M_2/2$ is even, this portion of the algorithm requires

$$\# \text{ Comp.} = P * \{2 * N_{M2} + 13 * M_2 - 18\}$$

In both cases, $N_{M2} =$ number of computations in the $M_2/2$-point FFT.

## Stage 3: Choose Outer Filter Transform Length

Choose a transform length $M_1$ for the outer summation in Equation 7-6 based on the criteria in Chapter 6. Using a larger number than $M_1 = N_1 + P - 1$ requires adding zeros (zero padding), which is equivalent to adding a border of zeros at the ends of the columns of the image.

## Stage 4: Perform Outer Filter Frequency Domain Processing

For each row $k_1 = 0, 1, 2, \ldots, (P - 1)$, compute either the overlap-and-add or overlap-and-save algorithm from Chapter 6 and replace the $X(j, k_2)$ with the results $y(k_1, k_2)$. This requires

$$\# \text{ Comp.} = Q * \{2 * N_{M1} + 13 * M_1 - 16\}$$

for real input sequences $x(j, i)$ and $M_1/2$ odd. If $M_1/2$ is even, this portion of the algorithm requires

$$\# \text{ Comp.} = Q * \{2 * N_{M1} + 13 * M_1 - 18\}$$

In both cases, $N_{M1} =$ number of computations in the $M_1/2$-point FFT. The total number of computations using the frequency domain approach is

$$\# \text{ Comp.} = Q * \{2 * N_{M1} + 13 * M_1 - 16\} + P * \{2 * N_{M2} + 13 * M_2 - 16\}$$

for $M/2$ odd, and for $M/2$ even

$$\# \text{ Comp.} = Q * \{2 * N_{M1} + 13 * M_1 - 18\} + P * \{2 * N_{M2} + 13 * M_2 - 18\}$$

### 7.2.3 Three and More Dimensions

Just as frequency analysis can be extended into more than two dimensions, the linear filtering equation can also be written in more than two dimensions. Again, the most common technique for reducing the computational load from multidimensional linear filtering is to restrict the unit pulse response to one that can be factored into functions of the individual dimensions, and then use frequency domain filtering on the resulting one-dimensional linear filters.

## 7.3 PATTERN MATCHING

One-dimensional pattern matching is defined in Chapter 6. Just as one-dimensional pattern matching can be performed in the time or frequency domain to find a pattern in a waveform, two-dimensional pattern matching can be performed in the spatial or frequency domain to find two-dimensional patterns in an image [1, 2, 3]. If $h(j, i)$ is the pattern to be located in an image $x(j, i)$, then the best match to that pattern is found when $y(k_1, k_2)$ is largest in the equation

$$y(k_1, k_2) = \sum_{j=0}^{N_1-1} \sum_{i=0}^{N_2-1} x(k_1 + j, k_2 + i) * h(j, i) \qquad (7\text{-}8)$$

For a general unit pulse response this equation requires an enormous number of computations. Suppose the image has $P$ rows and $Q$ columns of pixels, and the two-dimensional unit pulse response has $N_1$ rows and $N_2$ columns. Generally, $N_1$ and $N_2$ are much smaller than $P$ and $Q$.

Equation 7-8 is computed for each value of $k_1 = 0, 1, 2, \ldots, (P - 1)$ and $k_2 = 0, 1, 2, \ldots, (Q - 1)$. Since $P \gg N_1$ and $Q \gg N_2$, almost all of the $P * Q$ computations of Equation 7-5 require the full $(N_1 * N_2)$ multiplies and $(N_1 * N_2 - 1)$ adds. Therefore, $P * Q * \{2 * N_1 * N_2 - 1\}$ computations is a good estimate for real input sequences and unit pulse responses. If the input sequence is complex and the unit pulse response remains real, these numbers double.

### 7.3.1 Separable Two-Dimensional Pattern Matching

One of the most popular techniques to reduce the computational requirements of the two-dimensional pattern matching is to require the two-dimensional unit pulse response to be the product of two one-dimensional unit pulse responses. This dramatically reduces the computational load because it allows Equation 7-8 to be rewritten as

$$y(k_1, k_2) = \sum_{j=0}^{N_1-1} \left\{ \sum_{i=0}^{N_2-1} x(k_1 + j, k_2 + i) * h(i) \right\} * h(j) \qquad (7\text{-}9)$$

The inner summation is a one-dimensional pattern matcher that is computed for each value of $j = 0, 1, 2, \ldots, (N_1 - 1)$ in each row $k_1 = 0, 1, 2, \ldots, (P - 1)$. Since each one-dimensional pattern matcher requires $N_2$ multiplies and $(N_2 - 1)$ adds, the inner summation requires $N_1 * P * [2 * N_2 - 1]$ arithmetic computations and produces the signal used by the outer summation which is now also only a one-dimensional pattern matcher. Similarly, the outer summation requires $N_2 * Q * [2 * N_1 - 1]$ arithmetic computations. The total computations for Equation 7-9 are then reduced to $N_1 * P * [2 * N_2 - 1] + N_2 * Q * [2 * N_1 - 1]$. This total can be roughly approximated as $2 * N_1 * N_2 * (P + Q)$. The ratio of the number of computations required for the two-dimensional approach to the separable one-dimensional approach is roughly

$$(P + Q)/(P * Q) \qquad (7\text{-}10)$$

For a 512 × 512 image, this ratio is $(512 + 512)/(512 * 512) = 1/256$, which is why this approach to the unit pulse response is commonly found in image processing. Note that Equation 7-10 is not dependent on the size of the unit pulse response. There actually is a weak dependence that has been lost in the equation because of the approximations made on the number of computations near the edge of the image.

## 7.3.2 Frequency Domain Approach

The frequency domain pattern matching algorithms in Chapter 6 can be used on Equation 7-9 to further reduce the computational requirements. Namely, each pattern matcher can be replaced by the three-step process in Chapter 6 for computing pattern matchers in the frequency domain. The frequency domain algorithm stages for computing the two-dimensional pattern matcher are as follows:

### Stage 1: Choose Inner Pattern Matcher Transform Length

Choose a transform length $M_2$ for the inner summation in Equation 7-9 based on the criteria in Chapter 6. Using a number larger than $M_2 = N_2 + Q - 1$ requires adding zeros (zero padding), which is equivalent to adding a border of zeros at the ends of the rows of the image.

### Stage 2: Perform Inner Pattern Matcher Frequency Domain Processing

For each row $k_1 = 0, 1, 2, \ldots, (P - 1)$, compute either the overlap-and-add or overlap-and-save algorithm from Chapter 6 and replace the $x(j, i)$ with the results $X(j, k_2)$. This approach requires

$$\# \text{ Comp.} = P * \{2 * N_{M2} + 13 * M_2 - 16\}$$

for real input sequences $x(j, i)$ and $M_2/2$ odd. If $M_2/2$ is even, this portion of the algorithm requires

$$\# \text{ Comp.} = P * \{2 * N_{M2} + 13 * M_2 - 18\}$$

### Stage 3: Choose Outer Pattern Matcher Transform Length

Choose a transform length $M_1$ for the outer summation in Equation 7-9 based on the criteria in Chapter 6. Using a number larger than $M_1 = N_1 + P - 1$ requires adding zeros (zero padding), which is equivalent to adding a border of zeros at the ends of the columns of the image.

### Stage 4: Perform Outer Pattern Matcher Frequency Domain Processing

For each row $k_1 = 0, 1, 2, \ldots, (P - 1)$, compute either the overlap-and-add or overlap-and-save algorithm from Chapter 6 and replace the $X(j, k_2)$ with the results $y(k_1, k_2)$. This requires roughly

$$\# \text{ Comp.} = Q * \{2 * N_{M1} + 13 * M_1 - 16\}$$

for real input sequences $x(j, i)$ and $M_1/2$ odd. If $M_1/2$ is even, this portion of the algorithm requires

$$\# \text{ Comp.} = Q * \{2 * N_{M1} + 13 * M_1 - 18\}$$

The total number of computations with the frequency domain approach is roughly

$$\# \text{ Comp.} = Q * \{2 * N_{M1} + 13 * M_1 - 16\} + P * \{2 * N_{M2} + 13 * M_2 - 16\}$$

for $M/2$ odd, and for $M/2$ even

$$\text{\# Comp.} = Q * \{2 * N_{M1} + 13 * M_1 - 18\} + P * \{2 * N_{M2} + 13 * M_2 - 18\}$$

### 7.3.3 Three and More Dimensions

Just as frequency analysis can be extended to more than two dimensions, the pattern matching equation can also be written in more than two dimensions. Again, the most common technique for reducing the computational load from multidimensional pattern matching is to restrict the unit pulse response to one that can be factored into functions of the individual dimensions, and then use frequency domain pattern matching on the resulting one-dimensional pattern matchers.

## 7.4 CONCLUSIONS

Having learned in this chapter how to break down multidimensional processing to more easily performed sequences of one-dimensional processing, we conclude the foundation portion of the book. Design Example 4 in Chapter 17, an image deblurrer, demonstrates two-dimensional processing. Now that what FFTs are and what they can do have been covered, the next two chapters show how to construct an FFT of any length.

## REFERENCES

[1] L. R. Rabiner and B. Gold, *Theory and Application of Digital Signal Processing*, Prentice-Hall, Englewood Cliffs, NJ, 1975.

[2] A. V. Oppenheim and R. W. Schafer, *Digital Signal Processing*, Prentice-Hall, Englewood Cliffs, NJ, 1975.

[3] E. Oran Brigham, *The Fast Fourier Transform and Its Applications*, Prentice-Hall, Englewood Cliffs, NJ, 1988.

# 8

# Building-Block Algorithms

## 8.0 INTRODUCTION

In this chapter the 2-, 3-, 4-, 5-, 7-, 8-, 9-, and 16-point FFT algorithms are presented because they are the most efficient and widely used FFT algorithm building blocks. The general-purpose FFT algorithms (Rader and Singleton) are included to provide the additional building blocks necessary to compute any transform length. This is because not all numbers have only 2, 3, 4, 5, 7, 8, 9, or 16 as factors, for example, $119 = 7 * 17$. More than one algorithm for computing a particular building block, except for 2 and 4, is given because each has different features that make it better suited to some applications than others. A unique feature of the book is the format in which they are all presented, with input adds, multiplies, and then output adds, so that all can be used with the Winograd algorithm in Chapter 9.

All of the building-block algorithms are FFTs, sometimes called small-point transforms. Since they are FFTs, they have all of the same properties, strengths, and weaknesses of the DFT described in Chapter 2.

## 8.1 FOUR PERFORMANCE MEASURES

The most common way to evaluate FFT algorithms is in terms of the number of computations and amount of memory required to compute them. The performance measures in this section quantify those computations and memory needs. The same four measures are used again in Chapter 9.

### 8.1.1 Number of Adds

This is the total number of real adds used for each building-block algorithm. It includes the two adds required as part of each of the complex multiplies.

### 8.1.2 Number of Multiplies

This is the total number of real multiplies for each building-block algorithm. Each complex multiply takes four real multiplies and two real adds (counted in the number of adds). The standard way of computing complex multiplies is as a sequence of four real multiplies and two real adds, as shown in Equation 8-1.

$$(a + jb) * (c + jd) = (ac - bd) + j(bc + ad) \tag{8-1}$$

However, it is possible to rewrite Equation 8-1 so that it is computed as three multiplies and three adds (Equation 8-2).

$$(a + jb) * (c + jd) = (a + b) * c - b * (c + d) + j[(a - b) * d + b * (c + d)] \tag{8-2}$$

This technique is not used in any of the building-block algorithms in this chapter. However, it could be used to modify the add and multiply count for a particular building block to satisfy the requirements of a particular application or arithmetic format. The drawback of this technique is that it introduces additional quantization noise into the FFT results, because of the way identical terms are added and then subtracted to form the results.

To understand how Equation 8-2 only requires three multiplies and three adds, consider $a + jb$, the FFT multiplier constant. Then $a + b$ and $a - b$ are constants that can be computed ahead of time and stored in memory. The sequence of computations is:

(a) Add $c$ and $d$ to form $(c + d)$.

(b) Multiply $(c + d)$ by $b$ to form $b * (c + d)$.

(c) Multiply $(a + b)$ by $c$ to form $(a + b) * c$.

(d) Multiply $(a - b)$ by $d$ to form $(a - b) * d$.

(e) Subtract the results of b and c to form the real part of the result.

(f) Add the results of d and b to form the imaginary part of the result.

Steps a, e, and f are additions (in one case a subtraction which is generally implemented as an addition of a negative number), and steps b–d are real multiplications.

### 8.1.3 Number of Memory Locations for Multiplier Constants

Each building-block algorithm requires a different number of multiplier constants. Each constant must be stored in data or program memory or computed as needed. The latter is seldom done any more because memory costs have been dramatically lowered. The number for this performance measure in the Comparison Matrix in Table 8-1 is the total of the different constants required by each algorithm. These include multiplication by 2 and 1/2, which can also be done by moving the binary point of fixed-point numbers or by changing the exponent of floating-point numbers.

### 8.1.4 Number of Data Memory Locations

Each algorithm begins and ends by using exactly $2 * N$ data memory locations to store the input data and output results, respectively. However, if no temporary registers are available for intermediate results, most of the algorithms in this chapter require additional data memory locations during the computations. In this chapter, Algorithm Steps and a Memory Map are given for each algorithm, and total data memory location requirements are listed in the Comparison Matrix, assuming the processor has no temporary registers. The difference between those numbers and $2 * N$ is the number of temporary registers needed to avoid using extra data memory locations for intermediate results.

## 8.2 TEN BUILDING-BLOCK ALGORITHM CONSTRAINTS

The following are the constraints the authors have used for the small-point transforms in this chapter:

1. The real and imaginary parts of the $i$-th input sample are $a_R(i)$ and $a_I(i)$. $A_R(i)$ and $A_I(i)$ are the real and imaginary parts of the $i$-th output frequency component.

2. All of the algorithms have been segmented to have all of the multiplications in the center so that they can be used by any of the FFT algorithms in Chapter 9 to form longer transform lengths. Chapter 9 explains the reasons for this constraint.

3. Intermediate results are labeled with sequential lowercase letters of the alphabet to indicate where they are located relative to other computational outputs. For example, the first set of intermediate computational results in each of the algorithm building blocks is labeled $b_R(i)$ and $b_I(i)$.

4. The sum and difference computations are performed by taking two pieces of data from data memory, performing the required computations, and returning the results to available data memory locations.

5. The multiply-accumulates are performed by sequentially pulling a data value from data memory, performing the multiplication, and adding the results to the processor's accumulator (Section 14.2.11). When the multiply-accumulate function is complete, the result is stored in a memory location, overwriting data that is no longer needed.

6. The sequence of computations shown for the first stage in each algorithm has been left the same as in its referenced article. The data labels have been changed to make them consistent for all the algorithms in the book.

7. The memory location (Memory Map) for intermediate results or output frequency components is shown next to each Algorithm Step.

8. For an $N$-point algorithm building block, the real input data, $a_R(i)$, is located in data memory locations $M(i)$, and the imaginary input data, $a_I(i)$, is located in data memory locations $M(N + i)$, where $i = 0, 1, 2, \ldots, (N - 1)$.

9. All of the multiplier constants are presented in their sine and cosine forms so that they may be computed in the arithmetic format (see Chapter 13) appropriate for the application.

10. All of the intermediate results and output frequency components are stored directly in data memory, rather than temporary storage locations, to ensure that the algorithm will work on all processors.

## 8.3 TWO-POINT FFT

The 2-point DFT is defined for $k = 0$ and 1 as

$$A(k) = \sum_{n=0}^{1} a(n) * e^{-j2\pi kn/2} \tag{8-3}$$

This simplest of DFTs and its FFT are the same. This algorithm requires four adds and no multiplies and its execution is straightforward. The strategy for converting these equations to code is to start at the top (compute $A_R(0)$) and identify the pair of inputs to be used first (in this case $a_R(0)$ and $a_R(1)$). Then look down the list to find the second (compute $A_R(1)$) place where these two inputs are used. Pull $a_R(0)$ and $a_R(1)$ from memory, compute $A_R(0)$ and $A_R(1)$, and store the results in data memory locations $M(0)$ and $M(1)$ previously occupied by $a_R(0)$ and $a_R(1)$. Next, repeat the same set of steps for $A_I(0)$ and $A_I(1)$.

| Algorithm Steps | Memory Map |
|---|---|
| $A_R(0) = a_R(0) + a_R(1)$ | $A_R(0) \Rightarrow M(0)$ |
| $A_I(0) = a_I(0) + a_I(1)$ | $A_I(0) \Rightarrow M(2)$ |
| $A_R(1) = a_R(0) - a_R(1)$ | $A_R(1) \Rightarrow M(1)$ |
| $A_R(1) = a_I(0) - a_I(1)$ | $A_I(1) \Rightarrow M(3)$ |

Since each set of results can be placed in the same data memory locations that the inputs were taken from, this algorithm requires only four data memory locations. The flowchart for the 2-point FFT is shown in Figure 8-1. Two inputs and two outputs are used to indicate that the same computational building block is used twice to compute the real and imaginary portions of the 2-point FFT output.

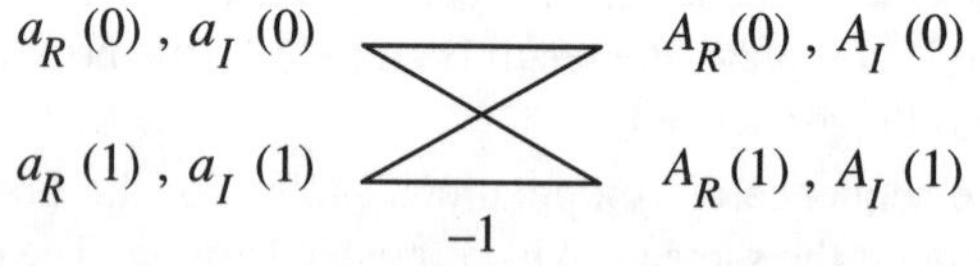

$a_R(0), a_I(0)$ ⟍⟋ $A_R(0), A_I(0)$

$a_R(1), a_I(1)$ ⟋⟍ $A_R(1), A_I(1)$
$-1$

**Figure 8-1** Two-point FFT algorithm flow graph.

Note that Figure 8-1 looks similar to the 2-point decimation-in-time (DIT) and decimation-in-frequency (DIF) figures in Section 10.4. The difference is the multiplier in the DIT and DIF flowcharts. When the 2-point transform is used in a larger power-of-two algorithm, it requires data reorganization as well as the complex multiplier to prepare the data for each succeeding stage of the algorithm. However, in the prime factor algorithm (Section 9.6), only data reorganization is required. Therefore, the universal building block is the 2-point FFT in Figure 8-1. Chapter 9 deals with how these algorithm building blocks are combined in different ways to form larger transform lengths, including power-of-two and prime factor algorithms.

## 8.4 THREE-POINT FFT

The 3-point DFT is defined for $k = 0$, 1, and 2 as

$$A(k) = \sum_{n=0}^{2} a(n) * e^{-j2\pi kn/3} \tag{8-4}$$

If the 3-point DFT is calculated directly from Equation 8-4, it requires four complex multiplies and six complex adds. Since a complex multiply uses 4 real multiplies and 2 real adds, and a complex add uses 2 real adds, the 3-point DFT requires 16 real multiplies and 20 real adds. The number of adds and multiplies for the two fast algorithms is significantly less than required for computing the DFT directly. However, if only a subset of the output frequency components is required, it may be more cost effective to compute the DFT equation directly for those terms. For example, if $A(0)$ is the only term needed, it can be computed with four adds and no multiplies by using the DFT directly. Each of the other two output frequencies requires two complex multiplies and two complex adds for a total of eight real adds and eight real multiplies. With this in mind the crossover point between using the DFT directly and one of the 3-point FFT algorithms can be determined based on the number of output frequency components that must be computed.

Since all of the input data is required for each of the output frequency component calculations, the direct DFT computations require six data memory locations for the input data and six more for the output frequency components. This is a total of 12 data memory locations, since the input and output are complex. Similarly, the DFT data addressing is sequential (i.e., 0 through 2 for each output frequency component), and the computational architecture is simple since they can all be performed by using a complex multiply accumulator (see Chapter 10 for details). Addressing the complex multiplier coefficients is sequential in two orders (1 and 2 or 2 and 1) or requires that the addresses be stored in program memory.

There are two common 3-point FFT algorithms. Both require 12 adds, 4 multiplies, and 2 memory locations for multiplier constants. The Winograd [1] algorithm is based on circular convolution properties and requires six data memory locations. The Singleton [2] algorithm is based on complex conjugate symmetry properties of the 3-point DFT and requires seven data memory locations.

### 8.4.1 Winograd 3-Point FFT

The strategy for converting these equations into code is to start at the top (compute $b_R(1)$) and identify the pair of inputs to be used first (in this case $a_R(1)$ and $a_R(2)$). Then look down the list to find the second (compute $b_R(2)$) place where these two inputs are used. Pull $a_R(1)$ and $a_R(2)$ from memory, compute $b_R(1)$ and $b_R(2)$, and store the results in data memory locations $M(1)$ and $M(2)$ previously occupied by $a_R(1)$ and $a_R(2)$.

Next, look for the computation for $b_I(1)$ on the list and repeat the same set of steps. Continue this process until all the Algorithm Steps have been computed and their results stored in the Memory Map addresses. Note that the algorithm steps for $A_R(0)$ and $A_I(0)$ only relabel the data values to their output labels once they have been used as required by other portions of the algorithm.

| Algorithm Steps | Memory Map |
|---|---|
| $b_R(1) = a_R(1) + a_R(2)$ | $b_R(1) \Rightarrow M(1)$ |
| $b_R(2) = a_R(1) - a_R(2)$ | $b_R(2) \Rightarrow M(2)$ |
| $b_I(1) = a_I(1) + a_I(2)$ | $b_I(1) \Rightarrow M(4)$ |
| $b_I(2) = a_I(1) - a_I(2)$ | $b_I(2) \Rightarrow M(5)$ |
| $b_R(0) = a_R(0) + b_R(1)$ | $b_R(0) \Rightarrow M(0)$ |
| $b_I(0) = a_I(0) + b_I(1)$ | $b_I(0) \Rightarrow M(3)$ |
| $c_R(1) = b_R(1) * [\cos(2\pi/3) - 1]$ | $c_R(1) \Rightarrow M(1)$ |
| $c_R(2) = b_I(2) * \sin(2\pi/3)$ | $c_R(2) \Rightarrow M(5)$ |
| $c_I(1) = b_I(1) * [\cos(2\pi/3) - 1]$ | $c_I(1) \Rightarrow M(4)$ |
| $c_I(2) = b_R(2) * \sin(2\pi/3)$ | $c_I(2) \Rightarrow M(2)$ |
| $d_R(0) = b_R(0) + c_R(1)$ | $d_R(0) \Rightarrow M(1)$ |
| $d_I(0) = b_I(0) + c_I(1)$ | $d_I(0) \Rightarrow M(4)$ |
| $A_R(0) = b_R(0)$ | $A_R(0) \Rightarrow M(0)$ |
| $A_I(0) = b_I(0)$ | $A_I(0) \Rightarrow M(3)$ |
| $A_R(1) = d_R(0) + c_R(2)$ | $A_R(1) \Rightarrow M(1)$ |
| $A_I(1) = d_I(0) - c_I(2)$ | $A_I(1) \Rightarrow M(4)$ |
| $A_R(2) = d_R(0) - c_R(2)$ | $A_R(2) \Rightarrow M(5)$ |
| $A_I(2) = d_I(0) + c_I(2)$ | $A_I(2) \Rightarrow M(2)$ |

This set of equations is shown pictorially with the flow graph in Figure 8-2.

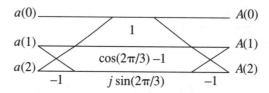

**Figure 8-2**  Winograd 3-point FFT flow graph.

## 8.4.2 Singleton 3-Point FFT

The strategy for converting these equations into code is to start at the top (compute $b_R(1)$) and identify the pair of inputs to be used first (in this case $a_R(1)$ and $a_R(2)$). Then look down the list to find the second (compute $b_R(2)$) place where these two inputs are used. Pull $a_R(1)$ and $a_R(2)$ from memory, compute $b_R(1)$ and $b_R(2)$, and store the results in data memory locations $M(1)$ and $M(2)$ previously occupied by $a_R(1)$ and $a_R(2)$.

Next, look for the computation for $b_I(1)$ on the list and repeat the same set of the steps. Continue this process until all the Algorithm Steps have been computed and their results stored in the Memory Map addresses.

| Algorithm Steps | Memory Map |
|---|---|
| $b_R(1) = a_R(1) + a_R(2)$ | $b_R(1) \Rightarrow M(1)$ |
| $b_R(2) = a_R(1) - a_R(2)$ | $b_R(2) \Rightarrow M(2)$ |

| Algorithm Steps | Memory Map |
|---|---|
| $b_I(1) = a_I(1) + a_I(2)$ | $b_I(1) \Rightarrow M(4)$ |
| $b_I(2) = a_I(1) - a_I(2)$ | $b_I(2) \Rightarrow M(5)$ |
| $c_R(1) = b_R(1) * \cos(2\pi/3) + a_R(0)$ | $c_R(1) \Rightarrow M(6)$ |
| $A_R(0) = a_R(0) + b_R(1)$ | $A_R(0) \Rightarrow M(0)$ |
| $c_R(2) = b_I(2) * \sin(2\pi/3)$ | $c_R(2) \Rightarrow M(5)$ |
| $c_I(1) = b_I(1) * \cos(2\pi/3) + a_I(0)$ | $c_I(1) \Rightarrow M(1)$ |
| $A_I(0) = a_I(0) + b_I(1)$ | $A_I(0) \Rightarrow M(3)$ |
| $c_I(2) = -b_R(2) * \sin(2\pi/3)$ | $c_I(2) \Rightarrow M(2)$ |
| $A_R(1) = c_R(1) + c_R(2)$ | $A_R(1) \Rightarrow M(5)$ |
| $A_I(1) = c_I(1) + c_I(2)$ | $A_I(1) \Rightarrow M(2)$ |
| $A_R(2) = c_R(1) - c_R(2)$ | $A_R(2) \Rightarrow M(4)$ |
| $A_I(2) = c_I(1) - c_I(1)$ | $A_I(2) \Rightarrow M(1)$ |

Figure 8-3 is a flow graph of these equations.

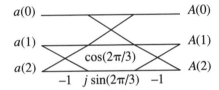

**Figure 8-3**    Singleton 3-point FFT flow graph.

## 8.5 FOUR-POINT FFT

The 4-point DFT is defined for $k = 0, 1, 2,$ and 3 as

$$A(k) = \sum_{n=0}^{3} a(n) * e^{-j2\pi kn/4} \tag{8-5}$$

If the 4-point DFT is computed directly from Equation 8-5, it requires no complex multiplies and 12 complex adds for a total of 24 real adds. The circular convolution, complex conjugate symmetry, and 90° and 180° symmetry approaches to a 4-point FFT all result in the same set of Algorithm Steps. The algorithm requires 16 adds, no multiplications, 8 data memory locations, and no memory locations for multiplier constants.

Since all of the input data is required for each output frequency component calculation, the direct DFT computations require eight data memory locations for the input data and eight more for the output frequency components. This is a total of 16 data memory locations, since the input and output are complex. Similarly, the DFT data addressing is sequential (i.e., 0 through 3 for each output frequency component), and the computational architecture is simple, since they can all be performed with additions.

The strategy for converting these equations into code is to start at the top (compute $b_R(0)$) and identify the pair of inputs to be used first (in this case $a_R(0)$ and $a_R(2)$). Then look down the list to find the second (compute $b_R(1)$) place where these two inputs are used. Pull $a_R(0)$ and $a_R(2)$ from memory, compute $b_R(0)$ and $b_R(1)$, and store the results in data memory locations $M(0)$ and $M(2)$ previously occupied by $a_R(0)$ and $a_R(2)$.

Next, look for the computation for $b_I(0)$ on the list and repeat the same set of steps. Continue this process until all the Algorithm Steps have been computed and their results stored in the Memory Map addresses.

| Algorithm Steps | Memory Map |
|---|---|
| $b_R(0) = a_R(0) + a_R(2)$ | $b_R(0) \Rightarrow M(0)$ |
| $b_R(1) = a_R(0) - a_R(2)$ | $b_R(1) \Rightarrow M(2)$ |
| $b_I(0) = a_I(0) + a_I(2)$ | $b_I(0) \Rightarrow M(4)$ |
| $b_I(1) = a_I(0) - a_I(2)$ | $b_I(1) \Rightarrow M(6)$ |
| $b_R(2) = a_R(1) + a_R(3)$ | $b_R(2) \Rightarrow M(1)$ |
| $b_R(3) = a_R(1) - a_R(3)$ | $b_R(3) \Rightarrow M(3)$ |
| $b_I(2) = a_I(1) + a_I(3)$ | $b_I(2) \Rightarrow M(5)$ |
| $b_I(3) = a_I(1) - a_I(3)$ | $b_I(3) \Rightarrow M(7)$ |
| $A_R(0) = b_R(0) + b_R(2)$ | $A_R(0) \Rightarrow M(0)$ |
| $A_I(0) = b_I(0) + b_I(2)$ | $A_I(0) \Rightarrow M(4)$ |
| $A_R(2) = b_R(0) - b_R(2)$ | $A_R(2) \Rightarrow M(1)$ |
| $A_I(2) = b_I(0) - b_I(2)$ | $A_I(2) \Rightarrow M(5)$ |
| $A_R(1) = b_R(1) + b_I(3)$ | $A_R(1) \Rightarrow M(2)$ |
| $A_R(3) = b_R(1) - b_I(3)$ | $A_R(3) \Rightarrow M(7)$ |
| $A_I(1) = b_I(1) - b_R(3)$ | $A_I(1) \Rightarrow M(3)$ |
| $A_I(3) = b_I(1) + b_R(3)$ | $A_I(3) \Rightarrow M(6)$ |

## 8.6 FIVE-POINT FFT

The 5-point DFT is defined for $k = 0, 1, 2, 3,$ and 4 as

$$A(k) = \sum_{n=0}^{4} a(n) * e^{-j2\pi kn/5} \qquad (8\text{-}6)$$

Three fast versions of the 5-point DFT are presented. The Winograd and Rader algorithms were developed by using a decomposition based on circular convolution properties. The Singleton algorithm was developed by using a decomposition based on the complex conjugate symmetry properties of the 5-point transform.

If the 5-point DFT is calculated directly from Equation 8-6, it requires 16 complex multiplies and 20 complex adds. Since a complex multiply uses 4 real multiplies and 2 real adds, and a complex add uses 2 real adds, the 5-point DFT requires 64 real multiplies and 72 real adds. The number of adds and multiplies for each of the building-block algorithms is significantly less than required for computing the DFT directly. However, if only a subset of the output frequency components is required, it may be more cost effective to compute the DFT equation directly for those terms. For example, if $A(0)$ is the only term needed, it can be computed with eight adds and no multiplies by using the DFT directly. Each of the other 4 output frequencies requires 4 complex multiplies and 4 complex adds for a total of

16 real adds and 16 real multiplies. With this in mind the crossover point between using the DFT directly and one of the 5-point FFT algorithms can be determined based on the number of output frequency components that must be computed.

Since all of the input data is required for each output frequency component calculation, the direct DFT computations require 10 data memory locations for the input data and 10 more for the output frequency components. This is a total of 20 data memory locations, since the input and output are complex. Similarly, the DFT data addressing is sequential (i.e., 0 through 4 for each output frequency component), and the computational architecture is simple, since they can all be performed with a complex multiply accumulator (see Chapter 10 for details). Addressing the complex multiplier coefficients requires either a modulo arithmetic scheme ($k * n$ mod 5) or that the addresses be stored in program memory.

Each of the three fast algorithms is presented, characterized, and summarized in the Comparison Matrix in Table 8-1. For example, the Rader algorithm has the simplest computational structure but requires the largest number of adds. The Singleton algorithm has the simplest memory mapping for the multiplier constants but requires more constants than the Winograd algorithm.

### 8.6.1  Winograd 5-Point FFT

The Winograd [1] 5-point FFT requires 10 multiplies, 34 adds, 12 data memory locations, and 5 multiplier constant memory locations. The four stages are as follows.

### Stage 1: Input Adds

This stage requires additional data memory locations to store intermediate results that reduce the total number of multiplications in the next stage. However, this stage does not require accessing any of the multiplier constants. The strategy for converting these equations to code is to start at the top (compute $b_R(1)$) and identify the pair of inputs to be used first (in this case $a_R(1)$ and $a_R(4)$). Then look down the list to find the second (compute $b_R(2)$) place where these two inputs are used. Pull $a_R(1)$ and $a_R(4)$ from memory, compute $b_R(1)$ and $b_R(2)$, and store the results in data memory locations $M(1)$ and $M(4)$ previously occupied by $a_R(1)$ and $a_R(4)$.

Next, look for the computation for $b_I(1)$ on the list and repeat the same set of steps. Continue this process until all the Algorithm Steps have been computed and their results stored in the Memory Map addresses. The computation of all the $b_R(j)$ and $b_I(j)$ terms are performed in-place by using the add-subtract butterfly algorithm. The computations of $c_R(1), c_R(3), c_I(1)$, and $c_I(3)$ use this same approach. However, the computations of $c_R(5)$ and $c_I(5)$ require additional data memory locations because $b_R(2), b_R(4), b_I(2)$, and $b_I(4)$ are also required in Stage 2.

| Algorithm Steps | Memory Map |
|---|---|
| $b_R(1) = a_R(1) + a_R(4)$ | $b_R(1) \Rightarrow M(1)$ |
| $b_I(1) = a_I(1) + a_I(4)$ | $b_I(1) \Rightarrow M(6)$ |
| $b_R(2) = a_R(1) - a_R(4)$ | $b_R(2) \Rightarrow M(4)$ |
| $b_I(2) = a_I(1) - a_I(4)$ | $b_I(2) \Rightarrow M(9)$ |
| $b_R(3) = a_R(2) + a_R(3)$ | $b_R(3) \Rightarrow M(2)$ |

| **Algorithm Steps** | **Memory Map** |
|---|---|
| $b_I(3) = a_I(2) + a_I(3)$ | $b_I(3) \Rightarrow M(7)$ |
| $b_R(4) = a_R(3) - a_R(2)$ | $b_R(4) \Rightarrow M(3)$ |
| $b_I(4) = a_I(3) - a_I(2)$ | $b_I(4) \Rightarrow M(8)$ |
| $c_R(1) = b_R(1) + b_R(3)$ | $c_R(1) \Rightarrow M(1)$ |
| $c_I(1) = b_I(1) + b_I(3)$ | $c_I(1) \Rightarrow M(6)$ |
| $c_R(3) = b_R(1) - b_R(3)$ | $c_R(3) \Rightarrow M(2)$ |
| $c_I(3) = b_I(1) - b_I(3)$ | $c_I(3) \Rightarrow M(7)$ |
| $c_R(5) = b_R(2) + b_R(4)$ | $c_R(5) \Rightarrow M(10)$ |
| $c_I(5) = b_I(2) + b_I(4)$ | $c_I(5) \Rightarrow M(11)$ |
| $d_R(0) = c_R(1) + a_R(0)$ | $d_R(0) \Rightarrow M(0)$ |
| $d_I(0) = c_I(1) + a_I(0)$ | $d_I(0) \Rightarrow M(5)$ |

## Stage 2: Multiplications

This stage contains all of the multiplications and requires additional data memory locations to store intermediate results. In all steps the multiplication is performed by pulling a data value from memory, multiplying it by the appropriate constant, and returning the result to the same data memory location. All these computations are performed in-place.

| **Algorithm Steps** | **Memory Map** |
|---|---|
| $d_R(1) = c_R(1) * [0.5 * \cos(2\pi/5) + 0.5 * \cos(4\pi/5) - 1]$ | $d_R(1) \Rightarrow M(1)$ |
| $d_I(1) = c_I(1) * [0.5 * \cos(2\pi/5) + 0.5 * \cos(4\pi/5) - 1]$ | $d_I(1) \Rightarrow M(6)$ |
| $e_R(3) = c_R(3) * [0.5 * \cos(2\pi/5) - 0.5 * \cos(4\pi/5)]$ | $e_R(3) \Rightarrow M(2)$ |
| $e_I(3) = c_I(3) * [0.5 * \cos(2\pi/5) - 0.5 * \cos(4\pi/5)]$ | $e_I(3) \Rightarrow M(7)$ |
| $e_R(5) = c_R(5) * \sin(4\pi/5)$ | $e_R(5) \Rightarrow M(10)$ |
| $e_I(5) = c_I(5) * \sin(4\pi/5)$ | $e_I(5) \Rightarrow M(11)$ |
| $d_R(2) = b_I(2) * [\sin(2\pi/5) + \sin(4\pi/5)]$ | $d_R(2) \Rightarrow M(9)$ |
| $d_I(2) = -b_R(2) * [\sin(2\pi/5) + \sin(4\pi/5)]$ | $d_I(2) \Rightarrow M(4)$ |
| $d_R(4) = -b_I(4) * [\sin(2\pi/5) - \sin(4\pi/5)]$ | $d_R(4) \Rightarrow M(8)$ |
| $d_I(4) = b_R(4) * [\sin(2\pi/5) - \sin(4\pi/5)]$ | $d_I(4) \Rightarrow M(3)$ |

## Stage 3: Postmultiply Adds

The output of this stage does not require additional data memory locations. The strategy for converting these equations to code is to start at the top (compute $e_R(1)$) and identify the pair of inputs to be used first (in this case $d_R(1)$ and $d_R(0)$). Pull $d_R(1)$ and $d_R(0)$ from memory, compute $e_R(1)$, and store the results in memory location $M(1)$ previously occupied by $d_R(1)$. This process is repeated until all the Algorithm Steps have been computed and their results stored in the Memory Map addresses.

| Algorithm Steps | Memory Map |
|---|---|
| $e_R(1) = d_R(1) + d_R(0)$ | $e_R(1) \Rightarrow M(1)$ |
| $e_I(1) = d_I(1) + d_I(0)$ | $e_I(1) \Rightarrow M(6)$ |
| $f_R(1) = e_R(1) + e_R(3)$ | $f_R(1) \Rightarrow M(1)$ |
| $f_I(1) = e_I(1) + e_I(3)$ | $f_I(1) \Rightarrow M(6)$ |
| $f_R(2) = d_R(2) - e_I(5)$ | $f_R(2) \Rightarrow M(9)$ |
| $f_I(2) = d_I(2) + e_R(5)$ | $f_I(2) \Rightarrow M(4)$ |
| $f_R(3) = e_R(1) - e_R(3)$ | $f_R(3) \Rightarrow M(2)$ |
| $f_I(3) = e_I(1) - e_I(3)$ | $f_I(3) \Rightarrow M(7)$ |
| $f_R(4) = d_R(4) - e_I(5)$ | $f_R(4) \Rightarrow M(8)$ |
| $f_I(4) = d_I(4) + e_R(5)$ | $f_I(4) \Rightarrow M(3)$ |

### Stage 4: Output Adds

The strategy for converting these equations to code is to start at the top (compute $A_R(1)$) and identify the pair of inputs to be used first (in this case $f_R(1)$ and $f_R(2)$). Then look down the list to find the second (compute $A_R(4)$) place where these two inputs are used. Pull $f_R(1)$ and $f_R(2)$ from memory, compute $A_R(1)$ and $A_R(4)$, and store the results in data memory locations $M(1)$ and $M(9)$ previously occupied by $f_R(1)$ and $f_R(2)$.

Next, look for the computation for $A_I(1)$ on the list and repeat the same set of steps. Continue this process until all the Algorithm Steps have been computed and their results stored in the Memory Map addresses. Note that the Algorithm Steps for $A_R(0)$ and $A_I(0)$ only relabel the data values to their output labels once they have been used as required by other portions of the algorithm.

| Algorithm Steps | Memory Map |
|---|---|
| $A_R(0) = d_R(0)$ | $A_R(0) \Rightarrow M(0)$ |
| $A_I(0) = d_I(0)$ | $A_I(0) \Rightarrow M(5)$ |
| $A_R(1) = f_R(1) + f_R(2)$ | $A_R(1) \Rightarrow M(1)$ |
| $A_I(1) = f_I(1) + f_I(2)$ | $A_I(1) \Rightarrow M(4)$ |
| $A_R(4) = f_R(1) - f_R(2)$ | $A_R(4) \Rightarrow M(9)$ |
| $A_I(4) = f_I(1) - f_I(2)$ | $A_I(4) \Rightarrow M(6)$ |
| $A_R(3) = f_R(3) + f_R(4)$ | $A_R(3) \Rightarrow M(2)$ |
| $A_I(3) = f_I(3) + f_I(4)$ | $A_I(3) \Rightarrow M(3)$ |
| $A_R(2) = f_R(3) - f_R(4)$ | $A_R(2) \Rightarrow M(8)$ |
| $A_I(2) = f_I(3) - f_I(4)$ | $A_I(2) \Rightarrow M(7)$ |

### 8.6.2 Singleton 5-Point FFT

The Singleton [2] 5-point FFT requires 32 adds, 16 multiplies, 12 data memory locations, and 4 multiplier constant memory locations. The method of computing the multiplier outputs in Stage 2 requires additional data memory locations. The three stages are as follows.

### Stage 1: Input Adds

This stage does not require additional data memory locations for accessing any of the multiplier constants. The strategy for converting these equations to code is to start at the top (compute $b_R(1)$) and identify the pair of inputs to be used first (in this case $a_R(1)$ and $a_R(4)$). Then look down the list to find the second (compute $b_R(2)$) place where these two inputs are used. Pull $a_R(1)$ and $a_R(4)$ from memory, compute $b_R(1)$ and $b_R(2)$, and store the results in data memory locations $M(1)$ and $M(4)$ previously occupied by $a_R(1)$ and $a_R(4)$.

Next, look for the computation for $b_I(1)$ on the list and repeat the same set of steps. Continue this process until all the Algorithm Steps have been computed and their results stored in the Memory Map addresses.

| Algorithm Steps | Memory Map |
|---|---|
| $b_R(1) = a_R(1) + a_R(4)$ | $b_R(1) \Rightarrow M(1)$ |
| $b_I(1) = a_I(1) + a_I(4)$ | $b_I(1) \Rightarrow M(6)$ |
| $b_R(2) = a_R(1) - a_R(4)$ | $b_R(2) \Rightarrow M(4)$ |
| $b_I(2) = a_I(1) - a_I(4)$ | $b_I(2) \Rightarrow M(9)$ |
| $b_R(3) = a_R(2) + a_R(3)$ | $b_R(3) \Rightarrow M(2)$ |
| $b_I(3) = a_I(2) + a_I(3)$ | $b_I(3) \Rightarrow M(7)$ |
| $b_R(4) = a_R(2) - a_R(3)$ | $b_R(4) \Rightarrow M(3)$ |
| $b_I(4) = a_I(2) - a_I(3)$ | $b_I(4) \Rightarrow M(8)$ |

### Stage 2: Multiply-Accumulates

This stage contains all of the multiplications and requires additional data memory locations to perform the sets of multiply-accumulate operations and store the intermediate results. The strategy for converting these steps into code is explained in Constraint 5 of Section 8.2.

| Algorithm Steps | Memory Map |
|---|---|
| $c_R(2) = b_R(2) * \sin(2\pi/5) + b_R(4) * \sin(4\pi/5)$ | $c_R(2) \Rightarrow M(10)$ |
| $c_I(2) = b_I(2) * \sin(2\pi/5) + b_I(4) * \sin(4\pi/5)$ | $c_I(2) \Rightarrow M(3)$ |
| $c_R(4) = b_R(2) * \sin(4\pi/5) - b_R(4) * \sin(2\pi/5)$ | $c_R(4) \Rightarrow M(11)$ |
| $c_I(4) = b_I(2) * \sin(4\pi/5) - b_I(4) * \sin(2\pi/5)$ | $c_I(4) \Rightarrow M(4)$ |
| $c_R(1) = b_R(1) * \cos(2\pi/5) + b_R(3) * \cos(4\pi/5) + a_R(0)$ | $c_R(1) \Rightarrow M(9)$ |
| $c_I(1) = b_I(1) * \cos(2\pi/5) + b_I(3) * \cos(4\pi/5) + a_I(0)$ | $c_I(1) \Rightarrow M(1)$ |
| $c_R(3) = b_R(1) * \cos(4\pi/5) + b_R(3) * \cos(2\pi/5) + a_R(0)$ | $c_R(3) \Rightarrow M(8)$ |
| $c_I(3) = b_I(1) * \cos(4\pi/5) + b_I(3) * \cos(2\pi/5) + a_I(0)$ | $c_I(3) \Rightarrow M(2)$ |
| $A_R(0) = a_R(0) + b_R(1) + b_R(3)$ | $A_R(0) \Rightarrow M(0)$ |
| $A_I(0) = a_I(0) + b_I(1) + b_I(3)$ | $A_I(0) \Rightarrow M(5)$ |

### Stage 3: Output Adds

The strategy for converting these equations to code is to start at the top (compute $A_R(1)$) and identify the pair of inputs to be used first (in this case $c_R(1)$ and $c_I(2)$). Then look down the list to find the second (compute $A_R(4)$) place where these two inputs are used. Pull $c_R(1)$ and $c_I(2)$ from memory, compute $A_R(1)$ and $A_R(4)$, and store the results in data memory locations $M(9)$ and $M(3)$ previously occupied by $c_R(1)$ and $c_I(2)$.

Next, look for the computation for $A_I(1)$ on the list and repeat the same set of steps. Continue this process until all the Algorithm Steps have been computed and their results stored in the Memory Map addresses.

| Algorithm Steps | Memory Map |
|---|---|
| $A_R(1) = c_R(1) + c_I(2)$ | $A_R(1) \Rightarrow M(9)$ |
| $A_I(1) = c_I(1) - c_R(2)$ | $A_I(1) \Rightarrow M(6)$ |
| $A_R(2) = c_R(3) + c_I(4)$ | $A_R(2) \Rightarrow M(8)$ |
| $A_I(2) = c_I(3) - c_R(4)$ | $A_I(2) \Rightarrow M(2)$ |
| $A_R(3) = c_R(3) - c_I(4)$ | $A_R(3) \Rightarrow M(4)$ |
| $A_I(3) = c_I(3) + c_R(4)$ | $A_I(3) \Rightarrow M(1)$ |
| $A_R(4) = c_R(1) - c_I(2)$ | $A_R(4) \Rightarrow M(3)$ |
| $A_I(4) = c_I(1) + c_R(2)$ | $A_I(4) \Rightarrow M(7)$ |

### 8.6.3 Rader 5-Point FFT

The Rader [3] 5-point FFT requires 42 adds, 12 multiplies, 12 data memory locations, and 4 multiplier constant memory locations. The structure of this algorithm is very similar to the 4-point transform because the 4-point transform is used twice in the computations. The first time is Stages 1 and 2. After these stages, complex multiplications are required to prepare the data for the output computations. Finally, the three stages after the multiplications are an inverse 4-point transform plus the computations required to include the fifth input data point in the output frequency components. Stage 4 is the first stage of a 4-point IFFT. Stage 5 is where the fifth input data point is added, and the final stage is the second stage of a 4-point IFFT. Section 8.11.1 provides more detail on the Rader algorithm, and Section 2.3 gives additional information on how the 4-point FFT algorithm is converted to a 4-point IFFT. The six stages are as follows.

### Stage 1: Input Adds

This stage does not require additional data memory locations or accessing of multiplier constants. The strategy for converting these equations to code is to start at the top (compute $b_R(1)$) and identify the pair of inputs to be used first (in this case $a_R(3)$ and $a_R(2)$). Then look down the list to find the second (compute $b_R(2)$) place where these two inputs are used. Pull $a_R(3)$ and $a_R(2)$ from memory, compute $b_R(1)$ and $b_R(2)$, and store the results in data memory locations $M(2)$ and $M(3)$ previously occupied by $a_R(3)$ and $a_R(2)$.

Next, look for the computation for $b_I(1)$ on the list and repeat the same set of steps. Continue this process until all the Algorithm Steps have been computed and their results stored in the Memory Map addresses.

| Algorithm Steps | Memory Map |
|---|---|
| $b_R(1) = a_R(3) + a_R(2)$ | $b_R(1) \Rightarrow M(2)$ |
| $b_I(1) = a_I(3) + a_I(2)$ | $b_I(1) \Rightarrow M(7)$ |
| $b_R(2) = a_R(3) - a_R(2)$ | $b_R(2) \Rightarrow M(3)$ |

|                   Algorithm Steps | Memory Map |
|---|---|
| $b_I(2) = a_I(3) - a_I(2)$ | $b_I(2) \Rightarrow M(8)$ |
| $b_R(3) = a_R(4) + a_R(1)$ | $b_R(3) \Rightarrow M(1)$ |
| $b_I(3) = a_I(4) + a_I(1)$ | $b_I(3) \Rightarrow M(6)$ |
| $b_R(4) = a_R(4) - a_R(1)$ | $b_R(4) \Rightarrow M(4)$ |
| $b_I(4) = a_I(4) - a_I(1)$ | $b_I(4) \Rightarrow M(9)$ |

### Stage 2: Second Set of Input Adds

This stage also does not require additional data memory locations or accessing of multiplier constants. The strategy for converting these equations to code is to start at the top (compute $c_R(1)$) and identify the pair of inputs to be used first (in this case $b_R(1)$ and $b_R(3)$). Then look down the list to find the second (compute $c_R(3)$) place where these two inputs are used. Pull $b_R(1)$ and $b_R(3)$ from memory, compute $c_R(1)$ and $c_R(3)$, and store the results in data memory locations $M(1)$ and $M(2)$ previously occupied by $b_R(1)$ and $b_R(3)$.

Next, look for the computation for $c_I(1)$ on the list and repeat the same set of steps. Continue this process until all the Algorithm Steps have been computed and their results stored in the Memory Map addresses.

|                   Algorithm Steps | Memory Map |
|---|---|
| $c_R(1) = b_R(1) + b_R(3)$ | $c_R(1) \Rightarrow M(1)$ |
| $c_I(1) = b_I(1) + b_I(3)$ | $c_I(1) \Rightarrow M(6)$ |
| $c_R(2) = b_R(2) + b_I(4)$ | $c_R(2) \Rightarrow M(3)$ |
| $c_I(2) = b_I(2) - b_R(4)$ | $c_I(2) \Rightarrow M(8)$ |
| $c_R(3) = b_R(1) - b_R(3)$ | $c_R(3) \Rightarrow M(2)$ |
| $c_I(3) = b_I(1) - b_I(3)$ | $c_I(3) \Rightarrow M(7)$ |
| $c_R(4) = b_R(2) - b_I(4)$ | $c_R(4) \Rightarrow M(9)$ |
| $c_I(4) = b_I(2) + b_R(4)$ | $c_I(4) \Rightarrow M(4)$ |

### Stage 3: Multiplies

This stage contains all of the multiplications and also requires additional data memory locations to store intermediate results. In Steps 1 through 4, multiply accumulation requires additional data memory locations because the input data is multiplied by two different constants as part of two different outputs. In Steps 5 through 8, multiplication is performed by pulling a data value from memory, multiplying it by the appropriate constant, and returning the result to the same data memory location (in-place).

|                   Algorithm Steps | Memory Map |
|---|---|
| $d_R(3) = (1/2) * [c_R(2) * \sin(2\pi/5) + c_I(2) * \sin(4\pi/5)]$ | $d_R(3) \Rightarrow M(10)$ |
| $d_I(3) = (1/2) * [-c_R(2) * \sin(4\pi/5) + c_I(2) * \sin(2\pi/5)]$ | $d_I(3) \Rightarrow M(8)$ |
| $d_R(4) = (1/2) * [-c_R(4) * \sin(2\pi/5) + c_I(4) * \sin(4\pi/5)]$ | $d_R(4) \Rightarrow M(3)$ |
| $d_I(4) = (1/2) * [-c_R(4) * \sin(4\pi/5) - c_I(4) * \sin(2\pi/5)]$ | $d_I(4) \Rightarrow M(9)$ |
| $d_R(1) = (1/2) * [\cos(2\pi/5) + \cos(4\pi/5)] * c_R(1)$ | $d_R(1) \Rightarrow M(11)$ |
| $d_I(1) = (1/2) * [\cos(2\pi/5) + \cos(4\pi/5)] * c_I(1)$ | $d_I(1) \Rightarrow M(4)$ |
| $d_R(2) = (1/2) * [-\cos(2\pi/5) + \cos(4\pi/5)] * c_R(3)$ | $d_R(2) \Rightarrow M(2)$ |
| $d_I(2) = (1/2) * [-\cos(2\pi/5) + \cos(4\pi/5)] * c_I(3)$ | $d_I(2) \Rightarrow M(7)$ |

## Stage 4: First Stage of Postmultiply Adds

The strategy for converting these equations to code is to start at the top (compute $e_R(1)$) and identify the pair of inputs to be used first (in this case $d_R(1)$ and $d_R(2)$). Then look down the list to find the second (compute $e_R(2)$) place where these two inputs are used. Pull $d_R(1)$ and $d_R(2)$ from memory, compute $e_R(1)$ and $e_R(2)$, and store the results in data memory locations $M(1)$ and $M(2)$ previously occupied by $d_R(1)$ and $d_R(2)$.

Next, look for the computation for $e_I(1)$ on the list and repeat the same set of steps. Continue this process until all the Algorithm Steps have been computed and their results stored in the Memory Map addresses.

| Algorithm Steps | Memory Map |
|---|---|
| $e_R(1) = d_R(1) + d_R(2)$ | $e_R(1) \Rightarrow M(1)$ |
| $e_I(1) = d_I(1) + d_I(2)$ | $e_I(1) \Rightarrow M(6)$ |
| $e_R(2) = d_R(1) - d_R(2)$ | $e_R(2) \Rightarrow M(2)$ |
| $e_I(2) = d_I(1) - d_I(2)$ | $e_I(2) \Rightarrow M(7)$ |
| $e_R(3) = d_R(3) + d_R(4)$ | $e_R(3) \Rightarrow M(3)$ |
| $e_I(3) = d_I(3) + d_I(4)$ | $e_I(3) \Rightarrow M(8)$ |
| $e_R(4) = d_R(3) - d_R(4)$ | $e_R(4) \Rightarrow M(4)$ |
| $e_I(4) = d_I(3) - d_I(4)$ | $e_I(4) \Rightarrow M(9)$ |

## Stage 5: Second Stage of Postmultiply Adds

Since $a_R(0)$ and $a_I(0)$ are each used in three of the computational steps, their data memory locations cannot be modified until the last time they are used. Since each other input to this stage is used only once, and is not needed again, the results are placed back in their data memory locations.

The strategy for converting these equations to code is to start at the top (compute $f_R(1)$) and identify the pair of inputs to be used first (in this case $e_R(1)$ and $a_R(0)$). Pull $e_R(1)$ and $a_R(0)$ from memory, compute $f_R(1)$, and store the results in data memory location $M(1)$ previously occupied by $e_R(1)$.

Next, look for the computation for $f_I(1)$ on the list and repeat the same set of steps. Continue this process until all the Algorithm Steps have been computed and their results stored in the Memory Map addresses.

| Algorithm Steps | Memory Map |
|---|---|
| $f_R(1) = e_R(1) + a_R(0)$ | $f_R(1) \Rightarrow M(1)$ |
| $f_I(1) = e_I(1) + a_I(0)$ | $f_I(1) \Rightarrow M(6)$ |
| $f_R(2) = e_R(2) + a_R(0)$ | $f_R(2) \Rightarrow M(2)$ |
| $f_I(2) = e_I(2) + a_I(0)$ | $f_I(2) \Rightarrow M(7)$ |
| $A_R(0) = c_R(1) + a_R(0)$ | $A_R(0) \Rightarrow M(0)$ |
| $A_I(0) = c_I(1) + a_I(0)$ | $A_I(0) \Rightarrow M(5)$ |

## Stage 6: Output Adds

The strategy for converting these equations to code is to start at the top (compute $A_R(1)$) and identify the pair of inputs to be used first (in this case $f_R(1)$ and $e_R(3)$). Then look down the list to find the second (compute $A_R(4)$) place where these two inputs are used. Pull $f_R(1)$ and $e_R(3)$ from memory, compute $A_R(1)$ and $A_R(4)$, and store the results in data memory locations $M(3)$ and $M(1)$ previously occupied by $f_R(1)$ and $e_R(3)$.

Next, look for the computation for $A_I(1)$ and repeat the same set of steps. Continue this process until all the Algorithm Steps have been computed and all of the results are returned to the data memory locations.

| Algorithm Steps | Memory Map |
|---|---|
| $A_R(1) = f_R(1) - e_R(3)$ | $A_R(1) \Rightarrow M(3)$ |
| $A_I(1) = f_I(1) - e_I(3)$ | $A_I(1) \Rightarrow M(8)$ |
| $A_R(2) = f_R(2) + e_I(4)$ | $A_R(2) \Rightarrow M(2)$ |
| $A_I(2) = f_I(2) - e_R(4)$ | $A_I(2) \Rightarrow M(7)$ |
| $A_R(3) = f_R(2) - e_I(4)$ | $A_R(3) \Rightarrow M(9)$ |
| $A_I(3) = f_I(2) + e_R(4)$ | $A_I(3) \Rightarrow M(6)$ |
| $A_R(4) = f_R(1) + e_R(3)$ | $A_R(4) \Rightarrow M(1)$ |
| $A_I(4) = f_I(1) + e_I(3)$ | $A_I(4) \Rightarrow M(4)$ |

## 8.7 SEVEN-POINT FFT

The 7-point DFT is defined for $k = 0, 1, 2, 3, 4, 5,$ and 6 as

$$A(k) = \sum_{n=0}^{6} a(n) * e^{-j2\pi kn/7} \qquad (8\text{-}7)$$

If the 7-point DFT is calculated directly from Equation 8-7, it requires 36 complex multiplies and 42 complex adds. Since a complex multiply uses 4 real multiplies and 2 real adds, and a complex add uses 2 real adds, the 7-point DFT requires 144 real multiplies and 156 real adds. The number of adds and multiplies shown for each of the fast algorithms is significantly less than required for computing the DFT directly. However, if only a subset of the output frequency components is required, it may be more cost effective to compute the DFT equation directly for those terms. For example, if $A(0)$ is the only term needed, it can be computed with 12 adds and no multiplies by using the DFT directly. Each of the other six output frequencies requires 5 complex multiplies and 5 complex adds for a total of 20 real adds and 20 real multiplies. With this in mind the crossover point between using the DFT directly and one of the 7-point FFT algorithms can be determined based on the number of output frequency components that must be computed.

Since all of the input data is required for each output frequency component calculation, the direct DFT computations require 14 data memory locations for the input data and 14 more for the output frequency components. This is a total of 28 data memory locations, since the input and output are complex. Similarly, the DFT data addressing is sequential (i.e., 0 through 6 for each output frequency component), and the computational architecture is simple, since they can all be performed by using a complex multiply accumulator (see Chapter 10 for details). Addressing the complex multiplier coefficients requires either a modulo arithmetic scheme ($k * n$ mod 7) or that the addresses be stored in program memory.

Two fast versions of the 7-point DFT are presented. The Winograd [1] algorithm was developed by using a decomposition based on circular convolution properties. The

Singleton [2] algorithm was developed by using a decomposition based on the complex conjugate symmetry properties of the 7-point transform.

### 8.7.1 Winograd 7-Point FFT

The 7-point Winograd [1] transform algorithm requires 16 multiplies, 72 adds, 22 data memory locations, and 8 multiplier constant memory locations. The eight stages are as follows.

### Stage 1: Input Adds

This stage does not require additional data memory locations or accessing any of the multiplier constants. Further, the add/subtract process is the same for all of the real and imaginary pairs. The strategy for converting these equations to code is to start at the top (compute $b_R(1)$) and identify the pair of inputs to be used first (in this case $a_R(1)$ and $a_R(6)$). Then look down the list to find the second (compute $b_R(2)$) place where these two inputs are used. Pull $a_R(1)$ and $a_R(6)$ from memory, compute $b_R(1)$ and $b_R(2)$, and store the results in data memory locations $M(1)$ and $M(6)$ previously occupied by $a_R(1)$ and $a_R(6)$.

Next, look for the computation for $b_I(1)$ on the list and repeat the same set of steps. Continue this process until all the Algorithm Steps have been computed and their results stored in the Memory Map addresses.

| Algorithm Steps | Memory Map |
|---|---|
| $b_R(1) = a_R(1) + a_R(6)$ | $b_R(1) \Rightarrow M(1)$ |
| $b_I(1) = a_I(1) + a_I(6)$ | $b_I(1) \Rightarrow M(8)$ |
| $b_R(2) = a_R(1) - a_R(6)$ | $b_R(2) \Rightarrow M(6)$ |
| $b_I(2) = a_I(1) - a_I(6)$ | $b_I(2) \Rightarrow M(13)$ |
| $b_R(3) = a_R(4) + a_R(3)$ | $b_R(3) \Rightarrow M(3)$ |
| $b_I(3) = a_I(4) + a_I(3)$ | $b_I(3) \Rightarrow M(10)$ |
| $b_R(4) = a_R(4) - a_R(3)$ | $b_R(4) \Rightarrow M(4)$ |
| $b_I(4) = a_I(4) - a_I(3)$ | $b_I(4) \Rightarrow M(11)$ |
| $b_R(5) = a_R(2) + a_R(5)$ | $b_R(5) \Rightarrow M(2)$ |
| $b_I(5) = a_I(2) + a_I(5)$ | $b_I(5) \Rightarrow M(9)$ |
| $b_R(6) = a_R(2) - a_R(5)$ | $b_R(6) \Rightarrow M(5)$ |
| $b_I(6) = a_I(2) - a_I(5)$ | $b_I(6) \Rightarrow M(12)$ |

### Stage 2: Second Set of Input Adds

This stage requires additional data memory locations to store intermediate results. The strategy for converting these equations to code is to start at the top (compute $c_R(1)$) and identify the pair of inputs to be used first (in this case $b_R(1)$ and $b_R(3)$). Then look down the list to find the second (compute $c_R(3)$) place where these two inputs are used. Pull $b_R(1)$ and $b_R(3)$ from memory, compute $c_R(1)$ and $c_R(3)$, and store the results in data memory locations $M(14)$ and $M(15)$ different than previously occupied by $b_R(1)$ and $b_R(3)$. Different data memory locations are required because $b_R(1)$ and $b_R(3)$ are also used in computing $c_R(4)$ and $c_R(2)$, respectively.

Next, look for the computation for $c_I(1)$ on the list and repeat the same set of steps. Continue this process until all the Algorithm Steps have been computed and their results stored in the Memory Map addresses. Note that $b_R(5)$, $b_I(5)$, $b_R(6)$, and $b_I(6)$ are also used in Stage 3.

| Algorithm Steps | Memory Map |
|---|---|
| $c_R(1) = b_R(1) + b_R(3)$ | $c_R(1) \Rightarrow M(14)$ |
| $c_I(1) = b_I(1) + b_I(3)$ | $c_I(1) \Rightarrow M(18)$ |
| $c_R(2) = b_R(3) - b_R(5)$ | $c_R(2) \Rightarrow M(3)$ |
| $c_I(2) = b_I(3) - b_I(5)$ | $c_I(2) \Rightarrow M(10)$ |
| $c_R(3) = b_R(1) - b_R(3)$ | $c_R(3) \Rightarrow M(15)$ |
| $c_I(3) = b_I(1) - b_I(3)$ | $c_I(3) \Rightarrow M(19)$ |
| $c_R(4) = b_R(5) - b_R(1)$ | $c_R(4) \Rightarrow M(1)$ |
| $c_I(4) = b_I(5) - b_I(1)$ | $c_I(4) \Rightarrow M(8)$ |
| $c_R(5) = b_R(2) + b_R(4)$ | $c_R(5) \Rightarrow M(16)$ |
| $c_I(5) = b_I(2) + b_I(4)$ | $c_I(5) \Rightarrow M(20)$ |
| $c_R(6) = b_R(2) - b_R(4)$ | $c_R(6) \Rightarrow M(17)$ |
| $c_I(6) = b_I(2) - b_I(4)$ | $c_I(6) \Rightarrow M(21)$ |
| $c_R(7) = b_R(4) - b_R(6)$ | $c_R(7) \Rightarrow M(4)$ |
| $c_I(7) = b_I(4) - b_I(6)$ | $c_I(7) \Rightarrow M(11)$ |
| $c_R(8) = b_R(6) - b_R(2)$ | $c_R(8) \Rightarrow M(6)$ |
| $c_I(8) = b_I(6) - b_I(2)$ | $c_I(8) \Rightarrow M(13)$ |

## Stage 3: Third Set of Input Adds

The strategy for converting these equations to code is to start at the top (compute $d_R(1)$) and identify the pair of inputs to be used first (in this case $b_R(5)$ and $c_R(1)$). In this case there is only one result associated with these two input data values. Pull $b_R(5)$ and $c_R(1)$ from memory, compute $d_R(1)$, and store the result in data memory location $M(2)$ previously occupied by $b_R(5)$.

Next, look for the computation for $d_I(1)$ on the list and repeat the same set of steps. Continue this process until all the Algorithm Steps have been computed and their results stored in the Memory Map addresses.

| Algorithm Steps | Memory Map |
|---|---|
| $d_R(1) = b_R(5) + c_R(1)$ | $d_R(1) \Rightarrow M(2)$ |
| $d_I(1) = b_I(5) + c_I(1)$ | $d_I(1) \Rightarrow M(9)$ |
| $d_R(2) = b_R(6) + c_R(5)$ | $d_R(2) \Rightarrow M(5)$ |
| $d_I(2) = b_I(6) + c_I(5)$ | $d_I(2) \Rightarrow M(12)$ |
| $e_R(0) = a_R(0) + d_R(1)$ | $e_R(0) \Rightarrow M(0)$ |
| $e_I(0) = a_I(0) + d_I(1)$ | $e_I(0) \Rightarrow M(7)$ |

## Stage 4: Multiplications

This stage contains all of the multiplications and also requires additional data memory locations to store intermediate results. In all cases the multiplication is performed by pulling a data value from memory, multiplying it by the appropriate constant, and returning the result to the same data memory location.

| Algorithm Steps | Memory Map |
|---|---|
| $e_R(1) = \{-1 + [\cos(2\pi/7) + \cos(4\pi/7) + \cos(6\pi/7)]/3\} * d_R(1)$ | $e_R(1) \Rightarrow M(2)$ |
| $e_I(1) = \{-1 + [\cos(2\pi/7) + \cos(4\pi/7) + \cos(6\pi/7)]/3\} * d_I(1)$ | $e_I(1) \Rightarrow M(9)$ |
| $e_R(2) = \{[2 * \cos(2\pi/7) - \cos(4\pi/7) - \cos(6\pi/7)]/3\} * c_R(3)$ | $e_R(2) \Rightarrow M(15)$ |
| $e_I(2) = \{[2 * \cos(2\pi/7) - \cos(4\pi/7) - \cos(6\pi/7)]/3\} * c_I(3)$ | $e_I(2) \Rightarrow M(19)$ |
| $e_R(3) = \{[\cos(2\pi/7) - 2 * \cos(4\pi/7) + \cos(6\pi/7)]/3\} * c_R(2)$ | $e_R(3) \Rightarrow M(3)$ |
| $e_I(3) = \{[\cos(2\pi/7) - 2 * \cos(4\pi/7) + \cos(6\pi/7)]/3\} * c_I(2)$ | $e_I(3) \Rightarrow M(10)$ |
| $e_R(4) = \{[\cos(2\pi/7) + \cos(4\pi/7) - 2 * \cos(6\pi/7)]/3\} * c_R(4)$ | $e_R(4) \Rightarrow M(1)$ |
| $e_I(4) = \{[\cos(2\pi/7) + \cos(4\pi/7) - 2 * \cos(6\pi/7)]/3\} * c_I(4)$ | $e_I(4) \Rightarrow M(8)$ |
| $e_R(5) = -\{[\sin(2\pi/7) + \sin(4\pi/7) - \sin(6\pi/7)]/3\} * d_I(2)$ | $e_R(5) \Rightarrow M(12)$ |
| $e_I(5) = \{[\sin(2\pi/7) + \sin(4\pi/7) - \sin(6\pi/7)]/3\} * d_R(2)$ | $e_I(5) \Rightarrow M(5)$ |
| $e_R(6) = -\{[2 * \sin(2\pi/7) - \sin(4\pi/7) + \sin(6\pi/7)]/3\} * c_I(6)$ | $e_R(6) \Rightarrow M(21)$ |
| $e_I(6) = \{[2 * \sin(2\pi/7) - \sin(4\pi/7) + \sin(6\pi/7)]/3\} * c_R(6)$ | $e_I(6) \Rightarrow M(17)$ |
| $e_R(7) = -\{[\sin(2\pi/7) - 2 * \sin(4\pi/7) - \sin(6\pi/7)]/3\} * c_I(7)$ | $e_R(7) \Rightarrow M(11)$ |
| $e_I(7) = \{[\sin(2\pi/7) - 2 * \sin(4\pi/7) - \sin(6\pi/7)]/3\} * c_R(7)$ | $e_I(7) \Rightarrow M(4)$ |
| $e_R(8) = -\{[\sin(2\pi/7) + \sin(4\pi/7) + 2 * \sin(6\pi/7)]/3\} * c_I(8)$ | $e_R(8) \Rightarrow M(13)$ |
| $e_I(8) = \{[\sin(2\pi/7) + \sin(4\pi/7) + 2 * \sin(6\pi/7)]/3\} * c_R(8)$ | $e_I(8) \Rightarrow M(6)$ |

## Stage 5: First Postmultiply Adds

The strategy for converting these equations to code is to start at the top (compute $f_R(1)$) and identify the pair of inputs to be used first (in this case $e_R(0)$ and $e_R(1)$). In this case there is only one result associated with these two input data values. Pull $e_R(0)$ and $e_R(1)$ from memory, compute $f_R(1)$, and store the result in data memory location $M(2)$ previously occupied by $e_R(1)$.

Next, look for the computation for $f_I(1)$ on the list and repeat the same set of steps. The remaining adds and subtracts require additional data memory locations because $e_R(5)$ is used in three places. Therefore, its data memory location cannot be used for results until the last time it is used as the input to a set of computations. Continue this process until all the Algorithm Steps have been computed and their results stored in the Memory Map addresses.

| Algorithm Steps | Memory Map |
|---|---|
| $f_R(1) = e_R(0) + e_R(1)$ | $f_R(1) \Rightarrow M(2)$ |
| $f_I(1) = e_I(0) + e_I(1)$ | $f_I(1) \Rightarrow M(9)$ |
| $f_R(2) = e_R(5) + e_R(6)$ | $f_R(2) \Rightarrow M(20)$ |

| Algorithm Steps | Memory Map |
|---|---|
| $f_I(2) = e_I(5) + e_I(6)$ | $f_I(2) \Rightarrow M(16)$ |
| $f_R(3) = e_R(5) - e_R(6)$ | $f_R(3) \Rightarrow M(21)$ |
| $f_I(3) = e_I(5) - e_I(6)$ | $f_I(3) \Rightarrow M(17)$ |
| $f_R(4) = e_R(5) - e_R(7)$ | $f_R(4) \Rightarrow M(12)$ |
| $f_I(4) = e_I(5) - e_I(7)$ | $f_I(4) \Rightarrow M(5)$ |

## Stage 6: Second Postmultiply Adds

The strategy for converting these equations to code is to start at the top (compute $g_R(1)$) and identify the pair of inputs to be used first (in this case $f_R(1)$ and $e_R(2)$). Notice that the same set of inputs is used to compute $g_R(2)$. However, $f_R(1)$ is also used to compute $g_R(3)$. Its memory location cannot be used to store $g_R(1)$ or $g_R(2)$, but can be used to store $g_R(3)$. Therefore, the strategy is to pull $f_R(1)$ and $e_R(2)$ from memory, compute $g_R(1)$ and $g_R(2)$, and store the results in data memory locations $M(14)$ and $M(15)$ previously occupied by $c_R(1)$ and $e_R(2)$.

Next, look for the computation for $g_I(1)$ on the list and repeat the same set of steps. Continue this process until all the Algorithm Steps have been computed and all of the results are returned to the data memory locations.

| Algorithm Steps | Memory Map |
|---|---|
| $g_R(1) = f_R(1) + e_R(2)$ | $g_R(1) \Rightarrow M(14)$ |
| $g_I(1) = f_I(1) + e_I(2)$ | $g_I(1) \Rightarrow M(18)$ |
| $g_R(2) = f_R(1) - e_R(2)$ | $g_R(2) \Rightarrow M(15)$ |
| $g_I(2) = f_I(1) - e_I(2)$ | $g_I(2) \Rightarrow M(19)$ |
| $g_R(3) = f_R(1) - e_R(3)$ | $g_R(3) \Rightarrow M(2)$ |
| $g_I(3) = f_I(1) - e_I(3)$ | $g_I(3) \Rightarrow M(9)$ |
| $g_R(4) = f_R(2) + e_R(7)$ | $g_R(4) \Rightarrow M(11)$ |
| $g_I(4) = f_I(2) + e_I(7)$ | $g_I(4) \Rightarrow M(4)$ |
| $g_R(5) = f_R(3) - e_R(8)$ | $g_R(5) \Rightarrow M(21)$ |
| $g_I(5) = f_I(3) - e_I(8)$ | $g_I(5) \Rightarrow M(17)$ |
| $g_R(6) = f_R(4) + e_R(8)$ | $g_R(6) \Rightarrow M(13)$ |
| $g_I(6) = f_I(4) + e_I(8)$ | $g_I(6) \Rightarrow M(6)$ |

## Stage 7: Third Postmultiply Adds

The strategy for converting these equations to code is to start at the top (compute $h_R(1)$) and identify the pair of inputs to be used first (in this case $g_R(1)$ and $e_R(3)$). For this set of computations only $e_R(4)$ and $e_I(4)$ are used more than once. Therefore, pull $g_R(1)$ and $e_R(3)$ from memory, compute $h_R(1)$, and store the result in data memory location $M(3)$ previously occupied by $e_R(3)$.

Next, look for the computation for $h_I(1)$ on the list and repeat the same set of steps. Continue this process until all the Algorithm Steps have been computed and all of the results are returned to the data memory locations.

| Algorithm Steps | Memory Map |
|---|---|
| $h_R(1) = g_R(1) + e_R(3)$ | $h_R(1) \Rightarrow M(3)$ |
| $h_I(1) = g_I(1) + e_I(3)$ | $h_I(1) \Rightarrow M(10)$ |
| $h_R(2) = g_R(2) - e_R(4)$ | $h_R(2) \Rightarrow M(15)$ |
| $h_I(2) = g_I(2) - e_I(4)$ | $h_I(2) \Rightarrow M(19)$ |
| $h_R(3) = g_R(3) + e_R(4)$ | $h_R(3) \Rightarrow M(1)$ |
| $h_I(3) = g_I(3) + e_I(4)$ | $h_I(3) \Rightarrow M(8)$ |

## Stage 8: Output Adds

The strategy for converting these equations to code is to start at the top (compute $A_R(1)$) and identify the pair of inputs to be used first (in this case $h_R(1)$ and $g_R(4)$). Next identify the other computation, $A_R(6)$, in the list that uses these same two inputs. Therefore, pull $h_R(1)$ and $g_R(4)$ from memory, compute $A_R(1)$ and $A_R(6)$, and store the result in data memory locations $M(3)$ and $M(11)$ previously occupied by $h_R(1)$ and $g_R(4)$. Next, look for the computation for $A_I(1)$ on the list and repeat the same set of steps.

The output of this stage requires only 14 data memory locations. Therefore, the results of computing $A_R(2)$ and $A_R(5)$, using intermediate results located in the extra data memory locations, are placed in available locations within the original $M(0)$ to $M(13)$. Continue this process until all the Algorithm Steps have been computed and all of the results are returned to the data memory locations.

| Algorithm Steps | Memory Map |
|---|---|
| $A_R(0) = e_R(0)$ | $A_R(0) \Rightarrow M(0)$ |
| $A_I(0) = e_I(0)$ | $A_I(0) \Rightarrow M(7)$ |
| $A_R(1) = h_R(1) - g_R(4)$ | $A_R(1) \Rightarrow M(3)$ |
| $A_I(1) = h_I(1) - g_I(4)$ | $A_I(1) \Rightarrow M(10)$ |
| $A_R(2) = h_R(2) - g_R(5)$ | $A_R(2) \Rightarrow M(2)$ |
| $A_I(2) = h_I(2) - g_I(5)$ | $A_I(2) \Rightarrow M(9)$ |
| $A_R(3) = h_R(3) + g_R(6)$ | $A_R(3) \Rightarrow M(1)$ |
| $A_I(3) = h_I(3) + g_I(6)$ | $A_I(3) \Rightarrow M(8)$ |
| $A_R(4) = h_R(3) - g_R(6)$ | $A_R(4) \Rightarrow M(13)$ |
| $A_I(4) = h_I(3) - g_I(6)$ | $A_I(4) \Rightarrow M(6)$ |
| $A_R(5) = h_R(2) + g_R(5)$ | $A_R(5) \Rightarrow M(5)$ |
| $A_I(5) = h_I(2) + g_I(5)$ | $A_I(5) \Rightarrow M(12)$ |
| $A_R(6) = h_R(1) + g_R(4)$ | $A_R(6) \Rightarrow M(11)$ |
| $A_I(6) = h_I(1) + g_I(4)$ | $A_I(6) \Rightarrow M(4)$ |

## 8.7.2 Singleton 7-Point FFT

The Singleton [2] 7-point FFT requires 60 adds, 36 multiplies, 17 data memory locations, and 6 multiplier constant memory locations. The three stages are as follows.

### Stage 1: Input Adds

This stage does not require additional data memory locations or accessing any of the multiplier constants. Further, the add/subtract process is the same for all of the real and imaginary pairs. The strategy for converting these equations to code is to start at the top (compute $b_R(1)$) and identify the pair of inputs to be used first (in this case $a_R(1)$ and $a_R(6)$). Then look down the list to find the second (compute $b_R(2)$) place where these two inputs are used. Pull $a_R(1)$ and $a_R(6)$ from memory, compute $b_R(1)$ and $b_R(2)$, and store the results in data memory locations $M(1)$ and $M(6)$ previously occupied by $a_R(1)$ and $a_R(6)$.

Next, look for the computation for $b_I(1)$ on the list and repeat the same set of steps. Continue this process until all the Algorithm Steps have been computed and their results stored in the Memory Map addresses.

| Algorithm Steps | Memory Map |
|---|---|
| $b_R(1) = a_R(1) + a_R(6)$ | $b_R(1) \Rightarrow M(1)$ |
| $b_I(1) = a_I(1) + a_I(6)$ | $b_I(1) \Rightarrow M(8)$ |
| $b_R(2) = a_R(1) - a_R(6)$ | $b_R(2) \Rightarrow M(6)$ |
| $b_I(2) = a_I(1) - a_I(6)$ | $b_I(2) \Rightarrow M(13)$ |
| $b_R(3) = a_R(2) + a_R(5)$ | $b_R(3) \Rightarrow M(2)$ |
| $b_I(3) = a_I(2) + a_I(5)$ | $b_I(3) \Rightarrow M(9)$ |
| $b_R(4) = a_R(2) - a_R(5)$ | $b_R(4) \Rightarrow M(5)$ |
| $b_I(4) = a_I(2) - a_I(5)$ | $b_I(4) \Rightarrow M(12)$ |
| $b_R(5) = a_R(3) + a_R(4)$ | $b_R(5) \Rightarrow M(3)$ |
| $b_I(5) = a_I(3) + a_I(4)$ | $b_I(5) \Rightarrow M(10)$ |
| $b_R(6) = a_R(3) - a_R(4)$ | $b_R(6) \Rightarrow M(4)$ |
| $b_I(6) = a_I(3) - a_I(4)$ | $b_I(6) \Rightarrow M(11)$ |

### Stage 2: Multiply-Accumulates

This stage contains all of the multiplications and also requires additional data memory locations to store intermediate results because of multiple multiply-accumulate operations requiring the same input data. The terms with the sine multipliers are computed first to minimize required memory. The Memory Map is based on Constraint 5 of Section 8.2.

| Algorithm Steps | Memory Map |
|---|---|
| $c_R(2) = b_R(2) * \sin(2\pi/7) + b_R(4) * \sin(4\pi/7) + b_R(6) * \sin(6\pi/7)$ | $c_R(2) \Rightarrow M(14)$ |
| $c_R(4) = b_R(2) * \sin(4\pi/7) - b_R(4) * \sin(6\pi/7) - b_R(6) * \sin(2\pi/7)$ | $c_R(4) \Rightarrow M(15)$ |
| $c_R(6) = b_R(2) * \sin(6\pi/7) - b_R(4) * \sin(2\pi/7) + b_R(6) * \sin(4\pi/7)$ | $c_R(6) \Rightarrow M(16)$ |
| $c_R(1) = a_R(0) + b_R(1) * \cos(2\pi/7) + b_R(3) * \cos(4\pi/7) + b_R(5) * \cos(6\pi/7)$ | $c_R(1) \Rightarrow M(4)$ |
| $c_R(3) = a_R(0) + b_R(1) * \cos(4\pi/7) + b_R(3) * \cos(6\pi/7) + b_R(5) * \cos(2\pi/7)$ | $c_R(3) \Rightarrow M(5)$ |
| $c_R(5) = a_R(0) + b_R(1) * \cos(6\pi/7) + b_R(3) * \cos(2\pi/7) + b_R(5) * \cos(4\pi/7)$ | $c_R(5) \Rightarrow M(6)$ |
| $A_R(0) = a_R(0) + b_R(1) + b_R(3) + b_R(5)$ | $A_R(0) \Rightarrow M(0)$ |
| $c_I(2) = b_I(2) * \sin(2\pi/7) + b_I(4) * \sin(4\pi/7) + b_I(6) * \sin(6\pi/7)$ | $c_I(2) \Rightarrow M(1)$ |

| **Algorithm Steps** | **Memory Map** |
|---|---|
| $c_I(4) = b_I(2) * \sin(4\pi/7) - b_I(4) * \sin(6\pi/7) - b_I(6) * \sin(2\pi/7)$ | $c_I(4) \Rightarrow M(2)$ |
| $c_I(6) = b_I(2) * \sin(6\pi/7) - b_I(4) * \sin(2\pi/7) + b_I(6) * \sin(4\pi/7)$ | $c_I(6) \Rightarrow M(3)$ |
| $c_I(1) = a_I(0) + b_I(1) * \cos(2\pi/7) + b_I(3) * \cos(4\pi/7) + b_I(5) * \cos(6\pi/7)$ | $c_I(1) \Rightarrow M(11)$ |
| $c_I(3) = a_I(0) + b_I(1) * \cos(4\pi/7) + b_I(3) * \cos(6\pi/7) + b_I(5) * \cos(2\pi/7)$ | $c_I(3) \Rightarrow M(12)$ |
| $c_I(5) = a_I(0) + b_I(1) * \cos(6\pi/7) + b_I(3) * \cos(2\pi/7) + b_I(5) * \cos(4\pi/7)$ | $c_I(5) \Rightarrow M(13)$ |
| $A_I(0) = a_I(0) + b_I(1) + b_I(3) + b_I(5)$ | $A_I(0) \Rightarrow M(7)$ |

### Stage 3: Output Adds

The strategy for converting these equations to code is to start at the top (compute $A_R(1)$) and identify the pair of inputs to be used first (in this case $c_R(1)$ and $c_I(2)$). Next identify the other computation, $A_R(6)$, in the list that uses these same two inputs. Therefore, pull $c_R(1)$ and $c_I(2)$ from memory, compute $A_R(1)$ and $A_R(6)$, and store the result in data memory locations $M(1)$ and $M(6)$ previously occupied by $c_R(1)$ and $c_I(2)$.

Next, look for the computation for $A_I(1)$ on the list and repeat the same set of steps. Continue this process until all the Algorithm Steps have been computed and all of the results are returned to the data memory locations.

| **Algorithm Steps** | **Memory Map** |
|---|---|
| $A_R(1) = c_R(1) + c_I(2)$ | $A_R(1) \Rightarrow M(1)$ |
| $A_I(1) = c_I(1) - c_R(2)$ | $A_I(1) \Rightarrow M(8)$ |
| $A_R(6) = c_R(1) - c_I(2)$ | $A_R(6) \Rightarrow M(4)$ |
| $A_I(6) = c_I(1) + c_R(2)$ | $A_I(6) \Rightarrow M(11)$ |
| $A_R(2) = c_R(3) + c_I(4)$ | $A_R(2) \Rightarrow M(2)$ |
| $A_I(2) = c_I(3) - c_R(4)$ | $A_I(2) \Rightarrow M(9)$ |
| $A_R(5) = c_R(3) - c_I(4)$ | $A_R(5) \Rightarrow M(5)$ |
| $A_I(5) = c_I(3) + c_R(4)$ | $A_I(5) \Rightarrow M(12)$ |
| $A_R(3) = c_R(5) + c_I(6)$ | $A_R(3) \Rightarrow M(3)$ |
| $A_I(3) = c_I(5) - c_R(6)$ | $A_I(3) \Rightarrow M(10)$ |
| $A_R(4) = c_R(5) - c_I(6)$ | $A_R(4) \Rightarrow M(6)$ |
| $A_I(4) = c_I(5) + c_R(6)$ | $A_I(4) \Rightarrow M(13)$ |

## 8.8 EIGHT-POINT FFT

The 8-point DFT is defined for $k = 0, 1, 2, 3, 4, 5, 6,$ and 7, as

$$A(k) = \sum_{n=0}^{7} a(n) * e^{-j2\pi kn/8} \tag{8-8}$$

Four fast versions of the 8-point DFT are presented. The Winograd [1] algorithm was developed by using a decomposition based on circular convolution properties. The radix-4 and -2 [4] and radix-2 [5] algorithms were developed based on 90° and 180° symmetries. The Practical Transform Length (PTL) [6] algorithm was developed using a decomposition based on complex conjugate symmetry properties.

If the 8-point DFT is calculated directly using Equation 8-8, it would require 16 complex multiplies and 56 complex adds. The number of complex multiplies is lower than expected (seven for each of seven output frequency components) because many of the multiplier constants are $\pm 1$ or $\pm j$ (see Figure 3-1). Since a complex multiply uses 4 real multiplies and 2 real adds, and a complex add uses 2 real adds, the 8-point DFT would require 64 real multiplies and 144 real adds. The number of adds and multiplies shown for each of the fast algorithms is significantly less than required for computing the DFT directly. However, if only a subset of the output frequency components is required, it may be more cost effective to compute the DFT equation directly for those terms. For example, if $A(0)$ is the only term needed, it can be computed with 14 adds and no multiplies using the DFT directly. Each of the other 7 output frequencies requires 6 complex multiplies and 6 complex adds for a total of 24 real adds and 24 real multiplies. With this in mind the crossover point between using the DFT directly and one of the 8-point FFT algorithms can be determined based on the number of output frequency components that must be computed.

Since all of the input data is required for each output frequency component calculation, the direct DFT computations require 16 memory locations for the input data and 16 more for the output frequency components. This is a total of 32 data memory locations, since the input and output are complex. Similarly, the DFT data addressing is sequential (i.e., 0 through 7 for each output frequency component), and the computational architecture is simple since they can all be performed with a complex multiply accumulator (see Chapter 10 for details). Addressing the complex multiplier coefficients requires either a modulo arithmetic scheme ($k * n$ mod 8) or that the addresses be stored in program memory.

### 8.8.1 Winograd 8-Point FFT

The Winograd [1] 8-point FFT requires 52 adds, 4 multiplies, 16 data memory locations, and one multiplier constant memory location. The four stages are as follows.

### Stage 1: Input Adds

This stage does not require any of the multiplier constants. Further, the add/subtract process is the same for all of the real and imaginary pairs. The strategy for converting these equations to code is to start at the top (compute $b_R(0)$) and identify the pair of inputs to be used first (in this case $a_R(0)$ and $a_R(4)$). Then look down the list to find the second (compute $b_R(1)$) place where these two inputs are used. Pull $a_R(0)$ and $a_R(4)$ from memory, compute $b_R(0)$ and $b_R(1)$, and store the results in data memory locations $M(0)$ and $M(4)$ previously occupied by $a_R(0)$ and $a_R(4)$.

Next, look for the computation for $b_I(0)$ on the list and repeat the same set of steps. Continue this process until all the Algorithm Steps have been computed and their results stored in the Memory Map addresses.

| Algorithm Steps | Memory Map |
|---|---|
| $b_R(0) = a_R(0) + a_R(4)$ | $b_R(0) \Rightarrow M(0)$ |
| $b_R(1) = a_R(0) - a_R(4)$ | $b_R(1) \Rightarrow M(4)$ |
| $b_I(0) = a_I(0) + a_I(4)$ | $b_I(0) \Rightarrow M(8)$ |
| $b_I(1) = a_I(0) - a_I(4)$ | $b_I(1) \Rightarrow M(12)$ |
| $b_R(2) = a_R(1) + a_R(5)$ | $b_R(2) \Rightarrow M(1)$ |

| **Algorithm Steps** | **Memory Map** |
| --- | --- |
| $b_R(3) = a_R(1) - a_R(5)$ | $b_R(3) \Rightarrow M(5)$ |
| $b_I(2) = a_I(1) + a_I(5)$ | $b_I(2) \Rightarrow M(9)$ |
| $b_I(3) = a_I(1) - a_I(5)$ | $b_I(3) \Rightarrow M(13)$ |
| $b_R(4) = a_R(2) + a_R(6)$ | $b_R(4) \Rightarrow M(2)$ |
| $b_R(5) = a_R(2) - a_R(6)$ | $b_R(5) \Rightarrow M(6)$ |
| $b_I(4) = a_I(2) + a_I(6)$ | $b_I(4) \Rightarrow M(10)$ |
| $b_I(5) = a_I(2) - a_I(6)$ | $b_I(5) \Rightarrow M(14)$ |
| $b_R(6) = a_R(3) + a_R(7)$ | $b_R(6) \Rightarrow M(3)$ |
| $b_R(7) = a_R(3) - a_R(7)$ | $b_R(7) \Rightarrow M(7)$ |
| $b_I(6) = a_I(3) + a_I(7)$ | $b_I(6) \Rightarrow M(11)$ |
| $b_I(7) = a_I(3) - a_I(7)$ | $b_I(7) \Rightarrow M(15)$ |
| $c_R(0) = b_R(0) + b_R(4)$ | $c_R(0) \Rightarrow M(0)$ |
| $c_R(1) = b_R(0) - b_R(4)$ | $c_R(1) \Rightarrow M(2)$ |
| $c_I(0) = b_I(0) + b_I(4)$ | $c_I(0) \Rightarrow M(8)$ |
| $c_I(1) = b_I(0) - b_I(4)$ | $c_I(1) \Rightarrow M(10)$ |
| $c_R(2) = b_R(2) + b_R(6)$ | $c_R(2) \Rightarrow M(1)$ |
| $c_R(3) = b_R(2) - b_R(6)$ | $c_R(3) \Rightarrow M(3)$ |
| $c_I(2) = b_I(2) + b_I(6)$ | $c_I(2) \Rightarrow M(9)$ |
| $c_I(3) = b_I(2) - b_I(6)$ | $c_I(3) \Rightarrow M(11)$ |
| $c_R(4) = b_R(3) + b_R(7)$ | $c_R(4) \Rightarrow M(5)$ |
| $c_R(5) = b_R(3) - b_R(7)$ | $c_R(5) \Rightarrow M(7)$ |
| $c_I(4) = b_I(3) + b_I(7)$ | $c_I(4) \Rightarrow M(13)$ |
| $c_I(5) = b_I(3) - b_I(7)$ | $c_I(5) \Rightarrow M(15)$ |

## Stage 2: Multiplies

This stage contains all of the multiplications. In all cases the multiplication is performed by pulling a data value from memory, multiplying it by the appropriate constant, and returning the result to the same data memory location. Note that only one multiplier constant is required.

| **Algorithm Steps** | **Memory Map** |
| --- | --- |
| $c_R(4) = c_R(4) * \cos(\pi/4)$ | $c_R(4) \Rightarrow M(5)$ |
| $c_R(5) = c_R(5) * \cos(\pi/4)$ | $c_R(5) \Rightarrow M(7)$ |
| $c_I(4) = c_I(4) * \cos(\pi/4)$ | $c_I(4) \Rightarrow M(13)$ |
| $c_I(5) = c_I(5) * \cos(\pi/4)$ | $c_I(5) \Rightarrow M(15)$ |

## Stage 3: Postmultiply Adds

This stage also does not require any of the multiplier constants. Further, the add/subtract process is the same for all of the real and imaginary pairs. The strategy for converting these equations to code is to start at the top (compute $d_R(0)$) and identify the pair of inputs to be used first (in this case $c_R(0)$ and $c_R(2)$). Then look down the list to find the second (compute $d_R(4)$) place where these two inputs are used. Pull $c_R(0)$ and $c_R(2)$ from memory, compute $d_R(0)$ and $d_R(4)$, and store the results in data memory locations $M(0)$ and $M(1)$ previously occupied by $c_R(0)$ and $c_R(2)$.

Next, look for the computation for $b_I(0)$ on the list and repeat the same set of steps. Continue this process until all the Algorithm Steps have been computed and their results stored in the Memory Map addresses. Notice that some of these additions require one imaginary input and one real input. This approach to these additions implements the required multiplication by $j = \sqrt{-1}$, which converts real parts of data to imaginary parts and imaginary parts to real parts (with a sign change).

| Algorithm Steps | Memory Map |
|---|---|
| $d_R(0) = c_R(0) + c_R(2)$ | $d_R(0) \Rightarrow M(0)$ |
| $d_R(4) = c_R(0) - c_R(2)$ | $d_R(4) \Rightarrow M(1)$ |
| $d_I(0) = c_I(0) + c_I(2)$ | $d_I(0) \Rightarrow M(8)$ |
| $d_I(4) = c_I(0) - c_I(2)$ | $d_I(4) \Rightarrow M(9)$ |
| $d_R(2) = c_R(1) + c_I(3)$ | $d_R(2) \Rightarrow M(2)$ |
| $d_I(2) = c_I(1) - c_R(3)$ | $d_I(2) \Rightarrow M(3)$ |
| $d_R(6) = c_R(1) - c_I(3)$ | $d_R(6) \Rightarrow M(11)$ |
| $d_I(6) = c_I(1) + c_R(3)$ | $d_I(6) \Rightarrow M(10)$ |
| $d_R(1) = b_R(1) + c_R(5)$ | $d_R(1) \Rightarrow M(4)$ |
| $d_R(5) = b_R(1) - c_R(5)$ | $d_R(5) \Rightarrow M(7)$ |
| $d_I(1) = b_I(1) + c_I(5)$ | $d_I(1) \Rightarrow M(12)$ |
| $d_I(5) = b_I(1) - c_I(5)$ | $d_I(5) \Rightarrow M(15)$ |
| $d_R(3) = b_I(5) + c_I(4)$ | $d_R(3) \Rightarrow M(13)$ |
| $d_R(7) = -b_I(5) + c_I(4)$ | $d_R(7) \Rightarrow M(14)$ |
| $d_I(3) = b_R(5) + c_R(4)$ | $d_I(3) \Rightarrow M(5)$ |
| $d_I(7) = b_R(5) - c_R(4)$ | $d_I(7) \Rightarrow M(6)$ |

### Stage 4: Output Adds

This stage also does not require any multiplier constants. Further, the add/subtract process is the same for all of the real and imaginary pairs. The strategy for converting these equations to code is to start at the top (compute $A_R(1)$) and identify the pair of inputs to be used first (in this case $d_R(1)$ and $d_R(3)$). Then look down the list to find the second (compute $A_R(7)$) place where these two inputs are used. Pull $d_R(1)$ and $d_R(3)$ from memory, compute $A_R(1)$ and $A_R(7)$, and store the results in data memory locations $M(4)$ and $M(13)$ previously occupied by $d_R(1)$ and $d_R(3)$.

Next, look for the computation for $A_I(1)$ on the list and repeat the same set of steps. Continue this process until all the Algorithm Steps have been computed and their results stored in the Memory Map addresses.

| Algorithm Steps | Memory Map |
|---|---|
| $A_R(0) = d_R(0)$ | $A_R(0) \Rightarrow M(0)$ |
| $A_I(0) = d_I(0)$ | $A_I(0) \Rightarrow M(8)$ |

| **Algorithm Steps** | **Memory Map** |
|---|---|
| $A_R(4) = d_R(4)$ | $A_R(4) \Rightarrow M(1)$ |
| $A_I(4) = d_I(4)$ | $A_I(4) \Rightarrow M(9)$ |
| $A_R(2) = d_R(2)$ | $A_R(2) \Rightarrow M(2)$ |
| $A_I(2) = d_I(2)$ | $A_I(2) \Rightarrow M(3)$ |
| $A_R(6) = d_R(6)$ | $A_R(6) \Rightarrow M(11)$ |
| $A_I(6) = d_I(6)$ | $A_I(6) \Rightarrow M(10)$ |
| $A_R(1) = d_R(1) + d_R(3)$ | $A_R(1) \Rightarrow M(4)$ |
| $A_I(1) = d_I(1) - d_I(3)$ | $A_I(1) \Rightarrow M(5)$ |
| $A_R(3) = d_R(5) + d_R(7)$ | $A_R(3) \Rightarrow M(14)$ |
| $A_I(3) = d_I(5) + d_I(7)$ | $A_I(3) \Rightarrow M(15)$ |
| $A_R(5) = -d_R(7) + d_R(5)$ | $A_R(5) \Rightarrow M(7)$ |
| $A_I(5) = -d_I(7) + d_I(5)$ | $A_I(5) \Rightarrow M(6)$ |
| $A_R(7) = d_R(1) - d_R(3)$ | $A_R(7) \Rightarrow M(13)$ |
| $A_I(7) = d_I(1) + d_I(3)$ | $A_I(7) \Rightarrow M(12)$ |

### 8.8.2 Eight-Point Radix-4 and -2 Algorithm

The radix-4 and -2 [4] 8-point FFT requires 52 adds, 4 multiplies, 16 data memory locations, and one location for the multiplier constant. The four stages are as follows:

### Stage 1: Input Adds

This stage does not require any of the multiplier constants. Further, the add/subtract process is the same for all of the real and imaginary pairs. The strategy for converting these equations to code is to start at the top (compute $b_R(0)$) and identify the pair of inputs to be used first (in this case $a_R(0)$ and $a_R(4)$). Then look down the list to find the second (compute $b_R(1)$) place where these two inputs are used. Pull $a_R(0)$ and $a_R(4)$ from memory, compute $b_R(0)$ and $b_R(1)$, and store the results in data memory locations $M(0)$ and $M(4)$ previously occupied by $a_R(0)$ and $a_R(4)$.

Next, look for the computation for $b_I(0)$ on the list and repeat the same set of steps. Continue this process until all the Algorithm Steps have been computed and their results stored in the Memory Map addresses.

| **Algorithm Steps** | **Memory Map** |
|---|---|
| $b_R(0) = a_R(0) + a_R(4)$ | $b_R(0) \Rightarrow M(0)$ |
| $b_I(0) = a_I(0) + a_I(4)$ | $b_I(0) \Rightarrow M(8)$ |
| $b_R(1) = a_R(0) - a_R(4)$ | $b_R(1) \Rightarrow M(4)$ |
| $b_I(1) = a_I(0) - a_I(4)$ | $b_I(1) \Rightarrow M(12)$ |
| $b_R(2) = a_R(2) + a_R(6)$ | $b_R(2) \Rightarrow M(2)$ |
| $b_I(2) = a_I(2) + a_I(6)$ | $b_I(2) \Rightarrow M(10)$ |
| $b_R(3) = a_R(2) - a_R(6)$ | $b_R(3) \Rightarrow M(6)$ |
| $b_I(3) = a_I(2) - a_I(6)$ | $b_I(3) \Rightarrow M(14)$ |
| $b_R(4) = a_R(1) + a_R(5)$ | $b_R(4) \Rightarrow M(1)$ |

| Algorithm Steps | Memory Map |
|---|---|
| $b_I(4) = a_I(1) + a_I(5)$ | $b_I(4) \Rightarrow M(9)$ |
| $b_R(5) = a_R(1) - a_R(5)$ | $b_R(5) \Rightarrow M(5)$ |
| $b_I(5) = a_I(1) - a_I(5)$ | $b_I(5) \Rightarrow M(13)$ |
| $b_R(6) = a_R(3) + a_R(7)$ | $b_R(6) \Rightarrow M(3)$ |
| $b_I(6) = a_I(3) + a_I(7)$ | $b_I(6) \Rightarrow M(11)$ |
| $b_R(7) = a_R(3) - a_R(7)$ | $b_R(7) \Rightarrow M(7)$ |
| $b_I(7) = a_I(3) - a_I(7)$ | $b_I(7) \Rightarrow M(15)$ |

## Stage 2: Second Set of Input Adds

This stage also does not require any multiplier constants. Further, the add/subtract process is the same for all of the real and imaginary pairs. The strategy for converting these equations to code is to start at the top (compute $c_R(0)$) and identify the pair of inputs to be used first (in this case $b_R(0)$ and $b_R(2)$). Then look down the list to find the second (compute $c_R(2)$) place where these two inputs are used. Pull $b_R(0)$ and $b_R(2)$ from memory, compute $c_R(0)$ and $c_R(2)$, and store the results in data memory locations $M(0)$ and $M(2)$ previously occupied by $b_R(0)$ and $b_R(2)$.

Next, look for the computation for $c_I(0)$ on the list and repeat the same set of steps. Continue this process until all the Algorithm Steps have been computed and their results stored in the Memory Map addresses.

| Algorithm Steps | Memory Map |
|---|---|
| $c_R(0) = b_R(0) + b_R(2)$ | $c_R(0) \Rightarrow M(0)$ |
| $c_I(0) = b_I(0) + b_I(2)$ | $c_I(0) \Rightarrow M(8)$ |
| $c_R(2) = b_R(0) - b_R(2)$ | $c_R(2) \Rightarrow M(2)$ |
| $c_I(2) = b_I(0) - b_I(2)$ | $c_I(2) \Rightarrow M(10)$ |
| $c_R(1) = b_R(1) + b_I(3)$ | $c_R(1) \Rightarrow M(4)$ |
| $c_I(1) = b_I(1) - b_R(3)$ | $c_I(1) \Rightarrow M(6)$ |
| $c_R(3) = b_R(1) - b_I(3)$ | $c_R(3) \Rightarrow M(14)$ |
| $c_I(3) = b_I(1) + b_R(3)$ | $c_I(3) \Rightarrow M(12)$ |
| $c_R(4) = b_R(4) + b_R(6)$ | $c_R(4) \Rightarrow M(1)$ |
| $c_I(4) = b_I(4) + b_I(6)$ | $c_I(4) \Rightarrow M(9)$ |
| $c_R(6) = b_R(4) - b_R(6)$ | $c_R(6) \Rightarrow M(3)$ |
| $c_I(6) = b_I(4) - b_I(6)$ | $c_I(6) \Rightarrow M(11)$ |
| $c_R(5) = b_R(5) + b_I(7)$ | $c_R(5) \Rightarrow M(5)$ |
| $c_I(5) = b_I(5) - b_R(7)$ | $c_I(5) \Rightarrow M(7)$ |
| $c_R(7) = b_R(5) - b_I(7)$ | $c_R(7) \Rightarrow M(15)$ |
| $c_I(7) = b_I(5) + b_R(7)$ | $c_I(7) \Rightarrow M(13)$ |

## Stage 3: Multiplies

This stage contains all of the multiplications. In all cases, multiplication is performed by pulling a data value from memory, multiplying it by the appropriate constant, and returning the result to the same data memory location. Note that only one multiplier constant is required because $\cos(2\pi/8) = \sin(2\pi/8)$.

| Algorithm Steps | Memory Map |
|---|---|
| $d_R(5) = c_R(5) * \cos(2\pi/8)$ | $d_R(5) \Rightarrow M(5)$ |
| $d_I(5) = c_I(5) * \sin(2\pi/8)$ | $d_I(5) \Rightarrow M(7)$ |
| $d_R(7) = c_R(7) * \cos(2\pi/8)$ | $d_R(7) \Rightarrow M(15)$ |
| $d_I(7) = c_I(7) * \sin(2\pi/8)$ | $d_I(7) \Rightarrow M(13)$ |

## Stage 4: Output Adds

This stage also does not require any multiplier constants. Further, the add/subtract process is the same for all of the real and imaginary pairs. The strategy for converting these equations to code is to start at the top (compute $A_R(0)$) and identify the pair of inputs to be used first (in this case $c_R(0)$ and $c_R(4)$). Then look down the list to find the second (compute $A_R(4)$) place where these two inputs are used. Pull $c_R(0)$ and $c_R(4)$ from memory, compute $A_R(0)$ and $A_R(4)$, and store the results in data memory locations $M(0)$ and $M(1)$ previously occupied by $c_R(0)$ and $c_R(4)$.

Next, look for the computation for $A_I(0)$ on the list and repeat the same set of steps. Continue this process until all the Algorithm Steps have been computed and their results stored in the Memory Map addresses. Notice that some of these additions require one imaginary input and one real input. This approach to these additions implements the required multiplication by $j = \sqrt{-1}$, which converts real parts of data to imaginary parts and imaginary parts to real parts (with a sign change).

| Algorithm Steps | Memory Map |
|---|---|
| $A_R(0) = c_R(0) + c_R(4)$ | $A_R(0) \Rightarrow M(0)$ |
| $A_I(0) = c_I(0) + c_I(4)$ | $A_I(0) \Rightarrow M(8)$ |
| $e_R(5) = d_R(5) + d_I(5)$ | $e_R(5) \Rightarrow M(5)$ |
| $e_I(5) = -d_R(5) + d_I(5)$ | $e_I(5) \Rightarrow M(7)$ |
| $e_R(7) = -d_R(7) + d_I(7)$ | $e_R(7) \Rightarrow M(15)$ |
| $e_I(7) = -d_R(7) - d_I(7)$ | $e_I(7) \Rightarrow M(13)$ |
| $A_R(1) = c_R(1) + e_R(5)$ | $A_R(1) \Rightarrow M(4)$ |
| $A_I(1) = c_I(1) + e_I(5)$ | $A_I(1) \Rightarrow M(6)$ |
| $A_R(2) = c_R(2) + c_I(6)$ | $A_R(2) \Rightarrow M(2)$ |
| $A_I(2) = c_I(2) - c_R(6)$ | $A_I(2) \Rightarrow M(3)$ |
| $A_R(3) = c_R(3) + e_R(7)$ | $A_R(3) \Rightarrow M(14)$ |
| $A_I(3) = c_I(3) + e_I(7)$ | $A_I(3) \Rightarrow M(12)$ |
| $A_R(4) = c_R(0) - c_R(4)$ | $A_R(4) \Rightarrow M(1)$ |
| $A_I(4) = c_I(0) - c_I(4)$ | $A_I(4) \Rightarrow M(9)$ |
| $A_R(5) = c_R(1) - e_R(5)$ | $A_R(5) \Rightarrow M(5)$ |
| $A_I(5) = c_I(1) - e_I(5)$ | $A_I(5) \Rightarrow M(7)$ |
| $A_R(6) = c_R(2) - c_I(6)$ | $A_R(6) \Rightarrow M(11)$ |
| $A_I(6) = c_I(2) + c_R(6)$ | $A_I(6) \Rightarrow M(10)$ |
| $A_R(7) = c_R(3) - e_R(7)$ | $A_R(7) \Rightarrow M(15)$ |
| $A_I(7) = c_I(3) - e_I(7)$ | $A_I(7) \Rightarrow M(13)$ |

### 8.8.3 Eight-Point Radix-2 Algorithm

The radix-2 [5] 8-point FFT requires 52 adds, 4 multiplies, 16 data memory locations, and one location for the multiplier constant. The six stages are as follows:

### Stage 1: Input Adds

This stage does not require any multiplier constants. Further, the add/subtract process is the same for all of the real and imaginary pairs. The strategy for converting these equations to code is to start at the top (compute $b_R(0)$) and identify the pair of inputs to be used first (in this case $a_R(0)$ and $a_R(4)$). Then look down the list to find the second (compute $b_R(1)$) place where these two inputs are used. Pull $a_R(0)$ and $a_R(4)$ from memory, compute $b_R(0)$ and $b_R(1)$, and store the results in data memory locations $M(0)$ and $M(4)$ previously occupied by $a_R(0)$ and $a_R(4)$.

Next, look for the computation for $b_I(0)$ on the list and repeat the same set of steps. Continue this process until all the Algorithm Steps have been computed and their results stored in the Memory Map addresses.

| Algorithm Steps | Memory Map |
|---|---|
| $b_R(0) = a_R(0) + a_R(4)$ | $b_R(0) \Rightarrow M(0)$ |
| $b_I(0) = a_I(0) + a_I(4)$ | $b_I(0) \Rightarrow M(8)$ |
| $b_R(1) = a_R(0) - a_R(4)$ | $b_R(1) \Rightarrow M(4)$ |
| $b_I(1) = a_I(0) - a_I(4)$ | $b_I(1) \Rightarrow M(12)$ |
| $b_R(2) = a_R(2) + a_R(6)$ | $b_R(2) \Rightarrow M(2)$ |
| $b_I(2) = a_I(2) + a_I(6)$ | $b_I(2) \Rightarrow M(10)$ |
| $b_R(3) = a_R(2) - a_R(6)$ | $b_R(3) \Rightarrow M(6)$ |
| $b_I(3) = a_I(2) - a_I(6)$ | $b_I(3) \Rightarrow M(14)$ |
| $b_R(4) = a_R(1) + a_R(5)$ | $b_R(4) \Rightarrow M(1)$ |
| $b_I(4) = a_I(1) + a_I(5)$ | $b_I(4) \Rightarrow M(9)$ |
| $b_R(5) = a_R(1) - a_R(5)$ | $b_R(5) \Rightarrow M(5)$ |
| $b_I(5) = a_I(1) - a_I(5)$ | $b_I(5) \Rightarrow M(13)$ |
| $b_R(6) = a_R(3) + a_R(7)$ | $b_R(6) \Rightarrow M(3)$ |
| $b_I(6) = a_I(3) + a_I(7)$ | $b_I(6) \Rightarrow M(11)$ |
| $b_R(7) = a_R(3) - a_R(7)$ | $b_R(7) \Rightarrow M(7)$ |
| $b_I(7) = a_I(3) - a_I(7)$ | $b_I(7) \Rightarrow M(15)$ |

### Stage 2: Second Set of Input Adds

This stage also does not require any multiplier constants. Further, the add/subtract process is the same for all of the real and imaginary pairs. The strategy for converting these equations to code is to start at the top (compute $d_R(0)$) and identify the pair of inputs to be used first (in this case $b_R(0)$ and $b_R(2)$). Then look down the list to find the second (compute $d_R(2)$) place where these two inputs are used. Pull $b_R(0)$ and $b_R(2)$ from memory, compute $d_R(0)$ and $d_R(2)$, and store the results in data memory locations $M(0)$ and $M(2)$ previously occupied by $b_R(0)$ and $b_R(2)$.

Next, look for the computation for $d_I(0)$ on the list and repeat the same set of steps. Continue this process until all the Algorithm Steps have been computed and their results stored in the Memory Map addresses. Notice that some of these additions require one imaginary input and one real input. This approach to these additions implements the required multiplication by $j = \sqrt{-1}$, which converts real parts of data to imaginary parts and imaginary parts to real parts (with a sign change).

| Algorithm Steps | Memory Map |
|---|---|
| $d_R(0) = b_R(0) + b_R(2)$ | $d_R(0) \Rightarrow M(0)$ |
| $d_I(0) = b_I(0) + b_I(2)$ | $d_I(0) \Rightarrow M(8)$ |
| $d_R(2) = b_R(0) - b_R(2)$ | $d_R(2) \Rightarrow M(2)$ |
| $d_I(2) = b_I(0) - b_I(2)$ | $d_I(2) \Rightarrow M(10)$ |
| $d_R(1) = b_R(1) + b_I(3)$ | $d_R(1) \Rightarrow M(4)$ |
| $d_I(1) = b_I(1) - b_R(3)$ | $d_I(1) \Rightarrow M(6)$ |
| $d_R(3) = b_R(1) - b_I(3)$ | $d_R(3) \Rightarrow M(14)$ |
| $d_I(3) = b_I(1) + b_R(3)$ | $d_I(3) \Rightarrow M(12)$ |
| $d_R(4) = b_R(4) + b_R(6)$ | $d_R(4) \Rightarrow M(1)$ |
| $d_I(4) = b_I(4) + b_I(6)$ | $d_I(4) \Rightarrow M(9)$ |
| $d_R(6) = b_R(4) - b_R(6)$ | $d_R(6) \Rightarrow M(3)$ |
| $d_I(6) = b_I(4) - b_I(6)$ | $d_I(6) \Rightarrow M(11)$ |

### Stage 3: Third Set of Input Adds

This stage also does not require any of the multiplier constants. Further, the add/subtract process is the same for all of the real and imaginary pairs. The strategy for converting these equations into code is to start at the top (compute $b_R(5)$) and identify the pair of inputs to be used first (in this case $b_R(5)$ and $b_I(5)$). Then look down the list to find the second (compute $b_I(5)$) place where these two inputs are used. Pull $b_R(5)$ and $b_I(5)$ from memory, use them to compute new values for $b_R(5)$ and $b_I(5)$, and store the results in data memory locations $M(5)$ and $M(13)$ previously occupied by the original values of $b_R(5)$ and $b_I(5)$. Repeat the same set of steps for $b_R(7)$ and $b_I(7)$. The inputs and outputs of this stage have the same labels, so all the terms in Stage 6 have the same label.

| Algorithm Steps | Memory Map |
|---|---|
| $b_R(5) = b_R(5) + b_I(5)$ | $b_R(5) \Rightarrow M(5)$ |
| $b_I(5) = -b_R(5) + b_I(5)$ | $b_I(5) \Rightarrow M(13)$ |
| $b_R(7) = b_R(7) + b_I(7)$ | $b_R(7) \Rightarrow M(7)$ |
| $b_I(7) = -b_R(7) + b_I(7)$ | $b_I(7) \Rightarrow M(15)$ |

### Stage 4: Multiplies

This stage contains all of the multiplications. In all cases the multiplication is performed by pulling a data value from memory, multiplying it by the appropriate constant, and returning the result to the same data memory location. Note that only one multiplier constant is required because $\cos(2\pi/8) = \sin(2\pi/8)$.

| Algorithm Steps | Memory Map |
|---|---|
| $c_R(5) = b_R(5) * \cos(2\pi/8)$ | $c_R(5) \Rightarrow M(5)$ |
| $c_I(5) = b_I(5) * \sin(2\pi/8)$ | $c_I(5) \Rightarrow M(13)$ |
| $c_R(7) = b_R(7) * \cos(2\pi/8)$ | $c_R(7) \Rightarrow M(7)$ |
| $c_I(7) = b_I(7) * \sin(2\pi/8)$ | $c_I(7) \Rightarrow M(15)$ |

## Stage 5: Postmultiply Adds

This stage also does not require any of the multiplier constants. Further, the add/subtract process is the same for all of the real and imaginary pairs. The strategy for converting these equations to code is to start at the top (compute $d_R(5)$) and identify the pair of inputs to be used first (in this case $c_R(5)$ and $c_I(7)$). Then look down the list to find the second (compute $d_R(7)$) place where these two inputs are used. Pull $c_R(5)$ and $c_I(7)$ from memory, compute $d_R(5)$ and $d_R(7)$, and store the results in data memory locations $M(5)$ and $M(15)$ previously occupied by $c_R(5)$ and $c_I(7)$.

Next, look for the computation for $d_I(5)$ on the list and repeat the same set of steps. Continue this process until all the Algorithm Steps have been computed and their results stored in the Memory Map addresses. Notice that all of these additions require one imaginary input and one real input. This approach to these additions implements the required multiplication by $j = \sqrt{-1}$, which converts real parts of data to imaginary parts and imaginary parts to real parts (with a sign change).

| Algorithm Steps | Memory Map |
|---|---|
| $d_R(5) = c_R(5) + c_I(7)$ | $d_R(5) \Rightarrow M(5)$ |
| $d_I(5) = c_I(5) - c_R(7)$ | $d_I(5) \Rightarrow M(7)$ |
| $d_R(7) = c_R(5) - c_I(7)$ | $d_R(7) \Rightarrow M(15)$ |
| $d_I(7) = c_I(5) + c_R(7)$ | $d_I(7) \Rightarrow M(13)$ |

## Stage 6: Output Adds

This stage also does not require any multiplier constants. Further, the add/subtract process is the same for all of the real and imaginary pairs. The strategy for converting these equations to code is to start at the top (compute $A_R(0)$) and identify the pair of inputs to be used first (in this case $d_R(0)$ and $d_R(4)$). Then look down the list to find the second (compute $A_R(4)$) place where these two inputs are used. Pull $d_R(0)$ and $d_R(4)$ from memory, compute $A_R(0)$ and $A_R(4)$, and store the results in data memory locations $M(0)$ and $M(1)$ previously occupied by $d_R(0)$ and $d_R(4)$.

Next, look for the computation for $A_I(0)$ on the list and repeat the same set of steps. Continue this process until all the Algorithm Steps have been computed and their results stored in the Memory Map addresses. Notice that some of these additions require one imaginary input and one real input. This approach to these additions implements the required multiplication by $j = \sqrt{-1}$, which converts real parts of data to imaginary parts and imaginary parts to real parts (with a sign change).

| Algorithm Steps | Memory Map |
|---|---|
| $A_R(0) = d_R(0) + d_R(4)$ | $A_R(0) \Rightarrow M(0)$ |
| $A_I(0) = d_I(0) + d_I(4)$ | $A_I(0) \Rightarrow M(8)$ |

| **Algorithm Steps** | **Memory Map** |
|---|---|
| $A_R(1) = d_R(1) + d_R(5)$ | $A_R(1) \Rightarrow M(4)$ |
| $A_I(1) = d_I(1) + d_I(5)$ | $A_I(1) \Rightarrow M(6)$ |
| $A_R(2) = d_R(2) + d_I(6)$ | $A_R(2) \Rightarrow M(2)$ |
| $A_I(2) = d_I(2) - d_R(6)$ | $A_I(2) \Rightarrow M(3)$ |
| $A_R(3) = d_R(3) + d_I(7)$ | $A_R(3) \Rightarrow M(13)$ |
| $A_I(3) = d_I(3) - d_R(7)$ | $A_I(3) \Rightarrow M(12)$ |
| $A_R(4) = d_R(0) - d_R(4)$ | $A_R(4) \Rightarrow M(1)$ |
| $A_I(4) = d_I(0) - d_I(4)$ | $A_I(4) \Rightarrow M(9)$ |
| $A_R(5) = d_R(1) - d_R(5)$ | $A_R(5) \Rightarrow M(5)$ |
| $A_I(5) = d_I(1) - d_I(5)$ | $A_I(5) \Rightarrow M(7)$ |
| $A_R(6) = d_R(2) - d_I(6)$ | $A_R(6) \Rightarrow M(11)$ |
| $A_I(6) = d_I(2) + d_R(6)$ | $A_I(6) \Rightarrow M(10)$ |
| $A_R(7) = d_R(3) - d_I(7)$ | $A_R(7) \Rightarrow M(14)$ |
| $A_I(7) = d_I(3) + d_R(7)$ | $A_I(7) \Rightarrow M(15)$ |

### 8.8.4 PTL 8-Point FFT

The PTL [6] 8-point FFT is a four-stage process with 52 adds, 4 multiplies, 16 data memory locations, and one multiplier constant memory location. The five stages are as follows.

### Stage 1: Input Adds

This stage does not require any multiplier constants. Further, the add/subtract process is the same for all of the real and imaginary pairs. The strategy for converting these equations to code is to start at the top (compute $b_R(0)$) and identify the pair of inputs to be used first (in this case $a_R(0)$ and $a_R(4)$). Then look down the list to find the second (compute $b_R(1)$) place where these two inputs are used. Pull $a_R(0)$ and $a_R(4)$ from memory, compute $b_R(0)$ and $b_R(1)$, and store the results in data memory locations $M(0)$ and $M(4)$ previously occupied by $a_R(0)$ and $a_R(4)$.

Next, look for the computation for $b_I(0)$ on the list and repeat the same set of steps. Continue this process until all the Algorithm Steps have been computed and their results stored in the Memory Map addresses.

| **Algorithm Steps** | **Memory Map** |
|---|---|
| $b_R(0) = a_R(0) + a_R(4)$ | $b_R(0) \Rightarrow M(0)$ |
| $b_R(1) = a_R(0) - a_R(4)$ | $b_R(1) \Rightarrow M(4)$ |
| $b_I(0) = a_I(0) + a_I(4)$ | $b_I(0) \Rightarrow M(8)$ |
| $b_I(1) = a_I(0) - a_I(4)$ | $b_I(1) \Rightarrow M(12)$ |
| $b_R(2) = a_R(1) + a_R(5)$ | $b_R(2) \Rightarrow M(1)$ |
| $b_R(3) = a_R(1) - a_R(5)$ | $b_R(3) \Rightarrow M(5)$ |
| $b_I(2) = a_I(1) + a_I(5)$ | $b_I(2) \Rightarrow M(9)$ |
| $b_I(3) = a_I(1) - a_I(5)$ | $b_I(3) \Rightarrow M(13)$ |
| $b_R(4) = a_R(2) + a_R(6)$ | $b_R(4) \Rightarrow M(2)$ |

| Algorithm Steps | Memory Map |
|---|---|
| $b_R(5) = a_R(2) - a_R(6)$ | $b_R(5) \Rightarrow M(6)$ |
| $b_I(4) = a_I(2) + a_I(6)$ | $b_I(4) \Rightarrow M(10)$ |
| $b_I(5) = a_I(2) - a_I(6)$ | $b_I(5) \Rightarrow M(14)$ |
| $b_R(6) = a_R(3) + a_R(7)$ | $b_R(6) \Rightarrow M(3)$ |
| $b_R(7) = a_R(3) - a_R(7)$ | $b_R(7) \Rightarrow M(7)$ |
| $b_I(6) = a_I(3) + a_I(7)$ | $b_I(6) \Rightarrow M(11)$ |
| $b_I(7) = a_I(3) - a_I(7)$ | $b_I(7) \Rightarrow M(15)$ |

### Stage 2: Second Set of Input Adds

This stage also does not require any multiplier constants. Further, the add/subtract process is the same for all of the real and imaginary pairs. The strategy for converting these equations to code is to start at the top (compute $c_R(0)$) and identify the pair of inputs to be used first (in this case $b_R(0)$ and $b_R(4)$). Then look down the list to find the second (compute $c_R(2)$) place where these two inputs are used. Pull $b_R(0)$ and $b_R(4)$ from memory, compute $c_R(0)$ and $c_R(2)$, and store the results in data memory locations $M(0)$ and $M(2)$ previously occupied by $b_R(0)$ and $b_R(4)$.

Next, look for the computation for $c_I(0)$ on the list and repeat the same set of steps. Continue this process until all the Algorithm Steps have been computed and their results stored in the Memory Map addresses.

| Algorithm Steps | Memory Map |
|---|---|
| $c_R(0) = b_R(0) + b_R(4)$ | $c_R(0) \Rightarrow M(0)$ |
| $c_I(0) = b_I(0) + b_I(4)$ | $c_I(0) \Rightarrow M(8)$ |
| $c_R(1) = b_R(1) + b_I(5)$ | $c_R(1) \Rightarrow M(4)$ |
| $c_I(1) = b_I(1) + b_R(5)$ | $c_I(1) \Rightarrow M(12)$ |
| $c_R(2) = b_R(0) - b_R(4)$ | $c_R(2) \Rightarrow M(2)$ |
| $c_I(2) = b_I(0) - b_I(4)$ | $c_I(2) \Rightarrow M(10)$ |
| $c_R(3) = b_R(1) - b_I(5)$ | $c_R(3) \Rightarrow M(14)$ |
| $c_I(3) = b_I(1) - b_R(5)$ | $c_I(3) \Rightarrow M(6)$ |
| $c_R(4) = b_R(2) + b_R(6)$ | $c_R(4) \Rightarrow M(1)$ |
| $c_I(4) = b_I(2) + b_I(6)$ | $c_I(4) \Rightarrow M(9)$ |
| $c_R(5) = b_R(3) + b_R(7)$ | $c_R(5) \Rightarrow M(5)$ |
| $c_I(5) = b_I(3) + b_I(7)$ | $c_I(5) \Rightarrow M(13)$ |
| $c_R(6) = b_R(2) - b_R(6)$ | $c_R(6) \Rightarrow M(3)$ |
| $c_I(6) = b_I(2) - b_I(6)$ | $c_I(6) \Rightarrow M(11)$ |
| $c_R(7) = b_R(3) - b_R(7)$ | $c_R(7) \Rightarrow M(7)$ |
| $c_I(7) = b_I(3) - b_I(7)$ | $c_I(7) \Rightarrow M(15)$ |

### Stage 3: Third Stage of Input Adds

The strategy for converting these equations to code is to start at the top (compute $d_R(5)$ and $d_I(5)$) and identify the pair of inputs to be used (in this case $c_R(5)$ and $c_I(7)$).

Pull $c_R(5)$ and $c_I(7)$ from memory, compute $d_R(5)$ and $d_I(5)$, and store the results in data memory locations $M(5)$ and $M(13)$ previously occupied by $c_R(5)$ and $c_I(7)$. Perform the same set of steps for $d_R(7)$ and $d_I(7)$.

| **Algorithm Steps** | **Memory Map** |
|---|---|
| $d_R(5) = c_R(5) + c_I(7)$ | $d_R(5) \Rightarrow M(5)$ |
| $d_I(5) = c_I(5) + c_R(7)$ | $d_I(5) \Rightarrow M(13)$ |
| $d_R(7) = c_R(5) - c_I(7)$ | $d_R(7) \Rightarrow M(15)$ |
| $d_I(7) = c_I(5) - c_R(7)$ | $d_I(7) \Rightarrow M(7)$ |

## Stage 4: Multiplies

This stage contains all of the multiplications. In all cases the multiplication is performed by pulling a data value from memory, multiplying it by the appropriate constant, and returning the result to the same data memory location. Note that only one multiplier constant is required.

| **Algorithm Steps** | **Memory Map** |
|---|---|
| $d_R(5) = d_R(5) * \cos(2\pi/8)$ | $d_R(5) \Rightarrow M(5)$ |
| $d_I(5) = d_I(5) * \cos(2\pi/8)$ | $d_I(5) \Rightarrow M(13)$ |
| $d_R(7) = d_R(7) * \cos(2\pi/8)$ | $d_R(7) \Rightarrow M(15)$ |
| $d_I(7) = d_I(7) * \cos(2\pi/8)$ | $d_I(7) \Rightarrow M(7)$ |

## Stage 5: Output Adds

This stage also does not require any multiplier constants. Further, the add/subtract process is the same for all of the real and imaginary pairs. The strategy for converting these equations to code is to start at the top (compute $A_R(0)$) and identify the pair of inputs to be used first (in this case $c_R(0)$ and $c_R(4)$). Then look down the list to find the second (compute $A_R(4)$) place where these two inputs are used. Pull $c_R(0)$ and $c_R(4)$ from memory, compute $A_R(0)$ and $A_R(4)$, and store the results in data memory locations $M(0)$ and $M(1)$ previously occupied by $c_R(0)$ and $c_R(4)$.

Next, look for the computation for $A_I(0)$ on the list and repeat the same set of steps. Continue this process until all the Algorithm Steps have been computed and their results stored in the Memory Map addresses.

| **Algorithm Steps** | **Memory Map** |
|---|---|
| $A_R(0) = c_R(0) + c_R(4)$ | $A_R(0) \Rightarrow M(0)$ |
| $A_I(0) = c_I(0) + c_I(4)$ | $A_I(0) \Rightarrow M(8)$ |
| $A_R(1) = c_R(1) + d_I(5)$ | $A_R(1) \Rightarrow M(4)$ |
| $A_I(1) = c_I(3) - d_R(7)$ | $A_I(1) \Rightarrow M(6)$ |
| $A_R(2) = c_R(2) + c_I(6)$ | $A_R(2) \Rightarrow M(2)$ |
| $A_I(2) = c_I(2) - c_R(6)$ | $A_I(2) \Rightarrow M(10)$ |
| $A_R(3) = c_R(3) - d_I(7)$ | $A_R(3) \Rightarrow M(14)$ |
| $A_I(3) = c_I(1) - d_R(5)$ | $A_I(3) \Rightarrow M(12)$ |
| $A_R(4) = c_R(0) - c_R(4)$ | $A_R(4) \Rightarrow M(1)$ |

| **Algorithm Steps** | **Memory Map** |
|---|---|
| $A_I(4) = c_I(0) - c_I(4)$ | $A_I(4) \Rightarrow M(9)$ |
| $A_R(5) = c_R(1) - d_I(5)$ | $A_R(5) \Rightarrow M(13)$ |
| $A_I(5) = c_I(3) + d_R(7)$ | $A_I(5) \Rightarrow M(15)$ |
| $A_R(6) = c_R(2) - c_I(6)$ | $A_R(6) \Rightarrow M(11)$ |
| $A_I(6) = c_I(2) + c_R(6)$ | $A_I(6) \Rightarrow M(3)$ |
| $A_R(7) = c_R(3) + d_I(7)$ | $A_R(7) \Rightarrow M(7)$ |
| $A_I(7) = c_I(1) + d_R(5)$ | $A_I(7) \Rightarrow M(5)$ |

## 8.9 NINE-POINT FFT

The 9-point DFT is defined for $k = 0, 1, 2, 3, 4, 5, 6, 7$, and 8, as

$$A(k) = \sum_{n=0}^{8} a(n) * e^{-j2\pi kn/9} \qquad (8\text{-}9)$$

If the 9-point DFT is calculated directly from Equation 8-9, it requires 64 complex multiplies and 72 complex adds. Since a complex multiply uses 4 real multiplies and 2 real adds, and a complex add uses 2 real adds, the 9-point DFT requires 256 real multiplies and 272 real adds. The number of adds and multiplies for each of the fast algorithms is significantly less than required for computing the DFT directly. However, if only a subset of the output frequency components is required, it may be more cost effective to compute the DFT equation directly for those terms. For example, if $A(0)$ is the only term needed, it can be computed with 16 adds and no multiplies by using the DFT directly. Each of the other eight output frequencies requires 8 complex multiplies and 8 complex adds for a total of 32 real adds and 32 real multiplies. With this in mind the crossover point between using the DFT directly and one of the 9-point FFT algorithms can be determined based on the number of output frequency components that must be computed.

Since all of the input data is required for each output frequency component calculation, the direct DFT computations require 18 data memory locations for the input data and 18 more for the output frequency components. This is a total of 36 data memory locations, since the input and output are complex. Similarly, the DFT data addressing is sequential (i.e., 0 through 8 for each output frequency component), and the computational architecture is simple, since they can all be performed by using a complex multiply accumulator (see Chapter 10 for details). Addressing the complex multiplier coefficients requires either a modulo arithmetic scheme ($k*n$ mod 9) or that the addresses be stored in program memory.

There have been a number of variations on the 9-point FFT, each having a different number of adds and multiplies. The reason for many algorithms is that the 9-point transform has the special property that it is $3 \times 3$ points. This results in some additional symmetries in the multiplier coefficients that have been exploited in various ways. Three variations are presented, characterized, and then summarized in the Comparison Matrix in Table 8-1.

### 8.9.1 Winograd 9-point FFT

The Winograd [1] 9-point FFT requires 90 adds, 20 multiplies, 26 data memory locations, and 10 multiplier constant memory locations (assuming the multiply by 0.5 is counted as one of the coefficients). The five stages are as follows.

### Stage 1: Input Adds

This stage does not require additional data memory or accessing any of the multiplier constants. Further, the add/subtract process is the same for all of the real and imaginary pairs. The strategy for converting these equations to code is to start at the top (compute $b_R(1)$) and identify the pair of inputs to be used first (in this case $a_R(1)$ and $a_R(8)$). Then look down the list to find the second (compute $b_R(2)$) place where these two inputs are used. Pull $a_R(1)$ and $a_R(8)$ from memory, compute $b_R(1)$ and $b_R(2)$, and store the results in data memory locations $M(1)$ and $M(8)$ previously occupied by $a_R(1)$ and $a_R(8)$.

Next, look for the computation for $b_I(1)$ on the list and repeat the same set of steps. Continue this process until all the Algorithm Steps have been computed and their results stored in the Memory Map addresses.

| Algorithm Steps | Memory Map |
|---|---|
| $b_R(1) = a_R(1) + a_R(8)$ | $b_R(1) \Rightarrow M(1)$ |
| $b_I(1) = a_I(1) + a_I(8)$ | $b_I(1) \Rightarrow M(10)$ |
| $b_R(2) = a_R(1) - a_R(8)$ | $b_R(2) \Rightarrow M(8)$ |
| $b_I(2) = a_I(1) - a_I(8)$ | $b_I(2) \Rightarrow M(17)$ |
| $b_R(3) = a_R(7) + a_R(2)$ | $b_R(3) \Rightarrow M(2)$ |
| $b_I(3) = a_I(7) + a_I(2)$ | $b_I(3) \Rightarrow M(11)$ |
| $b_R(4) = a_R(7) - a_R(2)$ | $b_R(4) \Rightarrow M(7)$ |
| $b_I(4) = a_I(7) - a_I(2)$ | $b_I(4) \Rightarrow M(16)$ |
| $b_R(5) = a_R(3) + a_R(6)$ | $b_R(5) \Rightarrow M(3)$ |
| $b_I(5) = a_I(3) + a_I(6)$ | $b_I(5) \Rightarrow M(12)$ |
| $b_R(6) = a_R(3) - a_R(6)$ | $b_R(6) \Rightarrow M(6)$ |
| $b_I(6) = a_I(3) - a_I(6)$ | $b_I(6) \Rightarrow M(15)$ |
| $b_R(7) = a_R(4) + a_R(5)$ | $b_R(7) \Rightarrow M(4)$ |
| $b_I(7) = a_I(4) + a_I(5)$ | $b_I(7) \Rightarrow M(13)$ |
| $b_R(8) = a_R(4) - a_R(5)$ | $b_R(8) \Rightarrow M(5)$ |
| $b_I(8) = a_I(4) - a_I(5)$ | $b_I(8) \Rightarrow M(14)$ |

### Stage 2: Second Set of Input Adds

This is the first stage that requires additional data memory locations to store computational results. The computational strategy is still the same as for the input adds. Start with the first computation on the list ($c_R(1)$) and find all of the other computations that involve the two input values $b_R(1)$ and $b_R(3)$. In this case there are two other computations that use $b_R(1)$, and two others that use $b_R(3)$. Therefore, when $c_R(1)$ and $c_R(2)$ are computed, their results must be placed in additional data memory locations $M(18)$ and $M(19)$ so that $b_R(1)$ and $b_R(3)$ are still available for the additional computations where they are used ($c_R(5)$ and $c_R(6)$).

This strategy is continued until all of the computations in this algorithm stage are completed. One caution is that some of the inputs to this stage are also needed in Stage 3. Therefore, all of the places where a data value is used in the algorithm must be taken into account.

| Algorithm Steps | Memory Map |
|---|---|
| $c_R(1) = b_R(1) + b_R(3)$ | $c_R(1) \Rightarrow M(18)$ |
| $c_I(1) = b_I(1) + b_I(3)$ | $c_I(1) \Rightarrow M(22)$ |
| $c_R(2) = b_R(1) - b_R(3)$ | $c_R(2) \Rightarrow M(19)$ |
| $c_I(2) = b_I(1) - b_I(3)$ | $c_I(2) \Rightarrow M(23)$ |
| $c_R(3) = b_R(2) + b_R(4)$ | $c_R(3) \Rightarrow M(20)$ |
| $c_I(3) = b_I(2) + b_I(4)$ | $c_I(3) \Rightarrow M(24)$ |
| $c_R(4) = b_R(2) - b_R(4)$ | $c_R(4) \Rightarrow M(21)$ |
| $c_I(4) = b_I(2) - b_I(4)$ | $c_I(4) \Rightarrow M(25)$ |
| $c_R(5) = b_R(3) - b_R(7)$ | $c_R(5) \Rightarrow M(2)$ |
| $c_I(5) = b_I(3) - b_I(7)$ | $c_I(5) \Rightarrow M(11)$ |
| $c_R(6) = b_R(7) - b_R(1)$ | $c_R(6) \Rightarrow M(1)$ |
| $c_I(6) = b_I(7) - b_I(1)$ | $c_I(6) \Rightarrow M(10)$ |
| $c_R(7) = b_R(4) - b_R(8)$ | $c_R(7) \Rightarrow M(7)$ |
| $c_I(7) = b_I(4) - b_I(8)$ | $c_I(7) \Rightarrow M(16)$ |
| $c_R(8) = b_R(8) - b_R(2)$ | $c_R(8) \Rightarrow M(8)$ |
| $c_I(8) = b_I(8) - b_I(2)$ | $c_I(8) \Rightarrow M(17)$ |
| $d_R(1) = c_R(1) + b_R(7)$ | $d_R(1) \Rightarrow M(4)$ |
| $d_I(1) = c_I(1) + b_I(7)$ | $d_I(1) \Rightarrow M(13)$ |
| $d_R(2) = c_R(3) + b_R(8)$ | $d_R(2) \Rightarrow M(5)$ |
| $d_I(2) = c_I(3) + b_I(8)$ | $d_I(2) \Rightarrow M(14)$ |
| $e_R(1) = d_R(1) + b_R(5)$ | $e_R(1) \Rightarrow M(18)$ |
| $e_I(1) = d_I(1) + b_I(5)$ | $e_I(1) \Rightarrow M(22)$ |
| $f_R(0) = e_R(1) + a_R(0)$ | $f_R(0) \Rightarrow M(0)$ |
| $f_I(0) = e_I(1) + a_I(0)$ | $f_I(0) \Rightarrow M(9)$ |

## Stage 3: Multiplies

This stage contains all of the multiplications. In all cases except $c_R(8)$ and $c_I(8)$, the multiplication is performed by pulling a data value from memory, multiplying it by the appropriate constant, and returning the result to the same data memory location. Since $c_{\bar{R}}(8)$ and $c_I(8)$ are multiplied during this stage as well as used in the next stage, the multiplied values $f_I(10)$ and $f_R(10)$, respectively, are stored in two of the additional data memory locations $M(20)$ and $M(24)$ used earlier.

| Algorithm Steps | Memory Map |
|---|---|
| $f_R(1) = -0.5 * d_R(1)$ | $f_R(1) \Rightarrow M(4)$ |
| $f_I(1) = -0.5 * d_I(1)$ | $f_I(1) \Rightarrow M(13)$ |
| $f_R(2) = \sin(6\pi/9) * d_I(2)$ | $f_R(2) \Rightarrow M(14)$ |
| $f_I(2) = -\sin(6\pi/9) * d_R(2)$ | $f_I(2) \Rightarrow M(5)$ |
| $f_R(3) = [\cos(6\pi/9) - 1] * b_R(5)$ | $f_R(3) \Rightarrow M(3)$ |
| $f_I(3) = [\cos(6\pi/9) - 1] * b_I(5)$ | $f_I(3) \Rightarrow M(12)$ |

| Algorithm Steps | Memory Map |
|---|---|
| $f_R(4) = \sin(6\pi/9) * b_I(6)$ | $f_R(4) \Rightarrow M(15)$ |
| $f_I(4) = -\sin(6\pi/9) * b_R(6)$ | $f_I(4) \Rightarrow M(6)$ |
| $f_R(5) = (1/3) * [2 * \cos(2\pi/9) - \cos(4\pi/9) - \cos(8\pi/9)] * c_R(2)$ | $f_R(5) \Rightarrow M(19)$ |
| $f_I(5) = (1/3) * [2 * \cos(2\pi/9) - \cos(4\pi/9) - \cos(8\pi/9)] * c_I(2)$ | $f_I(5) \Rightarrow M(23)$ |
| $f_R(6) = (1/3) * [\cos(2\pi/9) + \cos(4\pi/9) - 2 * \cos(8\pi/9)] * c_R(5)$ | $f_R(6) \Rightarrow M(2)$ |
| $f_I(6) = (1/3) * [\cos(2\pi/9) + \cos(4\pi/9) - 2 * \cos(8\pi/9)] * c_I(5)$ | $f_I(6) \Rightarrow M(11)$ |
| $f_R(7) = (1/3) * [\cos(2\pi/9) - 2 * \cos(4\pi/9) + \cos(8\pi/9)] * c_R(6)$ | $f_R(7) \Rightarrow M(1)$ |
| $f_I(7) = (1/3) * [\cos(2\pi/9) - 2 * \cos(4\pi/9) + \cos(8\pi/9)] * c_I(6)$ | $f_I(7) \Rightarrow M(10)$ |
| $f_R(8) = (1/3) * [2 * \sin(2\pi/9) + \sin(4\pi/9) - \sin(8\pi/9)] * c_I(4)$ | $f_R(8) \Rightarrow M(25)$ |
| $f_I(8) = -(1/3) * [2 * \sin(2\pi/9) + \sin(4\pi/9) - \sin(8\pi/9)] * c_R(4)$ | $f_I(8) \Rightarrow M(21)$ |
| $f_R(9) = (1/3) * [\sin(2\pi/9) - \sin(4\pi/9) - 2 * \sin(8\pi/9)] * c_I(7)$ | $f_R(9) \Rightarrow M(16)$ |
| $f_I(9) = -(1/3) * [\sin(2\pi/9) - \sin(4\pi/9) - 2 * \sin(8\pi/9)] * c_R(7)$ | $f_I(9) \Rightarrow M(7)$ |
| $f_R(10) = (1/3) * [\sin(2\pi/9) + 2 * \sin(4\pi/9) + \sin(8\pi/9)] * c_I(8)$ | $f_R(10) \Rightarrow M(24)$ |
| $f_I(10) = -(1/3) * [\sin(2\pi/9) + 2 * \sin(4\pi/9) + \sin(8\pi/9)] * c_R(8)$ | $f_I(10) \Rightarrow M(20)$ |

## Stage 4: Postmultiply Adds

Some of the computational results in this stage are given two labels (i.e., $h_R(1) = m_R(5)$). The first is the one in the derivation [1] of the algorithm, and the second is used to show the commonality of the output computations in all of the 9-point algorithms. The strategy for converting these equations to code is to start at the top (compute $g_R(1)$) and identify the pair of inputs to be used first (for the first Algorithm Step $f_R(1)$ is used for both inputs). Then look down the list to find the second (compute $g_R(2)$) place where this input is used. That Algorithm Step also uses $d_R(1)$. Pull $f_R(1)$ and $d_R(1)$ from memory, compute $g_R(1)$ and $g_R(2)$, and store the results in data memory locations $M(18)$ and $M(8)$ previously occupied by $e_R(1)$ and $c_R(8)$.

Next, look for the computation for $b_I(1)$ on the list and repeat the same set of steps. Continue this process until all the Algorithm Steps have been computed and their results stored in the Memory Map addresses. Note that the Algorithm Steps for $m_R(6)$ and $m_I(6)$ only relabel the data values once they have been used as required by other portions of the algorithm.

| Algorithm Steps | Memory Map |
|---|---|
| $g_R(1) = f_R(1) + f_R(1)$ | $g_R(1) \Rightarrow M(18)$ |
| $g_I(1) = f_I(1) + f_I(1)$ | $g_I(1) \Rightarrow M(22)$ |
| $g_R(2) = -d_R(1) + f_R(1)$ | $g_R(2) \Rightarrow M(8)$ |
| $g_I(2) = -d_I(1) + f_I(1)$ | $g_I(2) \Rightarrow M(17)$ |
| $g_R(3) = f_R(0) + f_R(3)$ | $g_R(3) \Rightarrow M(3)$ |
| $g_I(3) = f_I(0) + f_I(3)$ | $g_I(3) \Rightarrow M(12)$ |
| $g_R(4) = f_R(4) + f_R(8)$ | $g_R(4) \Rightarrow M(22)$ |
| $g_I(4) = f_I(4) + f_I(8)$ | $g_I(4) \Rightarrow M(18)$ |
| $g_R(5) = f_R(4) - f_R(9)$ | $g_R(5) \Rightarrow M(15)$ |

|  Algorithm Steps | Memory Map |
|---|---|
| $g_I(5) = f_I(4) - f_I(9)$ | $g_I(5) \Rightarrow M(6)$ |
| $g_R(6) = f_R(4) - f_R(8)$ | $g_R(6) \Rightarrow M(25)$ |
| $g_I(6) = f_I(4) - f_I(8)$ | $g_I(6) \Rightarrow M(21)$ |
| $h_R(1) = f_R(0) + g_R(2) = m_R(5)$ | $h_R(1) = m_R(5) \Rightarrow M(8)$ |
| $h_I(1) = f_I(0) + g_I(2) = m_I(5)$ | $h_I(1) = m_I(5) \Rightarrow M(17)$ |
| $h_R(2) = g_R(1) + g_R(3)$ | $h_R(2) \Rightarrow M(3)$ |
| $h_I(2) = g_I(1) + g_I(3)$ | $h_I(2) \Rightarrow M(12)$ |
| $h_R(3) = g_R(4) + f_R(9) = m_R(2)$ | $h_R(3) = m_R(2) \Rightarrow M(16)$ |
| $h_I(3) = g_I(4) + f_I(9) = -m_I(2)$ | $h_I(3) = -m_I(2) \Rightarrow M(7)$ |
| $h_R(4) = g_R(6) - f_R(10) = m_R(8)$ | $h_R(4) = m_R(8) \Rightarrow M(25)$ |
| $h_I(4) = g_I(6) - f_I(10) = -m_I(8)$ | $h_I(4) = -m_I(8) \Rightarrow M(21)$ |
| $h_R(5) = g_R(5) + f_R(10) = -m_R(4)$ | $h_R(5) = -m_R(4) \Rightarrow M(15)$ |
| $h_I(5) = g_I(5) + f_I(10) = m_I(4)$ | $h_I(5) = m_I(4) \Rightarrow M(6)$ |
| $k_R(1) = h_R(2) + f_R(5)$ | $k_R(1) \Rightarrow M(19)$ |
| $k_I(1) = h_I(2) + f_I(5)$ | $k_I(1) \Rightarrow M(23)$ |
| $k_R(2) = h_R(2) - f_R(6)$ | $k_R(2) \Rightarrow M(3)$ |
| $k_I(2) = h_I(2) - f_I(6)$ | $k_I(2) \Rightarrow M(12)$ |
| $k_R(3) = h_R(2) - f_R(5)$ | $k_R(3) \Rightarrow M(4)$ |
| $k_I(3) = h_I(2) - f_I(5)$ | $k_I(3) \Rightarrow M(13)$ |
| $l_R(1) = k_R(1) + f_R(6) = m_R(1)$ | $l_R(1) = m_R(1) \Rightarrow M(2)$ |
| $l_I(1) = k_I(1) + f_I(6) = m_I(1)$ | $l_I(1) = m_I(1) \Rightarrow M(11)$ |
| $l_R(2) = k_R(2) + f_R(7) = m_R(3)$ | $l_R(2) = m_R(3) \Rightarrow M(3)$ |
| $l_I(2) = k_I(2) + f_I(7) = m_I(3)$ | $l_I(2) = m_I(3) \Rightarrow M(12)$ |
| $l_R(3) = k_R(3) - f_R(7) = m_R(7)$ | $l_R(3) = m_R(7) \Rightarrow M(4)$ |
| $l_I(3) = k_I(3) - f_I(7) = m_I(7)$ | $l_I(3) = m_I(7) \Rightarrow M(13)$ |
| $m_R(6) = f_R(2)$ | $m_R(6) = f_R(2) \Rightarrow M(14)$ |
| $m_I(6) = -f_I(2)$ | $m_I(6) = -f_I(2) \Rightarrow M(5)$ |

## Stage 5: Output Adds

This stage also does not require any multiplier constants. The strategy for converting these equations to code is to start at the top (compute $A_R(1)$) and identify the pair of inputs to be used first (in this case $m_R(1)$ and $m_R(2)$). Then look down the column to find the second (compute $A_R(8)$) place where these two inputs are used. Pull $m_R(1)$ and $m_R(2)$ from memory, compute $A_R(1)$ and $A_R(8)$, and store the results in data memory locations $M(2)$ and $M(16)$ previously occupied by $m_R(1)$ and $m_R(2)$.

Next, look for the computation for $A_I(1)$ in the column and repeat the same set of steps. Continue this process until all of the computations are performed and all of the results are returned to the data memory locations. The $A_R(5)$ and $A_I(5)$ computations are placed in data memory locations different from where the inputs were taken. This is to meet the requirement that the output frequency components use the same locations as the input data sequence. Note that the Algorithm Steps for $A_R(0)$ and $A_I(0)$ only relabel the data values to their output labels once they have been used as required by other portions of the algorithm.

| Algorithm Steps | Memory Map |
|---|---|
| $A_R(0) = f_R(0)$ | $A_R(0) \Rightarrow M(0)$ |
| $A_I(0) = f_I(0)$ | $A_I(0) \Rightarrow M(9)$ |
| $A_R(1) = m_R(1) + m_R(2)$ | $A_R(1) \Rightarrow M(2)$ |
| $A_I(1) = m_I(1) - m_I(2)$ | $A_I(1) \Rightarrow M(7)$ |
| $A_R(2) = m_R(3) + m_R(4)$ | $A_R(2) \Rightarrow M(3)$ |
| $A_I(2) = m_I(3) - m_I(4)$ | $A_I(2) \Rightarrow M(6)$ |
| $A_R(3) = m_R(5) + m_R(6)$ | $A_R(3) \Rightarrow M(8)$ |
| $A_I(3) = m_I(5) - m_I(6)$ | $A_I(3) \Rightarrow M(5)$ |
| $A_R(4) = m_R(7) + m_R(8)$ | $A_R(4) \Rightarrow M(4)$ |
| $A_I(4) = m_I(7) - m_I(8)$ | $A_I(4) \Rightarrow M(13)$ |
| $A_R(5) = m_R(7) - m_R(8)$ | $A_R(5) \Rightarrow M(1)$ |
| $A_I(5) = m_I(7) + m_I(8)$ | $A_I(5) \Rightarrow M(10)$ |
| $A_R(6) = m_R(5) - m_R(6)$ | $A_R(6) \Rightarrow M(14)$ |
| $A_I(6) = m_I(5) + m_I(6)$ | $A_I(6) \Rightarrow M(17)$ |
| $A_R(7) = m_R(3) - m_R(4)$ | $A_R(7) \Rightarrow M(15)$ |
| $A_I(7) = m_I(3) + m_I(4)$ | $A_I(7) \Rightarrow M(12)$ |
| $A_R(8) = m_R(1) - m_R(2)$ | $A_R(8) \Rightarrow M(16)$ |
| $A_I(8) = m_I(1) + m_I(2)$ | $A_I(8) \Rightarrow M(11)$ |

### 8.9.2  PTL 9-point FFT

The PTL [6] 9-point FFT requires 94 adds, 52 multiplies, 22 data memory locations, and 8 multiplier constant locations. The three stages are as follows.

### Stage 1: Input Adds

This stage does not require additional data memory or accessing any of the multiplier constants. Further, the add/subtract process is the same for all of the real and imaginary pairs. The strategy for converting these equations to code is to start at the top (compute $b_R(1)$) and identify the pair of inputs to be used first (in this case $a_R(1)$ and $a_R(8)$). Then look down the list to find the second (compute $b_R(2)$) place where these two inputs are used. Pull $a_R(1)$ and $a_R(8)$ from memory, compute $b_R(1)$ and $b_R(2)$, and store the results in data memory locations $M(1)$ and $M(8)$ previously occupied by $a_R(1)$ and $a_R(8)$.

Next, look for the computation for $b_I(1)$ on the list and repeat the same set of steps. Continue this process until all the Algorithm Steps have been computed and their results stored in the Memory Map addresses.

| Algorithm Steps | Memory Map |
|---|---|
| $b_R(1) = a_R(1) + a_R(8)$ | $b_R(1) \Rightarrow M(1)$ |
| $b_I(1) = a_I(1) + a_I(8)$ | $b_I(1) \Rightarrow M(10)$ |
| $b_R(2) = a_R(1) - a_R(8)$ | $b_R(2) \Rightarrow M(8)$ |

| Algorithm Steps | Memory Map |
|---|---|
| $b_I(2) = a_I(1) - a_I(8)$ | $b_I(2) \Rightarrow M(17)$ |
| $b_R(3) = a_R(7) + a_R(2)$ | $b_R(3) \Rightarrow M(2)$ |
| $b_I(3) = a_I(7) + a_I(2)$ | $b_I(3) \Rightarrow M(11)$ |
| $b_R(4) = a_R(7) - a_R(2)$ | $b_R(4) \Rightarrow M(7)$ |
| $b_I(4) = a_I(7) - a_I(2)$ | $b_I(4) \Rightarrow M(16)$ |
| $b_R(5) = a_R(3) + a_R(6)$ | $b_R(5) \Rightarrow M(3)$ |
| $b_I(5) = a_I(3) + a_I(6)$ | $b_I(5) \Rightarrow M(12)$ |
| $b_R(6) = a_R(3) - a_R(6)$ | $b_R(6) \Rightarrow M(6)$ |
| $b_I(6) = a_I(3) - a_I(6)$ | $b_I(6) \Rightarrow M(15)$ |
| $b_R(7) = a_R(4) + a_R(5)$ | $b_R(7) \Rightarrow M(4)$ |
| $b_I(7) = a_I(4) + a_I(5)$ | $b_I(7) \Rightarrow M(13)$ |
| $b_R(8) = a_R(4) - a_R(5)$ | $b_R(8) \Rightarrow M(5)$ |
| $b_I(8) = a_I(4) - a_I(5)$ | $b_I(8) \Rightarrow M(14)$ |

## Stage 2: Multiply-Accumulates

This algorithm stage contains all of the multiplications and requires additional data memory locations to store the results because the input data is used for sets of computations. The data memory mapping assumes the multiply-accumulation process described as Constraint 5 in Section 8.2.

For example, consider the computation of $m_R(1)$, $m_R(3)$, $m_R(5)$, $m_R(7)$, and $f_R(0)$, which requires $b_R(1)$, $b_R(3)$, $b_R(5)$, $b_R(7)$, and $a_R(0)$. Because of the need for all five inputs to compute all five outputs, the first four outputs, say $m_R(1)$, $m_R(3)$, $m_R(5)$, and $m_R(7)$ are stored in additional data memory locations $M(21)$, $M(20)$, $M(19)$, and $M(18)$. Finally, $f_R(0)$ may be stored in one of the input data memory locations, say data memory location $M(0)$ occupied by $a_R(0)$. This leaves the four data memory locations $M(1)$, $M(2)$, $M(3)$, and $M(4)$, the ones used by $b_R(1)$, $b_R(3)$, $b_R(5)$, and $b_R(7)$, to be used for the extra locations required by other sets of multiply-accumulate operations. The extra locations are used for the imaginary equivalent of the real computations. This process is continued, always using leftover data memory locations, until all of the computations are performed.

| Algorithm Steps | Memory Map |
|---|---|
| $m_R(1) = b_R(1) * \cos(2\pi/9) + b_R(3) * \cos(4\pi/9) + b_R(5) * \cos(6\pi/9) + b_R(7) * \cos(8\pi/9) + a_R(0)$ | $m_R(1) \Rightarrow M(21)$ |
| $m_R(3) = b_R(1) * \cos(4\pi/9) + b_R(3) * \cos(8\pi/9) + b_R(5) * \cos(6\pi/9) + b_R(7) * \cos(2\pi/9) + a_R(0)$ | $m_R(3) \Rightarrow M(20)$ |
| $m_R(5) = [b_R(1) + b_R(3) + b_R(7)] * \cos(6\pi/9) + b_R(5) + a_R(0)$ | $m_R(5) \Rightarrow M(19)$ |
| $m_R(7) = b_R(1) * \cos(8\pi/9) + b_R(3) * \cos(2\pi/9) + b_R(5) * \cos(6\pi/9) + b_R(7) * \cos(4\pi/9) + a_R(0)$ | $m_R(7) \Rightarrow M(18)$ |
| $f_R(0) = b_R(1) + b_R(3) + b_R(5) + b_R(7) + a_R(0)$ | $f_R(0) \Rightarrow M(0)$ |
| $m_I(1) = b_I(1) * \cos(2\pi/9) + b_I(3) * \cos(4\pi/9) + b_I(5) * \cos(6\pi/9) + b_I(7) * \cos(8\pi/9) + a_I(0)$ | $m_I(1) \Rightarrow M(4)$ |
| $m_I(3) = b_I(1) * \cos(4\pi/9) + b_I(3) * \cos(8\pi/9) + b_I(5) * \cos(6\pi/9) + b_I(7) * \cos(2\pi/9) + a_I(0)$ | $m_I(3) \Rightarrow M(3)$ |
| $m_I(5) = [b_I(1) + b_I(3) + b_I(7)] * \cos(6\pi/9) + b_I(5) + a_I(0)$ | $m_I(5) \Rightarrow M(2)$ |
| $m_I(7) = b_I(1) * \cos(8\pi/9) + b_I(3) * \cos(2\pi/9) + b_I(5) * \cos(6\pi/9) + b_I(7) * \cos(4\pi/9) + a_I(0)$ | $m_I(7) \Rightarrow M(1)$ |
| $f_I(0) = b_I(1) + b_I(3) + b_I(5) + b_I(7) + a_I(0)$ | $f_I(0) \Rightarrow M(9)$ |
| $m_R(2) = b_I(2) * \sin(2\pi/9) - b_I(4) * \sin(4\pi/9) + b_I(6) * \sin(6\pi/9) + b_I(8) * \sin(8\pi/9)$ | $m_R(2) \Rightarrow M(11)$ |

| **Algorithm Steps** | **Memory Map** |
|---|---|
| $m_R(4) = b_I(2) * \sin(4\pi/9) - b_I(4) * \sin(8\pi/9) - b_I(6) * \sin(6\pi/9) - b_I(8) * \sin(2\pi/9)$ | $m_R(4) \Rightarrow M(12)$ |
| $m_R(6) = [b_I(2) + b_I(4) + b_I(8)] * \sin(6\pi/9)$ | $m_R(6) \Rightarrow M(13)$ |
| $m_R(8) = b_I(2) * \sin(8\pi/9) + b_I(4) * \sin(2\pi/9) + b_I(6) * \sin(6\pi/9) - b_I(8) * \sin(4\pi/9)$ | $m_R(8) \Rightarrow M(14)$ |
| $m_I(2) = b_R(2) * \sin(2\pi/9) - b_R(4) * \sin(4\pi/9) + b_R(6) * \sin(6\pi/9) + b_R(8) * \sin(8\pi/9)$ | $m_I(2) \Rightarrow M(17)$ |
| $m_I(4) = b_R(2) * \sin(4\pi/9) - b_R(4) * \sin(8\pi/9) - b_R(6) * \sin(6\pi/9) - b_R(8) * \sin(2\pi/9)$ | $m_I(4) \Rightarrow M(16)$ |
| $m_I(6) = [b_R(2) + b_R(4) + b_R(8)] * \sin(6\pi/9)$ | $m_I(6) \Rightarrow M(15)$ |
| $m_I(8) = b_R(2) * \sin(8\pi/9) + b_R(4) * \sin(2\pi/9) + b_R(6) * \sin(6\pi/9) - b_R(8) * \sin(4\pi/9)$ | $m_I(8) \Rightarrow M(5)$ |

### Stage 3: Output Adds

This stage also does not require any of the multiplier constants. The strategy for converting these equations to code is to start at the top (compute $A_R(1)$) and identify the pair of inputs to be used first (in this case $m_R(1)$ and $m_R(2)$). Then look down the column to find the second (compute $A_R(8)$) place where these two inputs are used. Pull $m_R(1)$ and $m_R(2)$ from memory, compute $A_R(1)$ and $A_R(8)$, and store the results in data memory locations $M(11)$ and $M(6)$ previously occupied by $m_R(1)$ and $m_R(2)$.

Next, look for the computation for $A_I(1)$ in the column and repeat the same set of steps. Continue this process until all of the computations are performed and all of the results are returned to the data memory locations. The $A_R(5)$, $A_R(6)$, $A_R(7)$, and $A_R(8)$ computations are placed in data memory locations different from where the inputs were taken. This is to meet the requirement that the output frequency components use the same locations as the input data sequence. Note that the Algorithm Steps for $A_R(0)$ and $A_I(0)$ only relabel the data values to their output labels once they have been used as required by other portions of the algorithm.

| **Algorithm Steps** | **Memory Map** |
|---|---|
| $A_R(0) = f_R(0)$ | $A_R(0) \Rightarrow M(0)$ |
| $A_I(0) = f_I(0)$ | $A_I(0) \Rightarrow M(9)$ |
| $A_R(1) = m_R(1) + m_R(2)$ | $A_R(1) \Rightarrow M(11)$ |
| $A_I(1) = m_I(1) - m_I(2)$ | $A_I(1) \Rightarrow M(4)$ |
| $A_R(2) = m_R(3) + m_R(4)$ | $A_R(2) \Rightarrow M(12)$ |
| $A_I(2) = m_I(3) - m_I(4)$ | $A_I(2) \Rightarrow M(3)$ |
| $A_R(3) = m_R(5) + m_R(6)$ | $A_R(3) \Rightarrow M(13)$ |
| $A_I(3) = m_I(5) - m_I(6)$ | $A_I(3) \Rightarrow M(2)$ |
| $A_R(4) = m_R(7) + m_R(8)$ | $A_R(4) \Rightarrow M(14)$ |
| $A_I(4) = m_I(7) - m_I(8)$ | $A_I(4) \Rightarrow M(1)$ |
| $A_R(5) = m_R(7) - m_R(8)$ | $A_R(5) \Rightarrow M(10)$ |
| $A_I(5) = m_I(7) + m_I(8)$ | $A_I(5) \Rightarrow M(5)$ |
| $A_R(6) = m_R(5) - m_R(6)$ | $A_R(6) \Rightarrow M(8)$ |
| $A_I(6) = m_I(5) + m_I(6)$ | $A_I(6) \Rightarrow M(15)$ |
| $A_R(7) = m_R(3) - m_R(4)$ | $A_R(7) \Rightarrow M(7)$ |
| $A_I(7) = m_I(3) + m_I(4)$ | $A_I(7) \Rightarrow M(16)$ |
| $A_R(8) = m_R(1) - m_R(2)$ | $A_R(8) \Rightarrow M(6)$ |
| $A_I(8) = m_I(1) + m_I(2)$ | $A_I(8) \Rightarrow M(17)$ |

### 8.9.3 Burrus and Eschenbacher 9-point FFT

The Burrus and Eschenbacher [7] 9-point FFT requires 84 adds, 20 multiplies, 26 data memory locations, and 8 multiplier constant memory locations. The five stages are as follows.

### Stage 1: Input Adds

This stage does not require additional data memory or accessing any of the multiplier constants. Further, the add/subtract process is the same for all of the real and imaginary pairs. The strategy for converting these equations to code is to start at the top (compute $b_R(1)$) and identify the pair of inputs to be used first (in this case $a_R(1)$ and $a_R(8)$). Look down the list for the second (compute $b_R(2)$) place where these two inputs are used. Pull $a_R(1)$ and $a_R(8)$ from memory, compute $b_R(1)$ and $b_R(2)$, and store the results in data memory locations $M(1)$ and $M(8)$ previously occupied by $a_R(1)$ and $a_R(8)$.

Next, look for the computation for $b_I(1)$ on the list and repeat the same set of steps. Continue this process until all the Algorithm Steps have been computed and their results stored in the Memory Map addresses.

| Algorithm Steps | Memory Map |
|---|---|
| $b_R(1) = a_R(1) + a_R(8)$ | $b_R(1) \Rightarrow M(1)$ |
| $b_I(1) = a_I(1) + a_I(8)$ | $b_I(1) \Rightarrow M(10)$ |
| $b_R(2) = a_R(1) - a_R(8)$ | $b_R(2) \Rightarrow M(8)$ |
| $b_I(2) = a_I(1) - a_I(8)$ | $b_I(2) \Rightarrow M(17)$ |
| $b_R(3) = a_R(7) + a_R(2)$ | $b_R(3) \Rightarrow M(2)$ |
| $b_I(3) = a_I(7) + a_I(2)$ | $b_I(3) \Rightarrow M(11)$ |
| $b_R(4) = a_R(7) - a_R(2)$ | $b_R(4) \Rightarrow M(7)$ |
| $b_I(4) = a_I(7) - a_I(2)$ | $b_I(4) \Rightarrow M(16)$ |
| $b_R(5) = a_R(3) + a_R(6)$ | $b_R(5) \Rightarrow M(3)$ |
| $b_I(5) = a_I(3) + a_I(6)$ | $b_I(5) \Rightarrow M(12)$ |
| $b_R(6) = a_R(3) - a_R(6)$ | $b_R(6) \Rightarrow M(6)$ |
| $b_I(6) = a_I(3) - a_I(6)$ | $b_I(6) \Rightarrow M(15)$ |
| $b_R(7) = a_R(4) + a_R(5)$ | $b_R(7) \Rightarrow M(4)$ |
| $b_I(7) = a_I(4) + a_I(5)$ | $b_I(7) \Rightarrow M(13)$ |
| $b_R(8) = a_R(4) - a_R(5)$ | $b_R(8) \Rightarrow M(5)$ |
| $b_I(8) = a_I(4) - a_I(5)$ | $b_I(8) \Rightarrow M(14)$ |

### Stage 2: Second Set of Input Adds

This is the first stage that requires additional data memory locations to store computational results. The computational strategy is still the same as for the input adds. Start with the first computation on the list ($c_R(1)$). In this case there are two other computations that use $a_R(0)$ and two others that use $b_R(5)$. Therefore, when $c_R(1)$ is computed, the result must be placed in the additional data memory location $M(18)$ so that $a_R(0)$ and $b_R(5)$ are still available for the additional computations.

This strategy is continued until all of the computations and all the results are stored in the data memory locations. One caution is that some of the inputs to this stage are needed in Stage 3.

| Algorithm Steps | Memory Map |
|---|---|
| $c_R(1) = a_R(0) + b_R(5)$ | $c_R(1) \Rightarrow M(18)$ |
| $c_I(1) = a_I(0) + b_I(5)$ | $c_I(1) \Rightarrow M(22)$ |
| $c_R(2) = b_R(1) + b_R(3) + b_R(7)$ | $c_R(2) \Rightarrow M(19)$ |
| $c_I(2) = b_I(1) + b_I(3) + b_I(7)$ | $c_I(2) \Rightarrow M(23)$ |
| $c_R(3) = b_R(3) - b_R(7)$ | $c_R(3) \Rightarrow M(20)$ |
| $c_I(3) = b_I(3) - b_I(7)$ | $c_I(3) \Rightarrow M(24)$ |
| $c_R(4) = b_R(1) - b_R(7)$ | $c_R(4) \Rightarrow M(4)$ |
| $c_I(4) = b_I(1) - b_I(7)$ | $c_I(4) \Rightarrow M(13)$ |
| $c_R(5) = b_R(1) - b_R(3)$ | $c_R(5) \Rightarrow M(1)$ |
| $c_I(5) = b_I(1) - b_I(3)$ | $c_I(5) \Rightarrow M(10)$ |
| $c_R(6) = b_R(2) + b_R(4) + b_R(8)$ | $c_R(6) \Rightarrow M(2)$ |
| $c_I(6) = b_I(2) + b_I(4) + b_I(8)$ | $c_I(6) \Rightarrow M(11)$ |
| $c_R(7) = b_R(4) - b_R(8)$ | $c_R(7) \Rightarrow M(21)$ |
| $c_I(7) = b_I(4) - b_I(8)$ | $c_I(7) \Rightarrow M(25)$ |
| $c_R(8) = b_R(8) - b_R(2)$ | $c_R(8) \Rightarrow M(5)$ |
| $c_I(8) = b_I(8) - b_I(2)$ | $c_I(8) \Rightarrow M(14)$ |
| $c_R(9) = b_R(4) - b_R(2)$ | $c_R(9) \Rightarrow M(8)$ |
| $c_I(9) = b_I(4) - b_I(2)$ | $c_I(9) \Rightarrow M(17)$ |
| $f_R(0) = c_R(1) + c_R(2)$ | $f_R(0) \Rightarrow M(7)$ |
| $f_I(0) = c_I(1) + c_I(2)$ | $f_I(0) \Rightarrow M(16)$ |

## Stage 3: Multiplies

This stage contains all of the multiplications. The individual data values are pulled from memory, multiplied by the appropriate constant, and stored in the same data memory location.

| Algorithm Steps | Memory Map |
|---|---|
| $d_R(1) = -b_R(6) * \sin(6\pi/9)$ | $d_R(1) \Rightarrow M(6)$ |
| $d_I(1) = -b_I(6) * \sin(6\pi/9)$ | $d_I(1) \Rightarrow M(15)$ |
| $d_R(2) = b_R(5) * \cos(6\pi/9)$ | $d_R(2) \Rightarrow M(3)$ |
| $d_I(2) = b_I(5) * \cos(6\pi/9)$ | $d_I(2) \Rightarrow M(12)$ |
| $d_R(3) = -c_R(3) * \cos(8\pi/9)$ | $d_R(3) \Rightarrow M(20)$ |
| $d_I(3) = -c_I(3) * \cos(8\pi/9)$ | $d_I(3) \Rightarrow M(24)$ |
| $d_R(4) = -c_R(4) * \cos(4\pi/9)$ | $d_R(4) \Rightarrow M(4)$ |
| $d_I(4) = -c_I(4) * \cos(4\pi/9)$ | $d_I(4) \Rightarrow M(13)$ |
| $d_R(5) = c_R(5) * \cos(2\pi/9)$ | $d_R(5) \Rightarrow M(1)$ |

| Algorithm Steps | Memory Map |
|---|---|
| $d_I(5) = c_I(5) * \cos(2\pi/9)$ | $d_I(5) \Rightarrow M(10)$ |
| $d_R(6) = -c_R(6) * \sin(6\pi/9)$ | $d_R(6) \Rightarrow M(2)$ |
| $d_I(6) = -c_I(6) * \sin(6\pi/9)$ | $d_I(6) \Rightarrow M(11)$ |
| $d_R(7) = c_R(7) * \sin(8\pi/9)$ | $d_R(7) \Rightarrow M(21)$ |
| $d_I(7) = c_I(7) * \sin(8\pi/9)$ | $d_I(7) \Rightarrow M(25)$ |
| $d_R(8) = c_R(8) * \sin(4\pi/9)$ | $d_R(8) \Rightarrow M(5)$ |
| $d_I(8) = c_I(8) * \sin(4\pi/9)$ | $d_I(8) \Rightarrow M(14)$ |
| $d_R(9) = c_R(9) * \sin(2\pi/9)$ | $d_R(9) \Rightarrow M(8)$ |
| $d_I(9) = c_I(9) * \sin(2\pi/9)$ | $d_I(9) \Rightarrow M(17)$ |
| $d_R(10) = c_R(2) * \cos(6\pi/9)$ | $d_R(10) \Rightarrow M(19)$ |
| $d_I(10) = c_I(2) * \cos(6\pi/9)$ | $d_I(10) \Rightarrow M(23)$ |

## Stage 4: Postmultiply Adds

This stage also requires additional data memory locations to store computational results. The strategy for converting these equations to code is to start at the top (compute $e_R(2)$) and identify the pair of inputs to be used first (in this case $d_R(2)$ and $a_R(0)$). Then look down the list to find the second (for this Algorithm Step there is none) place where these two inputs are used. Pull $d_R(2)$ and $a_R(0)$ from memory, compute $e_R(2)$, and store the results in data memory location $M(0)$ previously occupied by $a_R(0)$.

Next, look for the computation for $e_I(2)$ on the list and repeat the same set of steps. The calculations for $e_R(2)$ and $e_I(2)$ use inputs that are not used elsewhere. However, computing $m_R(1)$, $m_R(3)$, and $m_R(7)$ all require $e_R(2)$. This forces additional data memory locations to be used to ensure that $e_R(2)$ is not overwritten prior to using it all three places. Continue this process until all the Algorithm Steps have been computed and their results stored in the Memory Map addresses. Note that the Algorithm Steps for $m_R(6)$ and $m_I(6)$ only relabel the data values once they have been used as required by other portions of the algorithm.

| Algorithm Steps | Memory Map |
|---|---|
| $e_R(2) = d_R(2) + a_R(0)$ | $e_R(2) \Rightarrow M(0)$ |
| $e_I(2) = d_I(2) + a_I(0)$ | $e_I(2) \Rightarrow M(9)$ |
| $m_R(1) = e_R(2) + d_R(3) + d_R(5)$ | $m_R(1) \Rightarrow M(3)$ |
| $m_I(1) = e_I(2) + d_I(3) + d_I(5)$ | $m_I(1) \Rightarrow M(12)$ |
| $m_R(3) = e_R(2) - d_R(3) - d_R(4)$ | $m_R(3) \Rightarrow M(20)$ |
| $m_I(3) = e_I(2) - d_I(3) - d_I(4)$ | $m_I(3) \Rightarrow M(24)$ |
| $m_R(7) = e_R(2) + d_R(4) - d_R(5)$ | $m_R(7) \Rightarrow M(1)$ |
| $m_I(7) = e_I(2) + d_I(4) - d_I(5)$ | $m_I(7) \Rightarrow M(10)$ |
| $m_R(2) = -d_I(1) - d_I(7) - d_I(9)$ | $m_R(2) \Rightarrow M(9)$ |
| $m_I(2) = -d_R(1) - d_R(7) - d_R(9)$ | $m_I(2) \Rightarrow M(0)$ |
| $m_R(4) = d_I(1) - d_I(7) - d_I(8)$ | $m_R(4) \Rightarrow M(25)$ |
| $m_I(4) = d_R(1) - d_R(7) - d_R(8)$ | $m_I(4) \Rightarrow M(21)$ |

| Algorithm Steps | Memory Map |
|---|---|
| $m_R(8) = -d_I(1) - d_I(8) + d_I(9)$ | $m_R(8) \Rightarrow M(14)$ |
| $m_I(8) = -d_R(1) - d_R(8) + d_R(9)$ | $m_I(8) \Rightarrow M(5)$ |
| $m_R(5) = d_R(10) + c_R(1)$ | $m_R(5) \Rightarrow M(18)$ |
| $m_I(5) = d_I(10) + c_I(1)$ | $m_I(5) \Rightarrow M(22)$ |
| $m_R(6) = -d_I(6)$ | $m_R(6) \Rightarrow M(11)$ |
| $m_I(6) = -d_R(6)$ | $m_I(6) \Rightarrow M(2)$ |

## Stage 5: Output Adds

This stage also does not require any multiplier constants. The strategy for converting these equations to code is to start at the top (compute $A_R(1)$) and identify the pair of inputs to be used first (in this case $m_R(1)$ and $m_R(2)$). Then look down the column to find the second (compute $A_R(8)$) place where these two inputs are used. Pull $m_R(1)$ and $m_R(2)$ from memory, compute $A_R(1)$ and $A_R(8)$, and store the results in data memory locations $M(3)$ and $M(9)$ previously occupied by $m_R(1)$ and $m_R(2)$.

Next, look for the computation for $A_I(1)$ in the column and repeat the same set of steps. Continue this process until all of the computations are performed and all of the results are returned to the data memory locations. Note that the $A_R(2)$, $A_R(6)$, $A_R(7)$, $A_I(2)$, $A_I(6)$, and $A_I(7)$ computations are placed in data memory locations different from where the inputs were taken. This is to satisfy the constraint that the output frequency components are stored in the same locations as the input data sequence. Note that the Algorithm Steps for $A_R(0)$ and $A_I(0)$ only relabel the data values to their output labels once they have been used as required by other portions of the algorithm.

| Algorithm Steps | Memory Map |
|---|---|
| $A_R(0) = f_R(0)$ | $A_R(0) \Rightarrow M(7)$ |
| $A_I(0) = f_I(0)$ | $A_I(0) \Rightarrow M(16)$ |
| $A_R(1) = m_R(1) + m_R(2)$ | $A_R(1) \Rightarrow M(3)$ |
| $A_I(1) = m_I(1) - m_I(2)$ | $A_I(1) \Rightarrow M(0)$ |
| $A_R(2) = m_R(3) + m_R(4)$ | $A_R(2) \Rightarrow M(6)$ |
| $A_I(2) = m_I(3) - m_I(4)$ | $A_I(2) \Rightarrow M(8)$ |
| $A_R(3) = m_R(5) + m_R(6)$ | $A_R(3) \Rightarrow M(11)$ |
| $A_I(3) = m_I(5) - m_I(6)$ | $A_I(3) \Rightarrow M(2)$ |
| $A_R(4) = m_R(7) + m_R(8)$ | $A_R(4) \Rightarrow M(1)$ |
| $A_I(4) = m_I(7) - m_I(8)$ | $A_I(4) \Rightarrow M(5)$ |
| $A_R(5) = m_R(7) - m_R(8)$ | $A_R(5) \Rightarrow M(14)$ |
| $A_I(5) = m_I(7) + m_I(8)$ | $A_I(5) \Rightarrow M(10)$ |
| $A_R(6) = m_R(5) - m_R(6)$ | $A_R(6) \Rightarrow M(13)$ |
| $A_I(6) = m_I(5) + m_I(6)$ | $A_I(6) \Rightarrow M(17)$ |
| $A_R(7) = m_R(3) - m_R(4)$ | $A_R(7) \Rightarrow M(4)$ |
| $A_I(7) = m_I(3) + m_I(4)$ | $A_I(7) \Rightarrow M(15)$ |
| $A_R(8) = m_R(1) - m_R(2)$ | $A_R(8) \Rightarrow M(9)$ |
| $A_I(8) = m_I(1) + m_I(2)$ | $A_I(8) \Rightarrow M(12)$ |

## 8.10 SIXTEEN-POINT FFT

The 16-point DFT is defined for $k = 0, 1, 2, 3, 4, 5, 6, 7, 8, 9, 10, 11, 12, 13, 14,$ and 15 as

$$A(k) = \sum_{n=0}^{15} a(n) * e^{-j2\pi k*n/16} \qquad (8\text{-}10)$$

The Winograd [1] 16-point DFT was developed by using a decomposition based on circular convolution properties. Other popular 16-point FFTs are based on mixed-radix combinations of the 2-, 4-, and 8-point building-block algorithms and are presented in Chapter 9.

If the 16-point DFT is calculated directly from Equation 8-10, it requires 225 complex multiplies and 240 complex adds. Since a complex multiply uses 4 real multiplies and 2 real adds, and a complex add uses 2 real adds, the 16-point DFT requires 900 real multiplies and 930 real adds. The number of adds and multiplies for the fast algorithm is significantly less than required for computing the DFT directly. However, if only a subset of the output frequency components is required, it may be more cost effective to compute the DFT equation directly for those terms. For example, if $A(0)$ is the only term needed, it can be computed with 30 adds and no multiplies by using the DFT directly. Each of the other 15 output frequencies requires 15 complex multiplies and 15 complex adds for a total of 60 real adds and 60 real multiplies. With this in mind, the crossover point between using the DFT directly and the 16-point FFT algorithm can be determined based on the number of output frequency components that must be computed.

Since all of the input data is required for each output frequency component calculation, the direct DFT computations require 32 data memory locations for the input data and 32 more for the output frequency components. This is a total of 64 data memory locations, since the input and output are complex. Similarly, the DFT data addressing is sequential (i.e., 0 through 15 for each output frequency component), and the computational architecture is simple, since they can all be performed by using a complex multiply accumulator (see Chapter 10 for details). Addressing the complex multiplier coefficients requires either a modulo arithmetic scheme ($k * n \bmod(16)$) or that the addresses be stored in program memory. The Winograd algorithm is presented, characterized, and then summarized in the Comparison Matrix in Table 8-10.

### 8.10.1 Winograd 16-point FFT

The Winograd [1] 16-point FFT requires 148 adds, 20 multiplies, 36 data memory locations, and 6 multiplier constant memory locations. The seven stages are as follows.

### Stage 1: Input Adds

This stage does not require additional data memory or accessing any of the multiplier constants. Further, the add/subtract process is the same for all of the real and imaginary pairs. The strategy for converting these equations to code is to start at the top (compute $b_R(1)$) and identify the pair of inputs to be used first (in this case $a_R(0)$ and $a_R(8)$). Then look down the list to find the second (compute $b_R(2)$) place where these two inputs are used. Pull $a_R(1)$ and $a_R(8)$ from memory, compute $b_R(1)$ and $b_R(2)$, and store the results in data memory locations $M(0)$ and $M(8)$ previously occupied by $a_R(0)$ and $a_R(8)$.

Next, look for the computation for $b_I(1)$ on the list and repeat the same set of steps. Continue this process until all of the computations are performed and all of the results returned to the data memory locations.

| Algorithm Steps | Memory Map |
|---|---|
| $b_R(1) = a_R(0) + a_R(8)$ | $b_R(1) \Rightarrow M(0)$ |
| $b_I(1) = a_I(0) + a_I(8)$ | $b_I(1) \Rightarrow M(16)$ |
| $b_R(2) = a_R(0) - a_R(8)$ | $b_R(2) \Rightarrow M(8)$ |
| $b_I(2) = a_I(0) - a_I(8)$ | $b_I(2) \Rightarrow M(24)$ |
| $b_R(3) = a_R(4) + a_R(12)$ | $b_R(3) \Rightarrow M(4)$ |
| $b_I(3) = a_I(4) + a_I(12)$ | $b_I(3) \Rightarrow M(20)$ |
| $b_R(4) = a_R(4) - a_R(12)$ | $b_R(4) \Rightarrow M(12)$ |
| $b_I(4) = a_I(4) - a_I(12)$ | $b_I(4) \Rightarrow M(28)$ |
| $b_R(5) = a_R(2) + a_R(10)$ | $b_R(5) \Rightarrow M(2)$ |
| $b_I(5) = a_I(2) + a_I(10)$ | $b_I(5) \Rightarrow M(18)$ |
| $b_R(6) = a_R(2) - a_R(10)$ | $b_R(6) \Rightarrow M(10)$ |
| $b_I(6) = a_I(2) - a_I(10)$ | $b_I(6) \Rightarrow M(26)$ |
| $b_R(7) = a_R(6) + a_R(14)$ | $b_R(7) \Rightarrow M(6)$ |
| $b_I(7) = a_I(6) + a_I(14)$ | $b_I(7) \Rightarrow M(22)$ |
| $b_R(8) = a_R(6) - a_R(14)$ | $b_R(8) \Rightarrow M(14)$ |
| $b_I(8) = a_I(6) - a_I(14)$ | $b_I(8) \Rightarrow M(30)$ |
| $b_R(9) = a_R(1) + a_R(9)$ | $b_R(9) \Rightarrow M(1)$ |
| $b_I(9) = a_I(1) + a_I(9)$ | $b_I(9) \Rightarrow M(17)$ |
| $b_R(10) = a_R(1) - a_R(9)$ | $b_R(10) \Rightarrow M(9)$ |
| $b_I(10) = a_I(1) - a_I(9)$ | $b_I(10) \Rightarrow M(25)$ |
| $b_R(11) = a_R(5) + a_R(13)$ | $b_R(11) \Rightarrow M(5)$ |
| $b_I(11) = a_I(5) + a_I(13)$ | $b_I(11) \Rightarrow M(21)$ |
| $b_R(12) = a_R(5) - a_R(13)$ | $b_R(12) \Rightarrow M(13)$ |
| $b_I(12) = a_I(5) - a_I(13)$ | $b_I(12) \Rightarrow M(29)$ |
| $b_R(13) = a_R(3) + a_R(11)$ | $b_R(13) \Rightarrow M(3)$ |
| $b_I(13) = a_I(3) + a_I(11)$ | $b_I(13) \Rightarrow M(19)$ |
| $b_R(14) = a_R(3) - a_R(11)$ | $b_R(14) \Rightarrow M(11)$ |
| $b_I(14) = a_I(3) - a_I(11)$ | $b_I(14) \Rightarrow M(27)$ |
| $b_R(15) = a_R(7) + a_R(15)$ | $b_R(15) \Rightarrow M(7)$ |
| $b_I(15) = a_I(7) + a_I(15)$ | $b_I(15) \Rightarrow M(23)$ |
| $b_R(16) = a_R(7) - a_R(15)$ | $b_R(16) \Rightarrow M(15)$ |
| $b_I(16) = a_I(7) - a_I(15)$ | $b_I(16) \Rightarrow M(31)$ |

## Stage 2: Second Set of Input Adds

This stage also does not require additional data memory or accessing any multiplier constants. Further, the add/subtract process is the same for all of the real and imaginary pairs. The strategy for converting these equations to code is to start at the top (compute $c_R(1)$) and identify the pair of inputs to be used first (in this case $b_R(1)$ and $b_R(3)$). Then look down the list to find the second (compute $c_R(2)$) place where these two inputs are used. Pull $b_R(1)$ and $b_R(3)$ from memory, compute $c_R(1)$ and $c_R(2)$, and store the results in data memory locations $M(0)$ and $M(4)$ previously occupied by $b_R(1)$ and $b_R(3)$.

Next, look for the computation for $c_I(1)$ on the list and repeat the same set of steps. Continue this process until all the Algorithm Steps have been computed and their results stored in the Memory Map addresses.

| Algorithm Steps | Memory Map |
|---|---|
| $c_R(1) = b_R(1) + b_R(3)$ | $c_R(1) \Rightarrow M(0)$ |
| $c_I(1) = b_I(1) + b_I(3)$ | $c_I(1) \Rightarrow M(16)$ |
| $c_R(2) = b_R(1) - b_R(3)$ | $c_R(2) \Rightarrow M(4)$ |
| $c_I(2) = b_I(1) - b_I(3)$ | $c_I(2) \Rightarrow M(20)$ |
| $c_R(3) = b_R(5) + b_R(7)$ | $c_R(3) \Rightarrow M(2)$ |
| $c_I(3) = b_I(5) + b_I(7)$ | $c_I(3) \Rightarrow M(18)$ |
| $c_R(4) = b_R(5) - b_R(7)$ | $c_R(4) \Rightarrow M(6)$ |
| $c_I(4) = b_I(5) - b_I(7)$ | $c_I(4) \Rightarrow M(22)$ |
| $c_R(5) = b_R(9) + b_R(11)$ | $c_R(5) \Rightarrow M(1)$ |
| $c_I(5) = b_I(9) + b_I(11)$ | $c_I(5) \Rightarrow M(17)$ |
| $c_R(6) = b_R(9) - b_R(11)$ | $c_R(6) \Rightarrow M(5)$ |
| $c_I(6) = b_I(9) - b_I(11)$ | $c_I(6) \Rightarrow M(21)$ |
| $c_R(7) = b_R(13) + b_R(15)$ | $c_R(7) \Rightarrow M(3)$ |
| $c_I(7) = b_I(13) + b_I(15)$ | $c_I(7) \Rightarrow M(19)$ |
| $c_R(8) = b_R(13) - b_R(15)$ | $c_R(8) \Rightarrow M(7)$ |
| $c_I(8) = b_I(13) - b_I(15)$ | $c_I(8) \Rightarrow M(23)$ |
| $c_R(9) = b_R(6) + b_R(8)$ | $c_R(9) \Rightarrow M(10)$ |
| $c_I(9) = b_I(6) + b_I(8)$ | $c_I(9) \Rightarrow M(26)$ |
| $c_R(10) = b_R(6) - b_R(8)$ | $c_R(10) \Rightarrow M(14)$ |
| $c_I(10) = b_I(6) - b_I(8)$ | $c_I(10) \Rightarrow M(30)$ |
| $c_R(11) = b_R(10) + b_R(16)$ | $c_R(11) \Rightarrow M(9)$ |
| $c_I(11) = b_I(10) + b_I(16)$ | $c_I(11) \Rightarrow M(25)$ |
| $c_R(12) = b_R(10) - b_R(16)$ | $c_R(12) \Rightarrow M(15)$ |
| $c_I(12) = b_I(10) - b_I(16)$ | $c_I(12) \Rightarrow M(31)$ |
| $c_R(13) = b_R(12) + b_R(14)$ | $c_R(13) \Rightarrow M(11)$ |
| $c_I(13) = b_I(12) + b_I(14)$ | $c_I(13) \Rightarrow M(27)$ |
| $c_R(14) = b_R(12) - b_R(14)$ | $c_R(14) \Rightarrow M(13)$ |
| $c_I(14) = b_I(12) - b_I(14)$ | $c_I(14) \Rightarrow M(29)$ |

### Stage 3: Third Set of Input Adds

This stage requires additional data memory locations but not accessing any multiplier constants. Further, the add/subtract process is the same for all of the real and imaginary pairs. The strategy for converting these equations to code is to start at the top (compute $d_R(1)$) and identify the pair of inputs to be used first (in this case $c_R(1)$ and $c_R(3)$). Then look down the list to find the second (compute $d_R(2)$) place where these two inputs are used. Pull $c_R(1)$ and $c_R(3)$ from memory, compute $d_R(1)$ and $d_R(2)$, and store the results in data memory locations $M(0)$ and $M(2)$ previously occupied by $c_R(1)$ and $c_R(3)$.

Next, look for the computation for $d_I(1)$ on the list and repeat the same set of steps. Continue this process until all the Algorithm Steps have been computed and their results stored in the Memory Map addresses. The additional data memory locations $M(32)$, $M(33)$, $M(34)$, and $M(35)$ are required for $d_R(7)$, $d_R(8)$, $d_I(7)$, and $d_I(8)$ because their input values, $c_R(11)$ through $c_R(14)$ and $c_I(11)$ through $c_I(14)$, are also needed in Stage 4.

| Algorithm Steps | Memory Map |
|---|---|
| $d_R(1) = c_R(1) + c_R(3)$ | $d_R(1) \Rightarrow M(0)$ |
| $d_I(1) = c_I(1) + c_I(3)$ | $d_I(1) \Rightarrow M(16)$ |
| $d_R(2) = c_R(1) - c_R(3)$ | $d_R(2) \Rightarrow M(2)$ |
| $d_I(2) = c_I(1) - c_I(3)$ | $d_I(2) \Rightarrow M(18)$ |
| $d_R(3) = c_R(5) + c_R(7)$ | $d_R(3) \Rightarrow M(1)$ |
| $d_I(3) = c_I(5) + c_I(7)$ | $d_I(3) \Rightarrow M(17)$ |
| $d_R(4) = c_R(5) - c_R(7)$ | $d_R(4) \Rightarrow M(3)$ |
| $d_I(4) = c_I(5) - c_I(7)$ | $d_I(4) \Rightarrow M(19)$ |
| $d_R(5) = c_R(6) + c_R(8)$ | $d_R(5) \Rightarrow M(5)$ |
| $d_I(5) = c_I(6) + c_I(8)$ | $d_I(5) \Rightarrow M(21)$ |
| $d_R(6) = c_R(6) - c_R(8)$ | $d_R(6) \Rightarrow M(7)$ |
| $d_I(6) = c_I(6) - c_I(8)$ | $d_I(6) \Rightarrow M(23)$ |
| $d_R(7) = c_R(11) + c_R(13)$ | $d_R(7) \Rightarrow M(32)$ |
| $d_I(7) = c_I(11) + c_I(13)$ | $d_I(7) \Rightarrow M(34)$ |
| $d_R(8) = c_R(12) + c_R(14)$ | $d_R(8) \Rightarrow M(33)$ |
| $d_I(8) = c_I(12) + c_I(14)$ | $d_I(8) \Rightarrow M(35)$ |
| $e_R(1) = d_R(1) + d_R(3)$ | $e_R(1) \Rightarrow M(0)$ |
| $e_I(1) = d_I(1) + d_I(3)$ | $e_I(1) \Rightarrow M(16)$ |
| $e_R(2) = d_R(1) - d_R(3)$ | $e_R(2) \Rightarrow M(1)$ |
| $e_I(2) = d_I(1) - d_I(3)$ | $e_I(2) \Rightarrow M(17)$ |

## Stage 4: Multiplies

This stage contains all of the multiplications. In all cases the multiplication is performed by pulling a data value from memory, multiplying it by the appropriate constant, and returning the result to the same data memory location. In some of the multiplications the real part of a complex data value is the input and the output has an imaginary label. This process provides the required multiplications by $j = \sqrt{-1}$. Also note that $\sin(4p/16) = \cos(4p/16)$, which reduces the number of constants to be stored to 6. Note that several of the Algorithm Steps, such as $e_R(3)$ and $e_I(3)$, just relabel the data values. This is to make intermediate results from several stages have the same small letter label prior to proceeding with Stage 5.

| Algorithm Steps | Memory Map |
|---|---|
| $e_R(3) = d_R(2)$ | $e_R(3) \Rightarrow M(2)$ |
| $e_I(3) = d_I(2)$ | $e_I(3) \Rightarrow M(18)$ |
| $e_R(4) = d_I(4)$ | $e_R(4) \Rightarrow M(19)$ |
| $e_I(4) = -d_R(4)$ | $e_I(4) \Rightarrow M(3)$ |
| $e_R(5) = c_R(2)$ | $e_R(5) \Rightarrow M(4)$ |
| $e_I(5) = c_I(2)$ | $e_I(5) \Rightarrow M(20)$ |
| $e_R(6) = c_I(4)$ | $e_R(6) \Rightarrow M(22)$ |
| $e_I(6) = -c_R(4)$ | $e_I(6) \Rightarrow M(6)$ |
| $e_R(7) = \sin(4\pi/16) * d_I(5)$ | $e_R(7) \Rightarrow M(21)$ |
| $e_I(7) = -\sin(4\pi/16) * d_R(5)$ | $e_I(7) \Rightarrow M(5)$ |
| $e_R(8) = \cos(4\pi/16) * d_R(6)$ | $e_R(8) \Rightarrow M(7)$ |
| $e_I(8) = \cos(4\pi/16) * d_I(6)$ | $e_I(8) \Rightarrow M(23)$ |
| $e_R(9) = b_R(2)$ | $e_R(9) \Rightarrow M(8)$ |
| $e_I(9) = b_I(2)$ | $e_I(9) \Rightarrow M(24)$ |
| $e_R(10) = b_I(4)$ | $e_R(10) \Rightarrow M(28)$ |
| $e_I(10) = -b_R(4)$ | $e_I(10) \Rightarrow M(12)$ |
| $e_R(11) = \sin(4\pi/16) * c_I(9)$ | $e_R(11) \Rightarrow M(26)$ |
| $e_I(11) = -\sin(4\pi/16) * c_R(9)$ | $e_I(11) \Rightarrow M(10)$ |
| $e_R(12) = \cos(4\pi/16) * c_R(10)$ | $e_R(12) \Rightarrow M(14)$ |
| $e_I(12) = \cos(4\pi/16) * c_I(10)$ | $e_I(12) \Rightarrow M(30)$ |
| $e_R(13) = \sin(6\pi/16) * d_I(7)$ | $e_R(13) \Rightarrow M(34)$ |
| $e_I(13) = -\sin(6\pi/16) * d_R(7)$ | $e_I(13) \Rightarrow M(32)$ |
| $e_R(14) = [\sin(2\pi/16) - \sin(6\pi/16)] * c_I(11)$ | $e_R(14) \Rightarrow M(25)$ |
| $e_I(14) = -[\sin(2\pi/16) - \sin(6\pi/16)] * c_R(11)$ | $e_I(14) \Rightarrow M(9)$ |
| $e_R(15) = [\sin(2\pi/16) + \sin(6\pi/16)] * c_I(13)$ | $e_R(15) \Rightarrow M(27)$ |
| $e_I(15) = -[\sin(2\pi/16) + \sin(6\pi/16)] * c_R(13)$ | $e_I(15) \Rightarrow M(11)$ |
| $e_R(16) = \cos(6\pi/16) * d_R(8)$ | $e_R(16) \Rightarrow M(33)$ |
| $e_I(16) = \cos(6\pi/16) * d_I(8)$ | $e_I(16) \Rightarrow M(35)$ |
| $e_R(17) = [\cos(2\pi/16) + \cos(6\pi/16)] * c_R(12)$ | $e_R(17) \Rightarrow M(15)$ |
| $e_I(17) = [\cos(2\pi/16) + \cos(6\pi/16)] * c_I(12)$ | $e_I(17) \Rightarrow M(31)$ |
| $e_R(18) = -[\cos(2\pi/16) - \cos(6\pi/16)] * c_R(14)$ | $e_R(18) \Rightarrow M(13)$ |
| $e_I(18) = -[\cos(2\pi/16) - \cos(6\pi/16)] * c_I(14)$ | $e_I(18) \Rightarrow M(29)$ |

## Stage 5: Postmultiplies

This stage also does not require accessing any multiplier constants. The strategy for converting these equations to code is to start at the top (compute $f_R(1)$) and identify the pair of inputs to be used first (in this case $e_R(3)$ and $e_R(4)$). Then look down the list to find the second (compute $f_R(2)$) place where these two inputs are used. Pull $e_R(3)$ and $e_R(4)$ from memory, compute $f_R(1)$ and $f_R(2)$, and store the results in data memory locations $M(2)$ and $M(19)$ previously occupied by $e_R(3)$ and $e_R(4)$.

Next, look for the computation for $f_I(1)$ on the list and repeat the same set of steps. Continue this process until all the Algorithm Steps have been computed and their results stored in the Memory Map addresses. This stage does not require additional data memory locations. However, all four additional data memory locations required for this algorithm are used during this stage to simplify the data addressing. This leaves input data memory locations $M(11)$, $M(13)$, $M(27)$, and $M(29)$ unused. They will be reused in Stage 7 to end the algorithm with the results in the same data memory locations that were occupied by the input data.

Additionally, note that this stage has data ($e_R(13)$, $e_R(16)$, $e_I(13)$, and $e_I(16)$) that are independently used to compute two results. The Memory Map strategy in this case is to use $e_R(13)$, $e_R(16)$, $e_I(13)$, and $e_I(16)$ data memory locations for the output of the second computation that required these data values. If those data memory locations were used for the output of the first computations, their values would be destroyed before being able to use them for the second computation.

| Algorithm Steps | Memory Map |
|---|---|
| $f_R(1) = e_R(3) + e_R(4)$ | $f_R(1) \Rightarrow M(2)$ |
| $f_I(1) = e_I(3) + e_I(4)$ | $f_I(1) \Rightarrow M(18)$ |
| $f_R(2) = e_R(3) - e_R(4)$ | $f_R(2) \Rightarrow M(19)$ |
| $f_I(2) = e_I(3) - e_I(4)$ | $f_I(2) \Rightarrow M(3)$ |
| $f_R(3) = e_R(5) + e_R(7)$ | $f_R(3) \Rightarrow M(4)$ |
| $f_I(3) = e_I(5) + e_I(7)$ | $f_I(3) \Rightarrow M(20)$ |
| $f_R(4) = e_R(5) - e_R(7)$ | $f_R(4) \Rightarrow M(21)$ |
| $f_I(4) = e_I(5) - e_I(7)$ | $f_I(4) \Rightarrow M(5)$ |
| $f_R(5) = e_R(6) + e_R(8)$ | $f_R(5) \Rightarrow M(22)$ |
| $f_I(5) = e_I(6) + e_I(8)$ | $f_I(5) \Rightarrow M(6)$ |
| $f_R(6) = e_R(6) - e_R(8)$ | $f_R(6) \Rightarrow M(7)$ |
| $f_I(6) = e_I(6) - e_I(8)$ | $f_I(6) \Rightarrow M(23)$ |
| $f_R(7) = e_R(9) + e_R(12)$ | $f_R(7) \Rightarrow M(8)$ |
| $f_I(7) = e_I(9) + e_I(12)$ | $f_I(7) \Rightarrow M(24)$ |
| $f_R(8) = e_R(9) - e_R(12)$ | $f_R(8) \Rightarrow M(14)$ |
| $f_I(8) = e_I(9) - e_I(12)$ | $f_I(8) \Rightarrow M(30)$ |
| $f_R(9) = e_R(10) + e_R(11)$ | $f_R(9) \Rightarrow M(28)$ |
| $f_I(9) = e_I(10) + e_I(11)$ | $f_I(9) \Rightarrow M(12)$ |
| $f_R(10) = e_R(10) - e_R(11)$ | $f_R(10) \Rightarrow M(26)$ |
| $f_I(10) = e_I(10) - e_I(11)$ | $f_I(10) \Rightarrow M(10)$ |
| $f_R(11) = e_R(13) + e_R(14)$ | $f_R(11) \Rightarrow M(25)$ |
| $f_I(11) = e_I(13) + e_I(14)$ | $f_I(11) \Rightarrow M(9)$ |
| $f_R(12) = e_R(13) - e_R(15)$ | $f_R(12) \Rightarrow M(34)$ |
| $f_I(12) = e_I(13) - e_I(15)$ | $f_I(12) \Rightarrow M(32)$ |
| $f_R(13) = e_R(17) - e_R(16)$ | $f_R(13) \Rightarrow M(15)$ |
| $f_I(13) = e_I(17) - e_I(16)$ | $f_I(13) \Rightarrow M(31)$ |
| $f_R(14) = e_R(18) - e_R(16)$ | $f_R(14) \Rightarrow M(33)$ |
| $f_I(14) = e_I(18) - e_I(16)$ | $f_I(14) \Rightarrow M(35)$ |

## Stage 6: Second Set of Postmultiply Adds

The strategy for converting these equations to code is to start at the top (compute $g_R(1)$) and identify the pair of inputs to be used first (in this case $f_R(3)$ and $f_R(5)$). Then look down the list to find the second (compute $g_R(2)$) place where these two inputs are used. Pull $f_R(3)$ and $f_R(5)$ from memory, compute $g_R(1)$ and $g_R(2)$, and store the results in data memory locations $M(4)$ and $M(22)$ previously occupied by $f_R(3)$ and $f_R(5)$.

Next, look for the computation for $g_I(1)$ on the list and repeat the same set of steps. Continue this process until all the Algorithm Steps have been computed and their results stored in the Memory Map addresses. This stage does not require additional data memory locations. However, all four additional data memory locations required for this algorithm are also used during this stage to simplify the data addressing. This continues to leave input data memory locations $M(11)$, $M(13)$, $M(27)$, and $M(29)$ unused.

| Algorithm Steps | Memory Map |
|---|---|
| $g_R(1) = f_R(3) + f_R(5)$ | $g_R(1) \Rightarrow M(4)$ |
| $g_I(1) = f_I(3) + f_I(5)$ | $g_I(1) \Rightarrow M(20)$ |
| $g_R(2) = f_R(3) - f_R(5)$ | $g_R(2) \Rightarrow M(22)$ |
| $g_I(2) = f_I(3) - f_I(5)$ | $g_I(2) \Rightarrow M(6)$ |
| $g_R(3) = f_R(4) + f_R(6)$ | $g_R(3) \Rightarrow M(21)$ |
| $g_I(3) = f_I(4) + f_I(6)$ | $g_I(3) \Rightarrow M(5)$ |
| $g_R(4) = f_R(4) - f_R(6)$ | $g_R(4) \Rightarrow M(7)$ |
| $g_I(4) = f_I(4) - f_I(6)$ | $g_I(4) \Rightarrow M(23)$ |
| $g_R(5) = f_R(7) + f_R(11)$ | $g_R(5) \Rightarrow M(8)$ |
| $g_I(5) = f_I(7) + f_I(11)$ | $g_I(5) \Rightarrow M(24)$ |
| $g_R(6) = f_R(7) - f_R(11)$ | $g_R(6) \Rightarrow M(25)$ |
| $g_I(6) = f_I(7) - f_I(11)$ | $g_I(6) \Rightarrow M(9)$ |
| $g_R(7) = f_R(8) + f_R(12)$ | $g_R(7) \Rightarrow M(14)$ |
| $g_I(7) = f_I(8) + f_I(12)$ | $g_I(7) \Rightarrow M(30)$ |
| $g_R(8) = f_R(8) - f_R(12)$ | $g_R(8) \Rightarrow M(34)$ |
| $g_I(8) = f_I(8) - f_I(12)$ | $g_I(8) \Rightarrow M(32)$ |
| $g_R(9) = f_R(9) + f_R(13)$ | $g_R(9) \Rightarrow M(28)$ |
| $g_I(9) = f_I(9) + f_I(13)$ | $g_I(9) \Rightarrow M(12)$ |
| $g_R(10) = f_R(9) - f_R(13)$ | $g_R(10) \Rightarrow M(15)$ |
| $g_I(10) = f_I(9) - f_I(13)$ | $g_I(10) \Rightarrow M(31)$ |
| $g_R(11) = f_R(10) + f_R(14)$ | $g_R(11) \Rightarrow M(26)$ |
| $g_I(11) = f_I(10) + f_I(14)$ | $g_I(11) \Rightarrow M(10)$ |
| $g_R(12) = f_R(10) - f_R(14)$ | $g_R(12) \Rightarrow M(33)$ |
| $g_I(12) = f_I(10) - f_I(14)$ | $g_I(12) \Rightarrow M(35)$ |

## Stage 7: Output Adds

This stage does not require additional data memory or accessing any multiplier constants. Further, the add/subtract process is the same for all of the real and imaginary pairs.

The strategy for converting these equations to code is to start at the top (compute $A_R(1)$) and identify the pair of inputs to be used first (in this case $g_R(5)$ and $g_R(9)$). Then look down the list to find the second (compute $A_R(7)$) place where these two inputs are used. Pull $g_R(5)$ and $g_R(9)$ from memory, compute $A_R(1)$ and $A_R(7)$, and store the results in data memory locations $M(8)$ and $M(28)$ previously occupied by $g_R(5)$ and $g_R(9)$.

Next, look for the computation for $A_I(1)$ on the list and repeat the same set of steps. Continue this process until all the Algorithm Steps have been computed and their results stored in the Memory Map addresses. The only variation in the standard pattern of data addressing is for computing $A_R(11)$, $A_I(11)$, $A_R(13)$, and $A_I(13)$. The inputs for these computations come from the additional data memory locations needed earlier in the algorithm. Since the additional data memory locations are no longer needed, these computed results for $A_R(11)$, $A_I(11)$, $A_R(13)$, and $A_I(13)$ are stored in $M(13)$, $M(29)$, $M(27)$, and $M(11)$ respectively. The final result is the output frequencies being located in the same data memory locations used for the input data. Note that several of the Algorithm Steps, such as $A_R(0)$ and $A_I(0)$, only relabel the data values to their output labels once they have been used as required by other portions of the algorithm.

| Algorithm Steps | Memory Map |
|---|---|
| $A_R(0) = e_R(1)$ | $A_R(0) \Rightarrow M(0)$ |
| $A_I(0) = e_I(1)$ | $A_I(0) \Rightarrow M(16)$ |
| $A_R(1) = g_R(5) + g_R(9)$ | $A_R(1) \Rightarrow M(8)$ |
| $A_I(1) = g_I(5) + g_I(9)$ | $A_I(1) \Rightarrow M(24)$ |
| $A_R(2) = g_R(1)$ | $A_R(2) \Rightarrow M(4)$ |
| $A_I(2) = g_I(1)$ | $A_I(2) \Rightarrow M(20)$ |
| $A_R(3) = g_R(7) - g_R(11)$ | $A_R(3) \Rightarrow M(14)$ |
| $A_I(3) = g_I(7) - g_I(11)$ | $A_I(3) \Rightarrow M(30)$ |
| $A_R(4) = f_R(1)$ | $A_R(4) \Rightarrow M(2)$ |
| $A_I(4) = f_I(1)$ | $A_I(4) \Rightarrow M(18)$ |
| $A_R(5) = g_R(7) + g_R(11)$ | $A_R(5) \Rightarrow M(26)$ |
| $A_I(5) = g_I(7) + g_I(11)$ | $A_I(5) \Rightarrow M(10)$ |
| $A_R(6) = g_R(2)$ | $A_R(6) \Rightarrow M(22)$ |
| $A_I(6) = g_I(2)$ | $A_I(6) \Rightarrow M(6)$ |
| $A_R(7) = g_R(5) - g_R(9)$ | $A_R(7) \Rightarrow M(28)$ |
| $A_I(7) = g_I(5) - g_I(9)$ | $A_I(7) \Rightarrow M(12)$ |
| $A_R(8) = e_R(2)$ | $A_R(8) \Rightarrow M(1)$ |
| $A_I(8) = e_I(2)$ | $A_I(8) \Rightarrow M(17)$ |
| $A_R(9) = g_R(6) + g_R(10)$ | $A_R(9) \Rightarrow M(25)$ |
| $A_I(9) = g_I(6) + g_I(10)$ | $A_I(9) \Rightarrow M(9)$ |
| $A_R(10) = g_R(3)$ | $A_R(10) \Rightarrow M(21)$ |
| $A_I(10) = g_I(3)$ | $A_I(10) \Rightarrow M(5)$ |
| $A_R(11) = g_R(8) - g_R(12)$ | $A_R(11) \Rightarrow M(13)$ |
| $A_I(11) = g_I(8) - g_I(12)$ | $A_I(11) \Rightarrow M(29)$ |
| $A_R(12) = f_R(2)$ | $A_R(12) \Rightarrow M(19)$ |

| Algorithm Steps | Memory Map |
|---|---|
| $A_I(12) = f_I(2)$ | $A_I(12) \Rightarrow M(3)$ |
| $A_R(13) = g_R(8) + g_R(12)$ | $A_R(13) \Rightarrow M(27)$ |
| $A_I(13) = g_I(8) + g_I(12)$ | $A_I(13) \Rightarrow M(11)$ |
| $A_R(14) = g_R(4)$ | $A_R(14) \Rightarrow M(7)$ |
| $A_I(14) = g_I(4)$ | $A_I(14) \Rightarrow M(23)$ |
| $A_R(15) = g_R(6) - g_R(10)$ | $A_R(15) \Rightarrow M(15)$ |
| $A_I(15) = g_I(6) - g_I(10)$ | $A_I(15) \Rightarrow M(31)$ |

## 8.11 GENERAL ALGORITHMS FOR ALL ODD NUMBERS

The preceding sections describe specific algorithm building blocks for 2-, 3-, 4-, 5-, 7-, 8-, 9-, and 16-point FFTs. Chapter 9 shows how these can be combined to form any transform length that can be factored into the product of these numbers. However, transform lengths such as 13, $143 = 13 \times 11$, and $117 = 9 \times 19$ are not the product of these building block lengths. To compute all transform lengths efficiently, a fast algorithm must exist for computing all prime number ($p$) length building blocks. The Rader [3] algorithm provides this capability by converting the $p$-point FFT to a series of ($p - 1$)-point FFTs. The 5-point Rader FFT given in Section 8.6.3 is a special case of this algorithm.

Since all prime numbers except 2 are odd (all even numbers have at least one factor of 2), ($p - 1$) is always even and therefore has at least one factor of 2. For example, if $p = 67$, then ($p - 1$) $= 66 = 11 \times 2 \times 3$. If all of the factors of 2 are grouped (in this case just one factor of 2), the remaining factors are now all odd (in this case 11 and 3). If the factors of ($p - 1$) are 2, 3, 4, 5, 7, 8, 9, or 16, the algorithms in this chapter, combined with those in Chapter 9, can be used to compute the p-point FFT.

If some of the factors are not among the building-block algorithms provided, they must be obtained from some other source. The power-of-primes algorithm from Chapter 9 can be used for factors of 2 larger than 16. The Singleton [2] or general SWIFT [8] odd-point algorithms can be used for any odd-numbered factor. Therefore, coupled with the building blocks presented in this chapter and the algorithms presented in Chapter 9, the Singleton and general SWIFT odd-point algorithms can be used to compute an FFT of any length.

### 8.11.1 General Rader Algorithm

The general Rader [3] algorithm uses the circular convolution properties of prime number DFTs, much like the Winograd algorithm [1]. The eight stages are as follows.

### Stage 1: Remove a(0)

Separate the first input sample, $a(0)$, from the others and prepare to compute the output frequency components minus $a(0)$, $A(i) - a(0)$, for $i = 0, 1, 2, \ldots, (N - 1)$. This stage requires no computations or data manipulation.

## Stage 2: Reorganize the Input Data

For all prime numbers $N$ there is at least one factor, called a primitive root [9], that can be used to reorganize the numbers from 1 to $N - 1$ to take advantage of the circular convolution properties of the prime DFT. If $g$ is that primitive root, then the way to find the reorganized sequence pi is to compute

$$p_i \equiv g^i \quad \text{modulo } N$$

for $i = 1, 2, \ldots, (N - 1)$ where "modulo $N$" means to take the number $g^i$ and subtract $N$ from it until the number is less than $N$ but greater than zero.

For example, 3 and 5 are the primitive roots of 7. Therefore, either can be used to reorganize the input data to a 7-point DFT to prepare it for the Rader computational algorithm. Namely, the sequences for $g = 3$ and $g = 5$ are

$$g = 3 \text{ sequence: } 3, 2, 6, 4, 5, \text{ and } 1$$
$$g = 5 \text{ sequence: } 5, 4, 6, 2, 3, \text{ and } 1$$

The result is new input data sequences:

$$g = 3 \text{ input data sequence: } a(3), a(2), a(6), a(4), a(5), \text{ and } a(1)$$
$$g = 5 \text{ input data sequence: } a(5), a(4), a(6), a(2), a(3), \text{ and } a(1)$$

With the use of the table of primitive roots, this process can be performed for any prime number up to 5003 [10].

This stage requires no computation or data manipulation during FFT computations. For a given $N$-point prime number DFT, this reorganized data sequence can be computed ahead of time and stored in data or program memory.

## Stage 3: Compute an $(N-1)$-Point FFT

Compute an $(N - 1)$-point FFT of this new sequence. In all cases, $(N - 1)$ is an even number and therefore has at least one factor. For the 5-point Rader transform, $(N - 1) = 4$. For the 7-point example, $(N - 1) = 6$. Therefore, the $(N - 1)$-point FFT can be computed by combining building blocks with one of the algorithms in Chapter 9. This stage requires the number of computations associated with the $(N - 1)$-point FFT algorithm chosen from Chapter 9 with the building blocks from this chapter.

## Stage 4: Reorganize the Complex Multiplier Coefficients

For every primitive root there is another primitive root so that the product of the two is 1 modulo $N$. For the 7-point example, 5 plays this role for the primitive root 3, and 3 plays this role for the primitive root 5 ($3 \times 5 = 15 \equiv 1$ modulo 7). This stage reorganizes the complex multiplier coefficients using this other factor. Namely, for the 7-point transform and the generator $g = 3$, reorganize the complex multiplier coefficients, using the $g = 5$ sequence for the exponents, to $W_7^5, W_7^4, W_7^6, W_7^2, W_7^3, W_7^1$. This stage requires no computation or data manipulation during FFT computations. For a given $N$-point prime number DFT, this reorganized complex multiplier coefficient sequence can be computed ahead of time and stored in data or program memory.

### Stage 5: Compute an ($N$–1)-Point FFT of the Reorganized $W_7$ Sequence

Pretending that the new sequence of complex multiplier coefficients are the in-order data input to an $(N - 1)$-point FFT, compute that FFT. Again, that FFT can be computed using the building blocks in this chapter and the algorithms in Chapter 9. This stage requires the number of computations associated with the $(N - 1)$-point FFT algorithm chosen from Chapter 9 with the building blocks from this chapter. However, all of these computations can be performed ahead of time and stored as multiplier coefficients in data or program memory.

### Stage 6: Perform Complex Multiplications of the Outputs of Stages 3 and 5

Take the in-order $(N - 1)$ output data values of Stages 3 and 5 and multiply them to obtain a new sequence of data values. This stage requires $(N - 1)$ complex multiplies. Since a complex multiply uses four real multiplies and two real adds, this stage requires $4 * (N - 1)$ real multiplies and $2 * (N - 1)$ real adds.

### Stage 7: Compute IFFT

Compute the $(N - 1)$-point IFFT of the output sequence from Stage 7. Again, this IFFT can be computed using the building blocks in this chapter, the algorithms in Chapter 9, and the facts from Section 2.3. The result is the required $A(i) - a(0)$ for the $N$-point FFT, reordered by using the same generator that was used to reorder the complex multiplier coefficients. For the 7-point FFT, the output of this stage is:

$$[A(5)-a(0)], [A(4)-a(0)], [A(6)-a(0)], [A(2)-a(0)], [A(3)-a(0)], \text{ and } [A(1)-a(0)]$$

From Chapters 2 and 3, the IFFT requires the same number of computations as the comparable FFT. In fact, it uses the same algorithm, with some of the multiplier coefficients changed. Therefore, this stage requires the number of computations associated with the $(N - 1)$-point FFT algorithm chosen from Chapter 9 with the building blocks from this chapter.

### Stage 8: Compute the Output Frequency Components

This stage has two steps. First, $a(0)$ is added to each of the $(N - 1)$ outputs from Stage 7. Then all of the input data is added to form $A(0)$. This stage requires, at worst, $2 * (N - 1)$ complex adds.

### 8.11.2 General Singleton Algorithm

The general Singleton [2] algorithm uses the complex conjugate symmetry of the $W_N^{kn}$ multipliers in the DFT (Equation 8-11) and works for all odd numbers.

$$a(k) = \sum_{n=0}^{N-1} a(n) * W_N^{kn} \tag{8-11}$$

The three stages are as follows.

### Stage 1: Input Adds

For $i = 1, 2, \ldots, (N-1)/2$, compute

$$b_R(2i - 1) = a_R(i) + a_R(N - i)$$
$$b_R(2i) = a_R(i) - a_R(N - i)$$
$$b_I(2i - 1) = a_I(i) + a_I(N - i)$$
$$b_I(2i) = a_I(i) - a_I(N - i)$$

For $i = 1, 2, \ldots, (N-1)/2$:

(a) Pull $a_R(i)$ and $a_R(N - i)$ from their data memory locations, perform the add and subtract operations, and return the results, $b_R(2i - 1)$ and $b_R(2i)$, to the data memory locations previously occupied by $a_R(i)$ and $a_R(N - i)$.

(b) Pull $a_I(i)$ and $a_I(N - i)$ from their data memory locations, perform the add and subtract operations, and return the results, $b_I(2i - 1)$ and $b_I(2i)$ to the data memory locations previously occupied by $a_I(i)$ and $a_I(N - i)$.

Since all of these computations can be performed in-place, no additional data memory is required.

### Stage 2: Multiply-Accumulates

For $i = 1, 2, \ldots, (N-1)/2$, compute:

$$c_R(2i - 1) = \sum_{n=1}^{(N-1)/2} b_R(2n - 1) * \cos(2\pi ni/N) + a_R(0)$$

$$c_I(2i - 1) = \sum_{n=1}^{(N-1)/2} b_I(2n - 1) * \cos(2\pi ni/N) + a_I(0)$$

$$c_I(2i) = \sum_{n=1}^{(N-1)/2} b_R(2n) * \sin(2\pi ni/N)$$

$$c_R(2i) = \sum_{n=1}^{(N-1)/2} b_I(2n) * \sin(2\pi ni/N)$$

$$A_R(0) = \sum_{n=1}^{(N-1)/2} b_R(2n - 1) + a_R(0)$$

$$A_I(0) = \sum_{n=1}^{(N-1)/2} b_I(2n - 1) + a_I(0)$$

This is a total of $(N - 1) * (N - 1)$ additions and $(N - 1) * (N - 1)$ multiplications. Since the computations are all multiply accumulations and the input values are used by all of the computed results, the most efficient use of data memory is to:

(a) Compute the $(N - 1)/2$ different $c_R(2i - 1)$ terms and store them in $(N - 1)/2$ new data memory locations.

(b) Compute $A_R(0)$ and store its result in the location previously occupied by $a_R(0)$.

(c) Compute the $(N-1)/2$ different $c_I(2i-1)$ terms and store them in $(N-1)/2$ data memory locations previously occupied by the $(N - 1)/2$ different $b_R(2n - 1)$.

(d) Compute $A_I(0)$ and store its result in the location previously occupied by $a_I(0)$.

(e) Compute the $(N - 1)/2$ different $c_I(2i)$ terms and store them in $(N - 1)/2$ data memory locations previously occupied by the $(N - 1)/2$ different $b_I(2n - 1)$.

(f) Compute the $(N - 1)/2$ different $c_R(2i)$ terms and store them in $(N - 1)/2$ data memory locations previously occupied by the $(N - 1)/2$ different $b_R(2n)$.

The result is the need for $(N - 1)/2$ additional data memory locations.

### Stage 3: Output Adds

For $i = 1, 2, \ldots, (N - 1)/2$, compute:

$$A_R(i) = c_R(2i - 1) + c_I(2i)$$
$$A_R(N - i) = c_R(2i - 1) - c_I(2i)$$
$$A_I(i) = c_I(2i - 1) - c_R(2i)$$
$$A_I(N - i) = c_I(2i - 1) + c_R(2i)$$

This is a total of $2 * (N - 1)$ adds. These computations are performed in pairs. For $i = 1, 2, \ldots, (N - 1)/2$:

(a) Pull $c_R(2i - 1)$ and $c_I(2i)$ from their data memory locations, perform the add and subtract operations, and return the results, $A_R(i)$ and $A_R(N - i)$, to the data memory locations previously occupied by $c_R(2i - 1)$ and $c_I(2i)$.

(b) Pull $c_I(2i - 1)$ and $c_R(2i)$ from their data memory locations, perform the add and subtract operations, and return the results, $A_I(i)$ and $A_I(N - i)$, to the data memory locations previously occupied by $c_I(2i - 1)$ and $c_R(2i)$.

The total number of computations is $(N + 3) * (N - 1)$ adds and $(N - 1) * (N - 1)$ multiplies. The algorithm requires $2 * N + (N - 1)/2$ data memory locations.

### 8.11.3 General SWIFT Odd-Point Algorithm

The general SWIFT odd-point algorithm also uses the complex conjugate symmetry of the $W_N^{kn}$ multipliers in the DFT (Equation 8-11). The only difference is how the first input sample and first output frequency component are treated. Depending on the approach, half of the multipliers are changed. The three stages are as follows.

### Stage 1: Input Adds

For $i = 1, 2, \ldots, (N - 1)/2$, compute

$$b_R(2i - 1) = a_R(i) + a_R(N - i)$$

$$b_R(2i) = a_R(i) - a_R(N - i)$$

$$b_I(2i - 1) = a_I(i) + a_I(N - i)$$

$$b_I(2i) = a_I(i) - a_I(N - i)$$

$$A_R(0) = \sum_{i=1}^{(N-1)/2} b_R(2i - 1) + a_R(0)$$

$$A_I(0) = \sum_{i=1}^{(N-1)/2} b_I(2i - 1) + a_I(0)$$

This is a total of $3 * (N - 1)$ additions. Since all of these computations can be performed in-place, no additional data memory is required. These computations are performed in pairs. For $i = 1, 2, \ldots, (N - 1)/2$:

(a) Pull $a_R(i)$ and $a_R(N - i)$ from their data memory locations, perform the add and subtract operations, and return the results, $b_R(2i - 1)$ and $b_R(2i)$, to the data memory locations previously occupied by $a_R(i)$ and $a_I(N - i)$.

(b) Pull $a_I(i)$ and $a_I(N - i)$ from their data memory locations, perform the add and subtract operations, and return the results, $b_I(2i - 1)$ and $b_I(2i)$, to the data memory locations previously occupied by $a_I(i)$ and $a_I(N - i)$.

Finally, $A_R(0)$ and $A_I(0)$ are computed and the results stored in the locations previously occupied by $a_R(0)$ and $a_I(0)$.

### Stage 2: Multiply-Accumulates

For $i = 1, 2, \ldots, (N - 1)/2$, compute:

$$c_R(2i - 1) = \sum_{n=1}^{(N-1)/2} b_R(2n - 1) * [\cos(2\pi ni/N) - 1] + A_R(0)$$

$$c_I(2i - 1) = \sum_{n=1}^{(N-1)/2} b_I(2n - 1) * [\cos(2\pi ni/N) - 1] + A_I(0)$$

$$c_I(2i) = \sum_{n=1}^{(N-1)/2} b_R(2n) * \sin(2\pi ni/N)$$

$$c_R(2i) = \sum_{n=1}^{(N-1)/2} b_I(2n) * \sin(2\pi ni/N)$$

This is a total of $(N - 2) * (N - 1)$ additions and $(N - 1) * (N - 1)$ multiplications. Just as in the Singleton algorithm case, the most efficient use of data memory is to:

(a) Compute the $(N - 1)/2$ different $c_R(2i - 1)$ terms and store them in $(N - 1)/2$ new data memory locations.

(b) Compute the $(N-1)/2$ different $c_I(2i-1)$ terms and store them in $(N-1)/2$ data memory locations previously occupied by the $(N - 1)/2$ different $b_R(2n - 1)$.

(c) Compute the $(N - 1)/2$ different $c_I(2i)$ terms and store them in $(N - 1)/2$ data memory locations previously occupied by the $(N - 1)/2$ different $b_I(2n - 1)$.

(d) Compute the $(N - 1)/2$ different $c_R(2i)$ terms and store them in $(N - 1)/2$ data memory locations previously occupied by the $(N - 1)/2$ different $b_R(2n)$.

The result is the need for $(N - 1)/2$ additional data memory locations, and all of the computations are performed for the same multiply-accumulate structure, not in-place.

### Stage 3: Output Adds

For $i = 1, 2, \ldots, (N - 1)/2$, compute:

$$A_R(i) = c_R(2i - 1) + c_I(2i)$$
$$A_R(N - i) = c_R(2i - 1) - c_I(2i)$$
$$A_I(i) = c_I(2i - 1) - c_R(2i)$$
$$A_I(N - i) = c_I(2i - 1) + c_R(2i)$$

This is a total of $2 * (N - 1)$ adds. These computations are performed in pairs. For $i = 1, 2, \ldots, (N - 1)/2$:

(a) Pull $c_R(2i - 1)$ and $c_I(2i)$ from their data memory locations, perform the add and subtract operations, and return the results, $A_R(i)$ and $A_R(N - i)$, to the data memory locations previously occupied by $c_R(2i - 1)$ and $c_I(2i)$.

(b) Pull $c_I(2i - 1)$ and $c_R(2i)$ from their data memory locations, perform the add and subtract operations, and return the results, $A_I(i)$ and $A_I(N - i)$, to the data memory locations previously occupied by $c_I(2i - 1)$ and $c_R(2i)$.

The combination of all of the computations requires $(N + 3) * (N - 1)$ adds and $(N - 1) * (N - 1)$ multiplies. The algorithm requires $2 * N + (N - 1)/2$ data memory locations.

## 8.12 BUILDING-BLOCK ALGORITHM COMPARISON MATRIX

The performance measures of the three general algorithms at the bottom of the Comparison Matrix in Table 8-1 (see page 143) are described as formulas, so the specific values can be computed for any building-block length. The last two columns refer to memory locations.

## 8.13 CONCLUSIONS

A lot of space is spent on examples in this chapter because they provide the clearest picture and instruction on how to implement the familiar and not so familiar small-point transforms. Multiple algorithms for each length, except 2 and 4, prove the versatility and flexibility of

**Table 8-1**    Building-Block Algorithm Comparison Matrix

| Algorithm | # of adds | # of multiplies | # of data locations | # of const. locations |
|---|---|---|---|---|
| **2-Point** | 4 | 0 | 4 | 0 |
| **3-Point** | | | | |
| Winograd | 12 | 4 | 6 | 2 |
| Singleton | 12 | 4 | 7 | 2 |
| **4-Point** | 16 | 0 | 8 | 0 |
| **5-Point** | | | | |
| Winograd | 34 | 10 | 12 | 5 |
| Singleton | 32 | 16 | 12 | 4 |
| Rader | 42 | 12 | 12 | 4 |
| **7-Point** | | | | |
| Winograd | 72 | 16 | 22 | 8 |
| Singleton | 60 | 36 | 17 | 6 |
| **8-Point** | | | | |
| Winograd | 52 | 4 | 16 | 1 |
| Split-Radix | 52 | 4 | 16 | 1 |
| Radix-2 | 52 | 4 | 16 | 1 |
| PTL | 52 | 4 | 16 | 1 |
| **9-Point** | | | | |
| Winograd | 90 | 20 | 26 | 10 |
| PTL | 94 | 52 | 22 | 8 |
| Burrus-Eschenbacher | 84 | 20 | 26 | 8 |
| **16-Point** | | | | |
| Winograd | 148 | 20 | 36 | 6 |
| **General $N$-Point** | | | | |
| Rader | $2*A_{N-1}+6*(N-1)$ | $2*M_{N-1}+4*(N-1)$ | $C_{N-1}+2$ | $D_{N-1}+2$ |
| Singleton | $(N+3)*(N-1)$ | $(N-1)^2$ | $(5*N-1)/2$ | $(N-1)$ |
| SWIFT | $(N+3)*(N-1)$ | $(N-1)^2$ | $(5*N-1)/2$ | $(N-1)$ |

*Key to Variables*

$N$ = Number of complex points in building-block algorithm

$A_{N-1}$ = Number of adds required for $(N-1)$-point FFT

$M_{N-1}$ = Number of multiplies required for $(N-1)$-point FFT

$D_{N-1}$ = Number of memory locations used for data in $(N-1)$-point FFT

$C_{N-1}$ = Number of memory locations used for constants in $N$-point FFT

FFTs to provide optimized and customized products. With the building-block algorithms here an FFT of any length can be created by using the algorithms in the next chapter.

Another unique feature of the book—mapping—was introduced in this chapter and is done on two higher levels in Chapters 9 and 12. Here, mapping the result of each algorithm step into a data memory location is the first step toward converting FFT algorithms to optimized assembly language code. The next chapter shows how to do the necessary relabeling of the mappings in this chapter, so these building blocks can be used in larger algorithms. In Chapter 12 the third level of mapping shows how to distribute data and algorithms among multiple processors.

If an application only needs a small-point transform on a single processor, the methods and steps detailed in the next four chapters are not needed. The reader can proceed to Chapter 13 to see how to select an arithmetic format for implementing the algorithm on one of the chips in Chapter 14.

# REFERENCES

[1] S. Winograd, "On Computing the Discrete Fourier Transform," *Mathematics of Computation*, Vol. 32, No. 141, pp. 175–199 (1978).

[2] R. C. Singleton, "An Algorithm for Computing the Mixed Radix Fast Fourier Transform," *IEEE Transactions on Audio and Electroacoustics*, Vol. AU-17, pp. 93–103 (1969).

[3] C. M. Rader, "Discrete Fourier Transforms When the Number of Data Samples Is Prime," *Proceedings of the IEEE*, Vol. 56, pp. 1107–1108 (1968).

[4] J. W. Cooley, "The Structure of FFT Algorithms," *IEEE International Conference on Acoustics, Speech and Signal Processing Tutorial Session*, pp. 12–14 (1990).

[5] J. W. Cooley and J. W. Tukey, "An Algorithm for the Machine Calculation of Complex Fourier Series," *Mathematics of Computation*, Vol. 19, p. 297 (1965).

[6] J. Smith, "Next-Generation FFT Quickly Calculates Odd Sample Sizes," *Personal Engineering & Instrumentation News*, pp. 21–24 (1984).

[7] C. S. Burrus and P. W. Eschenbacher, "An In-Place In-Order Prime Factor FFT Algorithm," *Acoustic Speech and Signal Processing*, Vol. 29, No. 4, pp. 806–817 (1981).

[8] Patent No. 4,293,921, October 6, 1981, *Method and Signal Processor for Frequency Analysis of Time Domain Signals*, Winthrop W. Smith, Jr.

[9] *CRC Standard Mathematical Tables and Formulae*, CRC Press, Boca Raton, FL, pp. 96–101, 1991.

# 9

# Algorithm Construction

## 9.0 INTRODUCTION

An FFT algorithm is a sequence of computational steps used to compute the DFT efficiently. The most popular of these algorithms work only for transform lengths that are powers-of-two (i.e., 2, 4, 8, 16, 32, 64, ... points). However, there are FFT algorithms for any number ($N$) of data points. This chapter describes the computational stages and lists the computational steps for seven FFT algorithms, including the memory maps for storing the intermediate and final results of each.

The answers to the following questions help determine which FFT algorithm to use:

- How many adds and multiplies are required?
- How much data and program memory are required?

The seven algorithms in this chapter are:

- Presented with a general two-block algorithm and then with a 15- or 16-point example
- Constructed in a uniform format
- Able to use any of the building-block algorithms from Chapter 8
- Able to be combined to form even larger FFT algorithms

## 9.1 FOUR PERFORMANCE MEASURES

The most common way to evaluate FFT algorithms is in terms of the number of computations and amount of memory required to compute them. The performance measures in this section quantify those computations and memory needs. The same four measures were used in Chapter 8.

### 9.1.1 Number of Adds

The number of adds is the total number of real adds used for each of the algorithms. It includes the two adds required as part of each of the complex multiplies.

### 9.1.2 Number of Multiplies

The number of multiplies is the total number of real for each multiples for each algorithm. Each complex multiply takes four real multiplies and two real adds (counted in the number of adds).

### 9.1.3 Number of Memory Locations for Multiplier Constants

Each building-block algorithm requires a different number of multiplier constants. Each constant must be stored in data or program memory or computed as needed. The latter is seldom done any more because memory costs have been dramatically lowered. The number for this performance measure in the Comparison Matrix is the total number of different constants required by each algorithm. These include multiplication by 2 and 1/2, which can also be done by moving the binary point of fixed point numbers or by changing the exponent of floating-point numbers.

### 9.1.4 Number of Data Memory Locations

Each algorithm begins and ends by using exactly $2 * N$ data memory locations to store the input data and output results, respectively. However, if no temporary registers are available for intermediate results, most of the algorithms in this chapter require additional data memory locations during the computations. In this chapter, Algorithm Steps and a Memory Map are given for each algorithm, and total data memory location requirements are listed in the Comparison Matrix, assuming the processor has no temporary registers. The difference between those numbers and $2 * N$ is the number of temporary registers needed to avoid using extra data memory locations for intermediate results.

## 9.2 NINE ALGORITHM CONSTRAINTS

The following are the constraints the authors have used for the transforms in this chapter:

1. The real and imaginary parts of the $i$-th input sample are $a_R(i)$ and $a_I(i)$; $A_R(i)$ and $A_I(i)$ are the real and imaginary parts of the $i$-th output frequency component.
2. Intermediate results are labeled with subsequent lowercase letters of the alphabet to indicate where they are located relative to other computational outputs. For example, the first set of intermediate computational results in each of the algorithm building blocks is labeled $b_R(i)$ and $b_I(i)$.
3. The sum and difference computations are performed by taking two pieces of data from data memory, performing the required computations, and returning the results to available memory locations.
4. The multiply-accumulates are performed by sequentially pulling a data value from data memory, performing the multiplication, and adding the results to the proces-

sor's accumulator (Section 14.2.11). When the multiply-accumulate function is complete, the result is stored in a memory location, overwriting data that is no longer needed.

5. The sequence of computations shown for the first stage in each of the algorithms has been left the same as in its referenced article. The data labels have been changed to make them consistent for all the algorithms in the book.

6. The memory location (Memory Map) for intermediate results or output frequency components is shown next to each Algorithm Step.

7. For an $N$-point algorithm, the real input data, $a_R(i)$, is located in memory locations $M(i)$, and the imaginary input data, $a_I(i)$, is located in memory locations $M(N+i)$, where $i = 0, 1, 2, \ldots, N-1$.

8. All of the intermediate results and output frequency components are stored directly in data memory, rather than temporary storage locations, to ensure the algorithm will work on all processors.

9. All of the multiplier constants are presented in their sine and cosine form so that they may be computed in the arithmetic format (see Chapter 13) appropriate for the application.

## 9.3  THREE CONSTRUCTION APPROACHES

The seven FFT algorithms presented in this chapter are divided into three approaches: convolution, prime factor, and mixed-radix. For each algorithm, the general form is presented and discussed first. Then a specific example is presented to illustrate the features of each of the seven algorithms more clearly. These examples are chosen to be 15- and 16-point transforms. These lengths are large enough to show the characteristics of the algorithms and yet small enough to be reasonably presented. Keeping the lengths of the different examples close to each other also allows the algorithms in the different approaches to be compared.

The first approach is convolution-based algorithms. The mathematical technique for obtaining these FFT algorithms is based on converting the DFT into a set of convolution equations that have special properties to reduce the number of computations. Two prime factor–based algorithms, due to Bluestein and Winograd, are presented in general and then illustrated with 15-point examples. Performance measures are used to describe the properties and limitations of the algorithms.

The second approach of FFT algorithms is commonly called prime factor algorithms. The mathematics for obtaining these algorithms is based on modulo arithmetic theory. Two prime factor–based algorithms are presented in general and then illustrated with 15-point examples. Performance measures are used to describe the properties and limitations of the algorithms.

The third approach of FFT algorithms is called mixed-radix algorithms. This approach can be used for all transform lengths and includes the power-of-two algorithms which have been the most popular, yet most restrictive. The algorithm takes advantage of the complex conjugate symmetry properties of the DFT. The general algorithm is presented first and is followed by three examples, two of 16 points and one of 15 points. Performance measures are used to describe the properties and limitations of the algorithms.

## 9.4 ALGORITHM DATA MAPPING RELABELING

The memory mappings in the algorithm examples in Chapters 8 and 9 only work directly if these exact transforms are being computed and memory locations 0 through $2N - 1$ are available. In general, the building blocks in Chapter 8 will be combined in different ways than the examples in Chapter 9 in order to implement different transform lengths. This leads to the need to use different memory locations than in the examples.

Rather than having to construct a new memory mapping, this section provides a straightforward set of steps for converting the memory mappings in the Chapters 8 and 9 examples to any random ordering of the input data that occurred because of where the data was stored from prior computations. Section 9.4.1 defines the relabeling steps in general, and Section 9.4.2 provides a specific example.

### 9.4.1 General Address Relabeling

Step 1: For all of the stages in the $N$-point FFT, relabel the input addresses for real data with letters. Start with $M(AR)$ for $M(0)$, proceed to use $M(BR)$ for $M(1)$, and so forth, until all of the real data is relabeled.

Step 2: Label all real parts of all intermediate and output results in the algorithm that correspond with the "letter pair" address from Step 1.

Step 3: Repeat Step 1 for the imaginary data, labeling the input address with the letter corresponding to its real-part equivalent. For example, the real part of the zero-th input sample is in location zero. In Step 1 this was assigned memory location $M(AR)$.

Step 4: Label all imaginary parts of all intermediate and output results in the algorithm that correspond with the "letter pair" address from Step 3.

Step 5: For each input address pair $M(AR)$, $M(AI)$, set the $AR$ and $AI$ equal to the actual data location of the data that will be input to the algorithm.

Step 6: For each place in the $N$-point FFT that has letter labels (constructed in Steps 1 through 4), replace the labels with the actual data location assigned it in Step 5.

### 9.4.2 Four-Point FFT Address Relabeling Example

The 4-point FFT from Chapter 8 can be used as a simple example to illustrate Steps 1 through 6. The columns in Table 9-1 show the mapping steps, as follows:

1. The first eight entries in column 1 are the 4-point building-block input data mapping from Chapter 8.
2. The second eight entries in column 1 are a random ordering of the input data memory locations that might be required because of previous computations.
3. The first eight entries in column 2 are the result of performing Steps 1 and 3 of Section 9.4.1.
4. The second eight entries in column 2 are the result of performing Step 5 of Section 9.4.1.
5. The entries in column 3 are the result of performing Steps 2 and 4 of Section 9.4.1.
6. The entries in column 4 are the result of performing Step 6 of Section 9.4.1.

Once this is accomplished, the modified building blocks from Chapter 8 can be used to construct the needed building block computations with the new input data ordering.

**Table 9-1**    Four-Point Algorithm Example Memory Map Relabeling

| Column 1 | Column 2 | Column 3 | Column 4 |
|---|---|---|---|
| $a_R(0) \Rightarrow M(0)$ | $a_R(0) \Rightarrow M(AR)$ | $b_R(0) \Rightarrow M(AR)$ | $b_R(0) \Rightarrow M(0)$ |
| $a_R(1) \Rightarrow M(1)$ | $a_R(1) \Rightarrow M(BR)$ | $b_R(1) \Rightarrow M(CR)$ | $b_R(1) \Rightarrow M(6)$ |
| $a_I(0) \Rightarrow M(4)$ | $a_I(0) \Rightarrow M(AI)$ | $b_I(0) \Rightarrow M(AI)$ | $b_I(0) \Rightarrow M(3)$ |
| $a_I(1) \Rightarrow M(5)$ | $a_I(1) \Rightarrow M(BI)$ | $b_I(1) \Rightarrow M(CI)$ | $b_I(1) \Rightarrow M(5)$ |
| $a_R(2) \Rightarrow M(2)$ | $a_R(2) \Rightarrow M(CR)$ | $b_R(2) \Rightarrow M(BR)$ | $b_R(2) \Rightarrow M(1)$ |
| $a_R(3) \Rightarrow M(3)$ | $a_R(3) \Rightarrow M(DR)$ | $b_R(3) \Rightarrow M(DR)$ | $b_R(3) \Rightarrow M(4)$ |
| $a_I(2) \Rightarrow M(6)$ | $a_I(2) \Rightarrow M(CI)$ | $b_I(2) \Rightarrow M(BI)$ | $b_I(2) \Rightarrow M(7)$ |
| $a_I(3) \Rightarrow M(7)$ | $a_I(3) \Rightarrow M(DI)$ | $b_I(3) \Rightarrow M(DI)$ | $b_I(3) \Rightarrow M(2)$ |
| | | | |
| $a_R(0) \Rightarrow M(0)$ | $M(0) \Rightarrow M(AR)$ | $A_R(0) \Rightarrow M(AR)$ | $A_R(0) \Rightarrow M(0)$ |
| $a_R(1) \Rightarrow M(1)$ | $M(1) \Rightarrow M(BR)$ | $A_I(0) \Rightarrow M(AI)$ | $A_I(0) \Rightarrow M(3)$ |
| $a_I(0) \Rightarrow M(3)$ | $M(3) \Rightarrow M(AI)$ | $A_R(2) \Rightarrow M(BR)$ | $A_R(2) \Rightarrow M(1)$ |
| $a_I(1) \Rightarrow M(7)$ | $M(7) \Rightarrow M(BI)$ | $A_I(2) \Rightarrow M(BI)$ | $A_I(2) \Rightarrow M(7)$ |
| $a_R(2) \Rightarrow M(6)$ | $M(6) \Rightarrow M(CR)$ | $A_R(1) \Rightarrow M(CR)$ | $A_R(1) \Rightarrow M(6)$ |
| $a_R(3) \Rightarrow M(4)$ | $M(4) \Rightarrow M(DR)$ | $A_R(3) \Rightarrow M(DI)$ | $A_R(3) \Rightarrow M(2)$ |
| $a_I(2) \Rightarrow M(5)$ | $M(5) \Rightarrow M(CI)$ | $A_I(1) \Rightarrow M(DR)$ | $A_I(1) \Rightarrow M(4)$ |
| $a_I(3) \Rightarrow M(2)$ | $M(2) \Rightarrow M(DI)$ | $A_I(3) \Rightarrow M(CI)$ | $A_I(3) \Rightarrow M(5)$ |

## 9.5 CONVOLUTION APPROACH

### 9.5.1 Bluestein Algorithm Introduction

In Chapter 2 the analogy was made between the DFT and a bank of narrowband filters. The Bluestein [1] algorithm takes advantage of this fact to implement a fast version of the DFT using a linear filter in combination with pre- and postmultiplications as shown in Figure 9-1.

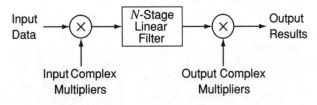

**Figure 9-1**    Bluestein algorithm block diagram.

In general, this algorithm only provides a speedup of $N^{1.5}$ rather than the $N * \log_2(N)$ computational speedup of other FFT algorithms. However, if the $N$-stage linear filter is implemented with the FFT techniques in Chapter 6, the Bluestein algorithm can provide computational performance that varies as $N * \log_2(N)$. Figure 9-2 shows the Bluestein algorithm with the $N$-stage linear filter replaced with its frequency domain processing equivalent from Chapter 6 (Figure 6-1). The $M$-point FFT that operates on the $N$-stage linear filter coefficients is used just once since the filter coefficients stay constant for a given transform length $N$.

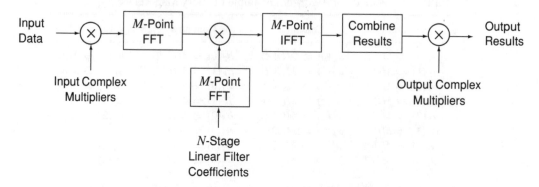

**Figure 9-2**    Frequency domain block diagram of Bluestein algorithm.

It seems logical that if an FFT is going to be used to compute the Bluestein algorithm for FFTs, the FFT might as well be used directly. The reason for the attractiveness of the Bluestein algorithm is that a standard power-of-two algorithm can be used to compute a non-power-of-two FFT. However, for the same non-power-of-two FFT length, the prime factor and Winograd implementations will require fewer multiplications than the Bluestein algorithm.

Once it has been decided that power-of-two algorithms provide the best approach for the $M$-point FFT needed in the Bluestein algorithm, the mixed-radix section of this chapter (Section 9.7) should be examined to see if other advantages can be taken to simplify the computations. The most useful simplification comes because of additional constraints the Bluestein algorithm puts on $M$. Namely, the algorithm requires that $M$, the FFT length, be at least twice $N$, the number of stages in the linear filter in Figure 9-1. This means that, for $N$ input samples, $M - N$ zeros (Section 2.3.10) are added to obtain the $M$ samples needed by the $M$-point FFT. Since $M \geq 2 * N$, it follows that $M - N \geq N$. Therefore, at least the second half of the inputs to the $M$-point FFT are zeros.

In Sections 9.7.5 and 9.7.6 the first input data samples are combined such that one comes from the first half of the data and one from the second half. This is the decimation-in-time decomposition in Section 10.4.1. In Stage 1 of the general mixed-radix algorithm in Section 9.7.4, if $P = 2$ and $Q = M/2$, then the samples ($k = 0$ and $k = 1$) that are combined in the $n$-th 2-point input building block are $a_R(k * N/2 + n)$ and $a_R(k * N/2 + n)$. This always puts one input ($k = 0$) in the first half of the data samples and the other ($k = 1$) in the second half. This means that if the first building block for the $M$-point FFT is two points ($P = 2$), one input to each 2-point FFT is always zero. Therefore, the 2-point FFTs require no computations. This replaces the single $M$-point mixed-radix FFT with two $M/2$-point FFTs, one of which requires a complex multiplier because of the details of the mixed-radix algorithm shown in Stage 2 of Section 9.7.4.

Since less than half of the outputs, $N$ to be exact, of the $M$-point IFFT are used, only half of its $M$ outputs need be computed. Similar to the input $M$-point FFT, if the

2-point IFFT is used as $Q$ rather than $P$, the output 2-point FFT is reduced to its subtract computation. Combining all of these facts to reduce the Bluestein computations results in converting the block diagram in Figure 9-2 to the one in Figure 9-3. Following the description of the general algorithm, a 15-point example is provided to concretely illustrate the algorithm and provide a direct comparison with the 15- and 16-point examples presented later in this chapter for other FFT algorithms.

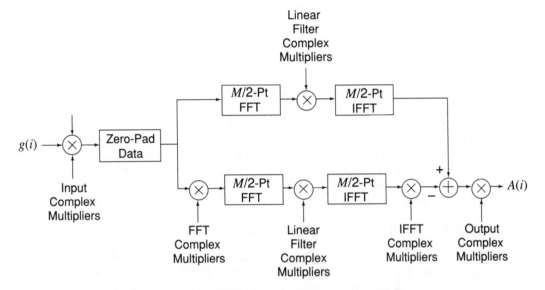

**Figure 9-3**    General Bluestein algorithm block diagram.

### 9.5.2 Number of Bluestein Algorithm Adds and Multiplies

The 10 stages required to implement the general Bluestein algorithm are presented and summarized in Figure 9-3. The total number of real adds required is $10 * N + 2 * M$ plus the number of real adds required for four $M/2$-point FFTs. Similarly, the required number of real multiplies is $4 * M + 16 * N$ plus the number of real multiplies required for four $M/2$-point FFTs.

### 9.5.3 Number of Bluestein Algorithm Memory Locations

Complex multipliers require two additional memory locations for temporary storage, and each $M/2$-point FFT requires some number of memory locations over and above the input and output data requirements. Since the FFT almost always requires at least two additional data memory locations, the data memory requirements are determined by the chosen $M/2$-point FFT. If the $M/2$-point FFTs are computed in sequence, not both at the same time, then the additional data memory required for the intermediate results of the first $M/2$-point FFT algorithm can also be used for the second $M/2$-point FFT. Therefore, the data memory requirement is $M$ (for the second $M/2$-point FFT) plus the requirements for the chosen $M/2$-point FFT.

There are $N$ complex multiplier constants on the input and the output, and $M$ complex multiplier constants in the center for the unit pulse response of the Bluestein filter. Additionally, there are $M/2$ complex constants at the input to the lower $M/2$-point FFT and at the output from the lower $M/2$-point IFFT. The $M/2$-point IFFT uses the same constants as the FFT with the sign of the imaginary parts changed. This is a total of $(4*N+3*M)$ memory locations plus those required for the chosen $M/2$-point FFT.

### 9.5.4 General Bluestein Algorithm

This sequence of stages assumes that the linear filter complex multipliers have been computed and stored in memory using the techniques in Chapter 6. The stages of the general Bluestein algorithm are as follows.

### Stage 1: Transform Length Selection

To perform an $N$-point FFT, select an $M$-point power-of-two algorithm, where $M$ is the smallest power-of-two greater than or equal to $(2*N-1)$. For example, if $N=15$, $M \geq 29$, which implies $M=32$. For the first and second stages in this algorithm, it makes no difference how the input data is stored in data memory. However, a strategy that will simplify subsequent stages is to store the real inputs in data memory locations 0 through $(N-1)$ and the imaginary inputs in locations $M$ through $(M+N-1)$.

### Stage 2: Multiplication by the Input Complex Multipliers

Modify the $N$-point complex input data sequence, $g(n) = g_R(n) + j*g_I(n)$ by multiplying it by $\exp(-j*\pi*n^2/N) = \cos(\pi*n^2/N) - j*\sin(\pi*n^2/N)$ to obtain $a(n) = a_R(n) + j*a_I(n)$. This requires $N$ complex multiplies, which is a total of $4*N$ real multiplies and $2*N$ real adds. The equations for $n = 0, 1, 2, \ldots, (N-1)$ are:

$$a_R(n) = g_R(n) * \cos(\pi*n^2/N) + g_I(n) * \sin(\pi*n^2/N)$$
$$a_I(n) = g_I(n) * \cos(\pi*n^2/N) - g_R(n) * \sin(\pi*n^2/N)$$

The complex data results are stored in the same locations from which the inputs were pulled. If no temporary registers are available, two additional memory locations, $M(2*M)$ and $M(2*M+1)$, are used to store the values computed from multiplying the sine constants by the input data, and the original data locations are used to store the values computed by multiplying the cosine constants by the input data. Those intermediate values are then pulled from the original and additional data memory locations and added to form the output values $a(n) = a_R(n) + j*a_I(n)$.

### Stage 3: Zero Padding

Append the $N$ input data points, $a(n) = a_R(n) + j*a_I(n)$, with $(M-N)$ zeros to obtain an $M$-point input sequence for the $M$-point FFT. The $(M-N)$ zeros are appended to the end of the actual data. The real zeros are stored in data memory locations $N$ through $(M-1)$, and the imaginary zeros in locations $(N+M)$ through $(2*M-1)$. The result is having all of the real input data to the $M$-point FFT stored in contiguous data memory locations 0 through $(M-1)$, and the imaginary data stored in data memory locations $M$ through $(2*M-1)$.

### Stage 4: FFT Input Stage Computation

*Step* 1: *Simulating the 2-Point Building-Block Computations*

Following the instructions in Step 1 of Stage 1 of the general mixed-radix algorithm in Section 9.7.4, the input data point groupings to the $n$-th 2-point building block are $a_R(k * M/2 + n)$ and $a_I(k * M/2 + n)$ (where $k = 0, 1$ and $n = 0, 1, \ldots, ((M/2) - 1)$). All of the inputs where $k = 1$ are zeros. Using the 2-point building-block equations from Chapter 8:

$$A_R(0) = a_R(0) + a_R(1) \qquad A_R(1) = a_R(0) - a_R(1)$$
$$A_I(0) = a_I(0) + a_I(1) \qquad A_I(1) = a_I(0) - a_I(1)$$

The $a_R(1)$ and $a_I(1)$ inputs to all $M/2$ of the required 2-point building blocks ($n = 0, 1, \ldots, ((M/2) - 1)$) are zero. Therefore, the outputs of all of those 2-point building blocks are just the input data:

$$A_R(0) = a_R(0) \qquad A_R(1) = a_R(0)$$
$$A_I(0) = a_I(0) \qquad A_I(1) = a_I(0)$$

If the labels from Step 2 of Stage 1 of the general mixed-radix algorithm in Section 9.7.4 are used, the $k$-th output ($k = 0, 1$) of the $n$-th 2-point building block ($n = 0, 1, \ldots, ((M/2) - 1)$)should be labeled $B_R(k * M/2 + n)$ and $B_I(k * M/2 + n)$ in preparation for input to the complex multiply portion of the mixed-radix algorithm. Specifically, the equations and their data memory map are:

$$B_R(k * M/2 + n) = a_R(n) \qquad B_R(k * M/2 + n) \Rightarrow M(k * M/2 + n)$$
$$B_I(k * M/2 + n) = a_I(n) \qquad B_I(k * M/2 + n) \Rightarrow M(k * M/2 + n + M)$$

The right column shows the resulting memory mapping, based on the locations of the input data and taking advantage of the initial data mapping that saved room for the added zeros.

*Step* 2: *Multiplication by FFT Complex Multipliers*

Each of the $B_R(k * M/2 + n)$ and $B_I(k * M/2 + n)$ needs to be multiplied by the specific complex number required by the general mixed-radix algorithm prior to entering the $M/2$-point portion of the $M$-point algorithm. The equations for this complex multiplication for $k = 0, 1$ and $n = 0, 1, \ldots, (M/2 - 1)$ are:

$$D_R(k * M/2 + n) = B_R(k * M/2 + n) * \cos(2\pi * kn/M)$$
$$+ B_I(k * M/2 + n) * \sin(2\pi * kn/M)$$
$$D_I(k * M/2 + n) = B_I(k * M/2 + n) * \cos(2\pi * kn/M)$$
$$- B_R(k * M/2 + n) * \sin(2\pi * kn/M)$$

If no temporary registers are assumed, each complex multiply required two additional data memory locations to store the results of multiplying each input value by two different constants prior to forming and storing the output results. Since the complex multiplications are computed sequentially, the same two additional memory locations can be used for each. The $D_R(k * M/2 + n)$ and $D_I(k * M/2 + n)$ are stored in the locations from which the $B_R(k * M/2 + n)$ and $B_I(k * M/2 + n)$ were pulled to perform the computations, specifically, in memory locations $M(k * M/2 + n)$ and $M(k * M/2 + n + M)$, respectively.

For $k = 0$, $\cos(2 * \pi * k * n/N) = 1$ and $\sin(2 * \pi * k * n/N) = 0$. Further, $B_R(k * M/2 + n) = B_I(k * M/2 + n) = 0$ for $n = N, N+1, \ldots, (M/2) - 1$. This reduces the number of complex multiplies to $N$, which is $4 * N$ real multiplies and $2 * N$ real adds. Figure 9-4 shows the locations of each of the results of the complex multiplies.

| Contents | Location |
|:---:|:---:|
| $D_R(0)$ | 0 |
| • | • |
| • | • |
| • | • |
| $D_R(M/2 - 1)$ | $M/2 - 1$ |
| $D_R(M/2)$ | $M/2$ |
| • | • |
| • | • |
| • | • |
| $D_R(M - 1)$ | $M - 1$ |
| $D_I(0)$ | $M$ |
| • | • |
| • | • |
| • | • |
| $D_I(M/2 - 1)$ | $3 * M/2 - 1$ |
| $D_I(M/2)$ | $3 * M/2$ |
| • | • |
| • | • |
| • | • |
| $D_I(M - 1)$ | $2 * M - 1$ |
| Temporary Data | $2 * M$ plus |

**Figure 9-4** Data memory map prior to the $M/2$-point FFT.

## Stage 5: Two *M*/2-Point FFT Computations

Again following the instructions in Step 1 of Stage 4 of the general mixed-radix algorithm, the $n$-th input to the $k$-th $M/2$-point algorithm is $D_R(k * M/2 + n)$ and $D_I(k * M/2 + n)$ (where $k = 0, 1$ and $n = 0, 1, \ldots, ((M/2) - 1)$).

For the first of the $M/2$-point FFTs ($k = 0$ is the top $M/2$-point FFT in Figure 9-3), the real data is located in the same place that was assumed in Chapter 8, namely, in locations $M(0)$ through $M(M/2 - 1)$, as shown in Figure 9-4. However, the corresponding imaginary data is offset in memory by $M$ locations rather than the $M/2$ locations from Chapter 8. Further, the addresses for the additional memory locations needed in the center of the computation must start after the end of the $M$ complex input data points, not after $M/2$ complex data points. For example, the first extra memory location in Chapter 8 comes at memory location $M(M)$. It must now be at $M(2 * M)$. Figure 9-4 summarizes these facts.

For the second $M/2$-point FFT ($k = 1$ is the bottom $M/2$-point FFT in Figure 9-3), the real input data addresses start at $M(M/2)$ and end at $M(M/2 + M/2 - 1)$. This makes them $M/2$ addresses higher than in the Chapter 8 building block. Similarly, the imaginary data addresses start at $M(M/2+M)$ and end at $M(M/2+M/2-1+M)$. This makes them $M$ addresses higher than in the Chapter 8 building block. This offset of the data locations makes it easy to directly use both the equations from Chapter 8 and their data memory map.

### Step 1: First M/2-point FFT Computations

The assumptions for the first $M/2$-point FFT ($k = 0$) are the following:

1. Use the $M/2$-point algorithm steps directly from Chapter 8 or from one of the mixed-radix algorithms in Section 9.7.
2. Use the memory addresses directly for all real data, except the additional memory locations required in the middle of the computations.
3. For the imaginary data, add $M/2$ to all of the memory locations, except for the additional memory locations required in the middle of the computations.
4. For the additional memory locations required in the middle of the computations, add $M$ to the memory location.
5. Relabel the output frequency components from $A_R(n)$ and $A_I(n)$ to $A_R(2*n)$ and $A_I(2*n)$.

### Step 2: Second M/2-Point FFT Computations

Similarly, the assumptions for the second $M/2$-point FFT ($k = 1$) are the following:

1. Use the $M/2$-point algorithm steps directly from Chapter 8 or Chapter 9, except modify all of the data labels by adding $M/2$ to them.
2. Add $M/2$ to the memory addresses for all real data, except the additional memory locations required in the middle of the computations.
3. Add $M$ to the memory addresses for all imaginary data, except for the additional memory locations required in the middle of the computations.
4. For the additional memory locations required in the middle of the computations, add $M$ to the memory location.
5. Relabel the output frequency components from $A_R(n)$ and $A_I(n)$ to $A_R(2*n+1)$ and $A_I(2*n+1)$.

The total number of computations required for this stage is twice the number of computations needed for the chosen $M/2$-point transform.

## Stage 6: Multiplication by Linear Filter Complex Multipliers

Multiply the $M$ relabeled complex outputs ($A_R(i)$, $A_I(i)$) of the two $M/2$-point FFTs by the $M$ complex outputs ($H_R(i)$, $H_I(i)$) of the unit pulse response FFT to obtain $C(n) = C_R(n) + j * C_I(n)$. In general, this requires $M$ complex multiplications, which is $4 * M$ real multiplies and $2 * M$ real adds. The equations are:

$$C_R(n) = A_R(n) * H_R(n) - A_I(n) * H_I(n)$$
$$C_I(n) = A_I(n) * H_R(n) + A_R(n) * H_I(n)$$

If no temporary registers are assumed, each complex multiply requires two additional data memory locations to store the results of multiplying each input value by two different constants prior to forming and storing the output results. Since the complex multiplies are computed sequentially, the same two additional memory locations can be used for each. The $C_R(n)$ and $C_I(n)$ are stored in the locations from which the $A_R(n)$ and $A_I(n)$ were pulled to perform the computations.

Some of the building-block algorithms in Chapter 8 and algorithms in Chapter 9 do not have all of their real outputs in the same data locations as the real inputs. Addressing convenience has resulted in some of the imaginary outputs being interspersed. It is convenient to correct this inconsistency during the complex multiply computations in this stage. Specifically, if the imaginary part of one of the $A_R(n)$ and $A_I(n)$ is stored in the lower portion of the data memory, change this when the complex multiply outputs are stored so that the real parts of all of the terms are stored together in the lower portion of the memory used for $C_R(n)$ and $C_I(n)$.

### Stage 7: Two *M*/2-Point IFFT Computations

Following the instructions in Step 1 of Stage 1 of the general mixed-radix algorithm, the input data point groupings to the $n$-th $M/2$-point algorithm are $C_R(k*2+n)$ and $C_I(k*2+n)$ (where $n = 0, 1$ and $k = 0, 1, \ldots, ((M/2)-1)$).

The inputs to the first $M/2$-point IFFT (upper IFFT in Figure 9-3) are $C_R(k*2)$ and $C_I(k*2)$ (where $k = 0, 1, \ldots, ((M/2)-1)$), and the inputs to the second $M/2$-point IFFT (lower IFFT in Figure 9-3) are $C_R(k*2+1)$ and $C_I(k*2+1)$ (where $k = 0, 1, \ldots, ((M/2)-1)$). These are the outputs of the two $M/2$-point FFTs, modified by complex multipliers. Therefore, these inputs occupy the same memory locations as the outputs of the $M/2$-point FFTs. In general, the Chapter 8 and Chapter 9 algorithms do not have their outputs in sequential memory addresses. Therefore, the inputs to the inverse $M/2$-point FFT will not be in sequential addresses, as was assumed in Chapters 8 and 9.

However, the first $M/2$-point IFFT does have all of its real inputs in the first $M/2$ memory locations and all of its imaginary inputs in memory locations $M$ through $(3*M/2-1)$ because they were put in these locations as part of Stage 6 of this algorithm. Likewise, the second $M/2$-point IFFT's real inputs are in memory locations $M/2$ through $(M-1)$, and imaginary inputs are in memory locations $(3*M/2)$ through $(2*M-1)$. The address relabeling in Section 9.4 is used to convert the memory mapping for the algorithms from Chapters 8 and 9 to a form that can be directly used here.

Each of the $e_R(k*2+n)$ and $e_I(k*2+n)$ needs to be multiplied by a specific complex number prior to entering the 2-point portion of the $M$-point algorithm. The equations for this complex multiplication for each $n = 0, 1$ and $k = 0, 1, \ldots, M/2 - 1$ are as follows:

$$f_R(k*2+n) = e_R(k*2+n)*\cos(2\pi*kn/M) - e_I(k*2+n)*\sin(2\pi*kn/M)$$
$$f_I(k*2+n) = e_I(k*2+n)*\cos(2\pi*kn/M) + e_R(k*2+n)*\sin(2\pi*kn/M)$$

If no temporary registers are assumed, each complex multiply required two additional data memory locations to store the results of multiplying each input value by two different constants prior to forming and storing the output results. However, if the complex multiplies are performed sequentially, the same two additional memory locations can be reused for all of the complex multiplies. The result is the need for only two additional memory locations. Store the results of the complex multiplies in the same locations from which the inputs to the complex multiplies were taken. For $n = 0$, these complex multiplies are just multiplies by 1. Therefore, one of the two $M/2$-point IFFTs does not have its outputs modified prior to computing the 2-point IFFTs. Since only $M/2 - 1$ of these $M/2$ complex outputs represent the needed result in Stage 8, only $M/2 - 1$ of the complex multiplies need be performed. The total number of computations for these $M/2 - 1$ complex multiplies is $4 * (M/2 - 1)$ real multiplies and $2 * (M/2 - 1)$ real adds.

### Stage 8: Computing the Output 2-Point Building Blocks

This stage has two steps. The first is to properly group the input data for each of the $M/2$ two-point algorithms. The second is to compute the appropriate part of each of the $M/2$ two-point algorithms.

#### Step 1: *Grouping the Input Data Points to the 2-Point Building Blocks*

For the $n$-th input to the $k$-th 2-point building block, choose $f_R(k * 2 + n)$ and $f_I(k * 2 + n)$ (where $k = 0, 1, \ldots, M/2 - 1$ and $n = 0, 1$) from the input data sequence. In terms of the input labels, $a_R(n)$ and $a_I(n)$, shown in Chapter 8, the inputs for the $k$-th 2-point building blocks are:

$$a_R(0) = f_R(2 * k) \qquad a_R(1) = f_R(2 * k + 1)$$
$$a_I(0) = f_I(2 * k) \qquad a_I(1) = f_I(2 * k + 1)$$

#### Step 2: *Computing a Portion of the Output 2-Point Building Blocks*

Using the 2-point building block from Chapter 8 gives:

$$A_R(0) = a_R(0) + a_R(1) \qquad A_R(1) = a_R(0) - a_R(1)$$
$$A_I(0) = a_I(0) + a_I(1) \qquad A_I(1) = a_I(0) - a_I(1)$$

The outputs of interest are the second pair of equations. Therefore, if the output frequency components of the $M$-point IFFT are $y_R(n * M/2 + k)$ and $y_I(n * M/2 + k)$, for the $n$-th output of the $k$-th 2-point building block, the outputs of interest are for $n = 1$. In terms of the output labels, $A_R(n)$ and $A_I(n)$, shown for the $M/2$-point radix-4 FFT, the outputs for the $k$-th 2-point building block are equated to the complete outputs, using the equations:

$$y_R(M/2 + k) = f_R(2 * k) - f_R(2 * k + 1)$$
$$y_I(M/2 + k) = f_I(2 * k) - f_I(2 * k + 1)$$

Since only $(M/2 - 1)$ of these $M/2$ complex outputs represent the needed result in Stage 10, only $(M/2 - 1)$ of the complex adds need be performed. The $(M/2 - 1)$ partial 2-point building block requires 30 real adds.

### Stage 9: Adjusting the Output Data

This stage has two steps:

1. For $n = N, (N + 1), \ldots, (2 * N - 1)$, multiply $y(n) = y_R(n) + j * y_I(n)$ by $\exp(-j * \pi * n^2/N) = \cos(\pi * n^2/N) - j * \sin(\pi * n^2/N)$ to obtain $z(n)$.
2. For $n = N, (N + 1), \ldots, (2 * N - 1)$, multiply $z(n)$ by $\exp(-j * \pi * N) = \cos(\pi * N) - j * \sin(\pi * N)$ to obtain $q(n)$.

These steps can be combined into a single complex multiply for each of the $N$ outputs. This is a total of $N$ complex multiplies, which is a total of $4 * N$ real multiplies and $2 * N$

real adds. If there are no temporary registers in the processor, then two additional memory locations are required to perform the complex computations. The outputs from this step are placed in the same locations from which the inputs to the step were pulled for each $n$. The equations are

$$q_R(n) = y_R(n) * \cos(\pi * N + \pi * n^2/N) + y_I(n) * \sin(\pi * N + \pi * n^2/N)$$
$$q_I(n) = y_I(n) * \cos(\pi * N + \pi * n^2/N) - y_R(n) * \sin(\pi * N + \pi * n^2/N)$$

### Stage 10: Extracting the N-Point FFT

The $N$-point FFT outputs, $G(n) = G_R(n) + j * G_I(n)$, are $q(N+n) = q_R(N+n) + j * q_I(N+n)$ where $n = 0, 1, \ldots, (N-1)$.

### 9.5.5 Fifteen-Point Bluestein Example

This 15-point example follows the general Bluestein algorithm for $M = 32 = 2 * 16$. It uses the mixed-radix algorithm for the 32-point transform and the 16-point radix-4 example from Section 9.7.4. Figure 9-5 is a block diagram of this example. Any of the mixed-radix 16-point examples in this chapter, or the 16-point Winograd building block from Chapter 8, could also have been used rather than the 16-point radix-4 algorithm. Following Section 9.4.4, the 15 complex input data samples are stored with the real parts in data memory locations 0 through 14, and the imaginary parts in data memory locations 32 through 46.

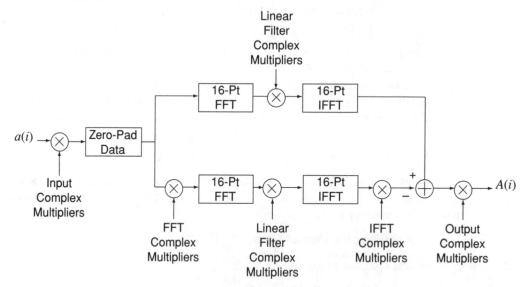

**Figure 9-5** Fifteen-point Bluestein algorithm block diagram.

This example requires 790 real adds and 464 real multiplies. This is about five times the number of computations needed for the other 15-point examples in this chapter. However, it can be computed using only power-of-two algorithms. This removes the need to develop special code or hardware and allows the application to take advantage of hardware and software refinements developed for the standard power-of-two FFTs. Further, the computational difference is not as great when unusual FFT lengths, such as prime numbers, are required.

The data memory required for this algorithm is the same as that required for two 16-point radix-4 mixed-radix algorithms. From the example in Section 9.7.5, this is 40 locations. Since the 16-point algorithms are computed sequentially, the additional eight ($40 - 32$) locations can be reused for the second 16-point FFT. The same is true for the IFFTs. Therefore, the total data memory required is $32 + 32 + 8 = 72$. The memory required for data constants is the sum of the requirements for the 16-point FFT plus those for each of the complex multiplies. For this example that is $4 * 15 + 3 * 32 + 6 = 162$. The complex multiply algorithm used here is the one used in the Singleton example in Section 9.7.7.

### Stage 1: Transform Length Selection

The 32-point FFT is chosen to execute the 15-point FFT because it is the smallest power-of-two greater than $2 * 15 = 30$ points.

### Stage 2: Modifying the Input Data

Modify the 15-point complex input data sequence, $g(n) = g_R(n) + j * g_I(n)$, by multiplying it by $\exp(-j * \pi * n^2/15) = \cos(\pi * n^2/15) - j * \sin(\pi * n^2/15)$ to obtain $a(n) = a_R(n) + j * a_I(n)$. This requires $4 * 15 = 60$ real multiplies and $2 * 15 = 30$ real adds. The equations are (for $n = 0, 1, \ldots, 15$):

$$a_R(n) = g_R(n) * \cos(\pi * n^2/15) + g_I(n) * \sin(\pi * n^2/15)$$
$$a_I(n) = g_I(n) * \cos(\pi * n^2/15) - g_R(n) * \sin(\pi * n^2/15)$$

The complex data results are stored in the same locations from which the inputs were pulled. If no temporary registers are available, two additional memory locations, $M(64)$ and $M(65)$ (Figure 9-4), are used to store the values computed from multiplying the sine term by the input data, and the original data locations are used to store the values computed by multiplying the cosine term by the input data. Those values are then pulled from memory and added to form the output values $a(n) = a_R(n) + j * a_I(n)$.

### Stage 3: Zero Padding

Append the 15 input data points, $a(n) = a_R(n) + j * a_I(n)$, with 17 complex zeros to obtain a 32-point input sequence for the 32-point FFT. The 17 complex zeros are appended to the end of the actual data (i.e., $n = 15, 16, \ldots, 31$). The real zeros are stored in data memory locations 15 through 31, and the imaginary zeros in locations 47 through 63.

### Stage 4: FFT Input Stage Computation

#### Step 1: Simulating the Input 2-Point Building-Block Computations

If the instructions in Step 1 of Stage 1 of the general mixed-radix algorithm in Section 9.7.4 are followed, the input data point groupings to the $n$-th 2-point building block are $a_R(k * 16 + n)$ and $a_I(k * 16 + n)$ (where $k = 0, 1,$ and $n = 0, 1, \ldots, 15$). All of the inputs where $k = 1$ are zeros. Using the 2-point building block from Chapter 8 gives:

$$A_R(0) = a_R(0) + a_R(1) \qquad A_R(1) = a_R(0) - a_R(1)$$
$$A_I(0) = a_I(0) + a_I(1) \qquad A_I(1) = a_I(0) - a_I(1)$$

The $a_R(1)$ and $a_I(1)$ inputs to all 16 of the required 2-point building blocks ($n = 0, 1, \ldots, 15$) are zero. Therefore, the outputs of all of those 2-point building blocks are just the input data:

$$A_R(0) = a_R(0) \qquad A_R(1) = a_R(0)$$
$$A_I(0) = a_I(0) \qquad A_I(1) = a_I(0)$$

Using the labels from Step 2 of Stage 1 of the general mixed-radix algorithm, the $k$-th output ($k = 0, 1$) of the $n$-th 2-point building block ($n = 0, 1, \ldots, 15$) should be labeled $B_R(k * 16 + n)$ and $B_I(k * 16 + n)$ in preparation for input to the complex multiply portion of the mixed-radix algorithm. Specifically,

$$B_R(k * 16 + n) = a_R(n) \qquad B_R(k * 16 + n) \Rightarrow M(k * 16 + n)$$
$$B_I(k * 16 + n) = a_I(n) \qquad B_I(k * 16 + n) \Rightarrow M(k * 16 + n + 32)$$

The right column shows the corresponding memory mapping, based on the locations of the input data and taking advantage of the initial data mapping that saved room for the added zeros. Each $a_R(n)$ and $a_I(n)$ is stored in two memory locations in preparation for subsequent steps.

### Step 2: Multiplication by FFT Complex Multipliers

Each $B_R(k*16+n)$ and $B_I(k*16+n)$ needs to be multiplied by the specific complex number required by the general mixed-radix algorithm prior to entering the 16-point portion of the 32-point algorithm. The equations for this complex multiplication for each $k = 0, 1$ and $n = 0, 1, \ldots, 15$ are:

$$D_R(k * 16 + n) = B_R(n) * \cos(2\pi * kn/32) + B_I(n) * \sin(2\pi * kn/32)$$
$$D_I(k * 16 + n) = B_I(n) * \cos(2\pi * kn/32) - B_R(n) * \sin(2\pi * kn/32)$$

If no temporary registers are assumed, each complex multiply required two additional data memory locations to store the results of multiplying each input value by two different constants prior to forming and storing the output results. However, if the complex multiplies are performed sequentially, the same two additional memory locations can be reused for all of the complex multiplies. The result is the need for only two additional memory locations. The $D_R(k * 16 + n)$ and $D_I(k * 16 + n)$ are stored in the locations from which the $B_R(k * 16 + n)$ and $B_I(k * 16 + n)$ were pulled to perform the computations. This step requires 15 complex multiplies, which is 60 real multiplies and 30 real adds.

### Stage 5: Two 16-Point FFT Computations

For the $n$-th input to the $k$-th 16-point algorithm, choose $D_R(k * 16 + n)$ and $D_I(k * 16 + n)$ (where $k = 0, 1$ and $n = 0, 1, \ldots, 15$) from the input data sequence. In terms of the input data labels, $a_R(n)$ and $a_I(n)$, shown in Chapter 8 for the 16-point radix-4 FFT, the inputs for the first 16-point FFTs and their data memory addresses are:

$$a_R(n) = D_R(n) \qquad a_R(n) \Rightarrow M(n)$$
$$a_I(n) = D_I(n) \qquad a_I(n) \Rightarrow M(n + 32)$$

For the second 16-point FFT they are:

$$a_R(n) = D_R(16 + n) \qquad a_R(n) \Rightarrow M(16 + n)$$
$$a_I(n) = D_I(16 + n) \qquad a_I(n) \Rightarrow M(48 + n)$$

Use the complex input data points, $a_R(n)$ and $a_I(n)$, defined in Step 1 to compute each of the two 16-point FFTs. The $n$-th output of the first 16-point FFT should be labeled $A_R(n*2)$ and $A_I(n*2)$. Similarly, the $n$-th output of the second 16-point FFT should be labeled $A_R(n*2+1)$ and $A_I(n*2+1)$. The $A_R(m)$ and $A_I(m)$, where $m = 0, 1, \ldots, 31$, are the final outputs of the 32-point FFT.

### Step 1: Computing the First of Two 16-Point Radix-4 FFTs

The approach for using the Algorithm Steps and Memory Map from Section 9.7.5 to compute the first of the two 16-point FFTs is as follows.

1. Use the 16-point radix-4 equations directly from Section 9.7.5.

2. Use the memory addresses in Section 9.7.5 for all real data, except the additional memory locations required in the middle of the computations.

3. For imaginary data, add 16 to all locations in Section 9.7.5, except for the additional memory locations required in the middle of the computations.

4. For the additional memory locations required in the middle of the computations in Section 9.7.5, add 32 to the memory location.

5. Relabel the output frequency components in Section 9.7.5 from $A_R(n)$ and $A_I(n)$ to $A_R(2*n)$ and $A_I(2*n)$.

### Step 2: Computing the Second of Two 16-Point Radix-4 FFTs

Similarly, the approach for using the Algorithm Steps and Memory Map in Section 9.7.5 for the second 16-point FFT is as follows.

1. Use the 16-point equations directly from Section 9.7.5, except modify all of the data labels $a_R(n)$ and $a_I(n)$ by adding 16 to them to obtain $a_R(n + 16)$ and $a_I(n + 16)$.

2. Add 16 to the memory addresses from Section 9.7.5 for all real data, except the additional memory locations required in the middle of the computations.

3. Add 32 to the memory addresses for all imaginary data in Section 9.7.5, except for the additional memory locations required in the middle of the computations.

4. For the additional memory locations in the middle of the computations in Section 9.7.5, add 32 to the memory location.

5. Relabel the output frequency components from Section 9.7.5 from $A_R(n)$ and $A_I(n)$ to $A_R(2*n + 1)$ and $A_I(2*n + 1)$.

Table 9-2 shows the output data addresses for the 16-point radix-4 FFT in Section 9.7.5 in column 1 and the offset addresses for the first and second 16-point FFTs in columns 2 and 3, based on following Steps 2 and 3 of this stage. The two 16-point FFTs require 288 real adds and 48 real multiplies.

### Stage 6: Multiplication by Linear Filter Complex Multipliers

Multiply the 32 complex outputs of the data FFT $(A_R(i), A_I(i))$ by the 32 complex outputs of the unit pulse response FFT $(H_R(i), H_I(i))$ to obtain $C(n) = C_R(n) + j*C_I(n)$.

**Table 9-2** Memory Maps for 15-Point Bluestein Algorithm Example

| Column 1 | Column 2 | Column 3 |
|----------|----------|----------|
| $A_R(0) \Rightarrow M(0)$ | $A_R(0) \Rightarrow M(0)$ | $A_R(1) \Rightarrow M(16)$ |
| $A_I(0) \Rightarrow M(16)$ | $A_I(0) \Rightarrow M(32)$ | $A_I(1) \Rightarrow M(48)$ |
| $A_R(1) \Rightarrow M(8)$ | $A_R(2) \Rightarrow M(8)$ | $A_R(3) \Rightarrow M(24)$ |
| $A_I(1) \Rightarrow M(24)$ | $A_I(2) \Rightarrow M(40)$ | $A_I(3) \Rightarrow M(56)$ |
| $A_R(2) \Rightarrow M(4)$ | $A_R(4) \Rightarrow M(4)$ | $A_R(5) \Rightarrow M(20)$ |
| $A_I(2) \Rightarrow M(20)$ | $A_I(4) \Rightarrow M(36)$ | $A_I(5) \Rightarrow M(52)$ |
| $A_R(3) \Rightarrow M(28)$ | $A_R(6) \Rightarrow M(44)$ | $A_R(7) \Rightarrow M(60)$ |
| $A_I(3) \Rightarrow M(12)$ | $A_I(6) \Rightarrow M(12)$ | $A_I(7) \Rightarrow M(28)$ |
| $A_R(4) \Rightarrow M(2)$ | $A_R(8) \Rightarrow M(2)$ | $A_R(9) \Rightarrow M(18)$ |
| $A_I(4) \Rightarrow M(18)$ | $A_I(8) \Rightarrow M(34)$ | $A_I(9) \Rightarrow M(50)$ |
| $A_R(5) \Rightarrow M(10)$ | $A_R(10) \Rightarrow M(10)$ | $A_R(11) \Rightarrow M(26)$ |
| $A_I(5) \Rightarrow M(26)$ | $A_I(10) \Rightarrow M(42)$ | $A_I(11) \Rightarrow M(58)$ |
| $A_R(6) \Rightarrow M(22)$ | $A_R(12) \Rightarrow M(38)$ | $A_R(13) \Rightarrow M(38)$ |
| $A_I(6) \Rightarrow M(6)$ | $A_I(12) \Rightarrow M(6)$ | $A_I(13) \Rightarrow M(6)$ |
| $A_R(7) \Rightarrow M(14)$ | $A_R(14) \Rightarrow M(14)$ | $A_R(15) \Rightarrow M(30)$ |
| $A_I(7) \Rightarrow M(30)$ | $A_I(14) \Rightarrow M(46)$ | $A_I(15) \Rightarrow M(62)$ |
| $A_R(8) \Rightarrow M(1)$ | $A_R(16) \Rightarrow M(1)$ | $A_R(17) \Rightarrow M(17)$ |
| $A_I(8) \Rightarrow M(17)$ | $A_I(16) \Rightarrow M(33)$ | $A_I(17) \Rightarrow M(49)$ |
| $A_R(9) \Rightarrow M(9)$ | $A_R(18) \Rightarrow M(9)$ | $A_R(19) \Rightarrow M(25)$ |
| $A_I(9) \Rightarrow M(25)$ | $A_I(18) \Rightarrow M(41)$ | $A_I(19) \Rightarrow M(57)$ |
| $A_R(10) \Rightarrow M(5)$ | $A_R(20) \Rightarrow M(5)$ | $A_R(21) \Rightarrow M(21)$ |
| $A_I(10) \Rightarrow M(21)$ | $A_I(20) \Rightarrow M(37)$ | $A_I(21) \Rightarrow M(53)$ |
| $A_R(11) \Rightarrow M(29)$ | $A_R(22) \Rightarrow M(45)$ | $A_R(23) \Rightarrow M(61)$ |
| $A_I(11) \Rightarrow M(13)$ | $A_I(22) \Rightarrow M(13)$ | $A_I(23) \Rightarrow M(29)$ |
| $A_R(12) \Rightarrow M(19)$ | $A_R(24) \Rightarrow M(35)$ | $A_R(25) \Rightarrow M(51)$ |
| $A_I(12) \Rightarrow M(3)$ | $A_I(24) \Rightarrow M(3)$ | $A_I(25) \Rightarrow M(19)$ |
| $A_R(13) \Rightarrow M(27)$ | $A_R(26) \Rightarrow M(43)$ | $A_R(27) \Rightarrow M(59)$ |
| $A_I(13) \Rightarrow M(11)$ | $A_I(26) \Rightarrow M(11)$ | $A_I(27) \Rightarrow M(27)$ |
| $A_R(14) \Rightarrow M(23)$ | $A_R(28) \Rightarrow M(39)$ | $A_R(29) \Rightarrow M(55)$ |
| $A_I(14) \Rightarrow M(7)$ | $A_I(28) \Rightarrow M(7)$ | $A_I(29) \Rightarrow M(39)$ |
| $A_R(15) \Rightarrow M(15)$ | $A_R(30) \Rightarrow M(15)$ | $A_R(31) \Rightarrow M(31)$ |
| $A_I(15) \Rightarrow M(31)$ | $A_I(30) \Rightarrow M(47)$ | $A_I(31) \Rightarrow M(63)$ |

In general, this requires 32 complex multiplications, which is $4 * 32 = 128$ real multiplies and $2 * 32 = 64$ real adds. The equations are (for $n = 0, 1, \ldots, 31$):

$$C_R(n) = A_R(n) * H_R(n) - A_I(n) * H_I(n)$$
$$C_I(n) = A_I(n) * H_R(n) + A_R(n) * H_I(n)$$

If no temporary registers are assumed, each complex multiply required two additional data memory locations to store the results of multiplying each input value by two different constants prior to forming and storing the output results. The $C_R(n)$ and $C_I(n)$ are stored in the locations the $A_R(n)$ and $A_I(n)$ were pulled from to perform the computations.

Addressing convenience has resulted in imaginary parts $A_I(6)$, $A_I(7)$, $A_I(12)$, $A_I(13)$, $A_I(22)$, $A_I(23)$, $A_I(24)$, $A_I(25)$, $A_I(26)$, $A_I(27)$, $A_I(28)$, and $A_I(29)$ being stored in the lower half of allotted data memory and their corresponding real parts stored in the upper half. It is convenient to correct this inconsistency during the complex multiply computations. Specifically, if the imaginary part of one of the $A_R(n)$ and $A_I(n)$ is stored in the lower portion of the data memory, change this when the complex multiply outputs are stored so that the real parts of all of the results are stored together in the lower portion of the memory used for $C_R(n)$ and $C_I(n)$. These 32 complex multiplies require 128 real multiplies and 64 real adds.

## Stage 7: Two 16-Point IFFT Computations

### Step 1: Organizing the Data for the 16-Point IFFTs

Following the instructions in Step 1 of Stage 1 of the general mixed-radix algorithm presented in Section 9.7.4, the $k$-th input data points to the $n$-th 16-point algorithm are $C_R(k*2+n)$ and $C_I(k*2+n)$ (where $n = 0, 1$ and $k = 0, 1, \ldots, 15$). In terms of the input labels, $a_R(n)$ and $a_I(n)$, for the 16-point FFT, the inputs for the first 16-point FFT are:

$$a_R(k) = C_R(2*k) \qquad a_I(k) = C_I(2*k)$$

and for the second 16-point FFT are:

$$a_R(k) = C_R(2*k+1) \qquad a_I(k) = C_I(2*k+1)$$

The inputs to the first 16-point IFFT are the outputs of the first 16-point FFT, modified by complex multipliers. Therefore, these inputs occupy the same memory locations as the outputs of the 16-point FFT. In general, the building-block algorithms do not have their outputs in sequential memory addresses. Therefore, the inputs to the inverse 16-point FFT will not be in sequential addresses, as was assumed in Chapter 8. However, the inputs to the first 16-point IFFT do have all of its real inputs in the first 16 memory locations and all of its imaginary outputs in memory locations 32 through 47. Likewise, the inputs to the second 16-point IFFT are in memory locations 16 through 31, and imaginary outputs are in memory locations 48 through 63. With this in mind, data address relabeling from Section 9.4 is applied to the 16-point radix-4 memory mapping in Section 9.7.5.

### Step 2: Computing the Two 16-Point IFFTs

If the labels from Step 2 of Stage 1 of the general mixed-radix algorithm are used, the $k$-th output ($k = 0, 1, \ldots, 15$) of the $n$-th 16-point transform ($n = 0, 1$) should be labeled $e_R(k*2+n)$ and $e_I(k*2+n)$ in preparation for input to the complex multiply portion of the 32-point mixed-radix algorithm. In terms of the output labels, $A_R(n)$ and $A_I(n)$, for the 16-point radix-4 FFT in Section 9.7.5, the outputs for the first 16-point FFT are:

$$e_R(k*2) = A_R(k) \qquad e_I(k*2) = A_I(k)$$

and for the second 16-point FFT are:

$$e_R(k*2+1) = A_R(k) \qquad e_I(k*2+1) = A_I(k)$$

The four columns in Table 9-3 are the remapping process for the first of the two 16-point radix-4 IFFTs.

**Table 9-3** Memory Maps for 15-Point Bluestein Algorithm Example

| Column 1 | Column 2 | Column 3 | Column 4 |
|---|---|---|---|
| $A_R(0) \Rightarrow M(0)$ | $C_R(0) \Rightarrow M(0)$ | $a_R(0) \Rightarrow M(0)$ | $A_R(0) = e_R(0) \Rightarrow M(0)$ |
| $A_I(0) \Rightarrow M(32)$ | $C_I(0) \Rightarrow M(32)$ | $a_I(0) \Rightarrow M(32)$ | $A_I(0) = e_I(0) \Rightarrow M(32)$ |
| $A_R(2) \Rightarrow M(8)$ | $C_R(2) \Rightarrow M(8)$ | $a_R(1) \Rightarrow M(8)$ | $A_R(1) = e_R(2) \Rightarrow M(1)$ |
| $A_I(2) \Rightarrow M(40)$ | $C_I(2) \Rightarrow M(40)$ | $a_I(1) \Rightarrow M(40)$ | $A_I(1) = e_I(2) \Rightarrow M(33)$ |
| $A_R(4) \Rightarrow M(4)$ | $C_R(4) \Rightarrow M(4)$ | $a_R(2) \Rightarrow M(4)$ | $A_R(2) = e_R(4) \Rightarrow M(2)$ |
| $A_I(4) \Rightarrow M(36)$ | $C_I(4) \Rightarrow M(36)$ | $a_I(2) \Rightarrow M(36)$ | $A_I(2) = e_I(4) \Rightarrow M(34)$ |
| $A_R(6) \Rightarrow M(44)$ | $C_R(6) \Rightarrow M(12)$ | $a_R(3) \Rightarrow M(12)$ | $A_R(3) = e_R(6) \Rightarrow M(35)$ |
| $A_I(6) \Rightarrow M(12)$ | $C_I(6) \Rightarrow M(44)$ | $a_I(3) \Rightarrow M(44)$ | $A_I(3) = e_I(6) \Rightarrow M(3)$ |
| $A_R(8) \Rightarrow M(2)$ | $C_R(8) \Rightarrow M(2)$ | $a_R(4) \Rightarrow M(2)$ | $A_R(4) = e_R(8) \Rightarrow M(4)$ |
| $A_I(8) \Rightarrow M(34)$ | $C_I(8) \Rightarrow M(34)$ | $a_I(4) \Rightarrow M(34)$ | $A_I(4) = e_I(8) \Rightarrow M(36)$ |
| $A_R(10) \Rightarrow M(10)$ | $C_R(10) \Rightarrow M(10)$ | $a_R(5) \Rightarrow M(10)$ | $A_R(5) = e_R(10) \Rightarrow M(5)$ |
| $A_I(10) \Rightarrow M(42)$ | $C_I(10) \Rightarrow M(42)$ | $a_I(5) \Rightarrow M(42)$ | $A_I(5) = e_I(10) \Rightarrow M(37)$ |
| $A_R(12) \Rightarrow M(38)$ | $C_R(12) \Rightarrow M(6)$ | $a_R(6) \Rightarrow M(6)$ | $A_R(6) = e_R(12) \Rightarrow M(38)$ |
| $A_I(12) \Rightarrow M(6)$ | $C_I(12) \Rightarrow M(38)$ | $a_I(6) \Rightarrow M(38)$ | $A_I(6) = e_I(12) \Rightarrow M(6)$ |
| $A_R(14) \Rightarrow M(14)$ | $C_R(14) \Rightarrow M(14)$ | $a_R(7) \Rightarrow M(14)$ | $A_R(7) = e_R(14) \Rightarrow M(7)$ |
| $A_I(14) \Rightarrow M(46)$ | $C_I(14) \Rightarrow M(46)$ | $a_I(7) \Rightarrow M(46)$ | $A_I(7) = e_I(14) \Rightarrow M(39)$ |
| $A_R(16) \Rightarrow M(1)$ | $C_R(16) \Rightarrow M(1)$ | $a_R(8) \Rightarrow M(1)$ | $A_R(8) = e_R(16) \Rightarrow M(8)$ |
| $A_I(16) \Rightarrow M(33)$ | $C_I(16) \Rightarrow M(33)$ | $a_I(8) \Rightarrow M(33)$ | $A_I(8) = e_I(16) \Rightarrow M(40)$ |
| $A_R(18) \Rightarrow M(9)$ | $C_R(18) \Rightarrow M(9)$ | $a_R(9) \Rightarrow M(9)$ | $A_R(9) = e_R(18) \Rightarrow M(9)$ |
| $A_I(18) \Rightarrow M(41)$ | $C_I(18) \Rightarrow M(41)$ | $a_I(9) \Rightarrow M(41)$ | $A_I(9) = e_I(18) \Rightarrow M(41)$ |
| $A_R(20) \Rightarrow M(5)$ | $C_R(20) \Rightarrow M(5)$ | $a_R(10) \Rightarrow M(5)$ | $A_R(10) = e_R(20) \Rightarrow M(10)$ |
| $A_I(20) \Rightarrow M(37)$ | $C_I(20) \Rightarrow M(37)$ | $a_I(10) \Rightarrow M(37)$ | $A_I(10) = e_I(20) \Rightarrow M(42)$ |
| $A_R(22) \Rightarrow M(45)$ | $C_R(22) \Rightarrow M(13)$ | $a_R(11) \Rightarrow M(13)$ | $A_R(11) = e_R(22) \Rightarrow M(43)$ |
| $A_I(22) \Rightarrow M(13)$ | $C_I(22) \Rightarrow M(45)$ | $a_I(11) \Rightarrow M(45)$ | $A_I(11) = e_I(22) \Rightarrow M(11)$ |
| $A_R(24) \Rightarrow M(35)$ | $C_R(24) \Rightarrow M(3)$ | $a_R(12) \Rightarrow M(3)$ | $A_R(12) = e_R(24) \Rightarrow M(44)$ |
| $A_I(24) \Rightarrow M(3)$ | $C_I(24) \Rightarrow M(35)$ | $a_I(12) \Rightarrow M(35)$ | $A_I(12) = e_I(24) \Rightarrow M(12)$ |
| $A_R(26) \Rightarrow M(43)$ | $C_R(26) \Rightarrow M(11)$ | $a_R(13) \Rightarrow M(11)$ | $A_R(13) = e_R(26) \Rightarrow M(45)$ |
| $A_I(26) \Rightarrow M(11)$ | $C_I(26) \Rightarrow M(43)$ | $a_I(13) \Rightarrow M(43)$ | $A_I(13) = e_I(26) \Rightarrow M(13)$ |
| $A_R(28) \Rightarrow M(39)$ | $C_R(28) \Rightarrow M(7)$ | $a_R(14) \Rightarrow M(7)$ | $A_R(14) = e_R(28) \Rightarrow M(46)$ |
| $A_I(28) \Rightarrow M(7)$ | $C_I(28) \Rightarrow M(39)$ | $a_I(14) \Rightarrow M(39)$ | $A_I(14) = e_I(28) \Rightarrow M(14)$ |
| $A_R(30) \Rightarrow M(15)$ | $C_R(30) \Rightarrow M(15)$ | $a_R(15) \Rightarrow M(15)$ | $A_R(15) = e_R(30) \Rightarrow M(15)$ |
| $A_I(30) \Rightarrow M(47)$ | $C_I(30) \Rightarrow M(47)$ | $a_I(15) \Rightarrow M(47)$ | $A_I(15) = e_I(30) \Rightarrow M(47)$ |

- Column 1 shows the data mapping out of the first 16-point input FFT.
- Column 2 shows the data mapping after the linear filter complex multiplications. The data addresses are identical to those in column 1 except for the terms where column 1 had the imaginary part at a lower address than the real part. In those cases, the real and imaginary addresses were swapped during the complex multiplication process.
- Column 3 shows the new memory addresses for each of the inputs to the first 16-point IFFT in terms of the data labeling found in Section 9.7.5.
- Column 4 shows the memory address for each of the first 16-point FFT's outputs, based on the memory relabeling technique, and the definition of how they are related to the actual output of the first stage of the required 32-point IFFT.

The four columns in Table 9-4 are the remapping process for the second of the two 16-point IFFTs.

**Table 9-4**   Output Memory Maps for 15-Point Bluestein Algorithm Example

| Column 1 | Column 2 | Column 3 | Column 4 |
|---|---|---|---|
| $A_R(1) \Rightarrow M(16)$ | $C_R(1) \Rightarrow M(16)$ | $a_R(0) \Rightarrow M(16)$ | $A_R(0) = e_R(1) \Rightarrow M(16)$ |
| $A_I(1) \Rightarrow M(48)$ | $C_I(1) \Rightarrow M(48)$ | $a_I(0) \Rightarrow M(48)$ | $A_I(0) = e_I(1) \Rightarrow M(48)$ |
| $A_R(3) \Rightarrow M(24)$ | $C_R(3) \Rightarrow M(24)$ | $a_R(1) \Rightarrow M(24)$ | $A_R(1) = e_R(3) \Rightarrow M(33)$ |
| $A_I(3) \Rightarrow M(56)$ | $C_I(3) \Rightarrow M(56)$ | $a_I(1) \Rightarrow M(56)$ | $A_I(1) = e_I(3) \Rightarrow M(49)$ |
| $A_R(5) \Rightarrow M(20)$ | $C_R(5) \Rightarrow M(20)$ | $a_R(2) \Rightarrow M(20)$ | $A_R(2) = e_R(5) \Rightarrow M(18)$ |
| $A_I(5) \Rightarrow M(52)$ | $C_I(5) \Rightarrow M(52)$ | $a_I(2) \Rightarrow M(52)$ | $A_I(2) = e_I(5) \Rightarrow M(50)$ |
| $A_R(7) \Rightarrow M(60)$ | $C_R(7) \Rightarrow M(28)$ | $a_R(3) \Rightarrow M(28)$ | $A_R(3) = e_R(7) \Rightarrow M(51)$ |
| $A_I(7) \Rightarrow M(28)$ | $C_I(7) \Rightarrow M(60)$ | $a_I(3) \Rightarrow M(60)$ | $A_I(3) = e_I(7) \Rightarrow M(19)$ |
| $A_R(9) \Rightarrow M(18)$ | $C_R(9) \Rightarrow M(18)$ | $a_R(4) \Rightarrow M(18)$ | $A_R(4) = e_R(9) \Rightarrow M(20)$ |
| $A_I(9) \Rightarrow M(50)$ | $C_I(9) \Rightarrow M(50)$ | $a_I(4) \Rightarrow M(50)$ | $A_I(4) = e_I(9) \Rightarrow M(52)$ |
| $A_R(11) \Rightarrow M(26)$ | $C_R(11) \Rightarrow M(26)$ | $a_R(5) \Rightarrow M(26)$ | $A_R(5) = e_R(11) \Rightarrow M(21)$ |
| $A_I(11) \Rightarrow M(58)$ | $C_I(11) \Rightarrow M(58)$ | $a_I(5) \Rightarrow M(58)$ | $A_I(5) = e_I(11) \Rightarrow M(53)$ |
| $A_R(13) \Rightarrow M(54)$ | $C_R(13) \Rightarrow M(22)$ | $a_R(6) \Rightarrow M(22)$ | $A_R(6) = e_R(13) \Rightarrow M(54)$ |
| $A_I(13) \Rightarrow M(22)$ | $C_I(13) \Rightarrow M(54)$ | $a_I(6) \Rightarrow M(54)$ | $A_I(6) = e_I(13) \Rightarrow M(22)$ |
| $A_R(15) \Rightarrow M(30)$ | $C_R(15) \Rightarrow M(30)$ | $a_R(7) \Rightarrow M(30)$ | $A_R(7) = e_R(15) \Rightarrow M(23)$ |
| $A_I(15) \Rightarrow M(62)$ | $C_I(15) \Rightarrow M(62)$ | $a_I(7) \Rightarrow M(62)$ | $A_I(7) = e_I(15) \Rightarrow M(55)$ |
| $A_R(17) \Rightarrow M(17)$ | $C_R(17) \Rightarrow M(17)$ | $a_R(8) \Rightarrow M(17)$ | $A_R(8) = e_R(17) \Rightarrow M(24)$ |
| $A_I(17) \Rightarrow M(49)$ | $C_I(17) \Rightarrow M(49)$ | $a_I(8) \Rightarrow M(49)$ | $A_I(8) = e_I(17) \Rightarrow M(56)$ |
| $A_R(19) \Rightarrow M(25)$ | $C_R(19) \Rightarrow M(25)$ | $a_R(9) \Rightarrow M(25)$ | $A_R(9) = e_R(19) \Rightarrow M(25)$ |
| $A_I(19) \Rightarrow M(57)$ | $C_I(19) \Rightarrow M(57)$ | $a_I(9) \Rightarrow M(57)$ | $A_I(9) = e_I(19) \Rightarrow M(57)$ |
| $A_R(21) \Rightarrow M(21)$ | $C_R(21) \Rightarrow M(21)$ | $a_R(10) \Rightarrow M(21)$ | $A_R(10) = e_R(21) \Rightarrow M(26)$ |
| $A_I(21) \Rightarrow M(53)$ | $C_I(21) \Rightarrow M(53)$ | $a_I(10) \Rightarrow M(53)$ | $A_I(10) = e_I(21) \Rightarrow M(58)$ |
| $A_R(23) \Rightarrow M(61)$ | $C_R(23) \Rightarrow M(29)$ | $a_R(11) \Rightarrow M(29)$ | $A_R(11) = e_R(23) \Rightarrow M(59)$ |
| $A_I(23) \Rightarrow M(29)$ | $C_I(23) \Rightarrow M(61)$ | $a_I(11) \Rightarrow M(61)$ | $A_I(11) = e_I(23) \Rightarrow M(27)$ |
| $A_R(25) \Rightarrow M(51)$ | $C_R(25) \Rightarrow M(19)$ | $a_R(12) \Rightarrow M(19)$ | $A_R(12) = e_R(25) \Rightarrow M(60)$ |
| $A_I(25) \Rightarrow M(19)$ | $C_I(25) \Rightarrow M(51)$ | $a_I(12) \Rightarrow M(51)$ | $A_I(12) = e_I(25) \Rightarrow M(28)$ |
| $A_R(27) \Rightarrow M(59)$ | $C_R(27) \Rightarrow M(27)$ | $a_R(13) \Rightarrow M(27)$ | $A_R(13) = e_R(27) \Rightarrow M(61)$ |
| $A_I(27) \Rightarrow M(27)$ | $C_I(27) \Rightarrow M(59)$ | $a_I(13) \Rightarrow M(59)$ | $A_I(13) = e_I(27) \Rightarrow M(29)$ |
| $A_R(29) \Rightarrow M(55)$ | $C_R(29) \Rightarrow M(23)$ | $a_R(14) \Rightarrow M(23)$ | $A_R(14) = e_R(29) \Rightarrow M(62)$ |
| $A_I(29) \Rightarrow M(23)$ | $C_I(29) \Rightarrow M(55)$ | $a_I(14) \Rightarrow M(55)$ | $A_I(14) = e_I(29) \Rightarrow M(30)$ |
| $A_R(31) \Rightarrow M(31)$ | $C_R(31) \Rightarrow M(31)$ | $a_R(15) \Rightarrow M(31)$ | $A_R(15) = e_R(31) \Rightarrow M(31)$ |
| $A_I(31) \Rightarrow M(63)$ | $C_I(31) \Rightarrow M(63)$ | $a_I(15) \Rightarrow M(63)$ | $A_I(15) = e_I(31) \Rightarrow M(63)$ |

- Column 1 shows the data mapping out of the second 16-point input FFT.

- Column 2 shows the data mapping after the linear filter complex multiplications. The data addresses are identical to those in column 1 except for the terms where column 1 had the imaginary part at a lower address than the real part. In those cases, the real and imaginary addresses were swapped during the complex multiplication process.

- Column 3 shows the new memory addresses for each of the inputs to the second 16-point IFFT, in terms of the data labeling found in Section 9.7.5.

- Column 4 shows the memory address for each of the second 16-point FFT's outputs, based on the memory relabeling technique, and the definition of how they are related to the actual output of the first stage of the required 32-point IFFT.

These two 16-point IFFTs require exactly the same number of computations as the 16-point FFTs in Stage 5. Therefore, Stage 7 requires 288 real adds and 48 real multiplies.

*Step 3: Performing Complex Multiplications*

Each of the $e_R(k*2+n)$ and $e_I(k*2+n)$ needs to be multiplied by a specific complex number prior to entering the 2-point portion of the 32-point algorithm. The equations for this complex multiplication for each $n = 0, 1$ and $k = 0, 1, \ldots, 15$ are:

$$f_R(k*2+n) = e_R(k*2+n) * \cos(2\pi * kn/32) - e_I(k*2+n) * \sin(2\pi * kn/32)$$
$$f_I(k*2+n) = e_I(k*2+n) * \cos(2\pi * kn/32) + e_R(k*2+n) * \sin(2\pi * kn/32)$$

If no temporary registers are assumed, each complex multiply required two additional data memory locations to store the results of multiplying each input value by two different constants prior to forming and storing the output results. However, if the complex multiplies are performed sequentially, the same two additional memory locations can be reused for all of the complex multiplies. The result is the need for only two additional memory locations. Store the results of the complex multiplies back in the same locations that the inputs to the complex multiplies were taken from. For $n = 0$, these complex multiplies are just multiplies by 1. Therefore, one of the two 16-point IFFTs does not have its outputs modified prior to computing the 2-point IFFTs. Since only 15 of these 16 complex outputs represent the needed result in Stage 8, only 15 of the complex multiplies need to be performed. The total number of computations for these 15 complex multiplies is 60 real multiplies and 30 real adds.

## Stage 8: Computing the Output 2-Point Building Blocks

This stage has two steps. The first is to properly group the input data for each of the 16 2-point building blocks. The second is to compute the appropriate part of each of the 16 2-point building blocks, based on the discussion in Stage 8 of the general Bluestein algorithm.

*Step 1: Grouping the Input Data Points to the 2-Point Building Blocks*

For the $n$-th input to the $k$-th 2-point building block, choose $f_R(k*2+n)$ and $f_I(k*2+n)$ (where $k = 0, 1, \ldots, 15$ and $n = 0, 1$) from the input data sequence. In terms of the input labels, $a_R(n)$ and $a_I(n)$, shown in Chapter 8, the inputs for the $k$-th 2-point building blocks are:

$$a_R(0) = f_R(2*k) \qquad a_R(1) = f_R(2*k+1)$$
$$a_I(0) = f_I(2*k) \qquad a_I(1) = f_I(2*k+1)$$

*Step 2: Computing a Portion of the Output 2-Point Building Blocks*

Using the 2-point building block from Chapter 8 yields:

$$A_R(0) = a_R(0) + a_R(1) \qquad A_R(1) = a_R(0) - a_R(1)$$
$$A_I(0) = a_I(0) + a_I(1) \qquad A_I(1) = a_I(0) - a_I(1)$$

The outputs of interest are the second pair of equations. Therefore, if the output frequency components of the 32-point IFFT are $y_R(n*16+k)$ and $y_I(n*16+k)$, for the $n$-th output of the $k$-th 2-point transform, the outputs of interest are for $n = 1$. In terms of the output labels, $A_R(n)$ and $A_I(n)$, shown for the 16-point radix-4 FFT, the outputs for the $k$-th 2-point building block are equated to the complete outputs by the equations:

$$y_R(16 + k) = f_R(2 * k) - f_R(2 * k + 1)$$
$$y_I(16 + k) = f_I(2 * k) - f_I(2 * k + 1)$$

Since only 15 of these 16 complex outputs represent the needed result in Stage 10, only 15 of the complex adds need to be performed. The 15 partial 2-point transform requires 30 real adds.

### Stage 9: Adjusting the Output Data

This stage has the following steps:

1. For $n = 15, 16, \ldots, 31$, multiply $y(n) = y_R(n) + j * y_I(n)$ by $\exp(-j * \pi * n^2/15) = \cos(\pi * n^2/15) - j * \sin(\pi * n^2/15)$ to obtain $z(n)$.
2. For $n = 15, 16, \ldots, 31$, multiply $z(n)$ by $\exp(-j * \pi * 15) = -1$ to obtain $q(n)$.

These two steps can be combined into a single complex multiply by multiplying the first complex multiplier by $-1$ to obtain:

$$q_R(n) = -y_R(n) * \cos(\pi * n^2/15) - y_I(n) * \sin(\pi * n^2/15)$$
$$q_I(n) = -y_I(n) * \cos(\pi * n^2/15) + y_R(n) * \sin(\pi * n^2/15)$$

Again, if there are no temporary registers in the processor, then two additional memory locations are required to perform the complex computations. However, if the complex multiplies are performed sequentially, the same two additional memory locations can be reused for all of the complex multiplies. The result is the need for only two additional memory locations. Store the results of the complex multiplies back in the same locations that the inputs to the complex multiplies were taken from.

Since only 15 of these 16 complex outputs represent the needed result in Stage 10, only 15 of the complex multiplies need be performed. This is a total of 60 real multiplies and 30 real adds.

### Stage 10: Extracting the 15-Point FFT

The 15-point FFT outputs, $G(n) = G_R(n) + j * G_I(n)$, are $q(15 + n) = q_R(15 + n) + j * q_I(15 + n)$ where $n = 0, 1, \ldots, 15$.

### 9.5.6  Winograd Algorithm Introduction

This algorithm was developed by mathematician Schmuel Winograd and originally published in 1976 [2]. The motivation for the development of this algorithm was that multiplication was extremely expensive in computation time, board area, and power. Thus the algorithm was designed to minimize the number of multiplications required to implement FFTs.

While Winograd succeeded in minimizing the number of multiplications, he also succeeded in complicating the computational building blocks and data memory mappings for his algorithm. The result was that the algorithm did not significantly decrease the cost of performing FFTs. In fact, in some cases the cost was increased over comparable power-of-two or Singleton algorithms presented in Section 9.7.

While advances in integrated circuit technology have lowered the cost of multiplication and complex data addressing, it has not improved the value of the Winograd transform, except when dedicated building blocks have been developed. The primary reason for this is that the multiply-accumulators (Chapter 10) used in DSP chips (Chapter 14) are all based on an architecture that does not allow the multiplier and accumulator to be used independently. Since the Winograd algorithm separates adds from multiplies, it is difficult to make efficient use of these computational building blocks to compute the Winograd algorithm.

The available Winograd building blocks (Chapter 8) are 2, 3, 4, 5, 7, 8, 9, and 16 points. Combining relatively prime sets of these allows the following 58 transform lengths:

$$N = 2, 3, 4, 5, 6, 7, 8, 9, 10, 12, 14, 15, 16, 18, 20, 21, 24, 28, 30, 35, 36, 40, 42, 45, 48,$$
$$56, 60, 63, 70, 72, 80, 84, 90, 105, 112, 120, 126, 140, 144, 168, 180, 210, 240, 252,$$
$$280, 315, 336, 360, 420, 504, 560, 630, 840, 1008, 1260, 1680, 2520, 5040$$

In the original derivation of the Winograd algorithm, the Winograd building blocks from Chapter 8 were combined to form these 58 different transform lengths. However, the technique can be extended to combining any building blocks that have all of their multiplies in the center and just adds and subtracts for the input and output computations. This is why the building blocks in Chapter 8 were configured in this format.

The general algorithm steps for computing the Winograd transform can be described completely with just two building blocks. The result is a larger transform that still has all of its multiplies in the center and only adds and subtracts on the input and output. The larger transform can now be combined into a larger transform with a third building block with the same technique for combining them that was used for the first two. This process can be continued as long as the add-multiply-add architecture is followed and all of the building blocks are relatively prime numbers. This process, using the general odd-number algorithms in Section 8.11, increases the number of transform lengths for the Winograd algorithm beyond the 58 listed. The only catch is that, since the non-Winograd building blocks do not have the minimum number of multiplies, their combination into larger FFTs does not result in a minimum number of multiplications.

Figure 9-6 is a Winograd algorithm block diagram for two factors, $P$ and $Q$. Since all of the $N$ input data points are processed by the $P$- and $Q$-point stages, the $N$ data points must be separated into sets of $P$ data points for the first input addition stage. There are $N/P = Q$ of these sets. Then the results from the first input addition stage must be divided into sets of $Q$ data points for processing by the second input addition stage.

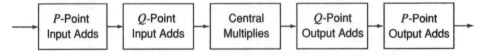

**Figure 9-6**   Top-level block diagram of two-factor Winograd algorithm.

In general, there are more outputs of the input adds than there are inputs. The result is that there are more than $N/Q = P$ sets of $Q$-point input adds to perform. If the order of $P$ and $Q$ is reversed, there are $P$ sets of $Q$-point input adds performed first, followed by more than $Q$ sets of $P$-point input adds. This implies that the total number of input adds (all of the $P$ and $Q$-point sets combined) changes as a function of which building block is implemented first.

### 9.5.7 Number of Winograd Algorithm Adds and Multiplies

The number of real adds is dependent on the order in which the building blocks are combined to form the larger transform. The equation for the number of real adds for a two-stage $N = P * Q$-point Winograd FFT in the order shown in Figure 9-6 is:

$$\# \text{ adds} = 2 * [Q * A_P + (M_P + 1) * A_Q]$$
$$\# \text{ multiplies} = 2 * (M_P + 1) * (M_Q + 1) - 1$$

where:   $A_P$ = number of real adds in $P$-point algorithm building block
$A_Q$ = number of real adds in $Q$-point algorithm building block
$M_P$ = number of real multiplies in $P$-point algorithm building block
$M_Q$ = number of real multiplies in $Q$-point algorithm building block

### 9.5.8 General Winograd Algorithm

The stages for combining two building blocks using the general Winograd [2] algorithm are as follows.

### Stage 1: Input Data Organization

If the complex input data sequence is $(a_R(n), a_I(n))$, the expression for the $k$-th input data value for the $m$-th $P$-point building block is $a_R((Q * k + P * m) \bmod N)$, $a_I((Q * k + P * m) \bmod N)$, where $k = 0, 1, \ldots, (P - 1)$ and $m = 0, 1, \ldots, (Q - 1)$. Specifically, the input samples to the first $(m = 0)P$-point input adds stage are $a_R(Q * k \bmod N)$ and $a_I(Q * k \bmod N)$, where $k = 0, 1, 2, \ldots, (P - 1)$. The input samples to the last $(m = Q - 1)P$-point input adds stage are $a_R(Q * k + P * (P - 1) \bmod N)$ and $a_I(Q * k + P * (P - 1) \bmod N)$, where $k = 0, 1, 2, \ldots, (P - 1)$.

### Stage 2: P-Point Building-Block Input Add Computations

Since there are $N/P = Q$ of the $P$-point input adds blocks, this stage requires $Q*$(number of $P$-point building block input adds) additions. There are $(M_P + 1)$ outputs from each of the $Q$ sets of input adds, for a total of $Q * (M_P + 1)$ outputs. Call the $k$-th

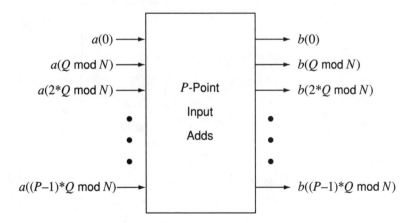

**Figure 9-7**   *P*-point input adds data configuration for $m = 0$.

complex output of the $m$-th $P$-point input adds building block $b_R((Q*k+P*m) \bmod N)$, $b_I((Q*k+P*m) \bmod N)$, where $k = 0, 1, \ldots, (M_P)$ and $m = 0, 1, \ldots, (Q-1)$. Specifically, the outputs from the first ($m = 0$) $P$-point input adds are $b_R(Q*k \bmod N)$ and $b_I(Q*k \bmod N)$, where $k = 0, 1, 2, \ldots, M_P$. The outputs from the last ($m = Q-1$) $P$-point input adds are $b_R(Q*k+Q*(P-1) \bmod N)$ and $b_I(Q*k+Q*(P-1) \bmod N)$, where $k = 0, 1, 2, \ldots, (P-1)$. Figure 9-7 shows the the input adds data ordering for the first ($m = 0$) of these $P$-point input adds.

### Stage 3: Q-Point Building-Block Input Add Data Organization

The outputs from Stage 2 are now regrouped to become input data for $(M_P + 1)$ replications of the $Q$-point building-block input add algorithm. With the labeling scheme from Stage 2, the $m$-th input to the $k$-th $Q$-point input adds is $b_R((Q*k+P*m) \bmod N)$, $b_I((Q*k+P*m) \bmod N)$, where $k = 0, 1, \ldots, (M_P)$ and $m = 0, 1, \ldots, (Q-1)$. Specifically, the inputs to the first ($k = 0$) $Q$-point input adds stage are $b_R(P*m \bmod N)$ and $b_I(P*m \bmod N)$, where $m = 0, 1, 2, \ldots, (Q-1)$. In Stage 3 these inputs are the first ($k = 0$) output of each of the $P$-point input adds. Similarly, the inputs to the $k$-th $Q$-point input adds are the $k$-th outputs of all of the $P$-point input adds. The arrow between blocks 1 and 2 in Figure 9-6 represents this data reorganization. This addressing is usually determined ahead of time and stored as a sequence of addresses or an addressing algorithm in program memory.

### Stage 4: Q-Point Building-Block Input Add Computations

Each group of $Q$ complex data points in Stage 3 becomes the input to a $Q$-point building-block's input adds. Since there are $(M_P + 1)$ of the $Q$-point input adds blocks, this stage requires $(M_P + 1) *$ (number of $Q$-point building-block input adds) additions. There are $(M_Q + 1)$ outputs from each $Q$-point input add, for a total of $(M_Q+1)*(M_P+1)$ outputs from the second block in Figure 9-6. Call the $m$-th complex output of the $k$-th $Q$-point transform $c_R(k*(M_Q+1)+m)$, $c_I(k*(M_Q+1)+m)$, where $k = 0, 1, \ldots, (M_P)$ and $m = 0, 1, \ldots, (M_Q)$. Figure 9-8 shows the input adds data ordering for the first ($k = 0$) of these $Q$-point input add stages.

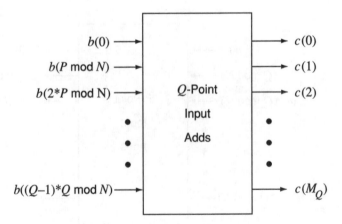

**Figure 9-8** $Q$-point input adds data configuration for $k = 0$.

## Stage 5: Central Multiplications

This stage contains all of the multiplications required for the Winograd transform. If the $k$-th multiplier constant for the $P$-point Winograd algorithm building block is $MP(k)$, and the $m$-th multiplier constant for the $Q$-point building block is $MQ(m)$, then the required multiplications are:

$$d_R(k * (M_Q + 1) + m) = MP(k) * MQ(m) * c_R(k * (M_Q + 1) + m)$$
$$d_I(k * (M_Q + 1) + m) = MP(k) * MQ(m) * c_I(k * (M_Q + 1) + m)$$

where $k = 0, 1, \ldots, (M_P)$ and $m = 0, 1, \ldots, (M_Q)$. Generally, the $MP(k) * MQ(m)$ multiplication is computed ahead of time and the constants stored in program or data memory. This requires $2 * (M_P * M_Q - 1)$ multiplications. This set of computations is represented in Figure 9-6 by the third block from the left. No data reorganization is required between the $Q$-point input adds and the central multiplications or between the central multiplications and the $Q$-point output adds as shown in Figure 9-9 for the first set $(k = 0)$ of multiplications.

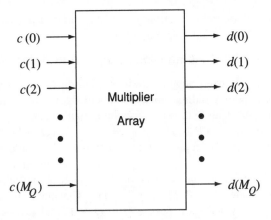

**Figure 9-9**    Central multiplication data configuration for $k = 0$.

## Stage 6: Q-Point Building-Block Output Add Data Organization

The outputs from Stage 5 become input data for $M_P + 1$ replications of the $Q$-point building-block output add algorithm. For the labeling scheme from Stage 5, the $m$-th input to the $k$-th $Q$-point output adds is $d_R(k * (M_Q + 1) + m)$, $d_I(k * (M_Q + 1) + m)$, where $k = 0, 1, \ldots, (M_P)$ and $m = 0, 1, \ldots, (M_Q)$. This set of operations is represented by the arrow between the third and fourth blocks from the left in Figure 9-6 and shown more explicitly for the first $(k = 0)$ of the $Q$-point output adds in Figure 9-10 (on page 172).

## Stage 7: Q-Point Building-Block Output Add Computations

Since there are $(M_P + 1)$ of the $Q$-point output adds blocks, this step requires $(M_P + 1) *$ (number of $Q$-point building-block output adds) additions. There are $Q$ outputs from each of the $Q$-point output adds, for a total of $Q * (M_P + 1)$ outputs. Call the $m$-th complex output of the $k$-th $Q$-point building block $e_R(k * Q + m)$, $e_I(k * Q + m)$, where $k = 0, 1, \ldots, (M_P)$ and $m = 0, 1, \ldots, (Q - 1)$. This set of computations is represented

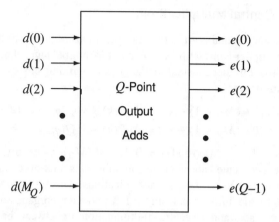

**Figure 9-10**   $Q$-point output adds data configuration for $k = 0$.

by the fourth block from the left in Figure 9-6 and shown in more detail in Figure 9-10 for the first ($k = 0$) of the $Q$-point output adds.

## Stage 8: *P*-Point Building-Block Output Add Data Organization

The outputs from Stage 7 are now regrouped to become input data for $Q$ replications of the $P$-point building-block output add algorithm. Using the labeling scheme from Stage 7, the $k$-th input to the $m$-th $P$-point output adds is $e_R(k * Q + m)$, $e_I(k * Q + m)$, where $k = 0, 1, \ldots, (P - 1)$ and $m = 0, 1, \ldots, (Q - 1)$. The arrow between blocks 4 and 5 in Figure 9-6 represents this operation. This addressing is determined ahead of time and stored as a sequence of addresses or an addressing algorithm in program memory. Specifically, the first ($m = 0$) $P$-point output adds stage are $e_R(k * Q)$ and $e_I(k * Q)$, where $k = 0, 1, \ldots, (P - 1)$. The inputs to the last ($m = Q - 1$) $P$-point output adds stage are $e_R(k * Q + Q - 1)$ and $e_I(k * Q + Q - 1)$, where $k = 0, 1, \ldots, (P - 1)$. Figure 9-11 shows this explictly for the first ($m = 0$) $P$-point output adds stage.

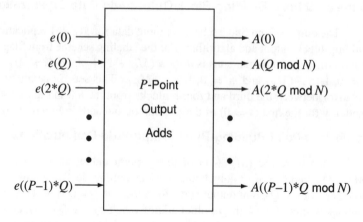

**Figure 9-11**   $P$-point output adds data configuration for $m = 0$.

### Stage 9: $P$-Point Building-Block Output Add Computations

Since there are $Q$ of the $P$-point output adds blocks, this step requires $Q *$ (number of $P$-point building-block output adds) additions. There are $P$ outputs from each of the $Q$ $P$-point output adds, for a total of $Q * P$ outputs. The $m$-th output of the $k$-th $P$-point building block is labeled $A_R[(Q * m + P * k) \bmod N]$ and $A_I[(Q * m + P * k) \bmod N]$, where $k = 0, 1, \ldots, (Q - 1)$ and $m = 0, 1, \ldots, (P - 1)$. This set of computations is represented by the fifth block from the left in Figure 9-6, and the results are shown more explicitly in Figure 9-11 for the first $(k = 0)$ $P$-point output adds stage.

### 9.5.9 Fifteen-Point Winograd Algorithm Example

The 15-point Winograd [2] algorithm can be implemented with either the 3-point or the 5-point building blocks first. Like the prime factor and mixed-radix algorithms in Sections 9.6 and 9.7, the order of the building blocks does not affect the number of multiplications. However, unlike the prime factor and mixed-radix algorithms, the order does affect the number of additions.

This example uses the Winograd 3- and 5-point building blocks. However, any of the 3- and 5-point building blocks from Chapter 8 can be used because they were designed to have an input add section, a central multiply section, and an output add section. From the Comparison Matrix in Chapter 8, the 3-point Winograd building block has six input adds, six output adds, and uses 3 for the number of multiply paths. The 5-point Winograd building block has 16 input adds, 18 output adds, and uses 6 for the number of multiply paths. Substituting these numbers into the equation for the number of computations gives that the total number of real multiplications is 34 and the total number of real adds is 174 if the input portion of the 5-point Winograd building block is computed first. The total number of real adds is 162 if the input add portion of the 3-point building block is computed first.

Figure 9-12 shows how the various portions of the 3- and 5-point Winograd building blocks are nested to form the 15-point Winograd FFT. The various 3- and 5-point input and output add blocks are labeled as they are below. The three distinct multiplier blocks are also shown explicitly in Figure 9-12. This 15-point example requires 36 data memory locations and 17 memory locations for multiplier constants.

The $b_R(i)$, $b_I(i)$, $c_R(i)$, $c_I(i)$, $d_R(i)$, $d_I(i)$, and $e_R(i)$, $e_I(i)$ used to label intermediate results in the description of the general Winograd algorithm in Section 9.5.8 are different from the intermediate result labels in this example. However, the computations and data reorganization are identical. The labels in Section 9.5.8 were chosen to show the interconnection pattern of the individual building blocks. The labels in this example were chosen to identify as closely as possible with the 3-point and 5-point Winograd building block labels in Chapter 8. The nonmodular nature of the different Winograd building blocks makes complete commonality between these descriptions impossible.

### Stage 1: Three-Point Input Adds

The 15 input data points must first be divided into five sets of 3 points to serve as inputs to each of the 3-point algorithms. Following the addressing in Section 9.5.8, this is done by starting with complex input data point $a_R(0)$, $a_I(0)$, and grouping it with complex

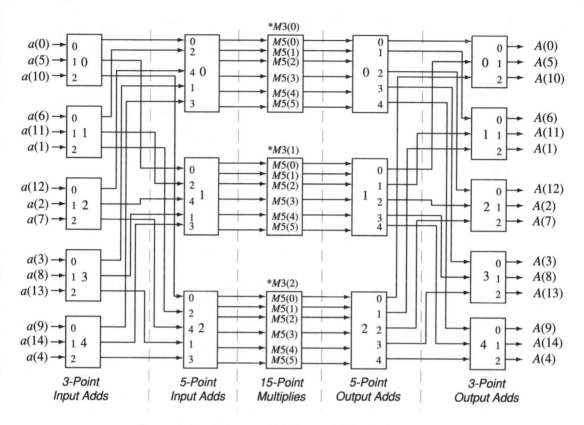

**Figure 9-12** Fifteen-point Winograd FFT block diagram.

input data point pairs $a_R(5)$, $a_I(5)$ and $a_R(10)$, $a_I(10)$. These provide the input to the top one of the five 3-point building blocks. This is followed by grouping the input data point pairs $a_R(1)$, $a_I(1)$, $a_R(6)$, $a_I(6)$, and $a_R(11)$, $a_I(11)$ to provide the input for the second of the five 3-point building blocks. The next grouping is data point pairs $a_R(2)$, $a_I(2)$, $a_R(7)$, $a_I(7)$, and $a_R(12)$, $a_I(12)$ for input into the third of the five 3-point building blocks. The next grouping is data point pairs $a_R(3)$, $a_I(3)$, $a_R(8)$, $a_I(8)$, and $a_R(13)$, $a_I(13)$ to provide input for the fourth of the five 3-point building blocks. The final grouping is data point pairs $a_R(4)$, $a_I(4)$, $a_R(9)$, $a_I(9)$, and $a_R(14)$, $a_I(14)$ for input into the fifth 3-point building block. The addressing in Section 9.5.8 determines the order in which these data points enter the 3-point input adds.

The strategy for converting these equations to code is to start at the top (compute $b_R(1)$) and identify the pair of inputs to be used first (in this case $a_R(5)$ and $a_R(10)$). Then look down the list to find the second (compute $b_R(2)$) place where these two inputs are used. Pull $a_R(5)$ and $a_R(10)$ from memory, compute $b_R(1)$ and $b_R(2)$, and store the results in memory locations $M(5)$ and $M(10)$, previously occupied by $a_R(5)$ and $a_R(10)$. The next step is to look at the next computation $b_I(1)$ on the list and repeat the same set of steps.

Continue this process until all the Algorithm Steps in Stage 1 have been computed and their results stored in the Memory Map addresses.

### First of Five 3-Point Algorithm Building-Block Input Adds

The inputs to these 3-point input adds are $a_R((5*k+3*m) \bmod 15)$, $a_I((5*k+3*m) \bmod 15)$ where $m = 0$. Performing the modulo arithmetic computations to determine the inputs results in the inputs being $a_R(0)$, $a_I(0)$, $a_R(5)$, $a_I(5)$, $a_R(10)$, and $a_I(10)$ for $k = 0, 1$, and 2. These input adds are represented in Figure 9-12 by the 3-point input adds block labeled 0. Further, the labels on the left of this input add block correspond to the input labels in the 3-point Winograd building block in Chapter 8.

| Algorithm Steps | Memory Map |
|---|---|
| $b_R(1) = a_R(5) + a_R(10)$ | $b_R(1) \Rightarrow M(5)$ |
| $b_R(2) = a_R(5) - a_R(10)$ | $b_R(2) \Rightarrow M(10)$ |
| $b_I(1) = a_I(5) + a_I(10)$ | $b_I(1) \Rightarrow M(20)$ |
| $b_I(2) = a_I(5) - a_I(10)$ | $b_I(2) \Rightarrow M(25)$ |
| $b_R(0) = a_R(0) + b_R(1)$ | $b_R(0) \Rightarrow M(0)$ |
| $b_I(0) = a_I(0) + b_I(1)$ | $b_I(0) \Rightarrow M(15)$ |

### Second of Five 3-Point Algorithm Building-Block Input Adds

The inputs to these 3-point input adds are $a_R((5*k+3*m) \bmod 15)$, $a_I((5*k+3*m) \bmod 15)$ where $m = 2$. Performing the modulo arithmetic computations to determine the inputs results in the inputs being $a_R(6)$, $a_I(6)$, $a_R(11)$, $a_I(11)$, $a_R(1)$, and $a_I(1)$ for $k = 0, 1$, and 2. These input adds are represented in Figure 9-12 by the 3-point input adds block labeled 1. Further, the labels on the left of this input add block correspond to the input labels in the 3-point Winograd building block in Chapter 8.

| Algorithm Steps | Memory Map |
|---|---|
| $b_R(4) = a_R(11) + a_R(1)$ | $b_R(4) \Rightarrow M(11)$ |
| $b_R(5) = a_R(11) - a_R(1)$ | $b_R(5) \Rightarrow M(1)$ |
| $b_I(4) = a_I(11) + a_I(1)$ | $b_I(4) \Rightarrow M(26)$ |
| $b_I(5) = a_I(11) - a_I(1)$ | $b_I(5) \Rightarrow M(16)$ |
| $b_R(3) = a_R(6) + b_R(4)$ | $b_R(3) \Rightarrow M(6)$ |
| $b_I(3) = a_I(6) + b_I(4)$ | $b_I(3) \Rightarrow M(21)$ |

### Third of Five 3-Point Algorithm Building-Block Input Adds

The inputs to these 3-point input adds are $a_R((5*k+3*m) \bmod 15)$, $a_I((5*k+3*m) \bmod 15)$ where $m = 4$. Performing the modulo arithmetic computations to determine the inputs results in the inputs being $a_R(12)$, $a_I(12)$, $a_R(2)$, $a_I(2)$, $a_R(7)$, and $a_I(7)$ for $k = 0, 1$, and 2. These input adds are represented in Figure 9-12 by the 3-point input adds block labeled 2. Further, the labels on the left of this input add block correspond to the input labels in the 3-point Winograd building block in Chapter 8.

| Algorithm Steps | Memory Map |
|---|---|
| $b_R(7) = a_R(2) + a_R(7)$ | $b_R(7) \Rightarrow M(2)$ |
| $b_R(8) = a_R(2) - a_R(7)$ | $b_R(8) \Rightarrow M(7)$ |
| $b_I(7) = a_I(2) + a_I(7)$ | $b_I(7) \Rightarrow M(17)$ |
| $b_I(8) = a_I(2) - a_I(7)$ | $b_I(8) \Rightarrow M(22)$ |
| $b_R(6) = a_R(12) + b_R(7)$ | $b_R(6) \Rightarrow M(12)$ |
| $b_I(6) = a_I(12) + b_I(7)$ | $b_I(6) \Rightarrow M(27)$ |

*Fourth of Five 3-Point Algorithm Building-Block Input Adds*

The inputs to these 3-point input adds are $a_R((5*k+3*m) \bmod 15)$, $a_I((5*k+3*m) \bmod 15)$ where $m = 1$. Performing the modulo arithmetic computations to determine the inputs results in the inputs being $a_R(3)$, $a_I(3)$, $a_R(8)$, $a_I(8)$, $a_R(13)$, and $a_I(13)$ for $k = 0, 1$, and 2. These input adds are represented in Figure 9-12 by the 3-point input adds block labeled 3. Further, the labels on the left of this input add block correspond to the input labels in the 3-point Winograd building block in Chapter 8.

| Algorithm Steps | Memory Map |
|---|---|
| $b_R(10) = a_R(8) + a_R(13)$ | $b_R(10) \Rightarrow M(8)$ |
| $b_R(11) = a_R(8) - a_R(13)$ | $b_R(11) \Rightarrow M(13)$ |
| $b_I(10) = a_I(8) + a_I(13)$ | $b_I(10) \Rightarrow M(23)$ |
| $b_I(11) = a_I(8) - a_I(13)$ | $b_I(11) \Rightarrow M(28)$ |
| $b_R(9) = a_R(3) + b_R(10)$ | $b_R(9) \Rightarrow M(3)$ |
| $b_I(9) = a_I(3) + b_I(10)$ | $b_I(9) \Rightarrow M(3)$ |

*Fifth of Five 3-Point Algorithm Building-Block Input Adds*

The inputs to these 3-point input adds are $a_R((5*k+3*m) \bmod 15)$, $a_I((5*k+3*m) \bmod 15)$ where $m = 3$. Performing the modulo arithmetic computations to determine the inputs results in the inputs being $a_R(9)$, $a_I(9)$, $a_R(14)$, $a_I(14)$, $a_R(4)$, and $a_I(4)$ for $k = 0, 1$, and 2. These input adds are represented in Figure 9-12 by the 3-point input adds block labeled 4. Further, the labels on the left of this input add block correspond to the input labels in the 3-point Winograd building block in Chapter 8.

| Algorithm Steps | Memory Map |
|---|---|
| $b_R(13) = a_R(14) + a_R(4)$ | $b_R(13) \Rightarrow M(14)$ |
| $b_R(14) = a_R(14) - a_R(4)$ | $b_R(14) \Rightarrow M(4)$ |
| $b_I(13) = a_I(14) + a_I(4)$ | $b_I(13) \Rightarrow M(29)$ |
| $b_I(14) = a_I(14) - a_I(4)$ | $b_I(14) \Rightarrow M(19)$ |
| $b_R(12) = a_R(9) + b_R(13)$ | $b_R(12) \Rightarrow M(9)$ |
| $b_I(12) = a_I(9) + b_I(13)$ | $b_I(12) \Rightarrow M(24)$ |

## Stage 2: Five-Point Input Adds

The outputs from the five sets of 3-point input adds must now be combined by using the input adds from the 5-point Winograd building block. The 5-point input adds are used three times ($15/3 = 3$), each using an input from the output of each of the 3-point input adds. The input combinations and their resulting outputs are listed below and are based on the addressing in Section 9.5.8.

The strategy for converting these equations to code is to start at the top (compute $t_R(1)$) and identify the pair of inputs to be used first (in this case $b_R(9)$ and $b_R(6)$). Then look down the list to find the second (compute $B_R(2)$) place where these two inputs are used. Pull $b_R(9)$ and $b_R(6)$ from memory, compute $t_R(1)$ and $B_R(2)$, and store the results in memory locations $M(12)$ and $M(3)$, previously occupied by $b_R(9)$ and $b_R(6)$. The next step is to look at the next computation $t_I(1)$ on the list and repeat the same set of steps. Continue this process until all the Algorithm Steps in Stage 2 have been computed and their results stored in the Memory Map addresses.

### First of Three 5-Point Winograd Building-Block Input Adds

The inputs are $b_R(0)$, $b_I(0)$, $b_R(6)$, $b_I(6)$, $b_R(12)$, $b_I(12)$, $b_R(3)$, $b_I(3)$, $b_R(9)$, and $b_I(9)$. They produce six complex outputs. There are many ways to allocate the additional memory locations, $t_R(i)$, $t_I(i)$ required to store this additional complex output data value. For this example they are located at $M(30)$ and $M(31)$. These input adds are represented in Figure 9-12 by the 5-point input adds block labeled 0. Further, the labels on the left of this input add block correspond to the input labels in the 5-point Winograd building block in Chapter 8.

| Algorithm Steps | Memory Map |
|:---:|:---:|
| $t_R(1) = b_R(9) + b_R(6)$ | $t_R(1) \Rightarrow M(12)$ |
| $t_I(1) = b_I(9) + b_I(6)$ | $t_I(1) \Rightarrow M(27)$ |
| $B_R(2) = b_R(9) - b_R(6)$ | $B_R(2) \Rightarrow M(3)$ |
| $B_I(2) = b_I(9) - b_I(6)$ | $B_I(2) \Rightarrow M(18)$ |
| $t_R(3) = b_R(12) + b_R(3)$ | $t_R(3) \Rightarrow M(6)$ |
| $t_I(3) = b_I(12) + b_I(3)$ | $t_I(3) \Rightarrow M(21)$ |
| $B_R(4) = b_R(12) - b_R(3)$ | $B_R(4) \Rightarrow M(9)$ |
| $B_I(4) = b_I(12) - b_I(3)$ | $B_I(4) \Rightarrow M(24)$ |
| $c_R(1) = t_R(1) + t_R(3)$ | $c_R(1) \Rightarrow M(12)$ |
| $c_I(1) = t_I(1) + t_I(3)$ | $c_I(1) \Rightarrow M(27)$ |
| $c_R(3) = t_R(1) - t_R(3)$ | $c_R(3) \Rightarrow M(16)$ |
| $c_I(3) = t_I(1) - t_I(3)$ | $c_I(3) \Rightarrow M(21)$ |
| $c_R(5) = B_R(2) + B_R(4)$ | $c_R(5) \Rightarrow M(30)$ |
| $c_I(5) = B_I(2) + B_I(4)$ | $c_I(5) \Rightarrow M(31)$ |
| $d_R(0) = c_R(1) + b_R(0)$ | $d_R(0) \Rightarrow M(0)$ |
| $d_I(0) = c_I(1) + b_I(0)$ | $d_I(0) \Rightarrow M(15)$ |

### Second of Three 5-Point Winograd Building-Block Input Adds

The inputs are $b_R(10)$, $b_I(10)$, $b_R(1)$, $b_I(1)$, $b_R(7)$, $b_I(7)$, $b_R(13)$, $b_I(13)$, $b_R(4)$, and $b_I(4)$. They produce six complex outputs. There are many ways to allocate the additional memory locations required to store this additional complex output data value. For this example, they are located at $M(34)$ and $M(35)$. These input adds are represented in Figure 9-12 by the 5-point input adds block labeled 1. Further, the labels on the left of this input add block correspond to the input labels in the 5-point Winograd building block in Chapter 8.

| Algorithm Steps | Memory Map |
|---|---|
| $t_R(6) = b_R(10) + b_R(7)$ | $t_R(6) \Rightarrow M(8)$ |
| $t_I(6) = b_I(10) + b_I(7)$ | $t_I(6) \Rightarrow M(23)$ |
| $B_R(7) = b_R(10) - b_R(7)$ | $B_R(7) \Rightarrow M(2)$ |
| $B_I(7) = b_I(10) - b_I(7)$ | $B_I(7) \Rightarrow M(17)$ |
| $t_R(8) = b_R(13) + b_R(4)$ | $t_R(8) \Rightarrow M(14)$ |
| $t_I(8) = b_I(13) + b_I(4)$ | $t_I(8) \Rightarrow M(29)$ |
| $B_R(9) = b_R(13) - b_R(4)$ | $B_R(9) \Rightarrow M(11)$ |
| $B_I(9) = b_I(13) - b_I(4)$ | $B_I(9) \Rightarrow M(26)$ |
| $c_R(6) = t_R(6) + t_R(8)$ | $c_R(6) \Rightarrow M(8)$ |
| $c_I(6) = t_I(6) + t_I(8)$ | $c_I(6) \Rightarrow M(23)$ |
| $c_R(8) = t_R(6) - t_R(8)$ | $c_R(8) \Rightarrow M(14)$ |
| $c_I(8) = t_I(6) - t_I(8)$ | $c_I(8) \Rightarrow M(29)$ |
| $c_R(10) = B_R(7) + B_R(9)$ | $c_R(10) \Rightarrow M(34)$ |
| $c_I(10) = B_I(7) + B_I(9)$ | $c_I(10) \Rightarrow M(35)$ |
| $d_R(5) = c_R(6) + b_R(1)$ | $d_R(5) \Rightarrow M(5)$ |
| $d_I(5) = c_I(6) + b_I(1)$ | $d_I(5) \Rightarrow M(20)$ |

### Third of Three 5-Point Winograd Building-Block Input Adds

The inputs are $b_R(5)$, $b_I(5)$, $b_R(11)$, $b_I(11)$, $b_R(2)$, $b_I(2)$, $b_R(8)$, $b_I(8)$, $b_R(14)$, and $b_I(14)$. They produce six complex outputs. There are many ways to allocate the additional memory locations required to store this additional complex output data value. For this example, they are located at $M(32)$ and $M(33)$. These input adds are represented in Figure 9-12 by the 5-point input adds block labeled 2. Further, the labels on the left of this input add block correspond to the input labels in the 5-point Winograd building block in Chapter 8.

| Algorithm Steps | Memory Map |
|---|---|
| $t_R(11) = b_R(11) + b_R(8)$ | $t_R(11) \Rightarrow M(7)$ |
| $t_I(11) = b_I(11) + b_I(8)$ | $t_I(11) \Rightarrow M(22)$ |
| $B_R(12) = b_R(11) - b_R(8)$ | $B_R(12) \Rightarrow M(13)$ |
| $B_I(12) = b_I(11) - b_I(8)$ | $B_I(12) \Rightarrow M(28)$ |
| $t_R(13) = b_R(14) + b_R(5)$ | $t_R(13) \Rightarrow M(1)$ |
| $t_I(13) = b_I(14) + b_I(5)$ | $t_I(13) \Rightarrow M(16)$ |
| $B_R(14) = b_R(14) - b_R(5)$ | $B_R(14) \Rightarrow M(4)$ |
| $B_I(14) = b_I(14) - b_I(5)$ | $B_I(14) \Rightarrow M(19)$ |
| $c_R(11) = t_R(11) + t_R(13)$ | $c_R(11) \Rightarrow M(7)$ |
| $c_I(11) = t_I(11) + t_I(13)$ | $c_I(11) \Rightarrow M(22)$ |
| $c_R(13) = t_R(11) - t_R(13)$ | $c_R(13) \Rightarrow M(1)$ |
| $c_I(13) = t_I(11) - t_I(13)$ | $c_I(13) \Rightarrow M(16)$ |
| $c_R(15) = B_R(12) + B_R(14)$ | $c_R(15) \Rightarrow M(32)$ |
| $c_I(15) = B_I(12) + B_I(14)$ | $c_I(15) \Rightarrow M(33)$ |
| $d_R(10) = c_R(11) + b_R(2)$ | $d_R(10) \Rightarrow M(10)$ |
| $d_I(10) = c_I(11) + b_I(2)$ | $d_I(10) \Rightarrow M(25)$ |

## Stage 3: Nested Multiplications

This stage performs all of the multiplications in the 15-point transform. It is composed of the product of multiplications from the 3- and 5-point building blocks as described in Section 9.5.8. The output from the first of the 5-point input add building blocks uses the normal 5-point transform multiplication constants. The outputs of the second of the 5-point building blocks also use these multiplication constants. However, these constants are multiplied by the 3-point building-block constant of $\cos(2 * \pi/3) - 1$. Likewise, the output of the third of the 5-point building blocks also uses the 5-point multiplication constants, multiplied by the 3-point building-block constant of $\sin(2 * \pi/3)$.

Since all of these computations are simple multiplications, the data addressing for this stage is to pull each of the data values from memory, perform the required multiplication, and return the results to the memory location occupied by the input data for the multiplication. The first set of multiplies requires 5 constants. Each of the other two sets of multiplications requires 6 constants for a total of 17 constants that are assumed to be stored in memory and 17 total multiplications.

*Multiplications for the Outputs of the First Set of 5-Point Building-Block Input Adds*

These multiplications are represented in Figure 9-12 by the top multiply block.

| **Algorithm Steps** | **Memory Map** |
|---|---|
| $M_R(0) = d_R(0) * 1$ | $M_R(0) \Rightarrow M(0)$ |
| $M_I(0) = d_I(0) * 1$ | $M_I(0) \Rightarrow M(15)$ |
| $M_R(1) = c_R(1) * [0.5 * \cos(2\pi/5) + 0.5 * \cos(4\pi/5) - 1]$ | $M_R(1) \Rightarrow M(12)$ |
| $M_I(1) = c_I(1) * [0.5 * \cos(2\pi/5) + 0.5 * \cos(4\pi/5) - 1]$ | $M_I(1) \Rightarrow M(27)$ |
| $M_R(3) = c_R(3) * [0.5 * \cos(2\pi/5) - 0.5 * \cos(4\pi/5)]$ | $M_R(3) \Rightarrow M(6)$ |
| $M_I(3) = c_I(3) * [0.5 * \cos(2\pi/5) - 0.5 * \cos(4\pi/5)]$ | $M_I(3) \Rightarrow M(21)$ |
| $M_R(15) = c_R(5) * \sin(4\pi/5)$ | $M_R(15) \Rightarrow M(30)$ |
| $M_I(15) = c_I(5) * \sin(4\pi/5)$ | $M_I(15) \Rightarrow M(31)$ |
| $M_R(2) = B_I(2) * [\sin(2\pi/5) + \sin(4\pi/5)]$ | $M_R(2) \Rightarrow M(18)$ |
| $M_I(2) = -B_R(2) * [\sin(2\pi/5) + \sin(4\pi/5)]$ | $M_I(2) \Rightarrow M(3)$ |
| $M_R(4) = -B_I(4) * [\sin(2\pi/5) - \sin(4\pi/5)]$ | $M_R(4) \Rightarrow M(24)$ |
| $M_I(4) = B_R(4) * [\sin(2\pi/5) - \sin(4\pi/5)]$ | $M_I(4) \Rightarrow M(9)$ |

*Multiplications for the Outputs of the Second Set of 5-Point Building-Block Input Adds*

These multiplications are represented in Figure 9-12 by the center multiply block.

| **Algorithm Steps** | **Memory Map** |
|---|---|
| $M_R(5) = d_R(5) * [\cos(2\pi/3) - 1]$ | $M_R(5) \Rightarrow M(5)$ |
| $M_I(5) = d_I(5) * [\cos(2\pi/3) - 1]$ | $M_I(5) \Rightarrow M(20)$ |
| $M_R(6) = c_R(6) * [0.5 * \cos(2\pi/5) + 0.5 * \cos(4\pi/5) - 1] * [\cos(2\pi/3) - 1]$ | $M_R(6) \Rightarrow M(8)$ |
| $M_I(6) = c_I(6) * [0.5 * \cos(2\pi/5) + 0.5 * \cos(4\pi/5) - 1] * [\cos(2\pi/3) - 1]$ | $M_I(6) \Rightarrow M(23)$ |
| $M_R(8) = c_R(8) * [0.5 * \cos(2\pi/5) - 0.5 * \cos(4\pi/5)] * [\cos(2\pi/3) - 1]$ | $M_R(8) \Rightarrow M(14)$ |
| $M_I(8) = c_I(8) * [0.5 * \cos(2\pi/5) - 0.5 * \cos(4\pi/5)] * [\cos(2\pi/3) - 1]$ | $M_I(8) \Rightarrow M(29)$ |
| $M_R(16) = c_R(10) * \sin(4\pi/5) * [\cos(2\pi/3) - 1]$ | $M_R(16) \Rightarrow M(34)$ |
| $M_I(16) = c_I(10) * \sin(4\pi/5) * [\cos(2\pi/3) - 1]$ | $M_I(16) \Rightarrow M(35)$ |

| **Algorithm Steps** | **Memory Map** |
|---|---|

$$M_R(7) = B_I(7) * [\sin(2\pi/5) + \sin(4\pi/5)] * [\cos(2\pi/3) - 1] \qquad M_R(7) \Rightarrow M(17)$$

$$M_I(7) = -B_R(7) * [\sin(2\pi/5) + \sin(4\pi/5)] * [\cos(2\pi/3) - 1] \qquad M_I(7) \Rightarrow M(2)$$

$$M_R(9) = -B_I(9) * [\sin(2\pi/5) - \sin(4\pi/5)] * [\cos(2\pi/3) - 1] \qquad M_R(9) \Rightarrow M(26)$$

$$M_I(9) = B_R(9) * [\sin(2\pi/5) - \sin(4\pi/5)] * [\cos(2\pi/3) - 1] \qquad M_I(9) \Rightarrow M(11)$$

*Multiplications for the Outputs of the Third Set of 5-Point Building-Block Input Adds*

These multiplications are represented in Figure 9-12 by the bottom multiply block.

| **Algorithm Steps** | **Memory Map** |
|---|---|

$$M_R(10) = -d_I(10) * \sin(2\pi/3) \qquad\qquad M_R(10) \Rightarrow M(25)$$

$$M_I(10) = -d_R(10) * \sin(2\pi/3) \qquad\qquad M_I(10) \Rightarrow M(10)$$

$$M_R(11) = -c_I(11) * [0.5 * \cos(2\pi/5) + 0.5 * \cos(4\pi/5) - 1] * \sin(2\pi/3) \qquad M_R(11) \Rightarrow M(22)$$

$$M_I(11) = -c_R(11) * [0.5 * \cos(2\pi/5) + 0.5 * \cos(4\pi/5) - 1] * \sin(2\pi/3) \qquad M_I(11) \Rightarrow M(7)$$

$$M_R(13) = -c_I(13) * [0.5 * \cos(2\pi/5) - 0.5 * \cos(4\pi/5)] * \sin(2\pi/3) \qquad M_R(13) \Rightarrow M(16)$$

$$M_I(13) = -c_R(13) * [0.5 * \cos(2\pi/5) - 0.5 * \cos(4\pi/5)] * \sin(2\pi/3) \qquad M_I(13) \Rightarrow M(1)$$

$$M_R(17) = c_I(15) * \sin(4\pi/5) * \sin(2\pi/3) \qquad\qquad M_R(17) \Rightarrow M(33)$$

$$M_I(17) = c_R(15) * \sin(4\pi/5) * \sin(2\pi/3) \qquad\qquad M_I(17) \Rightarrow M(32)$$

$$M_R(12) = B_R(12) * [\sin(2\pi/5) + \sin(4\pi/5)] * \sin(2\pi/3) \qquad M_R(12) \Rightarrow M(13)$$

$$M_I(12) = -B_I(12) * [\sin(2\pi/5) + \sin(4\pi/5)] * \sin(2\pi/3) \qquad M_I(12) \Rightarrow M(28)$$

$$M_R(14) = -B_R(14) * [\sin(2\pi/5) - \sin(4\pi/5)] * \sin(2\pi/3) \qquad M_R(14) \Rightarrow M(4)$$

$$M_I(14) = B_I(14) * [\sin(2\pi/5) - \sin(4\pi/5)] * \sin(2\pi/3) \qquad M_I(14) \Rightarrow M19)$$

## Stage 4: Output 5-Point Adds

This stage takes the outputs of each of the groups of multiplies in Stage 3 and performs adds and subtracts using the 5-point building block's output adds. The result is five complex outputs for each of the three sets of 5-point output adds. The inputs to each of these sets of computations is the outputs from the multiplications in Stage 3. Six complex input data values yields five complex output data values for each set of computations.

The strategy for converting these equations to code is to start at the top (compute $e_R(1)$) and identify the pair of inputs to be used first (in this case $M_R(1)$ and $M_R(0)$). Then look down the list to find the second place where these two inputs are used. In this case, $M_R(1)$ is not used again and $M_R(0)$ is only relabeled to become one of this stage's outputs. Therefore, pull $M_R(1)$ and $M_R(0)$ from memory, compute $e_R(1)$, relabel $M_R(0)$ as $N_R(0)$, and store the results in memory locations $M(12)$ and $M(0)$, previously occupied by $M_R(1)$ and $M_R(0)$. The next step is to look at the next computation $e_I(1)$ on the list and repeat the same set of steps. Continue this process until all the Algorithm Steps in Stage 4 have been computed and their results stored in the Memory Map addresses.

### First of Three Sets of 5-Point Building-Block Output Adds

These output adds are represented in Figure 9-12 by the 5-point output adds block labeled 0. Further, the labels on the right of this output adds block correspond to the output labels in the 5-point Winograd building block in Chapter 8.

| **Algorithm Steps** | **Memory Map** |
|---|---|
| $e_R(1) = M_R(1) + M_R(0)$ | $e_R(1) \Rightarrow M(12)$ |
| $e_I(1) = M_I(1) + M_I(0)$ | $e_I(1) \Rightarrow M(27)$ |
| $f_R(1) = e_R(1) + M_R(3)$ | $f_R(1) \Rightarrow M(6)$ |
| $f_I(1) = e_I(1) + M_I(3)$ | $f_I(1) \Rightarrow M(21)$ |
| $f_R(2) = M_R(2) - M_I(15)$ | $f_R(2) \Rightarrow M(18)$ |
| $f_I(2) = M_I(2) + M_R(15)$ | $f_I(2) \Rightarrow M(3)$ |
| $f_R(3) = e_R(1) - M_R(3)$ | $f_R(3) \Rightarrow M(12)$ |
| $f_I(3) = e_I(1) - M_I(3)$ | $f_I(3) \Rightarrow M(27)$ |
| $f_R(4) = M_R(4) - M_I(15)$ | $f_R(4) \Rightarrow M(24)$ |
| $f_I(4) = M_I(4) + M_R(15)$ | $f_I(4) \Rightarrow M(9)$ |
| $N_R(0) = M_R(0)$ | $N_R(0) \Rightarrow M(0)$ |
| $N_I(0) = M_I(0)$ | $N_I(0) \Rightarrow M(15)$ |
| $N_R(1) = f_R(1) + f_R(2)$ | $N_R(1) \Rightarrow M(6)$ |
| $N_I(1) = f_I(1) + f_I(2)$ | $N_I(1) \Rightarrow M(21)$ |
| $N_R(4) = f_R(1) - f_R(2)$ | $N_R(4) \Rightarrow M(18)$ |
| $N_I(4) = f_I(1) - f_I(2)$ | $N_I(4) \Rightarrow M(3)$ |
| $N_R(3) = f_R(3) + f_R(4)$ | $N_R(3) \Rightarrow M(12)$ |
| $N_I(3) = f_I(3) + f_I(4)$ | $N_I(3) \Rightarrow M(27)$ |
| $N_R(2) = f_R(3) - f_R(4)$ | $N_R(2) \Rightarrow M(24)$ |
| $N_I(2) = f_I(3) - f_I(4)$ | $N_I(2) \Rightarrow M(9)$ |

### *Second of Three Sets of 5-Point Building-Block Output Adds*

These output adds are represented in Figure 9-12 by the 5-point output adds block labeled 1. Further, the labels on the right of this output add block correspond to the output labels in the 5-point Winograd building block in Chapter 8.

| **Algorithm Steps** | **Memory Map** |
|---|---|
| $e_R(6) = M_R(6) + M_R(5)$ | $e_R(6) \Rightarrow M(8)$ |
| $e_I(6) = M_I(6) + M_I(5)$ | $e_I(6) \Rightarrow M(23)$ |
| $f_R(6) = e_R(6) + M_R(8)$ | $f_R(6) \Rightarrow M(14)$ |
| $f_I(6) = e_I(6) + M_I(8)$ | $f_I(6) \Rightarrow M(29)$ |
| $f_R(7) = M_R(7) - M_I(16)$ | $f_R(7) \Rightarrow M(17)$ |
| $f_I(7) = M_I(7) + M_R(16)$ | $f_I(7) \Rightarrow M(2)$ |
| $f_R(8) = e_R(6) - M_R(8)$ | $f_R(8) \Rightarrow M(8)$ |
| $f_I(8) = e_I(6) - M_I(8)$ | $f_I(8) \Rightarrow M(23)$ |
| $f_R(9) = M_R(9) - M_I(16)$ | $f_R(9) \Rightarrow M(26)$ |
| $f_I(9) = M_I(9) + M_R(16)$ | $f_I(9) \Rightarrow M(11)$ |
| $N_R(5) = M_R(5)$ | $N_R(5) \Rightarrow M(5)$ |
| $N_I(5) = M_I(5)$ | $N_I(5) \Rightarrow M(20)$ |

| Algorithm Steps | Memory Map |
|---|---|
| $N_R(6) = f_R(6) + f_R(7)$ | $N_R(6) \Rightarrow M(14)$ |
| $N_I(6) = f_I(6) + f_I(7)$ | $N_I(6) \Rightarrow M(29)$ |
| $N_R(9) = f_R(6) - f_R(7)$ | $N_R(9) \Rightarrow M(17)$ |
| $N_I(9) = f_I(6) - f_I(7)$ | $N_I(9) \Rightarrow M(2)$ |
| $N_R(8) = f_R(8) + f_R(9)$ | $N_R(8) \Rightarrow M(8)$ |
| $N_I(8) = f_I(8) + f_I(9)$ | $N_I(8) \Rightarrow M(23)$ |
| $N_R(7) = f_R(8) - f_R(9)$ | $N_R(7) \Rightarrow M(26)$ |
| $N_I(7) = f_I(8) - f_I(9)$ | $N_I(7) \Rightarrow M(11)$ |

### *Third of Three Sets of 5-Point Building-Block Output Adds*

These output adds are represented in Figure 9-12 by the 5-point output adds block labeled 2. Further, the labels on the right of this output add block correspond to the output labels in the 5-point Winograd building block in Chapter 8.

| Algorithm Steps | Memory Map |
|---|---|
| $e_R(11) = M_R(11) + M_R(10)$ | $e_R(11) \Rightarrow M(22)$ |
| $e_I(11) = M_I(11) + M_I(10)$ | $e_I(11) \Rightarrow M(7)$ |
| $f_R(11) = e_R(11) + M_R(13)$ | $f_R(11) \Rightarrow M(16)$ |
| $f_I(11) = e_I(11) + M_I(13)$ | $f_I(11) \Rightarrow M(1)$ |
| $f_R(12) = M_R(12) - M_I(17)$ | $f_R(12) \Rightarrow M(13)$ |
| $f_I(12) = M_I(12) + M_R(17)$ | $f_I(12) \Rightarrow M(28)$ |
| $f_R(13) = e_R(11) - M_R(13)$ | $f_R(13) \Rightarrow M(22)$ |
| $f_I(13) = e_I(11) - M_I(13)$ | $f_I(13) \Rightarrow M(7)$ |
| $f_R(14) = M_R(14) - M_I(17)$ | $f_R(14) \Rightarrow M(4)$ |
| $f_I(14) = M_I(14) + M_R(17)$ | $f_I(14) \Rightarrow M(19)$ |
| $N_R(10) = M_R(10)$ | $N_R(10) \Rightarrow M(25)$ |
| $N_I(10) = M_I(10)$ | $N_I(10) \Rightarrow M(10)$ |
| $N_R(11) = f_R(11) + f_R(12)$ | $N_R(11) \Rightarrow M(16)$ |
| $N_I(11) = f_I(11) + f_I(12)$ | $N_I(11) \Rightarrow M(1)$ |
| $N_R(14) = f_R(11) - f_R(12)$ | $N_R(14) \Rightarrow M(13)$ |
| $N_I(14) = f_I(11) - f_I(12)$ | $N_I(14) \Rightarrow M(28)$ |
| $N_R(13) = f_R(13) + f_R(14)$ | $N_R(13) \Rightarrow M(22)$ |
| $N_I(13) = f_I(13) + f_I(14)$ | $N_I(13) \Rightarrow M(7)$ |
| $N_R(12) = f_R(13) - f_R(14)$ | $N_R(12) \Rightarrow M(4)$ |
| $N_I(12) = f_I(13) - f_I(14)$ | $N_I(12) \Rightarrow M(19)$ |

### Stage 5: Three-Point Building-Block Output Adds

This is the final stage in the 15-point Winograd transform example. This stage performs five sets of 3-point building-block output adds, each using an input from each of the three 5-point output add computations in the previous stage. The $m$-th output of the $k$-th 3-point output add building block is $A_R((3*k+5*m) \bmod 15)$ and $A_I((3*k+5*m) \bmod 15)$.

The strategy for converting these equations into code is to start at the top (compute $d_R(1)$) and identify the pair of inputs to be used first (in this case $N_R(0)$ and $N_R(5)$). Then look down the list to find the second place where these two inputs are used. In this case, $N_R(5)$ is not used again and $N_R(0)$ is relabeled to become $A_R(0)$, one of the outputs. Therefore, pull $N_R(0)$ and $N_R(5)$ from memory, compute $d_R(1)$, relabel $N_R(0)$ as $A_R(0)$, and store the results in memory locations $M(5)$ and $M(0)$, previously occupied by $N_R(5)$ and $N_R(0)$. The next step is to look at the next computation $d_I(1)$ on the list and repeat the same set of steps. Continue this process until all the Algorithm Steps in Stage 5 have been computed and their results stored in the Memory Map addresses.

### First of Five 3-Point Building-Block Output Adds

These output adds are represented in Figure 9-12 by the 3-point output adds block labeled 0. Further, the labels on the right of this output add block correspond to the output labels in the 3-point Winograd building block in Chapter 8, for $k = 0$.

| Algorithm Steps | Memory Map |
|---|---|
| $d_R(1) = N_R(0) + N_R(5)$ | $d_R(1) \Rightarrow M(5)$ |
| $d_I(1) = N_I(0) + N_I(5)$ | $d_I(1) \Rightarrow M(20)$ |
| $A_R(0) = N_R(0)$ | $A_R(0) \Rightarrow M(0)$ |
| $A_I(0) = N_I(0)$ | $A_I(0) \Rightarrow M(15)$ |
| $A_R(5) = d_R(1) + N_R(10)$ | $A_R(5) \Rightarrow M(5)$ |
| $A_I(5) = d_I(1) - N_I(10)$ | $A_I(5) \Rightarrow M(20)$ |
| $A_R(10) = d_R(1) - N_R(10)$ | $A_R(10) \Rightarrow M(25)$ |
| $A_I(10) = d_I(1) + N_I(10)$ | $A_I(10) \Rightarrow M(10)$ |

### Second of Five 3-Point Building-Block Output Adds

These output adds are represented in Figure 9-12 by the 3-point output adds block labeled 1. Further, the labels on the right of this output add block correspond to the output labels in the 3-point Winograd building block in Chapter 8, for $k = 2$.

| Algorithm Steps | Memory Map |
|---|---|
| $d_R(4) = N_R(1) + N_R(6)$ | $d_R(4) \Rightarrow M(14)$ |
| $d_I(4) = N_I(1) + N_I(6)$ | $d_I(4) \Rightarrow M(29)$ |
| $A_R(6) = N_R(1)$ | $A_R(6) \Rightarrow M(6)$ |
| $A_I(6) = N_I(1)$ | $A_I(6) \Rightarrow M(21)$ |
| $A_R(11) = d_R(4) + N_R(11)$ | $A_R(11) \Rightarrow M(14)$ |
| $A_I(11) = d_I(4) - N_I(11)$ | $A_I(11) \Rightarrow M(29)$ |
| $A_R(1) = d_R(4) - N_R(11)$ | $A_R(1) \Rightarrow M(16)$ |
| $A_I(1) = d_I(4) + N_I(11)$ | $A_I(1) \Rightarrow M(1)$ |

### Third of Five 3-Point Building-Block Output Adds

These output adds are represented in Figure 9-12 by the 3-point output adds block labeled 2. Further, the labels on the right of this output add block correspond to the output labels in the 3-point Winograd building block in Chapter 8, for $k = 4$.

| Algorithm Steps | Memory Map |
|---|---|
| $d_R(7) = N_R(2) + N_R(7)$ | $d_R(7) \Rightarrow M(26)$ |
| $d_I(7) = N_I(2) + N_I(7)$ | $d_I(7) \Rightarrow M(11)$ |
| $A_R(12) = N_R(2)$ | $A_R(12) \Rightarrow M(24)$ |
| $A_I(12) = N_I(2)$ | $A_I(12) \Rightarrow M(9)$ |
| $A_R(2) = d_R(7) + N_R(12)$ | $A_R(2) \Rightarrow M(26)$ |
| $A_I(2) = d_I(7) - N_I(12)$ | $A_I(2) \Rightarrow M(11)$ |
| $A_R(7) = d_R(7) - N_R(12)$ | $A_R(7) \Rightarrow M(4)$ |
| $A_I(7) = d_I(7) + N_I(12)$ | $A_I(7) \Rightarrow M(19)$ |

*Fourth of Five 3-Point Building-Block Output Adds*

These output adds are represented in Figure 9-12 by the 3-point output adds block labeled 3. Further, the labels on the right of this output add block correspond to the output labels in the 3-point Winograd building block in Chapter 8, for $k = 1$.

| Algorithm Steps | Memory Map |
|---|---|
| $d_R(10) = N_R(3) + N_R(8)$ | $d_R(10) \Rightarrow M(8)$ |
| $d_I(10) = N_I(3) + N_I(8)$ | $d_I(10) \Rightarrow M(23)$ |
| $A_R(3) = N_R(3)$ | $A_R(3) \Rightarrow M(12)$ |
| $A_I(3) = N_I(3)$ | $A_I(3) \Rightarrow M(27)$ |
| $A_R(8) = d_R(10) + N_R(13)$ | $A_R(8) \Rightarrow M(8)$ |
| $A_I(8) = d_I(10) - N_I(13)$ | $A_I(8) \Rightarrow M(23)$ |
| $A_R(13) = d_R(10) - N_R(13)$ | $A_R(13) \Rightarrow M(22)$ |
| $A_I(13) = d_I(10) + N_I(13)$ | $A_I(13) \Rightarrow M(7)$ |

*Fifth of Five 3-Point Building-Block Output Adds*

These output adds are represented in Figure 9-12 by the 3-point output adds block labeled 4. Further, the labels on the right of this output add block correspond to the output labels in the 3-point Winograd building block in Chapter 8, for $k = 3$.

| Algorithm Steps | Memory Map |
|---|---|
| $d_R(13) = N_R(4) + N_R(9)$ | $d_R(13) \Rightarrow M(17)$ |
| $d_I(13) = N_I(4) + N_I(9)$ | $d_I(13) \Rightarrow M(2)$ |
| $A_R(9) = N_R(4)$ | $A_R(9) \Rightarrow M(18)$ |
| $A_I(9) = N_I(4)$ | $A_I(9) \Rightarrow M(3)$ |
| $A_R(14) = d_R(13) + N_R(14)$ | $A_R(14) \Rightarrow M(17)$ |
| $A_I(14) = d_I(13) - N_I(14)$ | $A_I(14) \Rightarrow M(2)$ |
| $A_R(4) = d_R(13) - N_R(14)$ | $A_R(4) \Rightarrow M(13)$ |
| $A_I(4) = d_I(13) + N_I(14)$ | $A_I(4) \Rightarrow M(28)$ |

## 9.6 PRIME FACTOR APPROACH

### 9.6.1 Prime Factor Algorithm Introduction

The prime factor [3] algorithm is a special form of the mixed-radix algorithm presented in Section 9.7. The major constraint on this algorithm is that the small-point building blocks must be relatively prime. This means that they cannot have any factors in common. For example, a 72-point transform can be implemented by using the prime factor algorithms because it can be decomposed into 8- and 9-point building blocks. While neither 8 nor 9 is a prime number, they have no factors in common and are therefore called relatively prime.

The drawback to this algorithm is that the relatively prime factors can get large and, therefore, cumbersome to implement. As an extreme example, the 256-point transform ($256 = 2^8$) cannot be factored and implemented using relatively prime factors. Transform lengths like 72 can only be implemented as $8 * 9$, not as $4 * 2 * 3 * 3$ or any of the other potential combinations of the factors of 72.

In exchange for these drawbacks, the prime number transform does not require any multiplications between the small-point transforms such as the mixed-radix algorithms in Section 9.7. These multiplications are replaced by reordering of the data, which can be performed at the beginning and end of the algorithm. This reduces the number of required computations and the corresponding quantization noise. It also reduces the number of multiplier constants required in the algorithm because the only ones to be stored are for the small-point building blocks themselves. Therefore, these algorithms are most likely to be used when quantization noise is critical, where data addressing is easier than multiplication, or where storage locations for multiplier constants are at a premium. Another important feature of this algorithm is that it can use any of the small-point building blocks from Chapter 8.

General Prime Factor Algorithm. Prime factor [3] algorithms are characterized by a sequence of small-point building blocks, from Chapter 8, without complex multipliers between. This sequence of building blocks is developed by factoring the transform length, $N$, into two numbers, $N = P * Q$, and computing the $N$-point transform based on $P$- and $Q$-point FFTs (Figure 9-13). Chapter 3 describes why that process works. If $P$ or $Q$ can be further factored, say $Q = R * S$, then the $Q$-point transform can be constructed from two building blocks ($R$- and $S$-point building blocks) with Figure 9-13 as a guide.

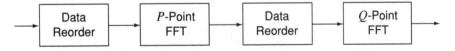

**Figure 9-13**    Top-level two-factor prime factor algorithm block diagram.

The result of factoring $N$ into $P * R * S$ is a block diagram that has a series of three building blocks without complex multipliers between them (Figure 9-14). The prime factor algorithm allows this factoring process to continue as long as the set of factors is relatively prime (i.e., they have no common factors). The extreme case is to factor $N$ until the building blocks are only primes and powers-of-primes. Even if $N$ is factored to this extreme, there are numerous orders in which those primes can be combined to form the complete transform. The order of the building blocks determines the data reordering used between the stages but does not affect the number of adds and multiplies.

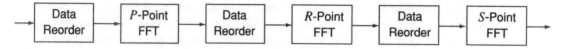

**Figure 9-14** Top-level three-factor prime factor algorithm block diagram.

**Thirty-Point Example.** There are three ways to factor 30 into two numbers ($2 * 15$, $3 * 10$, $5 * 6$). Therefore, the 30-point transform can be implemented, using the block diagram in Figure 9-13, as any one of these sequences of two building blocks. In fact, each of these choices can be implemented in two ways. The $2 * 15$ option can be implemented with either the 2- or 15-point transform first in Figure 9-13. However, in each case, one of the two factors can be factored further into two factors. The result in all three cases is three building blocks (2, 3, and 5 points). There are six ways of ordering these three numbers to implement the 30-point FFT. To summarize, there are 12 ways to implement the 30-point FFT independent of which algorithm is used for each building block. These are shown in Table 9-5. The first six sequence choices only have two building blocks, indicated by N/A in column $S$. The choice of building blocks from Chapter 8 for all but the 6-, 10-, and 15-point FFTs provides additional options to optimize the implementation for an application.

**Table 9-5** Thirty-Point Prime Factor Building-Block Sequences

| Sequence choices | $P$ | $R$ | $S$ |
|---|---|---|---|
| 1 | 2 | 15 | N/A |
| 2 | 15 | 2 | N/A |
| 3 | 5 | 6 | N/A |
| 4 | 6 | 5 | N/A |
| 5 | 3 | 10 | N/A |
| 6 | 10 | 3 | N/A |
| 7 | 2 | 3 | 5 |
| 8 | 2 | 5 | 3 |
| 9 | 3 | 2 | 5 |
| 10 | 3 | 5 | 2 |
| 11 | 5 | 2 | 3 |
| 12 | 5 | 3 | 2 |

Section 9.6.2 describes how to determine the number of adds and multiplies for the prime factor algorithm. Section 9.6.3 describes the general prime factor algorithm for two factors. Then the next two sections give two prime factor algorithms, Kolba-Parks and SWIFT, using 15-point transforms, so that their features can be most easily compared. The primary difference between the two algorithms is the strategy for organizing the data and then reorganizing it between the building blocks. The number of adds and multiplies, data

memory locations, and locations for multiplier constants is the same for both prime factor algorithms.

### 9.6.2  Number of Prime Factor Algorithm Adds and Multiplies

The number of real adds and multiplies is the sum of those required for the algorithm building blocks. Since there are $(N/P_i)P_i$-point transforms, the number of adds and multiplies contributed by these building blocks is just $(N/P_i)$ times the number of real adds and multiplies required by these algorithm building blocks. These numbers are listed in the Comparison Matrix in Chapter 8. If $N$ is factored into $n$ relatively prime factors, $P_i$, then:

$$\text{\# adds} = \sum_{i=1}^{n}(N/P_i) * A_i$$

$$\text{\# multiplies} = \sum_{i=1}^{n}(N/P_i) * M_i$$

(9-1)

where:   $A_i$ = number of real adds in $P_i$-point algorithm building block
         $M_i$ = number of real multiplies in $P_i$-point algorithm building block

### 9.6.3  General Prime Factor Algorithm for Two Factors

Since the prime factor algorithm is constructed by repeatedly factoring an integer into two other integers, it is completely described by the equations required to factor $N$ into two factors as depicted in Figure 9-13. To construct a prime factor algorithm for three factors $(P, R, S$, where $Q = R * S)$, first follow the two-step decomposition. Then for each of the $P$ $Q$-point transforms relabel its inputs as if they were $Q$ consecutive complex data points and reapply the two-step decomposition to split it into two factors. Each of those can be further subdivided by using the same approach if $R$ and $S$ can be factored.

The algorithm starts by properly grouping the complex data points from the total $N$-point input sequence for input to a set of $Q$ $P$-point algorithms. Once each of these $P$-point transforms is computed, their outputs are reorganized to provide the inputs to the $P$ $Q$-point algorithms. The $Q$-point algorithms are then computed and their outputs stored as the $N$ complex output frequency components.

### Stage 1: Input $P$-Point Building Blocks

This stage has two steps. The first is to properly group the input data for each of the $Q$ $P$-point building blocks. The second is to compute each of the $Q$ $P$-point building blocks. The number of adds and multiplies required for this stage is $Q$ times the number of adds and multiplies required for the chosen $P$-point algorithm. Since the $P$-point building blocks are computed sequentially, any additional memory required for the $P$-point building block is only needed once. This is because each $P$-point algorithm uses these additional locations, in sequence, not all at once. Therefore, the total memory required for this portion of the algorithm is $2 * N$ for the data plus the additional locations needed for one $P$-point building block.

*Step 1: Grouping the Input Data Points for the $P$-Point Building Blocks*

There are two strategies for grouping the input data to the $P$-point building blocks. Both result in the same groups of input data points. However, the order in which they are used as the $P$-point building block inputs is different for nearly all transform lengths. The equations for both input orderings are given. It is important to notice that the 15-point examples actually use the same ordering of the input data. This is an exception to the general rule.

For the Kolba-Parks [3] algorithm, the $k$-th input to the $n$-th $P$-point algorithm is $a_R((k*Q+P*n) \bmod N)$ and $a_I((k*Q+P*n) \bmod N)$, (where $k = 0, 1, \ldots, (P-1)$ and $n = 0, 1, \ldots, Q-1)$ from the input data sequence. Therefore, the zero-th ($k = 0$) input to the $n$-th $P$-point building block is $a_R(P*n)$ and $a_I(P*n)$, where $n = 0, 1, \ldots, (Q-1)$. Additionally, the subsequent inputs to the same $P$-point transform are separated by $Q$ samples because $k$ is incremented to determine the sample. Figure 9-15 shows the inputs for the second ($n = 1$) $P$-point building block.

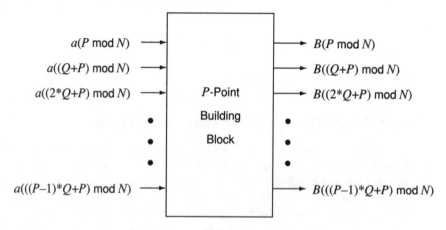

**Figure 9-15**   Kolba-Parks $P$-point building-block data configuration for $n = 1$.

For the SWIFT [4] algorithm, the $k$-th input to the $n$-th $P$-point building block is $a_R((k*Q+(Q*d+1)*n) \bmod N)$ and $a_I((k*Q+(Q*d+1)*n) \bmod N)$, where $c$ and $d$ are determined as the solution to the equation:

$$1 + P*c + Q*d = 0 \qquad (9\text{-}2)$$

and define the output sequence for the SWIFT algorithm. For the 15-point SWIFT example ($P = 3$ and $Q = 5$), the solution of Equation 9-2 is $c = -2$ and $d = 1$. Figure 9-16 shows these inputs for the second ($n = 1$) $P$-point building block.

### Step 2: *Computing the Q P-Point Building Blocks*

Use the complex input data points defined in Step 1 to compute each of the $Q$ $P$-point building blocks. Again, the two prime factor algorithms have different output data labeling. The simplest approach to output labeling is to use the same modulo arithmetic scheme as on the input. Therefore, for the Kolba-Parks algorithm, the $k$-th output of the $n$-th $P$-point building block is labeled $B_R((k*Q+P*n) \bmod N)$ and $B_I((k*Q+P*n) \bmod N)$, (where $k = 0, 1, \ldots, (P-1)$ and $n = 0, 1, \ldots, (Q-1)$). Similarly, for the SWIFT algorithm, the $k$-th output of the $n$-th $P$-point building block is $B_R((k*Q+(Q*d+1)*n) \bmod N)$ and $B_I((k*Q+(Q*d+1)*n) \bmod N)$, where $d$ is defined by Equation 9-2. Figures 9-15 and 9-16 show this labeling for the Kolba-Parks and SWIFT algorithms, respectively.

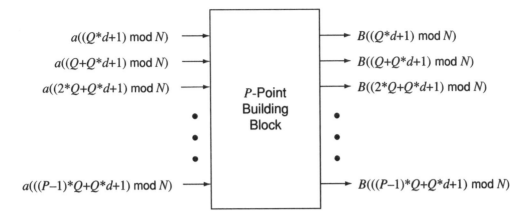

**Figure 9-16**   SWIFT $P$-point building-block data configuration for $n = 1$.

## Stage 2: Output Q-Point Building Blocks

This stage also has two steps. The first is to properly group the input data for each of the $P$ $Q$-point building blocks. The second is to compute each of the $P$ $Q$-point building blocks. The number of adds and multiplies required for this stage is $P$ times the number of adds and multiplies required for the chosen $Q$-point building block. Since the $Q$-point building blocks are performed sequentially, any additional memory required for the $Q$-point building block is only needed once. This is because each $Q$-point building block uses these additional locations, in sequence, not all at once. Therefore, the total memory required for this portion of the algorithm is $2 * N$ for the data plus the additional locations needed for one $Q$-point building block.

*Step* 1: *Grouping the Input Data Points to the Q-Point Building Blocks*

Again, the data ordering for this stage of the computations is different for the two prime factor algorithms. For the Kolba-Parks algorithm, the $n$-th input to the $k$-th $Q$- point building block is:

$$B_R((k * Q + P * n) \bmod (N)) \tag{9-3}$$

$$B_I((k * Q + P * n) \bmod (N)) \tag{9-4}$$

where $k = 0, 1, \ldots, (P-1)$ and $n = 0, 1, \ldots, (Q-1)$. Similarly, for the SWIFT algorithm, the $n$-th input to the $k$-th $Q$-point building block is $B_R((k * Q + (Q * d + 1) * n) \bmod N)$ and $B_I((k * Q + (Q * d + 1) * n) \bmod N)$, where $d$ is defined by Equation 9-2. Figures 9-17 and 9-18 show this labeling for the first $(k = 0)$ $Q$-point building block for the Kolba-Parks and SWIFT algorithms, respectively. In both algorithms, the inputs to the first $(k = 0)$ $Q$-point building block are the first outputs of each of the $P$-point building blocks. This pattern holds for each $Q$-point building block. Specifically, the inputs to the $k$-th $Q$-point building block are the $k$-th outputs of all of the $P$-point building blocks.

Each input data value to a $Q$-point building block comes from a different $P$-point building-block output. Therefore, the data memory locations where the required input data reside are not in the order assumed by the building-block $Q$-point building blocks in Chapter 8. To further complicate this, the output data memory map order for the

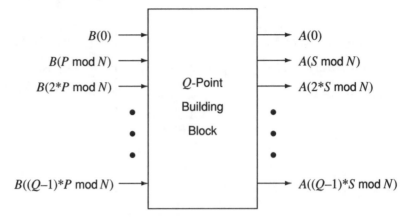

**Figure 9-17** Kolba-Parks $Q$-point building-block data configuration for $k = 0$.

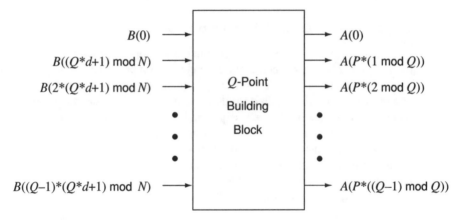

**Figure 9-18** SWIFT $Q$-point building-block data configuration for $k = 0$.

$P$-point building blocks in Chapter 8 is not in sequence. Therefore, to use the building-block algorithms from Chapter 8, the specified data memory locations must be relabeled. This process is straightforward and is completely described in Section 9.4.

*Step* 2: *Computing the P Q-Point Building Blocks*

Use the complex input data points defined in Step 1 to compute each of the $P$ $Q$-point building blocks. The output labeling is again different for the two prime factor algorithms. For the Kolba-Parks algorithm, the $n$-th output of the $k$-th $Q$-point building block should be labeled $A_R[(S*n+u*k) \bmod N]$ and $A_I[(S*n+u*k) \bmod N]$, where $S$ and $u$ are determined as solutions to the equations

$$S \equiv 1 \bmod(Q) \qquad u \equiv 1 \bmod(P)$$
$$S \equiv 0 \bmod(P) \qquad u \equiv 0 \bmod(Q)$$

For the 15-point Kolba-Parks example, $S = 6$ and $u = 10$. Figure 9-17 shows this labeling for the first $(k = 0)$ $Q$-point building block.

Similarly, for the SWIFT algorithm, the $n$-th output of the $k$-th $Q$-point building block is $A_R(P*[(n+c*k) \bmod Q]+k)$ and $A_I(P*[(n+c*k) \bmod Q]+k)$, where $c$

is defined by Equation 9-2, $n = 0, 1, \ldots, (Q-1)$, and $k = 0, 1, \ldots, (P-1)$. Figure 9-18 shows this labeling for the first ($k = 0$) $Q$-point building block.

### 9.6.4 Fifteen-Point Kolba-Parks FFT Example

The 15-point Kolba-Parks [3] algorithm can be implemented with either the 3-point or the 5-point building blocks first. If the 3-point transform is first, the 15 pieces of complex input data are divided into five sets of three complex points, one for each of the $15/3 = 5$ 3-point building blocks. Following the 3-point transforms, the intermediate results are reorganized into three sets of five pieces of complex data needed for input to the $15/5 = 3$ 5-point building-block computations. The order does not affect how many computations are required. This example uses the Singleton 3- and 5-point building blocks. A smaller number of adds and multiplies is required if the Winograd building blocks were used.

If the Comparison Matrix in Chapter 8 and Equation 9-1 are used, the total number of real adds required is $5 * 12 + 3 * 32 = 156$ and the total number of real multiplies is $5 * 4 + 3 * 16 = 68$. The total amount of data memory required is driven by the 5-point building block and is 32 locations. Explicitly, 30 locations are required for the 15 complex data points, plus 2 additional locations for the intermediate computations in the 5-point Singleton building block. Similarly, the 3-point Singleton building block has two multiplier constants and the 5-point Singleton building block has four for a total of six memory locations for multiplier constants. Figure 9-18 is a block diagram of this example. The stages are as follows.

### Stage 1: Three-Point Building Blocks

The 15 data points are divided into five sets of 3 points to serve as inputs to each of the 3-point building blocks. This is done by using the addressing from Section 9.6.3, starting with complex input data point pair $a_R(0)$, $a_I(0)$, and grouping it with complex input data point pairs $a_R(5)$, $a_I(5)$ and $a_R(10)$, $a_I(10)$. These provide the input to the top one of the five 3-point building blocks in Figure 9-19. This is followed by grouping the input data point pairs $a_R(3)$, $a_I(3)$, $a_R(8)$, $a_I(8)$, and $a_R(13)$, $a_I(13)$ to provide the input for the second of the five 3-point building blocks. The next grouping is data point pairs $a_R(6)$, $a_I(6)$, $a_R(11)$, $a_I(11)$, and $a_R(1)$, $a_I(1)$ for input into the third of the five 3-point building blocks. The next grouping is data point pairs $a_R(9)$, $a_I(9)$, $a_R(14)$, $a_I(14)$, and $a_R(4)$, $a_I(4)$ to provide input for the fourth of the five 3-point building blocks. The final grouping is data point pairs $a_R(12)$, $a_I(12)$, $a_R(2)$, $a_I(2)$, and $a_R(7)$, $a_I(7)$ for input into the fifth 3-point building block.

The order in which this data is used for inputs to the 3-point building blocks is the key point in removing the need for complex multipliers between the 3- and 5-point algorithms. From Section 9.6.3, the complex input data for the $k$-th input to the $m$-th 3-point building block is $a_R((5 * k + 3 * m) \bmod 15)$, $a_I((5 * k + 3 * m) \bmod 15)$, where $k = 0, 1,$ and 2, and $m = 0, 1, 2, 3,$ and 4.

The five groups of computations, listed as (a) through (e), each perform the 3-point building block. In this example, the Singleton 3-point algorithm building block from Chapter 8 is used. All of these 3-point building blocks could also have been the Winograd 3-point algorithm building block from Chapter 8. In fact, the five 3-point building blocks can be any combination of these two 3-point algorithm building blocks. The outputs of each of the

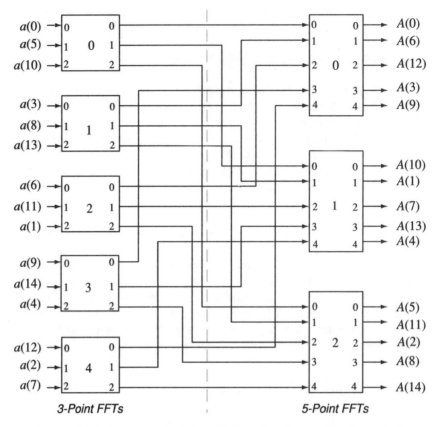

**Figure 9-19**  Fifteen-point Kolba-Parks prime factor algorithm block diagram.

3-point building blocks, labeled $B_R(i)$ and $B_I(i)$ for $i = 0, 5, 10$, are the equivalent of the $A_R(i)$ and $A_I(i)$ in the 3-point algorithm building block in Chapter 8.

The strategy for converting these equations to code is to start at the top (compute $b_R(5)$) and identify the pair of inputs to be used first (in this case $a_R(5)$ and $a_R(10)$). Then look down the list to find the second (compute $b_R(10)$) place where these two inputs are used. Pull $a_R(5)$ and $a_R(10)$ from memory, compute $b_R(5)$ and $b_R(10)$, and store the results in memory locations $M(5)$ and $M(10)$, previously occupied by $a_R(5)$ and $a_R(10)$. The next step is to look at the next computation $b_I(5)$ on the list and repeat the same set of steps. Continue this process until all the Algorithm Steps in Stage 1 have been computed and their results stored in the Memory Map addresses.

### First of Five 3-Point Algorithm Building Blocks

The inputs to this 3-point building block are $a_R((5*k + 3*m) \bmod 15)$, $a_I((5*k + 3*m) \bmod 15)$ where $m = 0$. Performing the modulo arithmetic computations results in the inputs being $a_R(0)$, $a_I(0)$, $a_R(5)$, $a_I(5)$, $a_R(10)$, and $a_I(10)$ for $k = 0, 1$, and 2. This set of computations is represented in Figure 9-19 by 3-point building block 0. Further, the labels on the left and right of this building block correspond to the input and output labels in the 3-point Singleton building block in Chapter 8.

| **Algorithm Steps** | **Memory Map** |
|---|---|
| $b_R(5) = a_R(5) + a_R(10)$ | $b_R(5) \Rightarrow M(5)$ |
| $b_R(10) = a_R(5) - a_R(10)$ | $b_R(10) \Rightarrow M(10)$ |
| $b_I(5) = a_I(5) + a_I(10)$ | $b_I(5) \Rightarrow M(20)$ |
| $b_I(10) = a_I(5) - a_I(10)$ | $b_I(10) \Rightarrow M(25)$ |
| $c_R(5) = b_R(5) * \cos(2\pi/3) + a_R(0)$ | $c_R(5) \Rightarrow M(30)$ |
| $B_R(0) = a_R(0) + b_R(5)$ | $B_R(0) \Rightarrow M(0)$ |
| $c_R(10) = b_I(10) * \sin(2\pi/3)$ | $c_R(10) \Rightarrow M(25)$ |
| $c_I(5) = b_I(5) * \cos(2\pi/3) + a_I(0)$ | $c_I(5) \Rightarrow M(5)$ |
| $B_I(0) = a_I(0) + b_I(5)$ | $B_I(0) \Rightarrow M(15)$ |
| $c_I(10) = -b_R(10) * \sin(2\pi/3)$ | $c_I(10) \Rightarrow M(10)$ |
| $B_R(5) = c_R(5) + c_R(10)$ | $B_R(5) \Rightarrow M(25)$ |
| $B_I(5) = c_I(5) + c_I(10)$ | $B_I(5) \Rightarrow M(10)$ |
| $B_R(10) = c_R(5) - c_R(10)$ | $B_R(10) \Rightarrow M(20)$ |
| $B_I(10) = c_I(5) - c_I(10)$ | $B_I(10) \Rightarrow M(5)$ |

### Second of Five 3-Point Algorithm Building Blocks

The inputs to this 3-point building block are $a_R((5 * k + 3 * m) \bmod 15)$, $a_I((5 * k + 3 * m) \bmod 15)$ where $m = 1$. Performing the modulo arithmetic computations results in the inputs being $a_R(3)$, $a_I(3)$, $a_R(8)$, $a_I(8)$, $a_R(13)$, and $a_I(13)$ for $k = 0, 1$, and $2$. This set of computations is represented in Figure 9-19 by 3-point building block 1. Further, the labels on the left and right of this building block correspond to the input and output labels in the 3-point Singleton building block in Chapter 8.

| **Algorithm Steps** | **Memory Map** |
|---|---|
| $b_R(8) = a_R(8) + a_R(13)$ | $b_R(8) \Rightarrow M(8)$ |
| $b_R(13) = a_R(8) - a_R(13)$ | $b_R(13) \Rightarrow M(13)$ |
| $b_I(8) = a_I(8) + a_I(13)$ | $b_I(8) \Rightarrow M(23)$ |
| $b_I(13) = a_I(8) - a_I(13)$ | $b_I(13) \Rightarrow M(28)$ |
| $c_R(8) = b_R(8) * \cos(2\pi/3) + a_R(3)$ | $c_R(8) \Rightarrow M(30)$ |
| $B_R(3) = a_R(3) + b_R(8)$ | $B_R(3) \Rightarrow M(3)$ |
| $c_R(13) = b_I(13) * \sin(2\pi/3)$ | $c_R(13) \Rightarrow M(28)$ |
| $c_I(8) = b_I(8) * \cos(2\pi/3) + a_I(3)$ | $c_I(8) \Rightarrow M(8)$ |
| $B_I(3) = a_I(3) + b_I(8)$ | $B_I(3) \Rightarrow M(18)$ |
| $c_I(13) = -b_R(13) * \sin(2\pi/3)$ | $c_I(13) \Rightarrow M(13)$ |
| $B_R(8) = c_R(8) + c_R(13)$ | $B_R(8) \Rightarrow M(28)$ |
| $B_I(8) = c_I(8) + c_I(13)$ | $B_I(8) \Rightarrow M(13)$ |
| $B_R(13) = c_R(8) - c_R(13)$ | $B_R(13) \Rightarrow M(23)$ |
| $B_I(13) = c_I(8) - c_I(13)$ | $B_I(13) \Rightarrow M(8)$ |

### Third of Five 3-Point Algorithm Building Blocks

The inputs to this 3-point building block are $a_R((5 * k + 3 * m) \bmod 15)$, $a_I((5 * k + 3 * m) \bmod 15)$ where $m = 2$. Performing the modulo arithmetic computations results in the inputs being $a_R(6)$, $a_I(6)$, $a_R(11)$, $a_I(11)$, $a_R(1)$, and $a_I(1)$ for $k = 0, 1$, and 2. This set of computations is represented in Figure 9-19 by 3-point building block 2. Further, the labels on the left and right of this building block correspond to the input and output labels in the 3-point Singleton building block in Chapter 8.

| **Algorithm Steps** | **Memory Map** |
|---|---|
| $b_R(6) = a_R(11) + a_R(1)$ | $b_R(6) \Rightarrow M(11)$ |
| $b_R(11) = a_R(11) - a_R(1)$ | $b_R(11) \Rightarrow M(1)$ |
| $b_I(6) = a_I(11) + a_I(1)$ | $b_I(6) \Rightarrow M(26)$ |
| $b_I(11) = a_I(11) - a_I(1)$ | $b_I(11) \Rightarrow M(16)$ |
| $c_R(6) = b_R(6) * \cos(2\pi/3) + a_R(6)$ | $c_R(6) \Rightarrow M(30)$ |
| $B_R(6) = a_R(6) + b_R(6)$ | $B_R(6) \Rightarrow M(6)$ |
| $c_R(11) = b_I(11) * \sin(2\pi/3)$ | $c_R(11) \Rightarrow M(16)$ |
| $c_I(6) = b_I(6) * \cos(2\pi/3) + a_I(6)$ | $c_I(6) \Rightarrow M(11)$ |
| $B_I(6) = a_I(6) + b_I(6)$ | $B_I(6) \Rightarrow M(21)$ |
| $c_I(11) = -b_R(11) * \sin(2\pi/3)$ | $c_I(11) \Rightarrow M(1)$ |
| $B_R(11) = c_R(6) + c_R(11)$ | $B_R(11) \Rightarrow M(16)$ |
| $B_I(11) = c_I(6) + c_I(11)$ | $B_I(11) \Rightarrow M(1)$ |
| $B_R(1) = c_R(6) - c_R(11)$ | $B_R(1) \Rightarrow M(26)$ |
| $B_I(1) = c_I(6) - c_I(11)$ | $B_I(1) \Rightarrow M(11)$ |

### Fourth of Five 3-Point Algorithm Building Blocks

The inputs to this 3-point building block are $a_R((5*k + 3*m) \bmod 15)$, $a_I((5*k + 3*m) \bmod 15)$ where $m = 3$. Performing the modulo arithmetic computations results in the inputs being $a_R(9)$, $a_I(9)$, $a_R(14)$, $a_I(14)$, $a_R(4)$, and $a_I(4)$ for $k = 0, 1$, and 2. This set of computations is represented in Figure 9-19 by 3-point building block 3. Further, the labels on the left and right of this building block correspond to the input and output labels in the 3-point Singleton building block in Chapter 8.

| **Algorithm Steps** | **Memory Map** |
|---|---|
| $b_R(9) = a_R(14) + a_R(4)$ | $b_R(9) \Rightarrow M(14)$ |
| $b_R(14) = a_R(14) - a_R(4)$ | $b_R(14) \Rightarrow M(4)$ |
| $b_I(9) = a_I(14) + a_I(4)$ | $b_I(9) \Rightarrow M(29)$ |
| $b_I(14) = a_I(14) - a_I(4)$ | $b_I(14) \Rightarrow M(19)$ |
| $c_R(9) = b_R(9) * \cos(2\pi/3) + a_R(9)$ | $c_R(9) \Rightarrow M(30)$ |
| $B_R(9) = a_R(9) + b_R(9)$ | $B_R(9) \Rightarrow M(9)$ |
| $c_R(14) = b_I(14) * \sin(2\pi/3)$ | $c_R(14) \Rightarrow M(19)$ |
| $c_I(9) = b_I(9) * \cos(2\pi/3) + a_I(9)$ | $c_I(9) \Rightarrow M(14)$ |
| $B_I(9) = a_I(9) + b_I(9)$ | $B_I(9) \Rightarrow M(24)$ |
| $c_I(14) = -b_R(14) * \sin(2\pi/3)$ | $c_I(14) \Rightarrow M(4)$ |
| $B_R(14) = c_R(9) + c_R(14)$ | $B_R(14) \Rightarrow M(19)$ |
| $B_I(14) = c_I(9) + c_I(14)$ | $B_I(14) \Rightarrow M(4)$ |
| $B_R(4) = c_R(9) - c_R(14)$ | $B_R(4) \Rightarrow M(29)$ |
| $B_I(4) = c_I(9) - c_I(14)$ | $B_I(4) \Rightarrow M(14)$ |

*Fifth of Five 3-Point Algorithm Building Blocks*

The inputs to this 3-point building block are $a_R((5*k+3*m) \bmod 15)$, $a_I((5*k+3*m) \bmod 15)$ where $m = 4$. Performing the modulo arithmetic computations results in the inputs being $a_R(12)$, $a_I(12)$, $a_R(2)$, $a_I(2)$, $a_R(7)$, and $a_I(7)$ for $k = 0, 1$, and 2. This set of computations is represented in Figure 9-19 by 3-point building block 4. Further, the labels on the left and right of this building block correspond to the input and output labels in the 3-point Singleton building block in Chapter 8.

|  Algorithm Steps | Memory Map |
|:---:|:---:|
| $b_R(7) = a_R(2) + a_R(7)$ | $b_R(7) \Rightarrow M(2)$ |
| $b_R(12) = a_R(2) - a_R(7)$ | $b_R(12) \Rightarrow M(7)$ |
| $b_I(7) = a_I(2) + a_I(7)$ | $b_I(7) \Rightarrow M(17)$ |
| $b_I(12) = a_I(2) - a_I(7)$ | $b_I(12) \Rightarrow M(22)$ |
| $c_R(7) = b_R(7) * \cos(2\pi/3) + a_R(12)$ | $c_R(7) \Rightarrow M(30)$ |
| $B_R(12) = a_R(12) + b_R(7)$ | $B_R(12) \Rightarrow M(12)$ |
| $c_R(12) = b_I(12) * \sin(2\pi/3)$ | $c_R(12) \Rightarrow M(22)$ |
| $c_I(7) = b_I(7) * \cos(2\pi/3) + a_I(12)$ | $c_I(7) \Rightarrow M(2)$ |
| $B_I(12) = a_I(12) + b_I(7)$ | $B_I(12) \Rightarrow M(27)$ |
| $c_I(12) = -b_R(12) * \sin(2\pi/3)$ | $c_I(12) \Rightarrow M(7)$ |
| $B_R(2) = c_R(7) + c_R(12)$ | $B_R(2) \Rightarrow M(22)$ |
| $B_I(2) = c_I(7) + c_I(12)$ | $B_I(2) \Rightarrow M(7)$ |
| $B_R(7) = c_R(7) - c_R(12)$ | $B_R(7) \Rightarrow M(17)$ |
| $B_I(7) = c_I(7) - c_I(12)$ | $B_I(7) \Rightarrow M(2)$ |

## Stage 2: Output 5-Point Building Blocks

For this example, the Singleton 5-point building block from Chapter 8 is used. Either of the two other 5-point building blocks could have been used without changing the rest of the structure of the algorithm. If the number of adds and multiplies is the overriding criterion, then the Winograd algorithm building block should be used in-place of the 5-point Singleton building block.

The three sets of 5-point algorithm building-block steps from Chapter 8 are listed as (a) through (c). In Chapter 8 the 5-point algorithm building block was presented as three stages. Since the individual stages of the 5-point building block are discussed in Chapter 8, they are not discussed again. The $m$-th input to the $k$-th 5-point building block is $B_R((5*k+3*m) \bmod 15)$ and $B_I((5*k+3*m) \bmod 15)$ from Stage 2, based on the addressing defined in Section 9.6.3.

The multiply stage of the 5-point Singleton algorithm required additional data memory locations. If the 15-point computations are performed in the order shown, the additional memory locations used by the first of the three 5-point building blocks can be reused by each of the other two 5-point building blocks.

The strategy for converting these equations to code is to start at the top (compute $b_R(1)$) and identify the pair of inputs to be used first (in this case $B_R(3)$ and $B_R(12)$). Then look down the list to find the second (compute $b_R(2)$) place where these two inputs are used. Pull $B_R(3)$ and $B_R(12)$ from memory, compute $b_R(1)$ and $b_R(2)$, and store the results in memory locations $M(3)$ and $M(12)$, previously occupied by $B_R(3)$ and $B_R(12)$. The next step is to look at the next computation $b_I(1)$ on the list and repeat the same set of steps. Continue this process until all the Algorithm Steps in Stage 2 have been computed and their results stored in the Memory Map addresses.

### First of Three 5-Point Building Blocks

This 5-point building block $(k = 0)$ has $B_R((5*k+3*m) \bmod 15)$ and $B_I((5*k+3*m) \bmod 15)$ $(m = 0, 1, 2, 3,$ and $4)$ as inputs and $A_R((10*k+6*m) \bmod 15)$ and $A_I((10*k+6*m) \bmod 15)$ $(m = 0, 1, 2, 3,$ and $4)$ as its output frequency components. Performing the modulo arithmetic computations results in the inputs being $B_R(0)$, $B_I(0)$, $B_R(3)$, $B_I(3)$, $B_R(6)$, $B_I(6)$, $B_R(9)$, $B_I(9)$, $B_R(12)$, and $B_I(12)$.

The multiplication portion of the algorithm requires two additional data memory locations because no temporary registers are assumed. The variables used for the intermediate computations were chosen to be the same as those used for the 5-point Singleton building block in Chapter 8 to make it easier to associate the computational steps with the discussion in Chapter 8. This set of computations is represented in Figure 9-19 by 5-point building block 0. Further, the labels on the left and right of this building block correspond to the input and output labels in the 5-point Singleton building block in Chapter 8.

| Algorithm Steps | Memory Map |
|---|---|
| $b_R(1) = B_R(3) + B_R(12)$ | $b_R(1) \Rightarrow M(3)$ |
| $b_I(1) = B_I(3) + B_I(12)$ | $b_I(1) \Rightarrow M(18)$ |
| $b_R(2) = B_R(3) - B_R(12)$ | $b_R(2) \Rightarrow M(12)$ |
| $b_I(2) = B_I(3) - B_I(12)$ | $b_I(2) \Rightarrow M(27)$ |
| $b_R(3) = B_R(6) + B_R(9)$ | $b_R(3) \Rightarrow M(6)$ |
| $b_I(3) = B_I(6) + B_I(9)$ | $b_I(3) \Rightarrow M(21)$ |
| $b_R(4) = B_R(6) - B_R(9)$ | $b_R(4) \Rightarrow M(9)$ |
| $b_I(4) = B_I(6) - B_I(9)$ | $b_I(4) \Rightarrow M(24)$ |
| $c_R(2) = b_R(2) * \sin(2\pi/5) + b_R(4) * \sin(4\pi/5)$ | $c_R(2) \Rightarrow M(30)$ |
| $c_I(2) = b_I(2) * \sin(2\pi/5) + b_I(4) * \sin(4\pi/5)$ | $c_I(2) \Rightarrow M(9)$ |
| $c_R(4) = b_R(2) * \sin(4\pi/5) - b_R(4) * \sin(2\pi/5)$ | $c_R(4) \Rightarrow M(31)$ |
| $c_I(4) = b_I(2) * \sin(4\pi/5) - b_I(4) * \sin(2\pi/5)$ | $c_I(4) \Rightarrow M(12)$ |
| $c_R(1) = b_R(1) * \cos(2\pi/5) + b_R(3) * \cos(4\pi/5) + B_R(0)$ | $c_R(1) \Rightarrow M(27)$ |
| $c_I(1) = b_I(1) * \cos(2\pi/5) + b_I(3) * \cos(4\pi/5) + B_I(0)$ | $c_I(1) \Rightarrow M(3)$ |
| $c_R(3) = b_R(1) * \cos(4\pi/5) + b_R(3) * \cos(2\pi/5) + B_R(0)$ | $c_R(3) \Rightarrow M(24)$ |
| $c_I(3) = b_I(1) * \cos(4\pi/5) + b_I(3) * \cos(2\pi/5) + B_I(0)$ | $c_I(3) \Rightarrow M(6)$ |
| $A_R(0) = B_R(0) + b_R(1) + b_R(3)$ | $A_R(0) \Rightarrow M(0)$ |
| $A_I(0) = B_I(0) + b_I(1) + b_I(3)$ | $A_I(0) \Rightarrow M(15)$ |
| $A_R(6) = c_R(1) + c_I(2)$ | $A_R(6) \Rightarrow M(27)$ |
| $A_I(6) = c_I(1) - c_R(2)$ | $A_I(6) \Rightarrow M(18)$ |
| $A_R(12) = c_R(3) + c_I(4)$ | $A_R(12) \Rightarrow M(24)$ |
| $A_I(12) = c_I(3) - c_R(4)$ | $A_I(12) \Rightarrow M(6)$ |
| $A_R(3) = c_R(3) - c_I(4)$ | $A_R(3) \Rightarrow M(12)$ |
| $A_I(3) = c_I(3) + c_R(4)$ | $A_I(3) \Rightarrow M(3)$ |
| $A_R(9) = c_R(1) - c_I(2)$ | $A_R(9) \Rightarrow M(9)$ |
| $A_I(9) = c_I(1) + c_R(2)$ | $A_I(9) \Rightarrow M(21)$ |

## Second of Three 5-Point Building Blocks

This 5-point building block ($k = 1$) has $B_R((5*k+3*m) \bmod 15)$ and $B_I((5*k+3*m) \bmod 15)(m = 0, 1, 2, 3,$ and $4)$ as inputs and $A_R((10*k+6*m) \bmod 15)$ and $A_I((10*k+6*m) \bmod 15)(m = 0, 1, 2, 3,$ and $4)$ as its output frequency components. Performing the modulo arithmetic computations results in the inputs being $B_R(5)$, $B_I(5)$, $B_R(8)$, $B_I(8)$, $B_R(11)$, $B_I(11)$, $B_R(14)$, $B_I(14)$, $B_R(2)$, and $B_I(2)$.

The multiplication portion of the algorithm requires two additional data memory locations because no temporary registers are assumed. The variables used for the intermediate computations were chosen to be the same as those used for the 5-point Singleton building block in Chapter 8 to make it easier to associate the computational steps with the discussion in Chapter 8. This set of computations is represented in Figure 9-19 by 5-point building block 1. Further, the labels on the left and right of this building block correspond to the input and output labels in the 5-point Singleton building block in Chapter 8.

| **Algorithm Steps** | **Memory Map** |
|---|---|
| $b_R(6) = B_R(8) + B_R(2)$ | $b_R(6) \Rightarrow M(28)$ |
| $b_I(6) = B_I(8) + B_I(2)$ | $b_I(6) \Rightarrow M(13)$ |
| $b_R(7) = B_R(8) - B_R(2)$ | $b_R(7) \Rightarrow M(22)$ |
| $b_I(7) = B_I(8) - B_I(2)$ | $b_I(7) \Rightarrow M(7)$ |
| $b_R(8) = B_R(11) + B_R(14)$ | $b_R(8) \Rightarrow M(16)$ |
| $b_I(8) = B_I(11) + B_I(14)$ | $b_I(8) \Rightarrow M(1)$ |
| $b_R(9) = B_R(11) - B_R(14)$ | $b_R(9) \Rightarrow M(19)$ |
| $b_I(9) = B_I(11) - B_I(14)$ | $b_I(9) \Rightarrow M(4)$ |
| $c_R(7) = b_R(7) * \sin(2\pi/5) + b_R(9) * \sin(4\pi/5)$ | $c_R(7) \Rightarrow M(30)$ |
| $c_I(7) = b_I(7) * \sin(2\pi/5) + b_I(9) * \sin(4\pi/5)$ | $c_I(7) \Rightarrow M(19)$ |
| $c_R(9) = b_R(7) * \sin(4\pi/5) - b_R(9) * \sin(2\pi/5)$ | $c_R(9) \Rightarrow M(31)$ |
| $c_I(9) = b_I(7) * \sin(4\pi/5) - b_I(9) * \sin(2\pi/5)$ | $c_I(9) \Rightarrow M(22)$ |
| $c_R(6) = b_R(6) * \cos(2\pi/5) + b_R(8) * \cos(4\pi/5) + B_R(5)$ | $c_R(6) \Rightarrow M(7)$ |
| $c_I(6) = b_I(6) * \cos(2\pi/5) + b_I(8) * \cos(4\pi/5) + B_I(5)$ | $c_I(6) \Rightarrow M(28)$ |
| $c_R(8) = b_R(6) * \cos(4\pi/5) + b_R(8) * \cos(2\pi/5) + B_R(5)$ | $c_R(8) \Rightarrow M(4)$ |
| $c_I(8) = b_I(6) * \cos(4\pi/5) + b_I(8) * \cos(2\pi/5) + B_I(5)$ | $c_I(8) \Rightarrow M(16)$ |
| $A_R(10) = B_R(5) + b_R(6) + b_R(8)$ | $A_R(10) \Rightarrow M(25)$ |
| $A_I(10) = B_I(5) + b_I(6) + b_I(8)$ | $A_I(10) \Rightarrow M(10)$ |
| $A_R(1) = c_R(6) + c_I(7)$ | $A_R(1) \Rightarrow M(7)$ |
| $A_I(1) = c_I(6) - c_R(7)$ | $A_I(1) \Rightarrow M(13)$ |
| $A_R(7) = c_R(8) + c_I(9)$ | $A_R(7) \Rightarrow M(4)$ |
| $A_I(7) = c_I(8) - c_R(9)$ | $A_I(7) \Rightarrow M(16)$ |
| $A_R(13) = c_R(8) - c_I(9)$ | $A_R(13) \Rightarrow M(22)$ |
| $A_I(13) = c_I(8) + c_R(9)$ | $A_I(13) \Rightarrow M(28)$ |
| $A_R(4) = c_R(6) - c_I(7)$ | $A_R(4) \Rightarrow M(19)$ |
| $A_I(4) = c_I(6) + c_R(7)$ | $A_I(4) \Rightarrow M(1)$ |

## Third of Three 5-Point Building Blocks

This 5-point building block ($k = 2$) has $B_R((5*k+3*m) \bmod 15)$ and $B_I((5*k+3*m) \bmod 15)(m = 0, 1, 2, 3,$ and $4)$ as inputs and $A_R((10*k+6*m) \bmod 15)$ and

$A_I((10 * k + 6 * m) \bmod 15)(m = 0, 1, 2, 3, \text{ and } 4)$ as its output frequency components. Performing the modulo arithmetic computations results in the inputs being $B_R(10)$, $B_I(10)$, $B_R(13)$, $B_I(13)$, $B_R(1)$, $B_I(1)$, $B_R(4)$, $B_I(4)$, $B_R(7)$, and $B_I(7)$.

The multiplication portion of the algorithm requires two additional data memory locations because no temporary registers are assumed. The variables used for the intermediate computations were chosen to be the same as those used for the 5-point Singleton building block in Chapter 8 to make it easier to associate the computational steps with the discussion in Chapter 8. This set of computations is represented in Figure 9-19 by 5-point building block 2. Further, the labels on the left and right of this building block correspond to the input and output labels in the 5-point Singleton building block in Chapter 8.

| **Algorithm Steps** | **Memory Map** |
|---|---|
| $b_R(11) = B_R(13) + B_R(7)$ | $b_R(11) \Rightarrow M(23)$ |
| $b_I(11) = B_I(13) + B_I(7)$ | $b_I(11) \Rightarrow M(8)$ |
| $b_R(12) = B_R(13) - B_R(7)$ | $b_R(12) \Rightarrow M(17)$ |
| $b_I(12) = B_I(13) - B_I(7)$ | $b_I(12) \Rightarrow M(2)$ |
| $b_R(13) = B_R(1) + B_R(4)$ | $b_R(13) \Rightarrow M(26)$ |
| $b_I(13) = B_I(1) + B_I(4)$ | $b_I(13) \Rightarrow M(11)$ |
| $b_R(14) = B_R(1) - B_R(4)$ | $b_R(14) \Rightarrow M(29)$ |
| $b_I(14) = B_I(1) - B_I(4)$ | $b_I(14) \Rightarrow M(14)$ |
| $c_R(12) = b_R(12) * \sin(2\pi/5) + b_R(14) * \sin(4\pi/5)$ | $c_R(12) \Rightarrow M(30)$ |
| $c_I(12) = b_I(12) * \sin(2\pi/5) + b_I(14) * \sin(4\pi/5)$ | $c_I(12) \Rightarrow M(29)$ |
| $c_R(14) = b_R(12) * \sin(4\pi/5) - b_R(14) * \sin(2\pi/5)$ | $c_R(14) \Rightarrow M(31)$ |
| $c_I(14) = b_I(12) * \sin(4\pi/5) - b_I(14) * \sin(2\pi/5)$ | $c_I(14) \Rightarrow M(17)$ |
| $c_R(11) = b_R(11) * \cos(2\pi/5) + b_R(13) * \cos(4\pi/5) + B_R(10)$ | $c_R(11) \Rightarrow M(2)$ |
| $c_I(11) = b_I(11) * \cos(2\pi/5) + b_I(13) * \cos(4\pi/5) + B_I(10)$ | $c_I(11) \Rightarrow M(23)$ |
| $c_R(13) = b_R(11) * \cos(4\pi/5) + b_R(13) * \cos(2\pi/5) + B_R(10)$ | $c_R(13) \Rightarrow M(14)$ |
| $c_I(13) = b_I(11) * \cos(4\pi/5) + b_I(13) * \cos(2\pi/5) + B_I(10)$ | $c_I(13) \Rightarrow M(26)$ |
| $A_R(5) = B_R(10) + b_R(11) + b_R(13)$ | $A_R(5) \Rightarrow M(20)$ |
| $A_I(5) = B_I(10) + b_I(11) + b_I(13)$ | $A_I(5) \Rightarrow M(5)$ |
| $A_R(11) = c_R(11) + c_I(12)$ | $A_R(11) \Rightarrow M(2)$ |
| $A_I(11) = c_I(11) - c_R(12)$ | $A_I(11) \Rightarrow M(8)$ |
| $A_R(2) = c_R(13) + c_I(14)$ | $A_R(2) \Rightarrow M(14)$ |
| $A_I(2) = c_I(13) - c_R(14)$ | $A_I(2) \Rightarrow M(26)$ |
| $A_R(8) = c_R(13) - c_I(14)$ | $A_R(8) \Rightarrow M(17)$ |
| $A_I(8) = c_I(13) + c_R(14)$ | $A_I(8) \Rightarrow M(23)$ |
| $A_R(14) = c_R(11) - c_I(12)$ | $A_R(14) \Rightarrow M(29)$ |
| $A_I(14) = c_I(11) + c_R(12)$ | $A_I(14) \Rightarrow M(11)$ |

### 9.6.5 Fifteen-Point SWIFT Example

The 15-point SWIFT [4] algorithm can be implemented with either the 3-point or the 5-point building blocks first. If the 3-point building block is first, the 15 pieces of complex input data are divided into five sets of three complex points, one for each of the $15/3 = 5$ 3-point building blocks. Following the 3-point building blocks, the intermediate results are divided into three sets of five pieces of complex data needed for input to the $15/5 = 3$ 5-point building-block computations. This algorithm is similar to the Kolba-Parks algorithm but uses a different data mapping strategy. The order does not affect how many computations are required.

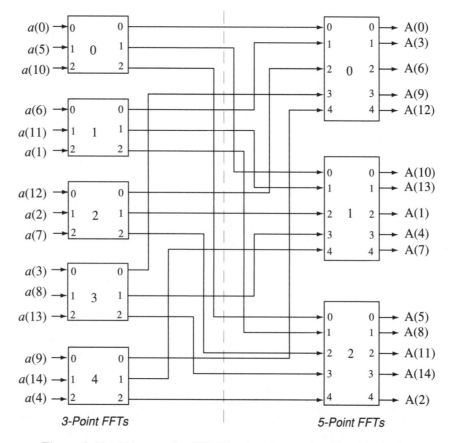

**Figure 9-20**    Fifteen-point SWIFT prime factor algorithm block diagram.

This example uses the Singleton 3- and 5-point building blocks. A smaller number of adds and multiplies would be needed if the Winograd building blocks were used. If the Comparison Matrix in Chapter 8 and the equation presented in the discussion of the performance features for the prime factor algorithm are used, the total number of real adds required is $5*12+3*32 = 156$, and the total number of real multiplies is $5*4+3*16 = 68$. The total amount of data memory required is driven by the 5-point algorithm and is 32 locations. Explicitly, 30 locations are required for the 15 complex data points, plus 2 additional locations for the intermediate computations in the 5-point Singleton building block. Similarly, the 3-point Singleton building block has two multiplier constants and the 5-point Singleton building block has four, for a total of six memory locations for multiplier constants. The stages are as follows.

### Stage 1: Three-Point Building Blocks

The 15 data points must first be divided into five sets of 3 points to serve as inputs to each of the 3-point building blocks. This is done by starting with complex input data point pair $a_R(0)$, $a_I(0)$, and grouping it with complex input data point pairs $a_R(5)$, $a_I(5)$ and $a_R(10)$, $a_I(10)$. These provide the input to the top one of the five 3-point transforms. This is followed by grouping the input data point pairs $a_R(1)$, $a_I(1)$, $a_R(6)$, $a_I(6)$, and $a_R(11)$, $a_I(11)$ to provide the input for the second of the five 3-point building blocks. The next grouping is data point pairs $a_R(2)$, $a_I(2)$, $a_R(7)$, $a_I(7)$, and $a_R(12)$, $a_I(12)$ for input into the third of the five 3-point building blocks. The next grouping is data point pairs $a_R(3)$, $a_I(3)$, $a_R(8)$, $a_I(8)$, and $a_R(13)$, $a_I(13)$ to provide input for the fourth of the five 3-point transforms. The final grouping is data point pairs $a_R(4)$, $a_I(4)$, $a_R(9)$, $a_I(9)$, and $a_R(14)$, $a_I(14)$ for input into the fifth 3-point building block.

The order in which this data is used for inputs to the 3-point building blocks is the key point in removing the need for complex multipliers between the 3- and 5-point building blocks. For the 15-point transform, the SWIFT algorithm requires the complex input data for the $k$-th input to the $m$-th 3-point transform to be $a_R((5*k+6*m) \bmod 15)$, $a_I((5*k+6*m) \bmod 15)$ where $k = 0, 1,$ and 2, and $m = 0, 1, 2, 3,$ and 4.

The five groups of computations, listed as (a) through (e), each perform a 3-point building block. In this example, the Singleton 3-point algorithm building block from Chapter 8 is used. All of these 3-point transforms could also have been the Winograd 3-point algorithm building block from Chapter 8. In fact, the five 3-point building blocks can be any combination of the two 3-point algorithm building blocks. The outputs of each of the 3-point building blocks, labeled $B_R(i)$ and $B_I(i)$ for $i = 0, 5, 10$, are the equivalent of the $A_R(i)$ and $A_I(i)$ in the 3-point algorithm building block in Chapter 8.

The strategy for converting these equations to code is to start at the top (compute $b_R(5)$) and identify the pair of inputs to be used first (in this case $a_R(5)$ and $a_R(10)$). Then look down the list to find the second (compute $b_R(10)$) place where these two inputs are used. Pull $a_R(5)$ and $a_R(10)$ from memory, compute $b_R(5)$ and $b_R(10)$, and store the results in memory locations $M(5)$ and $M(10)$, previously occupied by $a_R(5)$ and $a_R(10)$. The next step is to look at the next computation $b_I(1)$ on the list and repeat the same set of steps. Continue this process until all the Algorithm Steps in Stage 1 have been computed and their results stored in the Memory Map addresses.

### First of Five 3-Point Building Blocks

The inputs to this 3-point building block are $a_R((5 * k + 6 * m) \bmod 15)$, $a_I((5 * k + 6 * m) \bmod 15)$ where $m = 0$. Performing the modulo arithmetic computations to determine the inputs results in the inputs of $a_R(0)$, $a_I(0)$, $a_R(5)$, $a_I(5)$, $a_R(10)$, and $a_I(10)$ for $k = 0, 1,$ and 2. This set of computations is represented in Figure 9-20 by 3-point building block 0. Further, the labels on the left and right of this building block correspond to the input and output labels in the 3-point Singleton building block in Chapter 8.

<table>
<thead>
<tr><th colspan="2" style="text-align:center">Algorithm Steps</th><th style="text-align:center">Memory Map</th></tr>
</thead>
<tbody>
<tr><td style="text-align:right">$b_R(5) = a_R(5) + a_R(10)$</td><td></td><td style="text-align:center">$b_R(5) \Rightarrow M(5)$</td></tr>
<tr><td style="text-align:right">$b_R(10) = a_R(5) - a_R(10)$</td><td></td><td style="text-align:center">$b_R(10) \Rightarrow M(10)$</td></tr>
<tr><td style="text-align:right">$b_I(5) = a_I(5) + a_I(10)$</td><td></td><td style="text-align:center">$b_I(5) \Rightarrow M(20)$</td></tr>
<tr><td style="text-align:right">$b_I(10) = a_I(5) - a_I(10)$</td><td></td><td style="text-align:center">$b_I(10) \Rightarrow M(25)$</td></tr>
<tr><td style="text-align:right">$c_R(5) = b_R(5) * \cos(2\pi/3) + a_R(0)$</td><td></td><td style="text-align:center">$c_R(5) \Rightarrow M(30)$</td></tr>
<tr><td style="text-align:right">$B_R(0) = a_R(0) + b_R(5)$</td><td></td><td style="text-align:center">$B_R(0) \Rightarrow M(0)$</td></tr>
<tr><td style="text-align:right">$c_R(10) = b_I(10) * \sin(2\pi/3)$</td><td></td><td style="text-align:center">$c_R(10) \Rightarrow M(25)$</td></tr>
<tr><td style="text-align:right">$c_I(5) = b_I(5) * \cos(2\pi/3) + a_I(0)$</td><td></td><td style="text-align:center">$c_I(5) \Rightarrow M(5)$</td></tr>
<tr><td style="text-align:right">$B_I(0) = a_I(0) + b_I(5)$</td><td></td><td style="text-align:center">$B_I(0) \Rightarrow M(15)$</td></tr>
<tr><td style="text-align:right">$c_I(10) = -b_R(10) * \sin(2\pi/3)$</td><td></td><td style="text-align:center">$c_I(10) \Rightarrow M(10)$</td></tr>
<tr><td style="text-align:right">$B_R(5) = c_R(5) + c_R(10)$</td><td></td><td style="text-align:center">$B_R(5) \Rightarrow M(25)$</td></tr>
<tr><td style="text-align:right">$B_I(5) = c_I(5) + c_I(10)$</td><td></td><td style="text-align:center">$B_I(5) \Rightarrow M(10)$</td></tr>
<tr><td style="text-align:right">$B_R(10) = c_R(5) - c_R(10)$</td><td></td><td style="text-align:center">$B_R(10) \Rightarrow M(20)$</td></tr>
<tr><td style="text-align:right">$B_I(10) = c_I(5) - c_I(10)$</td><td></td><td style="text-align:center">$B_I(10) \Rightarrow M(5)$</td></tr>
</tbody>
</table>

### Second of Five 3-Point Building Blocks

The inputs to this 3-point building block are $a_R((5 * k + 6 * m) \bmod 15)$, $a_I((5 * k + 6 * m) \bmod 15)$ where $m = 1$. Performing the modulo arithmetic computations to determine the inputs results in the inputs being $a_R(6)$, $a_I(6)$, $a_R(11)$, $a_I(11)$, $a_R(1)$, and $a_I(1)$ for $k = 0, 1,$ and 2. This set of computations is represented in Figure 9-20 by 3-point building block 1. Further, the labels on the left and right of this building block correspond to the input and output labels in the 3-point Singleton building block in Chapter 8.

<table>
<thead>
<tr><th colspan="2" style="text-align:center">Algorithm Steps</th><th style="text-align:center">Memory Map</th></tr>
</thead>
<tbody>
<tr><td style="text-align:right">$b_R(6) = a_R(11) + a_R(1)$</td><td></td><td style="text-align:center">$b_R(6) \Rightarrow M(11)$</td></tr>
<tr><td style="text-align:right">$b_R(11) = a_R(11) - a_R(1)$</td><td></td><td style="text-align:center">$b_R(11) \Rightarrow M(1)$</td></tr>
<tr><td style="text-align:right">$b_I(6) = a_I(11) + a_I(1)$</td><td></td><td style="text-align:center">$b_I(6) \Rightarrow M(26)$</td></tr>
<tr><td style="text-align:right">$b_I(11) = a_I(11) - a_I(1)$</td><td></td><td style="text-align:center">$b_I(11) \Rightarrow M(16)$</td></tr>
<tr><td style="text-align:right">$c_R(6) = b_R(6) * \cos(2\pi/3) + a_R(6)$</td><td></td><td style="text-align:center">$c_R(6) \Rightarrow M(30)$</td></tr>
<tr><td style="text-align:right">$B_R(6) = a_R(6) + b_R(6)$</td><td></td><td style="text-align:center">$B_R(6) \Rightarrow M(6)$</td></tr>
<tr><td style="text-align:right">$c_R(11) = b_I(11) * \sin(2\pi/3)$</td><td></td><td style="text-align:center">$c_R(11) \Rightarrow M(16)$</td></tr>
<tr><td style="text-align:right">$c_I(6) = b_I(6) * \cos(2\pi/3) + a_I(6)$</td><td></td><td style="text-align:center">$c_I(6) \Rightarrow M(11)$</td></tr>
<tr><td style="text-align:right">$B_I(6) = a_I(6) + b_I(6)$</td><td></td><td style="text-align:center">$B_I(6) \Rightarrow M(21)$</td></tr>
<tr><td style="text-align:right">$c_I(11) = -b_R(11) * \sin(2\pi/3)$</td><td></td><td style="text-align:center">$c_I(11) \Rightarrow M(1)$</td></tr>
<tr><td style="text-align:right">$B_R(11) = c_R(6) + c_R(11)$</td><td></td><td style="text-align:center">$B_R(11) \Rightarrow M(16)$</td></tr>
<tr><td style="text-align:right">$B_I(11) = c_I(6) + c_I(11)$</td><td></td><td style="text-align:center">$B_I(11) \Rightarrow M(1)$</td></tr>
<tr><td style="text-align:right">$B_R(1) = c_R(6) - c_R(11)$</td><td></td><td style="text-align:center">$B_R(1) \Rightarrow M(26)$</td></tr>
<tr><td style="text-align:right">$B_I(1) = c_I(6) - c_I(11)$</td><td></td><td style="text-align:center">$B_I(1) \Rightarrow M(11)$</td></tr>
</tbody>
</table>

### Third of Five 3-Point Building Blocks

The inputs to this 3-point building block are $a_R((5 * k + 6 * m) \bmod 15)$, $a_I((5 * k + 6 * m) \bmod 15)$ where $m = 2$. Performing the modulo arithmetic computations to determine the inputs results in the inputs being $a_R(12)$, $a_I(12)$, $a_R(2)$, $a_I(2)$, $a_R(7)$, and $a_I(7)$ for $k = 0, 1$, and 2. This set of computations is represented in Figure 9-20 by 3-point building block 2. Further, the labels on the left and right of this building block correspond to the input and output labels in the 3-point Singleton building block in Chapter 8.

| **Algorithm Steps** | **Memory Map** |
|:---:|:---:|
| $b_R(7) = a_R(2) + a_R(7)$ | $b_R(7) \Rightarrow M(2)$ |
| $b_R(12) = a_R(2) - a_R(7)$ | $b_R(12) \Rightarrow M(7)$ |
| $b_I(7) = a_I(2) + a_I(7)$ | $b_I(7) \Rightarrow M(17)$ |
| $b_I(12) = a_I(2) - a_I(7)$ | $b_I(12) \Rightarrow M(22)$ |
| $c_R(7) = b_R(7) * \cos(2\pi/3) + a_R(12)$ | $c_R(7) \Rightarrow M(30)$ |
| $B_R(12) = a_R(12) + b_R(7)$ | $B_R(12) \Rightarrow M(12)$ |
| $c_R(12) = b_I(12) * \sin(2\pi/3)$ | $c_R(12) \Rightarrow M(22)$ |
| $c_I(7) = b_I(7) * \cos(2\pi/3) + a_I(12)$ | $c_I(7) \Rightarrow M(2)$ |
| $B_I(12) = a_I(12) + b_I(7)$ | $B_I(12) \Rightarrow M(27)$ |
| $c_I(12) = -b_R(12) * \sin(2\pi/3)$ | $c_I(12) \Rightarrow M(7)$ |
| $B_R(2) = c_R(7) + c_R(12)$ | $B_R(2) \Rightarrow M(22)$ |
| $B_I(2) = c_I(7) + c_I(12)$ | $B_I(2) \Rightarrow M(7)$ |
| $B_R(7) = c_R(7) - c_R(12)$ | $B_R(7) \Rightarrow M(17)$ |
| $B_I(7) = c_I(7) - c_I(12)$ | $B_I(7) \Rightarrow M(2)$ |

### Fourth of Five 3-Point Building Blocks

The inputs to this 3-point building block are $a_R((5 * k + 6 * m) \bmod 15)$, $a_I((5*6+3*m) \bmod 15)$ where $m = 3$. Performing the modulo arithmetic computations to determine the inputs results in the inputs being $a_R(3)$, $a_I(3)$, $a_R(8)$, $a_I(8)$, $a_R(13)$, and $a_I(13)$ for $k = 0, 1$, and 2. This set of computations is represented in Figure 9-20 by 3-point building block 3. Further, the labels on the left and right of this building block correspond to the input and output labels in the 3-point Singleton building block in Chapter 8.

| **Algorithm Steps** | **Memory Map** |
|:---:|:---:|
| $b_R(8) = a_R(8) + a_R(13)$ | $b_R(8) \Rightarrow M(8)$ |
| $b_R(13) = a_R(8) - a_R(13)$ | $b_R(13) \Rightarrow M(13)$ |
| $b_I(8) = a_I(8) + a_I(13)$ | $b_I(8) \Rightarrow M(23)$ |
| $b_I(13) = a_I(8) - a_I(13)$ | $b_I(13) \Rightarrow M(28)$ |
| $c_R(8) = b_R(8) * \cos(2\pi/3) + a_R(3)$ | $c_R(8) \Rightarrow M(30)$ |
| $B_R(3) = a_R(3) + b_R(8)$ | $B_R(3) \Rightarrow M(3)$ |

| Algorithm Steps | Memory Map |
|---|---|
| $c_R(13) = b_I(13) * \sin(2\pi/3)$ | $c_R(13) \Rightarrow M(28)$ |
| $c_I(8) = b_I(8) * \cos(2\pi/3) + a_I(3)$ | $c_I(8) \Rightarrow M(8)$ |
| $B_I(3) = a_I(3) + b_I(8)$ | $B_I(3) \Rightarrow M(18)$ |
| $c_I(13) = -b_R(13) * \sin(2\pi/3)$ | $c_I(13) \Rightarrow M(13)$ |
| $B_R(8) = c_R(8) + c_R(13)$ | $B_R(8) \Rightarrow M(28)$ |
| $B_I(8) = c_I(8) + c_I(13)$ | $B_I(8) \Rightarrow M(13)$ |
| $B_R(13) = c_R(8) - c_R(13)$ | $B_R(13) \Rightarrow M(23)$ |
| $B_I(13) = c_I(8) - c_I(13)$ | $B_I(13) \Rightarrow M(8)$ |

### Fifth of Five 3-Point Building Blocks

The inputs to this 3-point building block are $a_R((5 * k + 6 * m) \bmod 15)$, $a_I((5 * k + 6 * m) \bmod 15)$ where $m = 4$. Performing the modulo arithmetic computations to determine the inputs results in the inputs being $a_R(9)$, $a_I(9)$, $a_R(14)$, $a_I(14)$, $a_R(4)$, and $a_I(4)$ for $k = 0, 1$, and 2. This set of computations is represented in Figure 9-20 by 3-point building block 4. Further, the labels on the left and right of this building block correspond to the input and output labels in the 3-point Singleton building block in Chapter 8.

| Algorithm Steps | Memory Map |
|---|---|
| $b_R(9) = a_R(14) + a_R(4)$ | $b_R(9) \Rightarrow M(14)$ |
| $b_R(14) = a_R(14) - a_R(4)$ | $b_R(14) \Rightarrow M(4)$ |
| $b_I(9) = a_I(14) + a_I(4)$ | $b_I(9) \Rightarrow M(29)$ |
| $b_I(14) = a_I(14) - a_I(4)$ | $b_I(14) \Rightarrow M(19)$ |
| $c_R(9) = b_R(9) * \cos(2\pi/3) + a_R(9)$ | $c_R(9) \Rightarrow M(30)$ |
| $B_R(9) = a_R(9) + b_R(9)$ | $B_R(9) \Rightarrow M(9)$ |
| $c_R(14) = b_I(14) * \sin(2\pi/3)$ | $c_R(14) \Rightarrow M(19)$ |
| $c_I(9) = b_I(9) * \cos(2\pi/3) + a_I(9)$ | $c_I(9) \Rightarrow M(14)$ |
| $B_I(9) = a_I(9) + b_I(9)$ | $B_I(9) \Rightarrow M(24)$ |
| $c_I(14) = -b_R(14) * \sin(2\pi/3)$ | $c_I(14) \Rightarrow M(4)$ |
| $B_R(14) = c_R(9) + c_R(14)$ | $B_R(14) \Rightarrow M(19)$ |
| $B_I(14) = c_I(9) + c_I(14)$ | $B_I(14) \Rightarrow M(4)$ |
| $B_R(4) = c_R(9) - c_R(14)$ | $B_R(4) \Rightarrow M(29)$ |
| $B_I(4) = c_I(9) - c_I(14)$ | $B_I(4) \Rightarrow M(14)$ |

### Stage 2: Output 5-Point Building Blocks

For this example the Singleton 5-point building block from Chapter 8 is used. However, either of the two other 5-point building blocks could have been used without changing the rest of the structure of the building block. If the number of adds and multiplies is the overriding criterion, then the Winograd algorithm building block should be used in place of the 5-point Singleton building block.

Three sets of 5-point algorithm building-block Algorithm Steps from Chapter 8 are presented. In Chapter 8 the 5-point algorithm building block was presented as three stages. Since the features of the individual stages of the 5-point algorithm block are discussed in Chapter 8, they are not discussed again. The $m$-th input to the $k$-th 5-point building block is $B_R((5 * k + 6 * m) \bmod 15)$ and $B_I((5 * k + 6 * m) \bmod 15)$ from the previous stage.

The multiply stage of the 5-point Singleton building block required additional data memory locations under the set of constraints used in Chapter 8. If the 15-point computations are performed in the order shown, the additional memory locations used by the first of the three 5-point building blocks can be reused by each of the other two 5-point building blocks.

The strategy for converting these equations to code is to start at the top (compute $b_R(1)$) and identify the pair of inputs to be used first (in this case $B_R(6)$ and $B_R(9)$). Then look down the list to find the second (compute $b_R(2)$) place where these two inputs are used. Pull $B_R(6)$ and $B_R(9)$ from memory, compute $b_R(1)$ and $b_R(2)$, and store the results in memory locations $M(6)$ and $M(9)$, previously occupied by $B_R(6)$ and $B_R(9)$. The next step is to look at the next computation $b_I(1)$ on the list and repeat the same set of steps. Continue this process until all the Algorithm Steps in Stage 2 have been computed and their results stored in the Memory Map addresses.

### First of Three 5-Point Building Blocks

This 5-point building block ($k = 0$) has $B_R((5 * k + 6 * m) \bmod 15)$ and $B_I((5*k+6*m) \bmod 15)(m = 0, 1, 2, 3, \text{and } 4)$ as inputs and $A_R((10*k+3*m) \bmod 15)$ and $A_I((10 * k + 3 * m) \bmod 15)(m = 0, 1, 2, 3, \text{and } 4)$ as its output frequency components. Performing the modulo arithmetic computations to determine the inputs results in the inputs being $B_R(0)$, $B_I(0)$, $B_R(6)$, $B_I(6)$, $B_R(12)$, $B_I(12)$, $B_R(3)$, $B_I(3)$, $B_R(9)$, and $B_I(9)$.

The multiplication portion of the building block requires two additional data memory locations because no temporary registers are assumed. The variables used for the intermediate computations were chosen to be the same as those used for the 5-point Singleton building block in Chapter 8 to make it easier to associate the computational steps with the discussion of its features and memory mappings in Chapter 8. This set of computations is represented in Figure 9-20 by 5-point building block 0. Further, the labels on the left and right of this building block correspond to the input and output labels in the 5-point Singleton building block in Chapter 8.

| Algorithm Steps | Memory Map |
|---|---|
| $b_R(1) = B_R(6) + B_R(9)$ | $b_R(1) \Rightarrow M(6)$ |
| $b_I(1) = B_I(6) + B_I(9)$ | $b_I(1) \Rightarrow M(21)$ |
| $b_R(2) = B_R(6) - B_R(9)$ | $b_R(2) \Rightarrow M(9)$ |
| $b_I(2) = B_I(6) - B_I(9)$ | $b_I(2) \Rightarrow M(24)$ |

| **Algorithm Steps** | **Memory Map** |
|---|---|
| $b_R(3) = B_R(12) + B_R(3)$ | $b_R(3) \Rightarrow M(12)$ |
| $b_I(3) = B_I(12) + B_I(3)$ | $b_I(3) \Rightarrow M(27)$ |
| $b_R(4) = B_R(12) - B_R(3)$ | $b_R(4) \Rightarrow M(3)$ |
| $b_I(4) = B_I(12) - B_I(3)$ | $b_I(4) \Rightarrow M(18)$ |
| $c_R(2) = b_R(2) * \sin(2\pi/5) + b_R(4) * \sin(4\pi/5)$ | $c_R(2) \Rightarrow M(30)$ |
| $c_I(2) = b_I(2) * \sin(2\pi/5) + b_I(4) * \sin(4\pi/5)$ | $c_I(2) \Rightarrow M(3)$ |
| $c_R(4) = b_R(2) * \sin(4\pi/5) - b_R(4) * \sin(2\pi/5)$ | $c_R(4) \Rightarrow M(31)$ |
| $c_I(4) = b_I(2) * \sin(4\pi/5) - b_I(4) * \sin(2\pi/5)$ | $c_I(4) \Rightarrow M(9)$ |
| $c_R(1) = b_R(1) * \cos(2\pi/5) + b_R(3) * \cos(4\pi/5) + B_R(0)$ | $c_R(1) \Rightarrow M(24)$ |
| $c_I(1) = b_I(1) * \cos(2\pi/5) + b_I(3) * \cos(4\pi/5) + B_I(0)$ | $c_I(1) \Rightarrow M(6)$ |
| $c_R(3) = b_R(1) * \cos(4\pi/5) + b_R(3) * \cos(2\pi/5) + B_R(0)$ | $c_R(3) \Rightarrow M(18)$ |
| $c_I(3) = b_I(1) * \cos(4\pi/5) + b_I(3) * \cos(2\pi/5) + B_I(0)$ | $c_I(3) \Rightarrow M(12)$ |
| $A_R(0) = B_R(0) + b_R(1) + b_R(3)$ | $A_R(0) \Rightarrow M(0)$ |
| $A_I(0) = B_I(0) + b_I(1) + b_I(3)$ | $A_I(0) \Rightarrow M(15)$ |
| $A_R(3) = c_R(1) + c_I(2)$ | $A_R(3) \Rightarrow M(24)$ |
| $A_I(3) = c_I(1) - c_R(2)$ | $A_I(3) \Rightarrow M(21)$ |
| $A_R(6) = c_R(3) + c_I(4)$ | $A_R(6) \Rightarrow M(18)$ |
| $A_I(6) = c_I(3) - c_R(4)$ | $A_I(6) \Rightarrow M(12)$ |
| $A_R(9) = c_R(3) - c_I(4)$ | $A_R(9) \Rightarrow M(9)$ |
| $A_I(9) = c_I(3) + c_R(4)$ | $A_I(9) \Rightarrow M(6)$ |
| $A_R(12) = c_R(1) - c_I(2)$ | $A_R(12) \Rightarrow M(3)$ |
| $A_I(12) = c_I(1) + c_R(2)$ | $A_I(12) \Rightarrow M(27)$ |

### Second of Three 5-Point Building Blocks

This 5-point building block ($k = 1$) has $B_R((5 * k + 6 * m) \bmod 15)$ and $B_I((5*k+6*m) \bmod 15)(m = 0, 1, 2, 3, \text{and } 4)$ as inputs and $A_R((10*k+3*m) \bmod 15)$ and $A_I((10 * k + 3 * m) \bmod 15)(m = 0, 1, 2, 3, \text{and } 4)$ as its output frequency components. Performing the modulo arithmetic computations to determine the inputs results in the inputs being $B_R(5)$, $B_I(5)$, $B_R(11)$, $B_I(11)$, $B_R(2)$, $B_I(2)$, $B_R(8)$, $B_I(8)$, $B_R(14)$, and $B_I(14)$.

The multiplication portion of the building block requires two additional data memory locations because no temporary registers are assumed. The variables used for the intermediate computations were chosen to be the same as those used for the 5-point Singleton building block in Chapter 8 to make it easier to associate the computational steps with the discussion of its features and memory mappings in Chapter 8. This set of computations is represented in Figure 9-20 by 5-point building block 1. Further, the labels on the left and right of this building block correspond to the input and output labels in the 5-point Singleton building block in Chapter 8.

| **Algorithm Steps** | **Memory Map** |
|---|---|
| $b_R(6) = B_R(11) + B_R(14)$ | $b_R(6) \Rightarrow M(16)$ |
| $b_I(6) = B_I(11) + B_I(14)$ | $b_I(6) \Rightarrow M(1)$ |
| $b_R(7) = B_R(11) - B_R(14)$ | $b_R(7) \Rightarrow M(19)$ |
| $b_I(7) = B_I(11) - B_I(14)$ | $b_I(7) \Rightarrow M(4)$ |
| $b_R(8) = B_R(2) + B_R(8)$ | $b_R(8) \Rightarrow M(22)$ |
| $b_I(8) = B_I(2) + B_I(8)$ | $b_I(8) \Rightarrow M(7)$ |
| $b_R(9) = B_R(2) - B_R(8)$ | $b_R(9) \Rightarrow M(28)$ |
| $b_I(9) = B_I(2) - B_I(8)$ | $b_I(9) \Rightarrow M(13)$ |
| $c_R(7) = b_R(7) * \sin(2\pi/5) + b_R(9) * \sin(4\pi/5)$ | $c_R(7) \Rightarrow M(30)$ |
| $c_I(7) = b_I(7) * \sin(2\pi/5) + b_I(9) * \sin(4\pi/5)$ | $c_I(7) \Rightarrow M(28)$ |
| $c_R(9) = b_R(7) * \sin(4\pi/5) - b_R(9) * \sin(2\pi/5)$ | $c_R(9) \Rightarrow M(31)$ |
| $c_I(9) = b_I(7) * \sin(4\pi/5) - b_I(9) * \sin(2\pi/5)$ | $c_I(9) \Rightarrow M(19)$ |
| $c_R(6) = b_R(6) * \cos(2\pi/5) + b_R(8) * \cos(4\pi/5) + B_R(5)$ | $c_R(6) \Rightarrow M(4)$ |
| $c_I(6) = b_I(6) * \cos(2\pi/5) + b_I(8) * \cos(4\pi/5) + B_I(5)$ | $c_I(6) \Rightarrow M(16)$ |
| $c_R(8) = b_R(6) * \cos(4\pi/5) + b_R(8) * \cos(2\pi/5) + B_R(5)$ | $c_R(8) \Rightarrow M(13)$ |
| $c_I(8) = b_I(6) * \cos(4\pi/5) + b_I(8) * \cos(2\pi/5) + B_I(5)$ | $c_I(8) \Rightarrow M(22)$ |
| $A_R(10) = B_R(5) + b_R(6) + b_R(8)$ | $A_R(10) \Rightarrow M(25)$ |
| $A_I(10) = B_I(5) + b_I(6) + b_I(8)$ | $A_I(10) \Rightarrow M(10)$ |
| $A_R(13) = c_R(6) + c_I(7)$ | $A_R(13) \Rightarrow M(4)$ |
| $A_I(13) = c_I(6) - c_R(7)$ | $A_I(13) \Rightarrow M(1)$ |
| $A_R(1) = c_R(8) + c_I(9)$ | $A_R(1) \Rightarrow M(13)$ |
| $A_I(1) = c_I(8) - c_R(9)$ | $A_I(1) \Rightarrow M(22)$ |
| $A_R(4) = c_R(8) - c_I(9)$ | $A_R(4) \Rightarrow M(19)$ |
| $A_I(4) = c_I(8) + c_R(9)$ | $A_I(4) \Rightarrow M(16)$ |
| $A_R(7) = c_R(6) - c_I(7)$ | $A_R(7) \Rightarrow M(28)$ |
| $A_I(7) = c_I(6) + c_R(7)$ | $A_I(7) \Rightarrow M(7)$ |

### Third of Three 5-Point Building Blocks

This 5-point building block ($k = 2$) has $B_R((5 * k + 6 * m) \bmod 15)$ and $B_I((5*k+6*m) \bmod 15)(m = 0, 1, 2, 3,$ and 4) as inputs and $A_R((10*k+3*m) \bmod 15)$ and $A_I((10*k+3*m) \bmod 15)(m = 0, 1, 2, 3,$ and 4) as its output frequency components. Performing the modulo arithmetic computations to determine the inputs results in the inputs being $B_R(10)$, $B_I(10)$, $B_R(1)$, $B_I(1)$, $B_R(7)$, $B_I(7)$, $B_R(13)$, $B_I(13)$, $B_R(4)$, and $B_I(4)$.

The multiplication portion of the building block requires two additional data memory locations because no temporary registers are assumed. The variables used for the inter-mediate computations were chosen to be the same as those used for the 5-point Singleton building block in Chapter 8 to make it easier to associate the computational steps with the discussion of its features and memory mappings in Chapter 8. This set of computations is represented in Figure 9-20 by 5-point building block 2. Further, the labels on the left and right of this building block correspond to the input and output labels in the 5-point Singleton building block in Chapter 8.

| **Algorithm Steps** | **Memory Map** |
|---|---|

$$b_R(11) = B_R(1) + B_R(4)$$
$$b_I(11) = B_I(1) + B_I(4)$$
$$b_R(12) = B_R(1) - B_R(4)$$
$$b_I(12) = B_I(1) - B_I(4)$$
$$b_R(13) = B_R(7) + B_R(13)$$
$$b_I(13) = B_I(7) + B_I(13)$$
$$b_R(14) = B_R(7) - B_R(13)$$
$$b_I(14) = B_I(7) - B_I(13)$$
$$c_R(12) = b_R(12) * \sin(2\pi/5) + b_R(14) * \sin(4\pi/5)$$
$$c_I(12) = b_I(12) * \sin(2\pi/5) + b_I(14) * \sin(4\pi/5)$$
$$c_R(14) = b_R(12) * \sin(4\pi/5) - b_R(14) * \sin(2\pi/5)$$
$$c_I(14) = b_I(12) * \sin(4\pi/5) - b_I(14) * \sin(2\pi/5)$$
$$c_R(11) = b_R(11) * \cos(2\pi/5) + b_R(13) * \cos(4\pi/5) + B_R(10)$$
$$c_I(11) = b_I(11) * \cos(2\pi/5) + b_I(13) * \cos(4\pi/5) + B_I(10)$$
$$c_R(13) = b_R(11) * \cos(4\pi/5) + b_R(13) * \cos(2\pi/5) + B_R(10)$$
$$c_I(13) = b_I(11) * \cos(4\pi/5) + b_I(13) * \cos(2\pi/5) + B_I(10)$$
$$A_R(5) = B_R(10) + b_R(11) + b_R(13)$$
$$A_I(5) = B_I(10) + b_I(11) + b_I(13)$$
$$A_R(8) = c_R(11) + c_I(12)$$
$$A_I(8) = c_I(11) - c_R(12)$$
$$A_R(11) = c_R(13) + c_I(14)$$
$$A_I(11) = c_I(13) - c_R(14)$$
$$A_R(14) = c_R(13) - c_I(14)$$
$$A_I(14) = c_I(13) + c_R(14)$$
$$A_R(2) = c_R(11) - c_I(12)$$
$$A_I(2) = c_I(11) + c_R(12)$$

Memory Map:
$b_R(11) \Rightarrow M(26)$, $b_I(11) \Rightarrow M(11)$, $b_R(12) \Rightarrow M(29)$, $b_I(12) \Rightarrow M(14)$, $b_R(13) \Rightarrow M(17)$, $b_I(13) \Rightarrow M(2)$, $b_R(14) \Rightarrow M(23)$, $b_I(14) \Rightarrow M(8)$, $c_R(12) \Rightarrow M(30)$, $c_I(12) \Rightarrow M(23)$, $c_R(14) \Rightarrow M(31)$, $c_I(14) \Rightarrow M(29)$, $c_R(11) \Rightarrow M(14)$, $c_I(11) \Rightarrow M(26)$, $c_R(13) \Rightarrow M(8)$, $c_I(13) \Rightarrow M(17)$, $A_R(5) \Rightarrow M(20)$, $A_I(5) \Rightarrow M(5)$, $A_R(8) \Rightarrow M(14)$, $A_I(8) \Rightarrow M(11)$, $A_R(11) \Rightarrow M(8)$, $A_I(11) \Rightarrow M(17)$, $A_R(14) \Rightarrow M(29)$, $A_I(14) \Rightarrow M(26)$, $A_R(2) \Rightarrow M(23)$, $A_I(2) \Rightarrow M(2)$

## 9.7 MIXED-RADIX APPROACH

### 9.7.1 Mixed-Radix Algorithm Introduction

Mixed-radix [5, 6] algorithms are characterized by a sequence of small-point building blocks, from Chapter 8, with complex multipliers between. This sequence of building blocks is developed by factoring the transform length, $N$, into two numbers, $N = P * Q$, and computing the $N$-point transform based on $P$- and $Q$-point building blocks (See Figure 9-21). A description of why that process works can be found in Chapter 3. If $P$ or $Q$ can be further factored, say $Q = R * S$, then the $Q$-point transform can be constructed from two building blocks ($R$- and $S$-point building blocks) using Figure 9-21 as a guide.

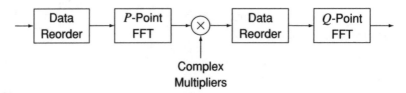

**Figure 9-21**   Top-level two-factor mixed-radix algorithm block diagram.

The result of factoring $N$ into $P * R * S$ is an algorithm that has a series of three building
blocks with complex multipliers between (Figure 9-22). The mixed-radix algorithm allows
this factoring process to stop at any point. The extreme case is to factor $N$ until the building
blocks are only prime numbers. Even if $N$ is factored to all prime numbers, there are
numerous orders in which those primes can be combined to form the complete transform.
The order of the building blocks determines the multiplier constants used between the stages
but does not affect the number of adds and multiplies.

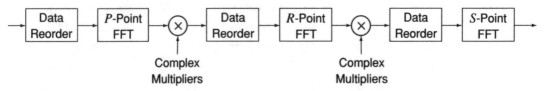

**Figure 9-22**   Top-level three-factor mixed-radix algorithm block diagram.

**Forty-five–Point Example.**   There are two ways to factor 45 into two numbers
($3 * 15$ and $5 * 9$). Therefore, the 45-point transform can be implemented by using the
block diagram in Figure 9-21. The $3 * 15$ option can be implemented with either the 3- or
15-point transform first in Figure 9-21. However, for either the $3 * 15$ or $5 * 9$ cases, the
second factor can be factored further. The result in all three cases is three building blocks
(3, 3, and 5 points). There are three ways of ordering these three numbers to implement the
45-point FFT. To summarize, there are seven ways to implement the 45-point FFT using the
mixed-radix algorithm, without having to choose which algorithm to use for each building
block. These are shown in Table 9-6.

**Table 9-6**   Forty-five–Point Mixed-Radix
Building-Block Sequences

| Sequence choices | $P$ | $R$ | $S$ |
|:---:|:---:|:---:|:---:|
| 1 | 3 | 15 | N/A |
| 2 | 15 | 3 | N/A |
| 3 | 5 | 9 | N/A |
| 4 | 9 | 5 | N/A |
| 5 | 3 | 3 | 5 |
| 6 | 3 | 5 | 3 |
| 7 | 5 | 3 | 3 |

The first four sequence choices only have two building blocks, indicated by N/A under column $S$. The choice of algorithm building blocks from Chapter 8, for all but the 15-point FFT, provides additional options to optimize the implementation for the application. The 15-point FFT can be implemented with any of the algorithms in this chapter.

A derivation of the mixed-radix algorithm shows that the complex multipliers between the $P$- and $Q$-point building blocks have a predictable pattern. If the complex multipliers are viewed as connected to the output of the $P$-point building block, then:

1. The zeroth $P$-point building block has all 1's as output multipliers.

2. The outputs of the other $(Q-1)$ $P$-point building blocks have complex multipliers for all but their top output $D(n)$, which has 1 as the multiplier, for a total of $P-1$ complex multiplies.

3. The complex multiplier at the $k$-th output, $B(k*Q+n)$, of the $n$-th $P$-point building block is $\cos(2*\pi*k*n/N) - j*\sin(2*\pi*k*n/N)$, as shown in Figure 9-23.

4. After multiplication, the $k$-th output, $D(k*Q+n)$, of the $n$-th $P$-point building block is connected to the $n$-th input of the $k$-th $Q$-point building block shown in Figure 9-24.

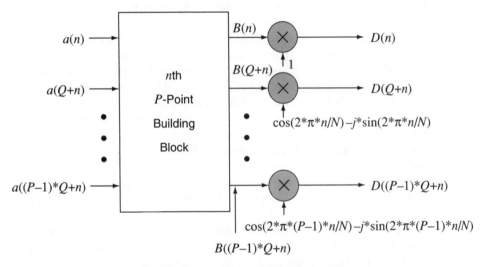

**Figure 9-23**    $n$-th $P$-point building-block output's complex multipliers.

Comments 1 and 2, combined with Figure 9-23, show that there are $Q-1$ of the $P$-point building blocks that each have $P-1$ complex multiplies on the output for a total of $(Q-1)*(P-1)$ complex multiplies.

If the $N$-point transform is further decomposed into three or more factors, say by factoring $Q$, these same four facts determine the number of building blocks and complex multiplier constants needed for each of the decomposed $Q$-point transforms. The only change is to replace $N$ with $Q$ and to replace $Q$ with $R$ and $S$, where $Q = R*S$. With this information and the algorithm building blocks from Chapter 8, a complete block diagram can be constructed for a transform of any length with several combinations of building blocks.

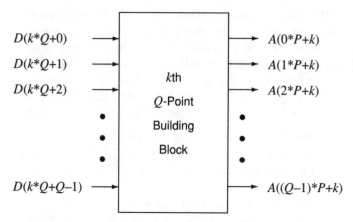

**Figure 9-24** $k$-th $Q$-point building-block input's origins.

### 9.7.2 Number of Mixed-Radix Algorithm Adds and Multiplies

The number of real adds and multiplies is the sum of those required for the algorithm building blocks and those required by the complex multiplies between the building blocks. This subsection develops these equations for the number of adds and multiplies for $N$-point transforms that have been decomposed into two or three algorithm building blocks. It also describes a straightforward procedure to use to determine the number of adds and multiplies for an $N$-point transform comprising any number of algorithm building blocks.

Since there are $(N/P_i)$ of the $P_i$-point building blocks, the number of adds and multiplies contributed by these building blocks is just $(N/P_i)$ times the number of real adds and multiplies required by the $P_i$-point algorithm building block. These numbers are listed explicitly in the Comparison Matrix in Chapter 8 for $P_i = 2, 3, 4, 5, 7, 8, 9$, and 16. An equation is also provided in that Comparison Matrix for computing the number of adds and multiplies for all other prime numbers.

To determine the number of complex multiplies required between the building blocks, start with the two building blocks $P$ and $Q$. From Section 9.7.1, the number of complex multiplies is $(Q - 1) * (P - 1)$, regardless of whether $P$ or $Q$ is first. Since each complex multiply has real and imaginary parts, they each require two memory locations for storing multiplier constants and $4 * (P - 1) * (Q - 1)$ real multiplies and $2 * (P - 1) * (Q - 1)$ real adds. In practice, this can be reduced because some of these constants will be the same. However, taking advantage of these symmetries usually requires a more complex memory mapping. Therefore, for the algorithms presented, assume this worst-case number of memory locations for constants and a simple memory mapping. The specific examples for each algorithm illustrate some of the symmetries of the complex multiplier coefficients that can be used to advantage.

If the $Q$-point building block is further decomposed into $R$- and $S$-point building blocks, then $(S - 1) * (R - 1)$ additional complex multiplies are required for each $Q$-point building block. Since there are $P$ of these $Q$-point building blocks, $P * (S - 1) * (R - 1)$ additional complex multiplies are required. There are $N/P$ $P$-point, $N/R$ $R$-point, and $N/S$ $S$-point building blocks to compute. This fact allows the number of complex multiplies to be easily determined if one of these three factors is further decomposed into two factors.

For $P$, $R$, and $S$, the total number of complex multiplies is $2 * P * R * S - R * S - P * S - P * R + 1$. This total does not change as the sequence of using $P$, $R$, and $S$ changes. Since the number of $P$-, $R$-, and $S$-point building blocks also does not depend on the order in which they are used, the total number of adds and multiplies does not depend on the order of the factors in the algorithm.

The add and multiply totals for the 2-, 3-, 4-, 5-, 7-, 8-, 9-, and 16-point building blocks are in the Chapter 8 Comparison Matrix. Together with four multiplies and two adds for each complex multiply between the building blocks, the total number of real adds and multiplies for an $N$-point transform, where $N$ is factored into two building blocks, $P$ and $Q$, is:

$$\# \, \text{adds} = P * A_Q + Q * A_P + 2 * (P - 1) * (Q - 1)$$
$$\# \, \text{multiplies} = P * M_Q + Q * M_P + 4 * (P - 1) * (Q - 1)$$

where:   $A_Q =$ number of real adds in $Q$-point algorithm building block
$A_P =$ number of real adds in $P$-point algorithm building block
$M_Q =$ number of real multiplies in $Q$-point algorithm building block
$M_P =$ number of real multiplies in $P$-point algorithm building block

If $N$ is factored into three building blocks ($P$, $R$, and $S$), the total number of real adds and multiplies for an $N$-point transform is:

$$\# \, \text{adds} = (N/P) * A_P + (N/R) * A_R + (N/S) * A_S$$
$$+ 2 * (2 * N - R * S - P * S - P * R + 1)$$
$$\# \, \text{multiplies} = (N/P) * M_P + (N/R) * M_R + (N/S) * M_S$$
$$+ 4 * (2 * N - R * S - P * S - P * R + 1)$$

where:   $A_P =$ number of real adds in $P$-point algorithm building block
$A_R =$ number of real adds in $R$-point algorithm building block
$A_S =$ number of real adds in $S$-point algorithm building block
$M_P =$ number of real multiplies in $P$-point algorithm building block
$M_R =$ number of real multiplies in $R$-point algorithm building block
$M_S =$ number of real multiplies in the $S$-point algorithm building block

### 9.7.3 Categories of the Mixed-Radix Algorithm

The mixed-radix algorithms fall into three categories but can all be described by the general mixed-radix algorithm in Section 9.7.4. The first has the same algorithm building block in each block in Figures 9-21 and 9-22. This is illustrated in Section 9.7.5 with a 16-point ($4 * 4$) power-of-primes example. The second category of mixed-radix algorithms has different powers of the same prime in the various building blocks. This category is illustrated in Section 9.7.6 with a 16-point ($8 * 2$) power-of-primes example. The third mixed-radix category allows any of the algorithm building blocks from Chapter 8 to be used. In Section 9.7.7, a 15-point example is used to illustrate this category.

### 9.7.4 General Mixed-Radix Algorithm for Two Factors

Since the mixed-radix algorithm is constructed by repeatedly factoring an integer into two other integers, the general mixed-radix algorithm is completely described by the equations required to factor $N$ into two factors as depicted in Figure 9-21. To construct a mixed-radix algorithm for three factors ($P$, $R$, $S$, where $Q = R * S$), follow the algorithm in Stages 1 through 6 to form a two-factor decomposition. Then, for each of the $P$ $Q$-point building blocks, relabel its inputs as if they were $Q$ consecutive complex data points and reapply the two-factor decomposition algorithm to split the $Q$-point building block into two factors. Each of those can be further subdivided with the same approach. The relabeling scheme is given in Section 9.4.

The algorithm starts by grouping the input data points for each of the $Q$ $P$-point building blocks (Stage 1, Step 1) and computing the $Q$ $P$-point building blocks with these data subsets as inputs (Stage 1, Step 2). Then the outputs of the $P$-point building blocks are multiplied by the proper complex numbers (Stage 2 and as shown in Figure 9-23). To complete the algorithm, the outputs of the complex multiplications are reorganized and fed to the $P$ $Q$-point building blocks (Stage 3, Step 1 as shown in Figure 9-24). Finally, the $P$ $Q$-point building blocks convert their input data to the output frequency components (Stage 3, Step 2).

## Stage 1: Input *P*-Point Building Blocks

This stage has two steps. The first is to properly group the input data for each of the $Q$ $P$-point building blocks. The second is to compute each of the $Q$ $P$-point building blocks. The number of adds and multiplies required for this stage is $Q$ times the number of adds and multiplies required for the chosen $P$-point building block. Since the $P$-point building blocks are performed sequentially, any additional memory required for the $P$-point building block is only needed once. The reason is that each $P$-point building block uses these additional locations in sequence, not all at once. Therefore, the total memory required for this portion of the algorithm is $2 * N$ for the data plus the additional locations needed for one $P$-point building block.

*Step* 1: *Grouping the Input Data Points for the P-Point Building Blocks*

For the $k$-th input to the $n$-th $P$-point building block, choose $a_R(k * Q + n)$ and $a_I(k * Q + n)$ (where $k = 0, 1, \ldots, (P - 1)$ and $n = 0, 1, \ldots, (Q - 1)$) from the input data sequence as shown in Figure 9-23.

*Step* 2: *Computing the Q P-Point Building Blocks*

Use the complex input data points defined in Step 1 to compute the outputs of each of the $Q$ $P$-point building blocks. The $k$-th output of the $n$-th $P$-point building block should be labeled $B_R(k * Q + n)$ and $B_I(k * Q + n)$ in preparation for input to the complex multiply portion of the algorithm.

## Stage 2: Complex Multiplications

Each output from the $P$-point building blocks is multiplied by a specific complex number prior to entering the $Q$-point portion of the overall algorithm. The equations for this complex multiplication for each $k = 0, 1, \ldots, (P - 1)$ and $n = 0, 1, \ldots, (Q - 1)$ are:

$$D_R(k * Q + n) = B_R(k * Q + n) * \cos(2\pi * kn/N) + B_I(k * Q + n) * \sin(2\pi * kn/N)$$
$$D_I(k * Q + n) = B_I(k * Q + n) * \cos(2\pi * kn/N) - B_R(k * Q + n) * \sin(2\pi * kn/N)$$

If no temporary registers are assumed in the processor performing the algorithm, each complex multiply required two additional data memory locations to store the results of multiplying each input value by two different constants prior to forming and storing the output results. Figure 9-23 illustrates this stage of the algorithm for the $n$-th $P$-point building block. Since the complex multiplies are performed one at a time, only two additional memory locations are required. In the 16-point radix-4 example (Section 9.7.5), the multiplies are all grouped together. This requires two additional memory locations for each of the complex multiplies. The 16-point radix-8 and -2 example (Section 9.7.6) and the 15-point Singleton example (Section 9.7.7) reduce the added memory locations required at the expense of interweaving adds with the multiplies. Details of the architectures in Chapters 11 and 12 determine which approach is best for an application.

### Stage 3: Output *Q*-Point Building Blocks

This stage has two steps. The first is to properly group the input data for each of the $P$ $Q$-point building blocks. The second is to compute each of the $P$ $Q$-point building blocks. The number of adds and multiplies required for this stage is $P$ times the number of adds and multiplies required for the chosen $Q$-point building block. Since the $Q$-point building blocks are performed sequentially, any additional memory required for the $Q$-point building block is only needed once. This is because each $Q$-point building block uses these additional locations in sequence, not all at once. Therefore, the total memory required for this portion of the algorithm is $2 * N$ for the data plus the additional locations needed for one $Q$-point building block.

### Step 1: *Grouping the Input Data Points to the Q-Point Building Blocks*

For the $n$-th input to the $k$-th $Q$-point building block, choose $D_R(k * Q + n)$ and $D_I(k * Q + n)$ (where $k = 0, 1, \ldots, (P - 1)$ and $n = 0, 1, \ldots, (Q - 1)$) from the input data sequence. Each input to a $Q$-point building block comes from a different $P$-point building-block output. Therefore, the data memory locations where the required input data reside are not in the order assumed by the $Q$-point building blocks in Chapter 8. To further complicate this, the output data memory address order for the $P$-point building blocks in Chapter 8 is not in order. Therefore, to use the building-block algorithms from Chapter 8, the specified data memory locations must be relabeled. This process is straightforward and completely described in Section 9.4.

### Step 2: *Computing the P Q-Point Building Blocks*

Use the complex input data points defined in Step 1 to compute each of the $P$ $Q$-point building blocks. The $n$-th output of the $k$-th $Q$-point building block should be labeled $A_R(n * P + k)$ and $A_I(n * P + k)$. These are the final outputs of the $N$-point FFT.

### 9.7.5 Sixteen-Point Radix-4 Primes-to-a-Power FFT Example

The primes-to-a-power [5, 6] algorithm requires each FFT building block in Figures 9-21 or 9-22 to have the same algorithm building block. The power-of-two algorithms, made popular by the 1965 Cooley and Tukey paper [6], are in this class. They are a set of algorithms for computing an $N$-point DFT, where $N = 2^P$, and $P$ is any positive integer. For example, $N = 64$ ($2^6$), $N = 256$ ($2^8$), and $N = 1024$ ($2^{10}$). Since 4, 8, and 16 are also powers-of-two, the 2-, 4-, 8-, or 16-point building blocks can be inserted into Figures 9-21 and 9-22 to produce a transform from this category. However, any of the other prime algorithm building blocks could also have been used. For example, an 81-point transform can be implemented by using four blocks with 3-point building blocks or two blocks with 9-point building blocks.

In Figure 9-21, the radix-4 16-point FFT has 4-point building blocks in each of two stages ($P = Q = 4$). It is a five-stage process with 144 adds and 24 multiplications. The equations for adds and multiplies in Section 9.7.2 imply the need for 146 real adds and 36 real multiplies, based on the 4-point building block having 16 real adds and no real multiplies. The actual numbers are reduced by taking advantage of some special-case multiplier constants. Specifically, multiplication by $\cos(8\pi/16) + j * \sin(8\pi/16) = j$ requires no multiplication or addition, and multiplication by $\cos(4\pi/16) + j * \sin(4\pi/16) = (\sqrt{2}) * (1 + j)$ requires only two multiplications.

The storage requirements are 40 locations for data memory and 6 locations for multiplier constants. This is larger than required by the other mixed-radix algorithms, because a different approach to complex multiplication was used in this example to illustrate the difference in storage requirements. Namely, the approach used in this example computed all of the multiplications required for the complex multiplies between the stages and stored the results. Then the adds needed to complete the complex multiplies were performed. It is the multiplies that cause the need for additional data memory locations. Each complex multiply only requires two additional memory locations. Therefore, if each complex multiply is completed before proceeding to the next one, only two additional memory locations are required, making the total 34 rather than 40 locations.

The data mapping shown next to the algorithm steps is an example. Specifically, Stage 1 is the four 4-point building blocks that must be performed on the input. The next two stages provide all of the complex multiplications required between Stages 1 and 3, and the final stage performs the four 4-point output building blocks.

Figure 9-25 is a block diagram of this example that shows the data memory mapping implemented in the detailed algorithm steps. Each 4-point building block is labeled to identify it with the steps of each stage of computation. The numbers inside the left and right edges of the 4-point building blocks are the corresponding input and output labels as defined in Chapter 8. For example, $a(12)$ is the complex input for the terms labeled $a_R(3)$ and $a_I(3)$ in the 4-point building-block description in Chapter 8.

The radix-4 power-of-primes algorithm stages for a 16-point radix-4 FFT are as follows.

### Stage 1: Input 4-Point Building Blocks

This stage does not require additional data memory or accessing any of the multiplier constants. Further, the add/subtract process is the same for all of the real and imaginary pairs. The strategy for converting these equations to code is to start at the top (compute $b_R(0)$) and identify the pair of inputs to be used first (in this case $a_R(0)$ and $a_R(8)$). Then look down the list to find the second (compute $b_R(1)$) place where these two inputs are used. Pull $a_R(1)$ and $a_R(8)$ from memory, compute $b_R(0)$ and $b_R(1)$, and store the results in memory locations $M(0)$ and $M(8)$, previously occupied by $a_R(0)$ and $a_R(8)$. The next step is to look at the next computation $b_I(0)$ on the list and repeat the same set of steps. Continue this process until all the Algorithm Steps in Stage 1 have been computed and their results stored in the Memory Map addresses.

### First of Four 4-Point Building Blocks

This set of computations is represented in Figure 9-25 by input 4-point building block 0. Further, the labels on the left and right of this building block correspond to the input and output labels in the 4-point building block in Chapter 8.

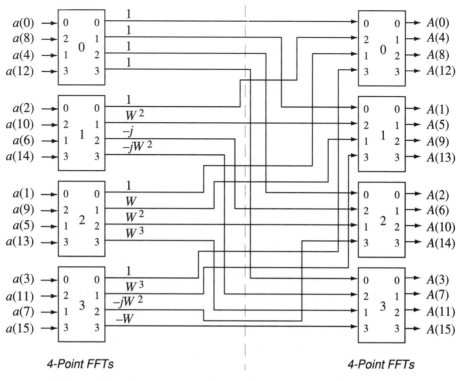

**Figure 9-25**    Sixteen-point radix-4 primes-to-a-power block diagram.

| **Algorithm Steps** | **Memory Map** |
|---|---|
| $b_R(0) = a_R(0) + a_R(8)$ | $b_R(0) \Rightarrow M(0)$ |
| $b_I(0) = a_I(0) + a_I(8)$ | $b_I(0) \Rightarrow M(16)$ |
| $b_R(1) = a_R(0) - a_R(8)$ | $b_R(1) \Rightarrow M(8)$ |
| $b_I(1) = a_I(0) - a_I(8)$ | $b_I(1) \Rightarrow M(24)$ |
| $b_R(2) = a_R(4) + a_R(12)$ | $b_R(2) \Rightarrow M(4)$ |
| $b_I(2) = a_I(4) + a_I(12)$ | $b_I(2) \Rightarrow M(20)$ |
| $b_R(3) = a_R(4) - a_R(12)$ | $b_R(3) \Rightarrow M(12)$ |
| $b_I(3) = a_I(4) - a_I(12)$ | $b_I(3) \Rightarrow M(28)$ |
| $c_R(0) = b_R(0) + b_R(2)$ | $c_R(0) \Rightarrow M(0)$ |
| $c_I(0) = b_I(0) + b_I(2)$ | $c_I(0) \Rightarrow M(16)$ |
| $c_R(1) = b_R(1) + b_I(3)$ | $c_R(1) \Rightarrow M(8)$ |
| $c_I(1) = b_I(1) - b_R(3)$ | $c_I(1) \Rightarrow M(24)$ |
| $c_R(2) = b_R(0) - b_R(2)$ | $c_R(2) \Rightarrow M(4)$ |
| $c_I(2) = b_I(0) - b_I(2)$ | $c_I(2) \Rightarrow M(20)$ |
| $c_R(3) = b_R(1) - b_I(3)$ | $c_R(3) \Rightarrow M(28)$ |
| $c_I(3) = b_I(1) + b_R(3)$ | $c_I(3) \Rightarrow M(12)$ |

### Second of Four 4-Point Building Blocks

This set of computations is represented in Figure 9-25 by input 4-point building block 1. Further, the labels on the left and right of this building block correspond to the input and output labels in the 4-point building block in Chapter 8.

| Algorithm Steps | Memory Map |
|---|---|
| $b_R(4) = a_R(2) + a_R(10)$ | $b_R(4) \Rightarrow M(2)$ |
| $b_I(4) = a_I(2) + a_I(10)$ | $b_I(4) \Rightarrow M(18)$ |
| $b_R(5) = a_R(2) - a_R(10)$ | $b_R(5) \Rightarrow M(10)$ |
| $b_I(5) = a_I(2) - a_I(10)$ | $b_I(5) \Rightarrow M(26)$ |
| $b_R(6) = a_R(6) + a_R(14)$ | $b_R(6) \Rightarrow M(6)$ |
| $b_I(6) = a_I(6) + a_I(14)$ | $b_I(6) \Rightarrow M(22)$ |
| $b_R(7) = a_R(6) - a_R(14)$ | $b_R(7) \Rightarrow M(14)$ |
| $b_I(7) = a_I(6) - a_I(14)$ | $b_I(7) \Rightarrow M(30)$ |
| $c_R(4) = b_R(4) + b_R(6)$ | $c_R(4) \Rightarrow M(2)$ |
| $c_I(4) = b_I(4) + b_I(6)$ | $c_I(4) \Rightarrow M(18)$ |
| $c_R(5) = b_R(5) + b_I(7)$ | $c_R(5) \Rightarrow M(10)$ |
| $c_I(5) = b_I(5) - b_R(7)$ | $c_I(5) \Rightarrow M(26)$ |
| $c_R(6) = b_R(4) - b_R(6)$ | $c_R(6) \Rightarrow M(6)$ |
| $c_I(6) = b_I(4) - b_I(6)$ | $c_I(6) \Rightarrow M(22)$ |
| $c_R(7) = b_R(5) - b_I(7)$ | $c_R(7) \Rightarrow M(30)$ |
| $c_I(7) = b_I(5) + b_R(7)$ | $c_I(7) \Rightarrow M(14)$ |

### Third of Four 4-Point Building Blocks

This set of computations is represented in Figure 9-25 by input 4-point building block 2. Further, the labels on the left and right of this building block correspond to the input and output labels in the 4-point building block in Chapter 8.

| Algorithm Steps | Memory Map |
|---|---|
| $b_R(8) = a_R(1) + a_R(9)$ | $b_R(8) \Rightarrow M(1)$ |
| $b_I(8) = a_I(1) + a_I(9)$ | $b_I(8) \Rightarrow M(17)$ |
| $b_R(9) = a_R(1) - a_R(9)$ | $b_R(9) \Rightarrow M(9)$ |
| $b_I(9) = a_I(1) - a_I(9)$ | $b_I(9) \Rightarrow M(25)$ |
| $b_R(10) = a_R(5) + a_R(13)$ | $b_R(10) \Rightarrow M(5)$ |
| $b_I(10) = a_I(5) + a_I(13)$ | $b_I(10) \Rightarrow M(21)$ |
| $b_R(11) = a_R(5) - a_R(13)$ | $b_R(11) \Rightarrow M(13)$ |
| $b_I(11) = a_I(5) - a_I(13)$ | $b_I(11) \Rightarrow M(29)$ |
| $c_R(8) = b_R(8) + b_R(10)$ | $c_R(8) \Rightarrow M(1)$ |
| $c_I(8) = b_I(8) + b_I(10)$ | $c_I(8) \Rightarrow M(17)$ |
| $c_R(9) = b_R(9) + b_I(11)$ | $c_R(9) \Rightarrow M(9)$ |
| $c_I(9) = b_I(9) - b_R(11)$ | $c_I(9) \Rightarrow M(25)$ |
| $c_R(10) = b_R(8) - b_R(10)$ | $c_R(10) \Rightarrow M(5)$ |
| $c_I(10) = b_I(8) - b_I(10)$ | $c_I(10) \Rightarrow M(21)$ |
| $c_R(11) = b_R(9) - b_I(11)$ | $c_R(11) \Rightarrow M(29)$ |
| $c_I(11) = b_I(9) + b_R(11)$ | $c_I(11) \Rightarrow M(13)$ |

*Fourth of Four 4-Point Building Blocks*

This set of computations is represented in Figure 9-25 by input 4-point building block 3. Further, the labels on the left and right of this building block correspond to the input and output labels in the 4-point building block in Chapter 8.

| Algorithm Steps | Memory Map |
|---|---|
| $b_R(12) = a_R(3) + a_R(11)$ | $b_R(12) \Rightarrow M(3)$ |
| $b_I(12) = a_I(3) + a_I(11)$ | $b_I(12) \Rightarrow M(19)$ |
| $b_R(13) = a_R(3) - a_R(11)$ | $b_R(13) \Rightarrow M(11)$ |
| $b_I(13) = a_I(3) - a_I(11)$ | $b_I(13) \Rightarrow M(27)$ |
| $b_R(14) = a_R(7) + a_R(15)$ | $b_R(14) \Rightarrow M(7)$ |
| $b_I(14) = a_I(7) + a_I(15)$ | $b_I(14) \Rightarrow M(23)$ |
| $b_R(15) = a_R(7) - a_R(15)$ | $b_R(15) \Rightarrow M(15)$ |
| $b_I(15) = a_I(7) - a_I(15)$ | $b_I(15) \Rightarrow M(31)$ |
| $c_R(12) = b_R(12) + b_R(14)$ | $c_R(12) \Rightarrow M(3)$ |
| $c_I(12) = b_I(12) + b_I(14)$ | $c_I(12) \Rightarrow M(19)$ |
| $c_R(13) = b_R(13) + b_I(15)$ | $c_R(13) \Rightarrow M(11)$ |
| $c_I(13) = b_I(13) - b_R(15)$ | $c_I(13) \Rightarrow M(27)$ |
| $c_R(14) = b_R(12) - b_R(14)$ | $c_R(14) \Rightarrow M(7)$ |
| $c_I(14) = b_I(12) - b_I(14)$ | $c_I(14) \Rightarrow M(23)$ |
| $c_R(15) = b_R(13) - b_I(15)$ | $c_R(15) \Rightarrow M(31)$ |
| $c_I(15) = b_I(13) + b_R(15)$ | $c_I(15) \Rightarrow M(15)$ |

## Stage 2: Complex Multiplies

This stage contains all of the multiplications. In all cases the multiplication is performed by pulling a data value from memory, multiplying it by the appropriate constant, and returning the result to data memory. The required multiplications are complex and therefore require four real multiplies. Therefore, each input data value gets multiplied twice. Since this algorithm assumes no temporary data locations, additional data memory locations are required. The complex multiplier to be applied to the $k$-th output of the $m$-th 4-point algorithm building block, $B_R(4*k+m) + B_I(4*k+m)$, is $\cos(2*\pi*k*m/16) + j*\sin(2*\pi*k*m/16)$.

In general, additional data locations are required for each of the complex multiplies. However, in the case of the complex multiplies for $c_R(5)$, $c_R(7)$, $c_R(10)$, $c_R(14)$, $c_I(5)$, $c_I(7)$, $c_I(10)$, and $c_I(14)$, the real and imaginary parts of the complex multiplier are equal $(\sin(\pi/16) = \cos(4\pi/16))$. This allows half the number of multiplications to be performed and removes the need for additional data storage locations. In some of the multiplications, the real part of a complex data value is the input and the output is the imaginary part of an intermediate result. This process provides the required multiplications by $j = \sqrt{-1}$. Also, $\sin(4\pi/16)$ equals $\cos(4\pi/16)$, which reduces the total number of constants to be stored to 6.

The approach used in this example is to perform all of the required multiplies and then combine these results with additions to complete the computation of the complex multiplies. This approach requires the most additional memory locations but does segregate the adds and multiplies. The approach used in the 16-point radix-8 and -2 and the 15-point Singleton examples completes each complex multiply before proceeding to the next to reduce the additional memory locations required to two. Hardware architectures, discussed in Chapters 11 and 12, will determine which of these two approaches is preferable.

*Complex Multiply Multiplications*

| **Algorithm Steps** | **Memory Map** |
|---|---|
| $d_R(5) = \cos(4\pi/16) * c_R(5)$ | $d_R(5) \Rightarrow M(10)$ |
| $d_I(5) = \sin(4\pi/16) * c_I(5)$ | $d_I(5) \Rightarrow M(26)$ |
| $d_R(7) = \cos(4\pi/16) * c_R(7)$ | $d_R(7) \Rightarrow M(30)$ |
| $d_I(7) = \sin(4\pi/16) * c_I(7)$ | $d_I(7) \Rightarrow M(14)$ |
| $d_R(17) = \sin(2\pi/16) * c_R(9)$ | $d_R(17) \Rightarrow M(32)$ |
| $d_R(9) = \cos(2\pi/16) * c_R(9)$ | $d_R(9) \Rightarrow M(9)$ |
| $d_I(17) = \cos(2\pi/16) * c_I(9)$ | $d_I(17) \Rightarrow M(36)$ |
| $d_I(9) = \sin(2\pi/16) * c_I(9)$ | $d_I(9) \Rightarrow M(25)$ |
| $d_R(10) = \cos(4\pi/16) * c_R(10)$ | $d_R(10) \Rightarrow M(5)$ |
| $d_I(10) = \cos(4\pi/16) * c_I(10)$ | $d_I(10) \Rightarrow M(21)$ |
| $d_R(18) = \sin(6\pi/16) * c_R(11)$ | $d_R(18) \Rightarrow M(33)$ |
| $d_I(18) = \sin(6\pi/16) * c_I(11)$ | $d_I(18) \Rightarrow M(37)$ |
| $d_R(11) = \cos(6\pi/16) * c_R(11)$ | $d_R(11) \Rightarrow M(29)$ |
| $d_I(11) = \cos(6\pi/16) * c_I(11)$ | $d_I(11) \Rightarrow M(13)$ |
| $d_R(19) = \sin(6\pi/16) * c_R(13)$ | $d_R(19) \Rightarrow M(34)$ |
| $d_I(19) = \sin(6\pi/16) * c_I(13)$ | $d_I(19) \Rightarrow M(38)$ |
| $d_R(13) = \cos(6\pi/16) * c_R(13)$ | $d_R(13) \Rightarrow M(11)$ |
| $d_I(13) = \cos(6\pi/16) * c_I(13)$ | $d_I(13) \Rightarrow M(27)$ |
| $d_R(14) = \cos(4\pi/16) * c_R(14)$ | $d_R(14) \Rightarrow M(7)$ |
| $d_I(14) = \cos(4\pi/16) * c_I(14)$ | $d_I(14) \Rightarrow M(23)$ |
| $d_R(20) = \sin(2\pi/16) * c_R(15)$ | $d_R(20) \Rightarrow M(35)$ |
| $d_I(20) = \sin(2\pi/16) * c_I(15)$ | $d_I(20) \Rightarrow M(39)$ |
| $d_R(15) = \cos(2\pi/16) * c_R(15)$ | $d_R(15) \Rightarrow M(31)$ |
| $d_I(15) = \cos(2\pi/16) * c_I(15)$ | $d_I(15) \Rightarrow M(15)$ |

*Complex Multiply Additions*

These steps combine the multiplications to form the complex multiplies required between the two sets of 4-point building blocks. Once these are combined there is no further need to use the additional data memory locations. Therefore, the addressing example for this step finishes with the output data being stored in the original 32 data memory locations.

Again, the strategy for converting these equations to code is to start at the top (compute $e_R(5)$) and identify the pair of inputs to be used first (in this case $d_R(5)$ and $d_I(5)$). Then look down the list to find the second (compute $e_I(5)$) place where these two inputs are used. Pull $d_R(5)$ and $d_I(5)$ from memory, compute $e_R(5)$ and $e_I(5)$, and store the results in memory locations $M(10)$ and $M(26)$, previously occupied by $d_R(5)$ and $d_I(5)$.

The next step is to swap the data memory locations for $c_R(6)$ and $c_I(6)$. This is accomplished by loading $c_R(6)$ and $c_I(6)$ into the computational unit and then storing them in the opposite memory locations from the ones they were taken from. Clearly this is not a requirement. It was done in this algorithm so that the output of each of the computational steps has the real part in the lower portion of data memory, and the imaginary part is in the upper portion of data memory. Continue this process until all the Algorithm Steps in Stage 2 have been computed and their results stored in the Memory Map addresses.

| Algorithm Steps | Memory Map |
|---|---|
| $e_R(5) = d_R(5) + d_I(5)$ | $e_R(5) \Rightarrow M(10)$ |
| $e_I(5) = -d_R(5) + d_I(5)$ | $e_I(5) \Rightarrow M(26)$ |
| $e_R(6) = c_I(6)$ | $e_R(6) \Rightarrow M(6)$ |
| $e_I(6) = -c_R(6)$ | $e_I(6) \Rightarrow M(22)$ |
| $e_R(7) = -d_R(7) + d_I(7)$ | $e_R(7) \Rightarrow M(14)$ |
| $e_I(7) = -d_R(7) - d_I(7)$ | $e_I(7) \Rightarrow M(30)$ |
| $e_R(9) = d_R(9) + d_I(9)$ | $e_R(9) \Rightarrow M(9)$ |
| $e_I(9) = -d_R(17) + d_I(17)$ | $e_I(9) \Rightarrow M(25)$ |
| $e_R(10) = d_R(10) + d_I(10)$ | $e_R(10) \Rightarrow M(5)$ |
| $e_I(10) = -d_R(10) + d_I(10)$ | $e_I(10) \Rightarrow M(21)$ |
| $e_R(11) = d_R(11) + d_I(18)$ | $e_R(11) \Rightarrow M(29)$ |
| $e_I(11) = -d_R(18) + d_I(11)$ | $e_I(11) \Rightarrow M(13)$ |
| $e_R(13) = d_R(13) + d_I(19)$ | $e_R(13) \Rightarrow M(11)$ |
| $e_I(13) = -d_R(19) + d_I(13)$ | $e_I(13) \Rightarrow M(27)$ |
| $e_R(14) = -d_R(14) + d_I(14)$ | $e_R(14) \Rightarrow M(7)$ |
| $e_I(14) = -d_R(14) - d_I(14)$ | $e_I(14) \Rightarrow M(23)$ |
| $e_R(15) = -d_R(15) - d_I(20)$ | $e_R(15) \Rightarrow M(31)$ |
| $e_I(15) = d_R(20) - d_I(15)$ | $e_I(15) \Rightarrow M(15)$ |

### Stage 3: Output 4-Point Building Blocks

This stage does not require additional memory locations. However, $f_R(8)$, $f_R(9)$, $f_I(8)$, and $f_I(9)$ use real and imaginary inputs to simulate multiplication by $j = \sqrt{-1}$. The result is that the real part of the output is stored in the upper half of the allotted data memory, and the imaginary part in the lower half.

The strategy for converting the equations to code is to start at the top (compute $f_R(0)$) and identify the pair of inputs to be used first (in this case $c_R(0)$ and $c_R(4)$). Then look down the list to find the second (compute $f_R(1)$) place where these two inputs are used. Pull $c_R(0)$ and $c_R(4)$ from memory, compute $f_R(0)$ and $f_R(1)$, and store the results in memory locations $M(0)$ and $M(2)$, previously occupied by $c_R(0)$ and $c_R(4)$. The next step is to look at the next computation $f_I(0)$ on the list and repeat the same set of steps. Continue this process until all the Algorithm Steps in Stage 3 have been computed and their results stored in the Memory Map addresses.

### First of Four 4-Point Building Blocks

This set of computations is represented in Figure 9-25 by output 4-point building block 0. Further, the labels on the left and right of this building block correspond to the input and output labels in the 4-point building block in Chapter 8.

| Algorithm Steps | Memory Map |
|---|---|
| $f_R(0) = c_R(0) + c_R(4)$ | $f_R(0) \Rightarrow M(0)$ |
| $f_I(0) = c_I(0) + c_I(4)$ | $f_I(0) \Rightarrow M(16)$ |
| $f_R(1) = c_R(0) - c_R(4)$ | $f_R(1) \Rightarrow M(2)$ |
| $f_I(1) = c_I(0) - c_I(4)$ | $f_I(1) \Rightarrow M(18)$ |
| $f_R(2) = c_R(8) + c_R(12)$ | $f_R(2) \Rightarrow M(1)$ |
| $f_I(2) = c_I(8) + c_I(12)$ | $f_I(2) \Rightarrow M(17)$ |
| $f_R(3) = c_R(8) - c_R(12)$ | $f_R(3) \Rightarrow M(3)$ |
| $f_I(3) = c_I(8) - c_I(12)$ | $f_I(3) \Rightarrow M(19)$ |
| $A_R(0) = f_R(0) + f_R(2)$ | $A_R(0) \Rightarrow M(0)$ |
| $A_I(0) = f_I(0) + f_I(2)$ | $A_I(0) \Rightarrow M(16)$ |
| $A_R(8) = f_R(0) - f_R(2)$ | $A_R(8) \Rightarrow M(1)$ |
| $A_I(8) = f_I(0) - f_I(2)$ | $A_I(8) \Rightarrow M(17)$ |
| $A_R(4) = f_R(1) + f_I(3)$ | $A_R(4) \Rightarrow M(2)$ |
| $A_I(4) = f_R(1) - f_I(3)$ | $A_I(4) \Rightarrow M(18)$ |
| $A_R(12) = f_R(1) - f_I(3)$ | $A_R(12) \Rightarrow M(19)$ |
| $A_I(12) = f_I(1) + f_R(3)$ | $A_I(12) \Rightarrow M(3)$ |

*Second of Four 4-Point Building Blocks*

This set of computations is represented in Figure 9-25 by output 4-point building block 1. Further, the labels on the left and right of this building block correspond to the input and output labels in the 4-point building block in Chapter 8.

| Algorithm Steps | Memory Map |
|---|---|
| $f_R(4) = c_R(1) + e_R(5)$ | $f_R(4) \Rightarrow M(8)$ |
| $f_I(4) = c_I(1) + e_I(5)$ | $f_I(4) \Rightarrow M(24)$ |
| $f_R(5) = c_R(1) - e_R(5)$ | $f_R(5) \Rightarrow M(10)$ |
| $f_I(5) = c_I(1) - e_I(5)$ | $f_I(5) \Rightarrow M(26)$ |
| $f_R(6) = e_R(9) + e_R(13)$ | $f_R(6) \Rightarrow M(9)$ |
| $f_I(6) = e_I(9) + e_I(13)$ | $f_I(6) \Rightarrow M(25)$ |
| $f_R(7) = e_R(9) - e_R(13)$ | $f_R(7) \Rightarrow M(11)$ |
| $f_I(7) = e_I(9) - e_I(13)$ | $f_I(7) \Rightarrow M(27)$ |
| $A_R(1) = f_R(4) + f_R(6)$ | $A_R(1) \Rightarrow M(8)$ |
| $A_I(1) = f_I(4) + f_I(6)$ | $A_I(1) \Rightarrow M(24)$ |
| $A_R(5) = f_R(5) + f_I(7)$ | $A_R(5) \Rightarrow M(10)$ |
| $A_I(5) = f_I(5) - f_R(7)$ | $A_I(5) \Rightarrow M(26)$ |
| $A_R(9) = f_R(4) - f_R(6)$ | $A_R(9) \Rightarrow M(9)$ |
| $A_I(9) = f_I(4) - f_I(6)$ | $A_I(9) \Rightarrow M(25)$ |
| $A_R(13) = f_R(5) - f_I(7)$ | $A_R(13) \Rightarrow M(27)$ |
| $A_I(13) = f_I(5) + f_R(7)$ | $A_I(13) \Rightarrow M(11)$ |

### Third of Four 4-Point Building Blocks

This set of computations is represented in Figure 9-25 by output 4-point building block 2. Further, the labels on the left and right of this building block correspond to the input and output labels in the 4-point building block in Chapter 8.

| Algorithm Steps | Memory Map |
|---|---|
| $f_R(8) = c_R(2) + c_I(6)$ | $f_R(8) \Rightarrow M(4)$ |
| $f_I(8) = c_I(2) - c_R(6)$ | $f_I(8) \Rightarrow M(20)$ |
| $f_R(9) = c_R(2) - c_I(6)$ | $f_R(9) \Rightarrow M(22)$ |
| $f_I(9) = c_I(2) + c_R(6)$ | $f_I(9) \Rightarrow M(6)$ |
| $f_R(10) = e_R(10) + e_R(14)$ | $f_R(10) \Rightarrow M(5)$ |
| $f_I(10) = e_I(10) + e_I(14)$ | $f_I(10) \Rightarrow M(21)$ |
| $f_R(11) = e_R(10) - e_R(14)$ | $f_R(11) \Rightarrow M(7)$ |
| $f_I(11) = e_I(10) - e_I(14)$ | $f_I(11) \Rightarrow M(23)$ |
| $A_R(2) = f_R(8) + f_R(10)$ | $A_R(2) \Rightarrow M(4)$ |
| $A_I(2) = f_I(8) + f_I(10)$ | $A_I(2) \Rightarrow M(20)$ |
| $A_R(6) = f_R(9) + f_I(11)$ | $A_R(6) \Rightarrow M(22)$ |
| $A_I(6) = f_I(9) - f_R(11)$ | $A_I(6) \Rightarrow M(6)$ |
| $A_R(10) = f_R(8) - f_R(10)$ | $A_R(10) \Rightarrow M(5)$ |
| $A_I(10) = f_I(8) - f_I(10)$ | $A_I(10) \Rightarrow M(21)$ |
| $A_R(14) = f_R(9) - f_I(11)$ | $A_R(14) \Rightarrow M(23)$ |
| $A_I(14) = f_I(9) + f_R(11)$ | $A_I(14) \Rightarrow M(7)$ |

### Fourth of Four 4-Point Building Blocks

This set of computations is represented in Figure 9-25 by output 4-point building block 3. Further, the labels on the left and right of this building block correspond to the input and output labels in the 4-point building block in Chapter 8.

| Algorithm Steps | Memory Map |
|---|---|
| $f_R(12) = c_R(3) + e_R(7)$ | $f_R(12) \Rightarrow M(28)$ |
| $f_I(12) = c_I(3) + e_I(7)$ | $f_I(12) \Rightarrow M(12)$ |
| $f_R(13) = c_R(3) - e_R(7)$ | $f_R(13) \Rightarrow M(14)$ |
| $f_I(13) = c_I(3) - e_I(7)$ | $f_I(13) \Rightarrow M(30)$ |
| $f_R(14) = e_R(11) + e_R(15)$ | $f_R(14) \Rightarrow M(29)$ |
| $f_I(14) = e_I(11) + e_I(15)$ | $f_I(14) \Rightarrow M(13)$ |
| $f_R(15) = e_R(11) - e_R(15)$ | $f_R(15) \Rightarrow M(31)$ |
| $f_I(15) = e_I(11) - e_I(15)$ | $f_I(15) \Rightarrow M(15)$ |
| $A_R(3) = f_R(12) + f_R(14)$ | $A_R(3) \Rightarrow M(28)$ |
| $A_I(3) = f_I(12) + f_I(14)$ | $A_I(3) \Rightarrow M(12)$ |
| $A_R(7) = f_R(13) + f_I(15)$ | $A_R(7) \Rightarrow M(14)$ |
| $A_I(7) = f_I(13) - f_R(15)$ | $A_I(7) \Rightarrow M(30)$ |

| Algorithm Steps | Memory Map |
|---|---|
| $A_R(11) = f_R(12) - f_R(14)$ | $A_R(11) \Rightarrow M(29)$ |
| $A_I(11) = f_I(12) - f_I(14)$ | $A_I(11) \Rightarrow M(13)$ |
| $A_R(15) = f_R(13) - f_I(15)$ | $A_R(15) \Rightarrow M(15)$ |
| $A_I(15) = f_I(13) + f_R(15)$ | $A_I(15) \Rightarrow M(31)$ |

### 9.7.6 Sixteen-Point Radix-8 and -2, Mixed Power-of-Primes Example

The mixed powers-of-primes [7] algorithm computes a transform length that can be written as one prime number raised to a power, but uses different algorithm building blocks in the blocks in Figure 9-21, as long as they are all powers of the same prime number. For example, an 81-point transform has five mixed power-of-primes implementations, namely $3*3*9, 3*9*3, 9*3*3, 3*27$, and $27*3$. The 16-point FFT can be implemented using 8-point and 2-point building blocks. Either the 2- or 8-point building blocks can be first, and any of the 8-point building blocks can be used. This example has the 8-point building blocks first.

The mixed power-of-primes 16-point FFT is a three-stage process with 148 adds and 28 multiplications. The reason these are lower than the general mixed-radix equation is that some of the complex multiplies can be performed with fewer computations because of their specific numerical values. Specifically, multiplication by $\cos(8\pi/16) + j*\sin(8\pi/16) = j$ requires no multiplication or addition, and multiplication by $\cos(4\pi/16) + j*\sin(4\pi/16) = (\sqrt{2})*(1 + j)$ requires only two multiplications.

The storage requirements are 34 locations for data memory and 6 locations for multiplier constants. The input stage implements the 8-point radix-4 and -2 building block from Section 8.8.2. Stage 2 implements the complex multiplications between Stages 1 and 3, and the output stage implements the eight 2-point building blocks from Section 8.3.

Figure 9-26 is a block diagram of this example. Each of the 8- and 2-point building blocks is labeled to identify it with the steps of each stage of computations. The numbers inside the left and right edges of the 8- and 2-point building blocks are the corresponding input and output labels as defined in Chapter 8. For example, $a(12)$ is the complex input for the terms labeled $a_R(6)$ and $a_I(6)$ in the 8-point radix-4 and -2 building-block description in Chapter 8.

The stages are described below.

### Stage 1: Input 8-Point Building Blocks

The strategy for converting these equations to code is to start at the top (compute $b_R(0)$) and identify the pair of inputs to be used first (in this case $a_R(0)$ and $a_R(8)$). Then look down the list to find the second (compute $b_R(1)$) place where these two inputs are used. Pull $a_R(0)$ and $a_R(8)$ from memory, compute $b_R(0)$ and $b_R(1)$, and store the results in memory locations $M(0)$ and $M(8)$, previously occupied by $a_R(0)$ and $a_R(8)$. The next step is to look at the next computation $b_I(0)$ on the list and repeat the same set of steps. Continue this process until all the Algorithm Steps in Stage 1 have been computed and their results stored in the Memory Map addresses.

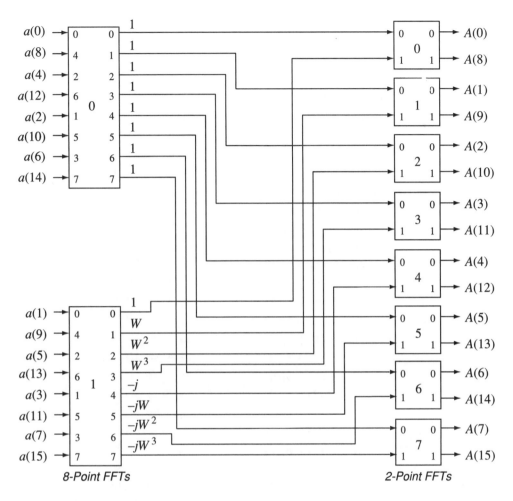

**Figure 9-26**    Sixteen-point radix-8 and -2 mixed power-of-primes block diagram.

### First of Two 8-Point Building Blocks

This set of computations is represented in Figure 9-26 by input 8-point radix-4 and -2 building block 0. Further, the labels on the left and right of this building block correspond to the input and output labels in the 8-point building block in Section 8.8.2.

| Algorithm Steps | Memory Map |
|---|---|
| $b_R(0) = a_R(0) + a_R(8)$ | $b_R(0) \Rightarrow M(0)$ |
| $b_I(0) = a_I(0) + a_I(8)$ | $b_I(0) \Rightarrow M(16)$ |
| $b_R(1) = a_R(0) - a_R(8)$ | $b_R(1) \Rightarrow M(8)$ |
| $b_I(1) = a_I(0) - a_I(8)$ | $b_I(1) \Rightarrow M(24)$ |
| $b_R(2) = a_R(4) + a_R(12)$ | $b_R(2) \Rightarrow M(4)$ |
| $b_I(2) = a_I(4) + a_I(12)$ | $b_I(2) \Rightarrow M(20)$ |
| $b_R(3) = a_R(4) - a_R(12)$ | $b_R(3) \Rightarrow M(12)$ |

| **Algorithm Steps** | **Memory Map** |
|---|---|
| $b_I(3) = a_I(4) - a_I(12)$ | $b_I(3) \Rightarrow M(28)$ |
| $b_R(4) = a_R(2) + a_R(10)$ | $b_R(4) \Rightarrow M(2)$ |
| $b_I(4) = a_I(2) + a_I(10)$ | $b_I(4) \Rightarrow M(18)$ |
| $b_R(5) = a_R(2) - a_R(10)$ | $b_R(5) \Rightarrow M(10)$ |
| $b_I(5) = a_I(2) - a_I(10)$ | $b_I(5) \Rightarrow M(26)$ |
| $b_R(6) = a_R(6) + a_R(14)$ | $b_R(6) \Rightarrow M(6)$ |
| $b_I(6) = a_I(6) + a_I(14)$ | $b_I(6) \Rightarrow M(22)$ |
| $b_R(7) = a_R(6) - a_R(14)$ | $b_R(7) \Rightarrow M(14)$ |
| $b_I(7) = a_I(6) - a_I(14)$ | $b_I(7) \Rightarrow M(30)$ |
| $c_R(0) = b_R(0) + b_R(2)$ | $c_R(0) \Rightarrow M(0)$ |
| $c_I(0) = b_I(0) + b_I(2)$ | $c_I(0) \Rightarrow M(16)$ |
| $c_R(1) = b_R(1) + b_I(3)$ | $c_R(1) \Rightarrow M(8)$ |
| $c_I(1) = b_I(1) - b_R(3)$ | $c_I(1) \Rightarrow M(24)$ |
| $c_R(2) = b_R(0) - b_R(2)$ | $c_R(2) \Rightarrow M(4)$ |
| $c_I(2) = b_I(0) - b_I(2)$ | $c_I(2) \Rightarrow M(20)$ |
| $c_R(3) = b_R(1) - b_I(3)$ | $c_R(3) \Rightarrow M(28)$ |
| $c_I(3) = b_I(1) + b_R(3)$ | $c_I(3) \Rightarrow M(12)$ |
| $c_R(4) = b_R(4) + b_R(6)$ | $c_R(4) \Rightarrow M(2)$ |
| $c_I(4) = b_I(4) + b_I(6)$ | $c_I(4) \Rightarrow M(18)$ |
| $c_R(5) = b_R(5) + b_I(7)$ | $c_R(5) \Rightarrow M(10)$ |
| $c_I(5) = b_I(5) - b_R(7)$ | $c_I(5) \Rightarrow M(26)$ |
| $c_R(6) = b_R(4) - b_R(6)$ | $c_R(6) \Rightarrow M(6)$ |
| $c_I(6) = b_I(4) - b_I(6)$ | $c_I(6) \Rightarrow M(22)$ |
| $c_R(7) = b_R(5) - b_I(7)$ | $c_R(7) \Rightarrow M(30)$ |
| $c_I(7) = b_I(5) + b_R(7)$ | $c_I(7) \Rightarrow M(14)$ |
| $d_R(5) = \cos(4\pi/16) * c_R(5)$ | $d_R(5) \Rightarrow M(10)$ |
| $d_I(5) = \cos(4\pi/16) * c_I(5)$ | $d_I(5) \Rightarrow M(26)$ |
| $d_R(7) = \cos(4\pi/16) * c_R(7)$ | $d_R(7) \Rightarrow M(30)$ |
| $d_I(7) = \cos(4\pi/16) * c_I(7)$ | $d_I(7) \Rightarrow M(14)$ |
| $e_R(5) = d_R(5) + d_I(5)$ | $e_R(5) \Rightarrow M(10)$ |
| $e_I(5) = -d_R(5) + d_I(5)$ | $e_I(5) \Rightarrow M(26)$ |
| $e_R(6) = c_I(6)$ | $e_R(6) \Rightarrow M(22)$ |
| $e_I(6) = -c_R(6)$ | $e_I(6) \Rightarrow M(6)$ |
| $e_R(7) = -d_R(7) + d_I(7)$ | $e_R(7) \Rightarrow M(14)$ |
| $e_I(7) = -d_R(7) - d_I(7)$ | $e_I(7) \Rightarrow M(30)$ |
| $f_R(0) = c_R(0) + c_R(4)$ | $f_R(0) \Rightarrow M(0)$ |
| $f_I(0) = c_I(0) + c_I(4)$ | $f_I(0) \Rightarrow M(16)$ |
| $f_R(1) = c_R(1) + e_R(5)$ | $f_R(1) \Rightarrow M(8)$ |

| Algorithm Steps | Memory Map |
|---|---|
| $f_I(1) = c_I(1) + e_I(5)$ | $f_I(1) \Rightarrow M(24)$ |
| $f_R(2) = c_R(2) + e_R(6)$ | $f_R(2) \Rightarrow M(4)$ |
| $f_I(2) = c_I(2) + e_I(6)$ | $f_I(2) \Rightarrow M(20)$ |
| $f_R(3) = c_R(3) + e_R(7)$ | $f_R(3) \Rightarrow M(28)$ |
| $f_I(3) = c_I(3) + e_I(7)$ | $f_I(3) \Rightarrow M(12)$ |
| $f_R(4) = c_R(0) - c_R(4)$ | $f_R(4) \Rightarrow M(2)$ |
| $f_I(4) = c_I(0) - c_I(4)$ | $f_I(4) \Rightarrow M(18)$ |
| $f_R(5) = c_R(1) - e_R(5)$ | $f_R(5) \Rightarrow M(10)$ |
| $f_I(5) = c_I(1) - e_I(5)$ | $f_I(5) \Rightarrow M(26)$ |
| $f_R(6) = c_R(2) - e_R(6)$ | $f_R(6) \Rightarrow M(22)$ |
| $f_I(6) = c_I(2) - e_I(6)$ | $f_I(6) \Rightarrow M(6)$ |
| $f_R(7) = c_R(3) - e_R(7)$ | $f_R(7) \Rightarrow M(14)$ |
| $f_I(7) = c_I(3) - e_I(7)$ | $f_I(7) \Rightarrow M(30)$ |

### Second of Two 8-Point Building Blocks

This set of computations is represented in Figure 9-26 by input 8-point radix-4 and -2 building block 1. Further, the labels on the left and right of this building block correspond to the input and output labels in the 8-point building block in Section 8.8.2.

| Algorithm Steps | Memory Map |
|---|---|
| $b_R(8) = a_R(1) + a_R(9)$ | $b_R(8) \Rightarrow M(1)$ |
| $b_I(8) = a_I(1) + a_I(9)$ | $b_I(8) \Rightarrow M(17)$ |
| $b_R(9) = a_R(1) - a_R(9)$ | $b_R(9) \Rightarrow M(9)$ |
| $b_I(9) = a_I(1) - a_I(9)$ | $b_I(9) \Rightarrow M(25)$ |
| $b_R(10) = a_R(5) + a_R(13)$ | $b_R(10) \Rightarrow M(5)$ |
| $b_I(10) = a_I(5) + a_I(13)$ | $b_I(10) \Rightarrow M(21)$ |
| $b_R(11) = a_R(5) - a_R(13)$ | $b_R(11) \Rightarrow M(13)$ |
| $b_I(11) = a_I(5) - a_I(13)$ | $b_I(11) \Rightarrow M(29)$ |
| $b_R(12) = a_R(3) + a_R(11)$ | $b_R(12) \Rightarrow M(3)$ |
| $b_I(12) = a_I(3) + a_I(11)$ | $b_I(12) \Rightarrow M(19)$ |
| $b_R(13) = a_R(3) - a_R(11)$ | $b_R(13) \Rightarrow M(11)$ |
| $b_I(13) = a_I(3) - a_I(11)$ | $b_I(13) \Rightarrow M(27)$ |
| $b_R(14) = a_R(7) + a_R(15)$ | $b_R(14) \Rightarrow M(7)$ |
| $b_I(14) = a_I(7) + a_I(15)$ | $b_I(14) \Rightarrow M(23)$ |
| $b_R(15) = a_R(7) - a_R(15)$ | $b_R(15) \Rightarrow M(15)$ |
| $b_I(15) = a_I(7) - a_I(15)$ | $b_I(15) \Rightarrow M(31)$ |
| $c_R(8) = b_R(8) + b_R(10)$ | $c_R(8) \Rightarrow M(1)$ |
| $c_I(8) = b_I(8) + b_I(10)$ | $c_I(8) \Rightarrow M(17)$ |
| $c_R(9) = b_R(9) + b_I(11)$ | $c_R(9) \Rightarrow M(9)$ |
| $c_I(9) = b_I(9) - b_R(11)$ | $c_I(9) \Rightarrow M(25)$ |

| Algorithm Steps | Memory Map |
|---|---|
| $c_R(10) = b_R(8) - b_R(10)$ | $c_R(10) \Rightarrow M(5)$ |
| $c_I(10) = b_I(8) - b_I(10)$ | $c_I(10) \Rightarrow M(21)$ |
| $c_R(11) = b_R(9) - b_I(11)$ | $c_R(11) \Rightarrow M(29)$ |
| $c_I(11) = b_I(9) + b_R(11)$ | $c_I(11) \Rightarrow M(13)$ |
| $c_R(12) = b_R(12) + b_R(14)$ | $c_R(12) \Rightarrow M(3)$ |
| $c_I(12) = b_I(12) + b_I(14)$ | $c_I(12) \Rightarrow M(19)$ |
| $c_R(13) = b_R(13) + b_I(15)$ | $c_R(13) \Rightarrow M(11)$ |
| $c_I(13) = b_I(13) - b_R(15)$ | $c_I(13) \Rightarrow M(27)$ |
| $c_R(14) = b_R(12) - b_R(14)$ | $c_R(14) \Rightarrow M(7)$ |
| $c_I(14) = b_I(12) - b_I(14)$ | $c_I(14) \Rightarrow M(23)$ |
| $c_R(15) = b_R(13) - b_I(15)$ | $c_R(15) \Rightarrow M(31)$ |
| $c_I(15) = b_I(13) + b_R(15)$ | $c_I(15) \Rightarrow M(15)$ |
| $d_R(13) = \cos(4\pi/16) * c_R(13)$ | $d_R(13) \Rightarrow M(11)$ |
| $d_I(13) = \cos(4\pi/16) * c_I(13)$ | $d_I(13) \Rightarrow M(27)$ |
| $d_R(15) = \cos(4\pi/16) * c_R(15)$ | $d_R(15) \Rightarrow M(31)$ |
| $d_I(15) = \cos(4\pi/16) * c_I(15)$ | $d_I(15) \Rightarrow M(15)$ |
| $e_R(13) = d_R(13) + d_I(13)$ | $e_R(13) \Rightarrow M(11)$ |
| $e_I(13) = -d_R(13) + d_I(13)$ | $e_I(13) \Rightarrow M(27)$ |
| $e_R(14) = c_I(14)$ | $e_R(14) \Rightarrow M(23)$ |
| $e_I(14) = -c_R(14)$ | $e_I(14) \Rightarrow M(7)$ |
| $e_R(15) = -d_R(15) + d_I(15)$ | $e_R(15) \Rightarrow M(15)$ |
| $e_I(15) = -d_R(15) - d_I(15)$ | $e_I(15) \Rightarrow M(31)$ |
| $f_R(8) = c_R(8) + c_R(12)$ | $f_R(8) \Rightarrow M(1)$ |
| $f_I(8) = c_I(8) + c_I(12)$ | $f_I(8) \Rightarrow M(17)$ |
| $f_R(9) = c_R(9) + e_R(13)$ | $f_R(9) \Rightarrow M(9)$ |
| $f_I(9) = c_I(9) + e_I(13)$ | $f_I(9) \Rightarrow M(25)$ |
| $f_R(10) = c_R(10) + e_R(14)$ | $f_R(10) \Rightarrow M(5)$ |
| $f_I(10) = c_I(10) + e_I(14)$ | $f_I(10) \Rightarrow M(21)$ |
| $f_R(11) = c_R(11) + e_R(15)$ | $f_R(11) \Rightarrow M(29)$ |
| $f_I(11) = c_I(11) + e_I(15)$ | $f_I(11) \Rightarrow M(13)$ |
| $f_R(12) = c_R(8) - c_R(12)$ | $f_R(12) \Rightarrow M(3)$ |
| $f_I(12) = c_I(8) - c_I(12)$ | $f_I(12) \Rightarrow M(19)$ |
| $f_R(13) = c_R(9) - e_R(13)$ | $f_R(13) \Rightarrow M(11)$ |
| $f_I(13) = c_I(9) - e_I(13)$ | $f_I(13) \Rightarrow M(27)$ |
| $f_R(14) = c_R(10) - e_R(14)$ | $f_R(14) \Rightarrow M(23)$ |
| $f_I(14) = c_I(10) - e_I(14)$ | $f_I(14) \Rightarrow M(7)$ |
| $f_R(15) = c_R(11) - e_R(15)$ | $f_R(15) \Rightarrow M(15)$ |
| $f_I(15) = c_I(11) - e_I(15)$ | $f_I(15) \Rightarrow M(31)$ |

### Stage 2: Interstage Complex Multiplies

This stage computes the complex multiplications required between the 8- and 2-point building-block stages. Since a complex multiplication requires four multiplies and two adds, each input data value is multiplied by two constants, and then these results are combined. Therefore, additional data memory locations are required to store the intermediate results of the multiplication portion of the complex multiplies. There is one exception to that in this example—the multiplication by $\sin(4\pi/16)$ and $\cos(4\pi/16)$, because these numbers are the same. Therefore, only one of the multiplications is required, and no additional data memory locations are needed to store the intermediate results.

The complex multiply computations are grouped to make them easier to see. For example, the first six computations are a complex multiply that requires two additional memory locations, $M(32)$ and $M(33)$. Each of the subsequent four sets of six computations is also a complex multiplication. In each case, $M(32)$ and $M(33)$ can be used for the required temporary results. After these computations, the next two sets of four computations are the multiplication by association with $\sin(4\pi/16)$ and $\cos(4\pi/16)$. Since these two constants are the same, the computations do not require additional memory locations. The last two lines simulate multiplication by $j$.

| Algorithm Steps | Memory Map |
|:---:|:---:|
| $g_R(9) = f_R(9) * \cos(2\pi/16)$ | $g_R(9) \Rightarrow M(32)$ |
| $g_R(17) = f_R(9) * \sin(2\pi/16)$ | $g_R(17) \Rightarrow M(9)$ |
| $g_I(9) = f_I(9) * \sin(2\pi/16)$ | $g_I(9) \Rightarrow M(33)$ |
| $g_I(17) = f_I(9) * \cos(2\pi/16)$ | $g_I(17) \Rightarrow M(25)$ |
| $h_I(9) = g_I(17) - g_R(17)$ | $h_I(9) \Rightarrow M(25)$ |
| $h_R(9) = g_R(9) + g_I(9)$ | $h_R(9) \Rightarrow M(9)$ |
| $g_R(11) = f_R(11) * \cos(6\pi/16)$ | $g_R(11) \Rightarrow M(32)$ |
| $g_R(18) = f_R(11) * \sin(6\pi/16)$ | $g_R(18) \Rightarrow M(29)$ |
| $g_I(11) = f_I(11) * \sin(6\pi/16)$ | $g_I(11) \Rightarrow M(33)$ |
| $g_I(18) = f_I(11) * \cos(6\pi/16)$ | $g_I(18) \Rightarrow M(13)$ |
| $h_I(11) = g_I(18) - g_R(18)$ | $h_I(11) \Rightarrow M(13)$ |
| $h_R(11) = g_R(11) + g_I(11)$ | $h_R(11) \Rightarrow M(29)$ |
| $g_R(13) = f_R(13) * \cos(2\pi/16)$ | $g_R(13) \Rightarrow M(32)$ |
| $g_R(19) = f_R(13) * \sin(2\pi/16)$ | $g_R(19) \Rightarrow M(11)$ |
| $g_I(13) = f_I(13) * \sin(2\pi/16)$ | $g_I(13) \Rightarrow M(33)$ |
| $g_I(19) = f_I(13) * \cos(2\pi/16)$ | $g_I(19) \Rightarrow M(27)$ |
| $h_R(13) = -g_R(19) + g_I(13)$ | $h_R(13) \Rightarrow M(11)$ |
| $h_I(13) = -g_I(13) - g_R(13)$ | $h_I(13) \Rightarrow M(27)$ |
| $g_R(15) = f_R(15) * \cos(6\pi/16)$ | $g_R(15) \Rightarrow M(32)$ |
| $g_R(20) = f_R(15) * \sin(6\pi/16)$ | $g_R(20) \Rightarrow M(15)$ |
| $g_I(15) = f_I(15) * \sin(6\pi/16)$ | $g_I(15) \Rightarrow M(33)$ |
| $g_I(20) = f_I(15) * \cos(6\pi/16)$ | $g_I(20) \Rightarrow M(31)$ |

| Algorithm Steps | Memory Map |
|---|---|
| $h_R(15) = -g_R(20) + g_I(20)$ | $h_R(15) \Rightarrow M(15)$ |
| $h_I(15) = -g_I(15) - g_R(15)$ | $h_I(15) \Rightarrow M(31)$ |
| $g_R(10) = f_R(10) * \cos(4\pi/16)$ | $g_R(10) \Rightarrow M(5)$ |
| $g_I(10) = f_I(10) * \cos(4\pi/16)$ | $g_I(10) \Rightarrow M(21)$ |
| $h_R(10) = g_R(10) + g_I(10)$ | $h_R(10) \Rightarrow M(5)$ |
| $h_I(10) = -g_R(10) + g_I(10)$ | $h_I(10) \Rightarrow M(21)$ |
| $g_R(14) = f_R(14) * \cos(4\pi/16)$ | $g_R(14) \Rightarrow M(23)$ |
| $g_I(14) = f_I(14) * \cos(4\pi/16)$ | $g_I(14) \Rightarrow M(7)$ |
| $h_R(14) = -g_R(14) + g_I(14)$ | $h_R(14) \Rightarrow M(23)$ |
| $h_I(14) = -g_R(14) - g_I(14)$ | $h_I(14) \Rightarrow M(7)$ |
| $g_R(12) = f_I(12)$ | $g_R(12) \Rightarrow M(19)$ |
| $g_I(12) = -f_R(12)$ | $g_I(12) \Rightarrow M(3)$ |

## Stage 3: Output 2-Point Building Blocks

This step is a sequence of eight 2-point algorithm building blocks and does not require additional data memory or accessing any of the multiplier constants. Further, the add/subtract process is the same for all of the real and imaginary pairs.

The strategy for converting these equations to code is to start at the top (compute $A_R(0)$) and identify the pair of inputs to be used first (in this case $f_R(0)$ and $f_R(8)$). Then look down the list to find the second (compute $A_R(8)$) place where these two inputs are used. Pull $f_R(0)$ and $f_R(8)$ from memory, compute $A_R(0)$ and $A_R(8)$, and store the results in memory locations $M(0)$ and $M(1)$, previously occupied by $f_R(0)$ and $f_R(8)$. The next step is to look at the next computation $A_I(0)$ on the list and repeat the same set of steps. Continue this process until all the Algorithm Steps in Stage 3 have been computed and their results stored in the Memory Map addresses.

### First of Eight 2-Point Building Blocks

This set of computations is represented in Figure 9-26 by output 2-point building block 0. Further, the labels on the left and right of this building block correspond to the input and output labels in the 2-point building block in Section 8.3.

| Algorithm Steps | Memory Map |
|---|---|
| $A_R(0) = f_R(0) + f_R(8)$ | $A_R(0) \Rightarrow M(0)$ |
| $A_I(0) = f_I(0) + f_I(8)$ | $A_I(0) \Rightarrow M(16)$ |
| $A_R(8) = f_R(0) - f_R(8)$ | $A_R(8) \Rightarrow M(1)$ |
| $A_I(8) = f_I(0) - f_I(8)$ | $A_I(8) \Rightarrow M(17)$ |

### Second of Eight 2-Point Building Blocks

This set of computations is represented in Figure 9-26 by output 2-point building block 1. Further, the labels on the left and right of this building block correspond to the input and output labels in the 2-point building block in Section 8.3.

| Algorithm Steps | Memory Map |
|---|---|
| $A_R(1) = f_R(1) + h_R(9)$ | $A_R(1) \Rightarrow M(8)$ |
| $A_I(1) = f_I(1) + h_I(9)$ | $A_I(1) \Rightarrow M(24)$ |
| $A_R(9) = f_R(1) - h_R(9)$ | $A_R(9) \Rightarrow M(9)$ |
| $A_I(9) = f_I(1) - h_I(9)$ | $A_I(9) \Rightarrow M(25)$ |

### Third of Eight 2-Point Building Blocks

This set of computations is represented in Figure 9-26 by output 2-point building block 2. Further, the labels on the left and right of this building block correspond to the input and output labels in the 2-point building block in Section 8.3.

| Algorithm Steps | Memory Map |
|---|---|
| $A_R(2) = f_R(2) + h_R(10)$ | $A_R(2) \Rightarrow M(4)$ |
| $A_I(2) = f_I(2) + h_I(10)$ | $A_I(2) \Rightarrow M(20)$ |
| $A_R(10) = f_R(2) - h_R(10)$ | $A_R(10) \Rightarrow M(5)$ |
| $A_I(10) = f_I(2) - h_I(10)$ | $A_I(10) \Rightarrow M(21)$ |

### Fourth of Eight 2-Point Building Blocks

This set of computations is represented in Figure 9-26 by output 2-point building block 3. Further, the labels on the left and right of this building block correspond to the input and output labels in the 2-point building block in Section 8.3.

| Algorithm Steps | Memory Map |
|---|---|
| $A_R(3) = f_R(3) + h_R(11)$ | $A_R(3) \Rightarrow M(28)$ |
| $A_I(3) = f_I(3) + h_I(11)$ | $A_I(3) \Rightarrow M(12)$ |
| $A_R(11) = f_R(3) - h_R(11)$ | $A_R(11) \Rightarrow M(29)$ |
| $A_I(11) = f_I(3) - h_I(11)$ | $A_I(11) \Rightarrow M(13)$ |

### Fifth of Eight 2-Point Building Blocks

This set of computations is represented in Figure 9-26 by output 2-point building block 5. Further, the labels on the left and right of this building block correspond to the input and output labels in the 2-point building block in Section 8.3.

| Algorithm Steps | Memory Map |
|---|---|
| $A_R(4) = f_R(4) + g_R(12)$ | $A_R(4) \Rightarrow M(2)$ |
| $A_I(4) = f_I(4) + g_I(12)$ | $A_I(4) \Rightarrow M(18)$ |
| $A_R(12) = f_R(4) - g_R(12)$ | $A_R(12) \Rightarrow M(19)$ |
| $A_I(12) = f_I(4) - g_I(12)$ | $A_I(12) \Rightarrow M(3)$ |

### Sixth of Eight 2-Point Building Blocks

This set of computations is represented in Figure 9-26 by output 2-point building block 5. Further, the labels on the left and right of this building block correspond to the input and output labels in the 2-point building block in Section 8.3.

| Algorithm Steps | Memory Map |
|---|---|
| $A_R(5) = f_R(5) + h_R(13)$ | $A_R(5) \Rightarrow M(10)$ |
| $A_I(5) = f_I(5) + h_I(13)$ | $A_I(5) \Rightarrow M(26)$ |
| $A_R(13) = f_R(5) - h_R(13)$ | $A_R(13) \Rightarrow M(11)$ |
| $A_I(13) = f_I(5) - h_I(13)$ | $A_I(13) \Rightarrow M(27)$ |

*Seventh of Eight 2-Point Building Blocks*

This set of computations is represented in Figure 9-26 by output 2-point building block 6. Further, the labels on the left and right of this building block correspond to the input and output labels in the 2-point building block in Section 8.3.

| Algorithm Steps | Memory Map |
|---|---|
| $A_R(6) = f_R(6) + h_R(14)$ | $A_R(6) \Rightarrow M(22)$ |
| $A_I(6) = f_I(6) + h_I(14)$ | $A_I(6) \Rightarrow M(6)$ |
| $A_R(14) = f_R(6) - h_R(14)$ | $A_R(14) \Rightarrow M(23)$ |
| $A_I(14) = f_I(6) - h_I(14)$ | $A_I(14) \Rightarrow M(7)$ |

*Eighth of Eight 2-Point Building Blocks*

This set of computations is represented in Figure 9-26 by output 2-point building block 2. Further, the labels on the left and right of this building block correspond to the input and output labels in the 2-point building block in Section 8.3.

| Algorithm Steps | Memory Map |
|---|---|
| $A_R(7) = f_R(7) + h_R(15)$ | $A_R(7) \Rightarrow M(14)$ |
| $A_I(7) = f_I(7) + h_I(15)$ | $A_I(7) \Rightarrow M(30)$ |
| $A_R(15) = f_R(7) - h_R(15)$ | $A_R(15) \Rightarrow M(15)$ |
| $A_I(15) = f_I(7) - h_I(15)$ | $A_I(15) \Rightarrow M(31)$ |

### 9.7.7 Fifteen-Point Singleton Mixed-Radix FFT Example

The Singleton mixed-radix [5] algorithm is the most general one. In Figure 9-21, any of the algorithm building blocks from Chapter 8 can be placed in the FFT stages.

The 15-point Singleton mixed-radix algorithm can be implemented with either the 3-point or the 5-point building blocks first. If the 3-point building block is first, the 15 pieces of complex input data are divided into five sets of three complex points, one for each of the $15/3 = 5$ 3-point transforms. Following the 3-point building blocks and complex multiplies, the intermediate results are divided into three sets of five pieces of complex data needed for input to the $15/5 = 3$ 5-point building-block computations. The order does not affect the number of computations required.

Figure 9-27 is a detailed block diagram of this example. At the block diagram level, any of the 3- and 5-point building blocks from Chapter 8 can be used. This example uses the Singleton 3-and 5-point building blocks. A smaller number of adds and multiplies would be needed if the Winograd building blocks were used.

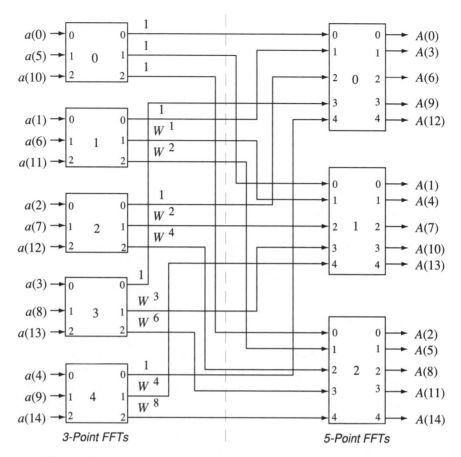

**Figure 9-27**    Fifteen-point Singleton mixed-radix algorithm block-diagram.

If the Comparison Matrix in Chapter 8 and the equation presented in Section 9.7.2 are used, the total number of real adds required is $5 * 12 + 3 * 32 + 2 * 2 * 4 = 172$, and the total number of real multiplies is $5 * 4 + 3 * 16 + 4 * 2 * 4 = 100$. The total amount of data memory required is driven by the 5-point building block and is $3 * 10$ basic complex data locations plus 2 temporary locations, for a total of 32 memory locations.

The 3-point Singleton building block has two multiplier constants ($\cos(2\pi/3)$ and $\sin(2\pi/3)$), the 5-point Singleton building block has four ($\cos(2\pi/5), \sin(2\pi/5), \cos(4\pi/5)$, and $\sin(4\pi/5)$), and the complex multiplies between the stages require eight constants that are not already required by the 3- and 5-point building blocks ($\cos(2\pi/15), \sin(2\pi/15), \cos(4\pi/15), \sin(4\pi/15), \cos(8\pi/15), \sin(8\pi/15), \cos(16\pi/15)$, and $\sin(16\pi/15)$). This is a total of 14 memory locations for multiplier constants.

### Stage 1: Three-Point Building Blocks

The 15 data points must first be divided into five sets of 3 points to serve as inputs to each of the 3-point building blocks. This is done by starting with complex input data point pair $a_R(0)$, $a_I(0)$ and grouping it with complex input data point pairs $a_R(5)$, $a_I(5)$ and $a_R(10)$, $a_I(10)$. These provide the input to the top one of the five 3-point building blocks. This is followed by grouping the input data point pairs $a_R(1)$, $a_I(1)$, $a_R(6)$, $a_I(6)$, and $a_R(11)$, $a_I(11)$ to provide the input for the second of the five 3-point building blocks. The next grouping is data point pairs $a_R(2)$, $a_I(2)$, $a_R(7)$, $a_I(7)$, and $a_R(12)$, $a_I(12)$ for input into the third of the five 3-point building blocks. The next grouping is data point pairs $a_R(3)$, $a_I(3)$, $a_R(8)$, $a_I(8)$, and $a_R(13)$, $a_I(13)$ to provide input for the fourth of the five 3-point building blocks. The final grouping is data point pairs $a_R(4)$, $a_I(4)$, $a_R(9)$, $a_I(9)$, and $a_R(14)$, $a_I(14)$ for input into the fifth 3-point building block. In general, the complex input data for the $k$-th input to the $m$-th 3-point building block are $a_R(5*k+m)$, $a_I(5*k+m)$ where $k=0$, 1, and 2, and $m=0$, 1, 2, 3, and 4.

The five groups of computations, listed as (a) through (e), each perform the 3-point building block. In this example, the Singleton 3-point algorithm building block from Section 8.4.2 is used. All of these 3-point transforms could also have been the Winograd 3-point algorithm building block from Chapter 8. In fact, the five 3-point transforms can be any combination of the two 3-point algorithm building blocks. The outputs of each of the 3-point building blocks, labeled $B_R(i)$ and $B_I(i)$ for $i=0, 5, 10$, are the equivalent of the $A_R(i)$ and $A_I(i)$ in the 3-point building block in Chapter 8. To translate these data addresses and data labels to each of the next four 3-point building blocks, add 1, 2, 3, and 4 to the addresses and data labels.

The strategy for converting these equations to code is to start at the top (compute $b_R(5)$) and identify the pair of inputs to be used first (in this case $a_R(5)$ and $a_R(10)$). Then look down the list to find the second (compute $b_R(10)$) place where these two inputs are used. Pull $a_R(0)$ and $a_R(10)$ from memory, compute $b_R(5)$ and $b_R(10)$ and store the results in memory locations $M(5)$ and $M(10)$, previously occupied by $a_R(5)$ and $a_R(10)$. The next step is to look at the next computation $b_I(5)$ on the list and repeat the same set of steps. Continue this process until all the Algorithm Steps in Stage 1 have been computed and their results stored in the Memory Map addresses.

### First of Five 3-Point Building Blocks

This set of computations is represented in Figure 9-27 by 3-point building block 0. Further, the labels on the left and right of this building block correspond to the input and output labels in the 3-point Singleton building block in Section 8.4.2.

| Algorithm Steps | Memory Map |
|---|---|
| $b_R(5) = a_R(5) + a_R(10)$ | $b_R(5) \Rightarrow M(5)$ |
| $b_R(10) = a_R(5) - a_R(10)$ | $b_R(10) \Rightarrow M(10)$ |
| $b_I(5) = a_I(5) + a_I(10)$ | $b_I(5) \Rightarrow M(20)$ |
| $b_I(10) = a_I(5) - a_I(10)$ | $b_I(10) \Rightarrow M(25)$ |
| $c_R(5) = b_R(5) * \cos(2\pi/3) + a_R(0)$ | $c_R(5) \Rightarrow M(30)$ |
| $B_R(0) = a_R(0) + b_R(5)$ | $B_R(0) \Rightarrow M(0)$ |
| $c_R(10) = b_I(10) * \sin(2\pi/3)$ | $c_R(10) \Rightarrow M(25)$ |
| $c_I(5) = b_I(5) * \cos(2\pi/3) + a_I(0)$ | $c_I(5) \Rightarrow M(5)$ |
| $B_I(0) = a_I(0) + b_I(5)$ | $B_I(0) \Rightarrow M(15)$ |
| $c_I(10) = -b_R(10) * \sin(2\pi/3)$ | $c_I(10) \Rightarrow M(10)$ |
| $B_R(5) = c_R(5) + c_R(10)$ | $B_R(5) \Rightarrow M(25)$ |

| Algorithm Steps | Memory Map |
|---|---|
| $B_I(5) = c_I(5) + c_I(10)$ | $B_I(5) \Rightarrow M(10)$ |
| $B_R(10) = c_R(5) - c_R(10)$ | $B_R(10) \Rightarrow M(20)$ |
| $B_I(10) = c_I(5) - c_I(10)$ | $B_I(10) \Rightarrow M(5)$ |

### Second of Five 3-Point Building Blocks

This set of computations is represented in Figure 9-27 by 3-point building block 1. Further, the labels on the left and right of this building block correspond to the input and output labels in the 3-point Singleton building block in Section 8.4.2.

| Algorithm Steps | Memory Map |
|---|---|
| $b_R(6) = a_R(6) + a_R(11)$ | $b_R(6) \Rightarrow M(6)$ |
| $b_R(11) = a_R(6) - a_R(11)$ | $b_R(11) \Rightarrow M(11)$ |
| $b_I(6) = a_I(6) + a_I(11)$ | $b_I(6) \Rightarrow M(21)$ |
| $b_I(11) = a_I(6) - a_I(11)$ | $b_I(11) \Rightarrow M(26)$ |
| $c_R(6) = b_R(6) * \cos(2\pi/3) + a_R(1)$ | $c_R(6) \Rightarrow M(30)$ |
| $B_R(1) = a_R(1) + b_R(6)$ | $B_R(1) \Rightarrow M(1)$ |
| $c_R(11) = b_I(11) * \sin(2\pi/3)$ | $c_R(11) \Rightarrow M(26)$ |
| $c_I(6) = b_I(6) * \cos(2\pi/3) + a_I(1)$ | $c_I(6) \Rightarrow M(6)$ |
| $B_I(1) = a_I(1) + b_I(6)$ | $B_I(1) \Rightarrow M(16)$ |
| $c_I(11) = -b_R(11) * \sin(2\pi/3)$ | $c_I(11) \Rightarrow M(11)$ |
| $B_R(6) = c_R(6) + c_R(11)$ | $B_R(6) \Rightarrow M(26)$ |
| $B_I(6) = c_I(6) + c_I(11)$ | $B_I(6) \Rightarrow M(11)$ |
| $B_R(11) = c_R(6) - c_R(11)$ | $B_R(11) \Rightarrow M(21)$ |
| $B_I(11) = c_I(6) - c_I(11)$ | $B_I(11) \Rightarrow M(6)$ |

### Third of Five 3-Point Building Blocks

This set of computations is represented in Figure 9-27 by 3-point building block 2. Further, the labels on the left and right of this building block correspond to the input and output labels in the 3-point Singleton building block in Section 8.4.2.

| Algorithm Steps | Memory Map |
|---|---|
| $b_R(7) = a_R(7) + a_R(12)$ | $b_R(7) \Rightarrow M(7)$ |
| $b_R(12) = a_R(7) - a_R(12)$ | $b_R(12) \Rightarrow M(12)$ |
| $b_I(7) = a_I(7) + a_I(12)$ | $b_I(7) \Rightarrow M(22)$ |
| $b_I(12) = a_I(7) - a_I(12)$ | $b_I(12) \Rightarrow M(27)$ |
| $c_R(7) = b_R(7) * \cos(2\pi/3) + a_R(2)$ | $c_R(7) \Rightarrow M(30)$ |
| $B_R(2) = a_R(2) + b_R(7)$ | $B_R(2) \Rightarrow M(2)$ |
| $c_R(12) = b_I(12) * \sin(2\pi/3)$ | $c_R(12) \Rightarrow M(27)$ |
| $c_I(7) = b_I(7) * \cos(2\pi/3) + a_I(2)$ | $c_I(7) \Rightarrow M(7)$ |
| $B_I(2) = a_I(2) + b_I(7)$ | $B_I(2) \Rightarrow M(17)$ |
| $c_I(12) = -b_R(12) * \sin(2\pi/3)$ | $c_I(12) \Rightarrow M(12)$ |

| Algorithm Steps | Memory Map |
|---|---|
| $B_R(7) = c_R(7) + c_R(12)$ | $B_R(7) \Rightarrow M(27)$ |
| $B_I(7) = c_I(7) + c_I(12)$ | $B_I(7) \Rightarrow M(12)$ |
| $B_R(12) = c_R(7) - c_R(12)$ | $B_R(12) \Rightarrow M(22)$ |
| $B_I(12) = c_I(7) - c_I(12)$ | $B_I(12) \Rightarrow M(7)$ |

### Fourth of Five 3-Point Building Blocks

This set of computations is represented in Figure 9-27 by 3-point building block 3. Further, the labels on the left and right of this building block correspond to the input and output labels in the 3-point Singleton building block in Section 8.4.2.

| Algorithm Steps | Memory Map |
|---|---|
| $b_R(8) = a_R(8) + a_R(13)$ | $b_R(8) \Rightarrow M(8)$ |
| $b_R(13) = a_R(8) - a_R(13)$ | $b_R(13) \Rightarrow M(13)$ |
| $b_I(8) = a_I(8) + a_I(13)$ | $b_I(8) \Rightarrow M(23)$ |
| $b_I(13) = a_I(8) - a_I(13)$ | $b_I(13) \Rightarrow M(28)$ |
| $c_R(8) = b_R(8) * \cos(2\pi/3) + a_R(3)$ | $c_R(8) \Rightarrow M(30)$ |
| $B_R(3) = a_R(3) + b_R(8)$ | $B_R(3) \Rightarrow M(3)$ |
| $c_R(13) = b_I(13) * \sin(2\pi/3)$ | $c_R(13) \Rightarrow M(28)$ |
| $c_I(8) = b_I(8) * \cos(2\pi/3) + a_I(3)$ | $c_I(8) \Rightarrow M(8)$ |
| $B_I(3) = a_I(3) + b_I(8)$ | $B_I(3) \Rightarrow M(18)$ |
| $c_I(13) = -b_R(13) * \sin(2\pi/3)$ | $c_I(13) \Rightarrow M(13)$ |
| $B_R(8) = c_R(8) + c_R(13)$ | $B_R(8) \Rightarrow M(28)$ |
| $B_I(8) = c_I(8) + c_I(13)$ | $B_I(8) \Rightarrow M(13)$ |
| $B_R(13) = c_R(8) - c_R(13)$ | $B_R(13) \Rightarrow M(23)$ |
| $B_I(13) = c_I(8) - c_I(13)$ | $B_I(13) \Rightarrow M(8)$ |

### Fifth of Five 3-Point Building Blocks

This set of computations is represented in Figure 9-27 by 3-point building block 4. Further, the labels on the left and right of this building block correspond to the input and output labels in the 3-point Singleton building block in Section 8.4.2.

| Algorithm Steps | Memory Map |
|---|---|
| $b_R(9) = a_R(9) + a_R(14)$ | $b_R(9) \Rightarrow M(9)$ |
| $b_R(14) = a_R(9) - a_R(14)$ | $b_R(14) \Rightarrow M(14)$ |
| $b_I(9) = a_I(9) + a_I(14)$ | $b_I(9) \Rightarrow M(24)$ |
| $b_I(14) = a_I(9) - a_I(14)$ | $b_I(14) \Rightarrow M(29)$ |
| $c_R(9) = b_R(9) * \cos(2\pi/3) + a_R(4)$ | $c_R(9) \Rightarrow M(6)$ |
| $B_R(4) = a_R(4) + b_R(9)$ | $B_R(4) \Rightarrow M(4)$ |
| $c_R(14) = b_I(14) * \sin(2\pi/3)$ | $c_R(14) \Rightarrow M(29)$ |
| $c_I(9) = b_I(9) * \cos(2\pi/3) + a_I(4)$ | $c_I(9) \Rightarrow M(9)$ |
| $B_I(4) = a_I(4) + b_I(9)$ | $B_I(4) \Rightarrow M(19)$ |

| Algorithm Steps | Memory Map |
|---|---|
| $c_I(14) = -b_R(14) * \sin(2\pi/3)$ | $c_I(14) \Rightarrow M(14)$ |
| $B_R(9) = c_R(9) + c_R(14)$ | $B_R(9) \Rightarrow M(29)$ |
| $B_I(9) = c_I(9) + c_I(14)$ | $B_I(9) \Rightarrow M(14)$ |
| $B_R(14) = c_R(9) - c_R(14)$ | $B_R(14) \Rightarrow M(24)$ |
| $B_I(14) = c_I(9) - c_I(14)$ | $B_I(14) \Rightarrow M(9)$ |

## Stage 2: Complex Multiplies

The complex multiplier to be applied to the $k$-th output of the $m$-th 3-point building block, $B_R(5*k+m) + j*B_I(5*k+m)$, is $\cos(2*\pi*k*m/15) - j*\sin(2*\pi*k*m/15)$ as shown in Figure 9-23. Assuming no temporary storage registers, the complex multiply requires two additional data memory locations ($M(30)$ and $M(31)$) if the results are to be placed back in the same memory locations where the $B_R(5*k+m)$ and $B_I(5*k+m)$ were accessed. The reason is that the real and imaginary parts, $B_R(5*k+m)$ and $B_I(5*k+m)$, are multiplied by different constants and both results are used twice. Once one complex multiply is performed, the two additional data memory locations ($M(30)$ and $M(31)$) are free to be used as the extra memory locations for the next complex multiply. Therefore, only two additional data memory locations are required.

Many of the Algorithm Steps in this stage are just renaming the intermediate results. This is done to make all of the intermediate results labels into the next stage have the same letter, $D$. For those Algorithm Steps that perform multiplication, the data is pulled from memory, the computation performed, and the results stored back in the same location. This stage's computations are as follows.

### First 3-Point Building-Block Output Complex Multiplies

When $m = 0$, the complex multiplier is 1, which requires no multiplication. The first four lines are a redefinition of the data variables so that the inputs to the output 5-point building blocks all use the same variable names. The final three lines are used to reverse the data memory locations of the real and imaginary parts of the last output of the zero-th 3-point building block. This rearrangement is not required. However, for this example, all of the real and imaginary parts that will be inputs to the 5-point building blocks are reordered so that the real part appears in the lower half of data memory and the imaginary parts appear in the upper half of data memory.

| Algorithm Steps | Memory Map |
|---|---|
| $D_R(0) = B_R(0)$ | $D_R(0) \Rightarrow M(0)$ |
| $D_I(0) = B_I(0)$ | $D_I(0) \Rightarrow M(15)$ |
| $D_R(5) = B_R(5)$ | $D_R(5) \Rightarrow M(25)$ |
| $D_I(5) = B_I(5)$ | $D_I(5) \Rightarrow M(10)$ |
| $T_R = B_I(10)$ | $T_R \Rightarrow M(30)$ |
| $D_R(10) = B_R(10)$ | $D_R(10) \Rightarrow M(20)$ |
| $D_I(10) = T_R$ | $D_I(10) \Rightarrow M(5)$ |

*Second 3-Point Building-Block Output Complex Multiplies*

The computations in this set perform the complex multiplies required at the output of the second of the five 3-point building blocks ($m = 1$). Additionally, the first two lines are used to redefine the data variables so that the inputs to the output 5-point building blocks all use the same variable names.

| **Algorithm Steps** | **Memory Map** |
|---|---|
| $D_R(1) = B_R(1)$ | $D_R(1) \Rightarrow M(1)$ |
| $D_I(1) = B_I(1)$ | $D_I(1) \Rightarrow M(16)$ |
| $T_R = B_R(6) * \cos(2\pi/15)$ | $T_R \Rightarrow M(30)$ |
| $T_I = B_I(6) * \sin(2\pi/15)$ | $T_I \Rightarrow M(31)$ |
| $C_R(6) = B_R(6) * \sin(2\pi/15)$ | $C_R(6) \Rightarrow M(26)$ |
| $C_I(6) = B_I(6) * \cos(2\pi/15)$ | $C_I(6) \Rightarrow M(11)$ |
| $D_I(6) = -C_R(6) + C_I(6)$ | $D_I(6) \Rightarrow M(26)$ |
| $D_R(6) = T_R + T_I$ | $D_R(6) \Rightarrow M(11)$ |
| $T_R = B_R(11) * \cos(4\pi/15)$ | $T_R \Rightarrow M(30)$ |
| $T_I = B_I(11) * \sin(4\pi/15)$ | $T_I \Rightarrow M(31)$ |
| $C_R(11) = B_R(11) * \sin(4\pi/15)$ | $C_R(11) \Rightarrow M(21)$ |
| $C_I(11) = B_I(11) * \cos(4\pi/15)$ | $C_I(11) \Rightarrow M(6)$ |
| $D_I(11) = -C_R(11) + C_I(11)$ | $D_I(11) \Rightarrow M(21)$ |
| $D_R(11) = T_R + T_I$ | $D_R(11) \Rightarrow M(6)$ |

*Third 3-Point Building-Block Output Complex Multiplies*

The computations in this set perform the complex multiplies required at the output of the third set of the five 3-point building blocks ($m = 2$). Again, all of the real and imaginary parts have been reordered after multiplication so that the inputs to the 5-point building blocks have their real part appearing in the bottom half of data memory, and the imaginary parts appear in the upper half of data memory. Additionally, the first two lines are used to redefine the data variables so that the inputs to the output 5-point building blocks all use the same variable names.

| **Algorithm Steps** | **Memory Map** |
|---|---|
| $D_R(2) = B_R(2)$ | $D_R(2) \Rightarrow M(2)$ |
| $D_I(2) = B_I(2)$ | $D_I(2) \Rightarrow M(17)$ |
| $T_R = B_R(7) * \cos(4\pi/15)$ | $T_R \Rightarrow M(30)$ |
| $T_I = B_I(7) * \sin(4\pi/15)$ | $T_I \Rightarrow M(31)$ |
| $C_R(7) = B_R(7) * \sin(4\pi/15)$ | $C_R(7) \Rightarrow M(27)$ |
| $C_I(7) = B_I(7) * \cos(4\pi/15)$ | $C_I(7) \Rightarrow M(12)$ |
| $D_I(7) = -C_R(7) + C_I(7)$ | $D_I(7) \Rightarrow M(27)$ |
| $D_R(7) = T_R + T_I$ | $D_R(7) \Rightarrow M(12)$ |

| Algorithm Steps | Memory Map |
|---|---|
| $T_R = B_R(12) * \cos(8\pi/15)$ | $T_R \Rightarrow M(30)$ |
| $T_I = B_I(12) * \sin(8\pi/15)$ | $T_I \Rightarrow M(31)$ |
| $C_R(12) = B_R(12) * \sin(8\pi/15)$ | $C_R(12) \Rightarrow M(22)$ |
| $C_I(12) = B_I(12) * \cos(8\pi/15)$ | $C_I(12) \Rightarrow M(7)$ |
| $D_I(12) = -C_R(12) + C_I(12)$ | $D_I(12) \Rightarrow M(22)$ |
| $D_R(12) = T_R + T_I$ | $D_R(12) \Rightarrow M(7)$ |

### Fourth 3-Point Building-Block Output Complex Multiplies

The computations in this set perform the complex multiplies required at the output of the fourth set of the five 3-point building blocks ($m = 3$). Again, all of the real and imaginary parts have been reordered after multiplication so that the inputs to the 5-point building blocks have their real part appearing in the bottom half of data memory, and the imaginary parts appear in the upper half of data memory. Additionally, the first two lines are used to redefine the data variables so that the inputs to the output 5-point building blocks all use the same variable names.

| Algorithm Steps | Memory Map |
|---|---|
| $D_R(3) = B_R(3)$ | $D_R(3) \Rightarrow M(3)$ |
| $D_I(3) = B_I(3)$ | $D_I(3) \Rightarrow M(18)$ |
| $T_R = B_R(8) * \cos(6\pi/15)$ | $T_R \Rightarrow M(30)$ |
| $T_I = B_I(8) * \sin(6\pi/15)$ | $T_I \Rightarrow M(31)$ |
| $C_R(8) = B_R(8) * \sin(6\pi/15)$ | $C_R(8) \Rightarrow M(28)$ |
| $C_I(8) = B_I(8) * \cos(6\pi/15)$ | $C_I(8) \Rightarrow M(13)$ |
| $D_I(8) = -C_R(8) + C_I(8)$ | $D_I(8) \Rightarrow M(28)$ |
| $D_R(8) = T_R + T_I$ | $D_R(8) \Rightarrow M(13)$ |
| $T_R = B_R(13) * \cos(12\pi/15)$ | $T_R \Rightarrow M(30)$ |
| $T_I = B_I(13) * \sin(12\pi/15)$ | $T_I \Rightarrow M(31)$ |
| $C_R(13) = B_R(13) * \sin(12\pi/15)$ | $C_R(13) \Rightarrow M(23)$ |
| $C_I(13) = B_I(13) * \cos(12\pi/15)$ | $C_I(13) \Rightarrow M(8)$ |
| $D_I(13) = -C_R(13) + C_I(13)$ | $D_I(13) \Rightarrow M(23)$ |
| $D_R(13) = T_R + T_I$ | $D_R(13) \Rightarrow M(8)$ |

### Fifth 3-Point Building-Block Output Complex Multiplies

The computations in this set perform the complex multiplies required at the output of the fifth set of the five 3-point building blocks ($m = 4$). Again, all of the real and imaginary parts have been reordered after multiplication so that the inputs to the 5-point building blocks have their real part appearing in the bottom half of data memory, and the imaginary parts appear in the upper half of data memory. Additionally, the first two lines are used to redefine the data variables so that the inputs to the output 5-point building blocks all use the same variable names.

| Algorithm Steps | Memory Map |
|---|---|
| $D_R(4) = B_R(4)$ | $D_R(4) \Rightarrow M(4)$ |
| $D_I(4) = B_I(4)$ | $D_I(4) \Rightarrow M(19)$ |
| $T_R = B_R(9) * \cos(8\pi/15)$ | $T_R \Rightarrow M(30)$ |
| $T_I = B_I(9) * \sin(8\pi/15)$ | $T_I \Rightarrow M(31)$ |
| $C_R(9) = B_R(9) * \sin(8\pi/15)$ | $C_R(9) \Rightarrow M(29)$ |
| $C_I(9) = B_I(9) * \cos(8\pi/15)$ | $C_I(9) \Rightarrow M(14)$ |
| $D_I(9) = -C_R(9) + C_I(9)$ | $D_I(9) \Rightarrow M(29)$ |
| $D_R(9) = T_R + T_I$ | $D_R(9) \Rightarrow M(14)$ |
| $T_R = B_R(14) * \cos(16\pi/15)$ | $T_R \Rightarrow M(30)$ |
| $T_I = B_I(14) * \sin(16\pi/15)$ | $T_I \Rightarrow M(31)$ |
| $C_R(14) = B_R(14) * \sin(16\pi/15)$ | $C_R(14) \Rightarrow M(24)$ |
| $C_I(14) = B_I(14) * \cos(16\pi/15)$ | $C_I(14) \Rightarrow M(9)$ |
| $D_I(14) = -C_R(14) + C_I(14)$ | $D_I(14) \Rightarrow M(24)$ |
| $D_R(14) = T_R + T_I$ | $D_R(14) \Rightarrow M(9)$ |

### Stage 3: Output 5-Point Building Blocks

For this example, the Singleton 5-point building block from Chapter 8 is used. However, either of the two other 5-point building blocks could have been used without changing the rest of the structure of the algorithm. If the number of adds and multiplies is the overriding criterion, then the Winograd algorithm building block should be used in place of the 5-point Singleton algorithm.

The three sets of 5-point algorithm building-block algorithm steps from Section 8.6.2 are listed as (a) through (c). In Chapter 8 this 5-point algorithm building block is presented as three stages. Since the features of the individual stages of the 5-point algorithm block are discussed in Chapter 8, they are not discussed again. The input data into the $m$-th input port of the $k$-th 5-point building block are the $D_R(5 * k + m)$ and $D_I(5 * k + m)$ from Stage 2.

The multiply stage of the 5-point Singleton building block requires additional data memory locations under the set of constraints used in Chapter 8. If the 15-point computations are performed in the order shown, the additional memory locations used by the first of the three 5-point building blocks can be reused by each of the other two 5-point building blocks.

The strategy for converting these equations into code is to start at the top (compute $b_R(1)$) and identify the pair of inputs to be used first (in this case $D_R(1)$ and $D_R(4)$). Then look down the list to find the second (compute $b_R(2)$) place where these two inputs are used. Pull $D_R(1)$ and $D_R(4)$ from memory, compute $b_R(1)$ and $b_R(2)$, and store the results in memory locations $M(1)$ and $M(4)$, previously occupied by $D_R(1)$ and $D_R(4)$. The next step is to look at the next computation $b_I(1)$ on the list and repeat the same set of steps. Continue this process until all the Algorithm Steps in Stage 3 have been computed and their results stored in the Memory Map addresses.

### First of Three 5-Point Building Blocks

This 5-point building block ($k = 0$) has $D_R(5 * k + m)$ and $D_I(5 * k + m)(m = 0, 1, 2, 3,$ and 4) as inputs and $A_R(3 * m + k)$ and $A_I(3 * m + k)(m = 0, 1, 2, 3,$ and 4) as its output frequency components. The multiplication portion of the building block requires two additional data memory locations because no temporary registers are assumed. The variables used for the intermediate computations were chosen to be the same as those used for the 5-point Singleton building block in Chapter 8 to make it easier to associate the computational steps with the discussion of its features and memory mappings in Chapter 8. This set of computations is represented in Figure 9-27 by 5-point building block 0. Further, the labels on the left and right of this building block correspond to the input and output labels in the 5-point Singleton building block in Section 8.6.2.

| **Algorithm Steps** | **Memory Map** |
|---|---|
| $b_R(1) = D_R(1) + D_R(4)$ | $b_R(1) \Rightarrow M(1)$ |
| $b_I(1) = D_I(1) + D_I(4)$ | $b_I(1) \Rightarrow M(16)$ |
| $b_R(2) = D_R(1) - D_R(4)$ | $b_R(2) \Rightarrow M(4)$ |
| $b_I(2) = D_I(1) - D_I(4)$ | $b_I(2) \Rightarrow M(19)$ |
| $b_R(3) = D_R(2) + D_R(3)$ | $b_R(3) \Rightarrow M(2)$ |
| $b_I(3) = D_I(2) + D_I(3)$ | $b_I(3) \Rightarrow M(17)$ |
| $b_R(4) = D_R(2) - D_R(3)$ | $b_R(4) \Rightarrow M(3)$ |
| $b_I(4) = D_I(2) - D_I(3)$ | $b_I(4) \Rightarrow M(18)$ |
| $c_R(2) = b_R(2) * \sin(2\pi/5) + b_R(4) * \sin(4\pi/5)$ | $c_R(2) \Rightarrow M(30)$ |
| $c_I(2) = b_I(2) * \sin(2\pi/5) + b_I(4) * \sin(4\pi/5)$ | $c_I(2) \Rightarrow M(3)$ |
| $c_R(4) = b_R(2) * \sin(4\pi/5) - b_R(4) * \sin(2\pi/5)$ | $c_R(4) \Rightarrow M(31)$ |
| $c_I(4) = b_I(2) * \sin(4\pi/5) - b_I(4) * \sin(2\pi/5)$ | $c_I(4) \Rightarrow M(4)$ |
| $c_R(1) = b_R(1) * \cos(2\pi/5) + b_R(3) * \cos(4\pi/5) + D_R(0)$ | $c_R(1) \Rightarrow M(19)$ |
| $c_I(1) = b_I(1) * \cos(2\pi/5) + b_I(3) * \cos(4\pi/5) + D_I(0)$ | $c_I(1) \Rightarrow M(1)$ |
| $c_R(3) = b_R(1) * \cos(4\pi/5) + b_R(3) * \cos(2\pi/5) + D_R(0)$ | $c_R(3) \Rightarrow M(18)$ |
| $c_I(3) = b_I(1) * \cos(4\pi/5) + b_I(3) * \cos(2\pi/5) + D_I(0)$ | $c_I(3) \Rightarrow M(2)$ |
| $A_R(0) = D_R(0) + b_R(1) + b_R(3)$ | $A_R(0) \Rightarrow M(0)$ |
| $A_I(0) = D_I(0) + b_I(1) + b_I(3)$ | $A_I(0) \Rightarrow M(15)$ |
| $A_R(3) = c_R(1) + c_I(2)$ | $A_R(3) \Rightarrow M(19)$ |
| $A_I(3) = c_I(1) - c_R(2)$ | $A_I(3) \Rightarrow M(16)$ |
| $A_R(6) = c_R(3) + c_I(4)$ | $A_R(6) \Rightarrow M(18)$ |
| $A_I(6) = c_I(3) - c_R(4)$ | $A_I(6) \Rightarrow M(2)$ |
| $A_R(9) = c_R(3) - c_I(4)$ | $A_R(9) \Rightarrow M(4)$ |
| $A_I(9) = c_I(3) + c_R(4)$ | $A_I(9) \Rightarrow M(1)$ |
| $A_R(12) = c_R(1) - c_I(2)$ | $A_R(12) \Rightarrow M(3)$ |
| $A_I(12) = c_I(1) + c_R(2)$ | $A_I(12) \Rightarrow M(17)$ |

### Second of Three 5-Point Building Blocks

This 5-point building block ($k = 1$) has $D_R(5 * k + m)$ and $D_I(5 * k + m)(m = 0, 1, 2, 3,$ and $4)$ as inputs and $A_R(3 * m + k)$ and $A_I(3 * m + k)(m = 0, 1, 2, 3,$ and $4)$ as its output frequency components. The multiplication portion of the algorithm requires two additional data memory locations because no temporary registers are assumed. This set of computations is represented in Figure 9-27 by 5-point building block 1. Further, the labels on the left and right of this building block correspond to the input and output labels in the 5-point Singleton building block in Section 8.6.2.

| **Algorithm Steps** | **Memory Map** |
|---|---|
| $b_R(6) = D_R(6) + D_R(9)$ | $b_R(6) \Rightarrow M(11)$ |
| $b_I(6) = D_I(6) + D_I(9)$ | $b_I(6) \Rightarrow M(26)$ |
| $b_R(7) = D_R(6) - D_R(9)$ | $b_R(7) \Rightarrow M(14)$ |
| $b_I(7) = D_I(6) - D_I(9)$ | $b_I(7) \Rightarrow M(29)$ |
| $b_R(8) = D_R(7) + D_R(8)$ | $b_R(8) \Rightarrow M(12)$ |
| $b_I(8) = D_I(7) + D_I(8)$ | $b_I(8) \Rightarrow M(27)$ |
| $b_R(9) = D_R(7) - D_R(8)$ | $b_R(9) \Rightarrow M(13)$ |
| $b_I(9) = D_I(7) - D_I(8)$ | $b_I(9) \Rightarrow M(28)$ |
| $c_R(7) = b_R(7) * \sin(2\pi/5) + b_R(9) * \sin(4\pi/5)$ | $c_R(7) \Rightarrow M(30)$ |
| $c_I(7) = b_I(7) * \sin(2\pi/5) + b_I(9) * \sin(4\pi/5)$ | $c_I(7) \Rightarrow M(13)$ |
| $c_R(9) = b_R(7) * \sin(4\pi/5) - b_R(9) * \sin(2\pi/5)$ | $c_R(9) \Rightarrow M(31)$ |
| $c_I(9) = b_I(7) * \sin(4\pi/5) - b_I(9) * \sin(2\pi/5)$ | $c_I(9) \Rightarrow M(14)$ |
| $c_R(6) = b_R(6) * \cos(2\pi/5) + b_R(8) * \cos(4\pi/5) + D_R(5)$ | $c_R(6) \Rightarrow M(29)$ |
| $c_I(6) = b_I(6) * \cos(2\pi/5) + b_I(8) * \cos(4\pi/5) + D_I(5)$ | $c_I(6) \Rightarrow M(11)$ |
| $c_R(8) = b_R(6) * \cos(4\pi/5) + b_R(8) * \cos(2\pi/5) + D_R(5)$ | $c_R(8) \Rightarrow M(28)$ |
| $c_I(8) = b_I(6) * \cos(4\pi/5) + b_I(8) * \cos(2\pi/5) + D_I(5)$ | $c_I(8) \Rightarrow M(12)$ |
| $A_R(1) = D_R(5) + b_R(6) + b_R(8)$ | $A_R(1) \Rightarrow M(25)$ |
| $A_I(1) = D_I(5) + b_I(6) + b_I(8)$ | $A_I(1) \Rightarrow M(10)$ |
| $A_R(4) = c_R(6) + c_I(7)$ | $A_R(4) \Rightarrow M(29)$ |
| $A_I(4) = c_I(6) - c_R(7)$ | $A_I(4) \Rightarrow M(26)$ |
| $A_R(7) = c_R(8) + c_I(9)$ | $A_R(7) \Rightarrow M(28)$ |
| $A_I(7) = c_I(8) - c_R(9)$ | $A_I(7) \Rightarrow M(12)$ |
| $A_R(10) = c_R(8) - c_I(9)$ | $A_R(10) \Rightarrow M(14)$ |
| $A_I(10) = c_I(8) + c_R(9)$ | $A_I(10) \Rightarrow M(11)$ |
| $A_R(13) = c_R(6) - c_I(7)$ | $A_R(13) \Rightarrow M(13)$ |
| $A_I(13) = c_I(6) + c_R(7)$ | $A_I(13) \Rightarrow M(27)$ |

### Third of Three 5-Point Building Blocks

This 5-point building block ($k = 2$) has $D_R(5 * k + m)$ and $D_I(5 * k + m)(m = 0, 1, 2, 3,$ and $4)$ as inputs and $A_R(3 * m + k)$ and $A_I(3 * m + k)(m = 0, 1, 2, 3,$ and $4)$ as its output frequency components. The multiplication portion of the algorithm requires two additional data memory locations because no temporary registers are assumed. This set of computations is represented in Figure 9-27 by 5-point building block 2. Further, the labels on the left and right of this building block correspond to the input and output labels in the 5-point Singleton building block in Section 8.6.2.

| Algorithm Steps | Memory Map |
|---|---|
| $b_R(11) = D_R(11) + D_R(14)$ | $b_R(11) \Rightarrow M(6)$ |
| $b_I(11) = D_I(11) + D_I(14)$ | $b_I(11) \Rightarrow M(21)$ |
| $b_R(12) = D_R(11) - D_R(14)$ | $b_R(12) \Rightarrow M(9)$ |
| $b_I(12) = D_I(11) - D_I(14)$ | $b_I(12) \Rightarrow M(24)$ |
| $b_R(13) = D_R(12) + D_R(13)$ | $b_R(13) \Rightarrow M(7)$ |
| $b_I(13) = D_I(12) + D_I(13)$ | $b_I(13) \Rightarrow M(22)$ |
| $b_R(14) = D_R(12) - D_R(13)$ | $b_R(14) \Rightarrow M(8)$ |
| $b_I(14) = D_I(12) - D_I(13)$ | $b_I(14) \Rightarrow M(23)$ |
| $c_R(12) = b_R(12) * \sin(2\pi/5) + b_R(14) * \sin(4\pi/5)$ | $c_R(12) \Rightarrow M(30)$ |
| $c_I(12) = b_I(12) * \sin(2\pi/5) + b_I(14) * \sin(4\pi/5)$ | $c_I(12) \Rightarrow M(8)$ |
| $c_R(14) = b_R(12) * \sin(4\pi/5) - b_R(14) * \sin(2\pi/5)$ | $c_R(14) \Rightarrow M(31)$ |
| $c_I(14) = b_I(12) * \sin(4\pi/5) - b_I(14) * \sin(2\pi/5)$ | $c_I(14) \Rightarrow M(9)$ |
| $c_R(11) = b_R(11) * \cos(2\pi/5) + b_R(13) * \cos(4\pi/5) + D_R(10)$ | $c_R(11) \Rightarrow M(24)$ |
| $c_I(11) = b_I(11) * \cos(2\pi/5) + b_I(13) * \cos(4\pi/5) + D_I(10)$ | $c_I(11) \Rightarrow M(6)$ |
| $c_R(13) = b_R(11) * \cos(4\pi/5) + b_R(13) * \cos(2\pi/5) + D_R(10)$ | $c_R(13) \Rightarrow M(23)$ |
| $c_I(13) = b_I(11) * \cos(4\pi/5) + b_I(13) * \cos(2\pi/5) + D_I(10)$ | $c_I(13) \Rightarrow M(7)$ |
| $A_R(2) = D_R(10) + b_R(11) + b_R(13)$ | $A_R(2) \Rightarrow M(20)$ |
| $A_I(2) = D_I(10) + b_I(11) + b_I(13)$ | $A_I(2) \Rightarrow M(5)$ |
| $A_R(5) = c_R(11) + c_I(12)$ | $A_R(5) \Rightarrow M(24)$ |
| $A_I(5) = c_I(11) - c_R(12)$ | $A_I(5) \Rightarrow M(21)$ |
| $A_R(8) = c_R(13) + c_I(14)$ | $A_R(8) \Rightarrow M(23)$ |
| $A_I(8) = c_I(13) - c_R(14)$ | $A_I(8) \Rightarrow M(7)$ |
| $A_R(11) = c_R(13) - c_I(14)$ | $A_R(11) \Rightarrow M(9)$ |
| $A_I(11) = c_I(13) + c_R(14)$ | $A_I(11) \Rightarrow M(6)$ |
| $A_R(14) = c_R(11) - c_I(12)$ | $A_R(14) \Rightarrow M(8)$ |
| $A_I(14) = c_I(11) + c_R(12)$ | $A_I(14) \Rightarrow M(22)$ |

## 9.8 COMPARISON MATRICES

**Table 9-7** Two-Building-Block FFT Algorithms Comparison Matrix

| Algorithm | # of adds | # of multiplies | # of data locations | # of const. locations |
|---|---|---|---|---|
| **Convolution** | | | | |
| Bluestein | $2*M+10*N$ $+4*A_{M/2}$ | $4*M+16*N$ $+4*M_{M/2}$ | $M+D_{M/2}$ | $4*N+3*M+C_{M/2}$ |
| Winograd | $Q*A_P$ $+(M_P+1)*A_Q$ | $-1+(M_P+1)*$ $(M_Q+1)$ | $D_P*D_Q$ | $(M_P+1)*(M_Q+1)-1$ |
| **Prime Factor** | $Q*A_P+P*A_Q$ | $Q*M_P+P*M_Q$ | $2*P*Q+\text{greatest}$ | $C_P+C_Q$ of $D_Q-2*Q$ and $D_P-2*P$ |
| **Mixed-Radix** | | | | |
| Primes-to-a-power | $2*(P-1)*(P-1)$ $+2*P*A_P$ | $4*(P-1)*(P-1)$ $+2*P*M_P$ | $2*P*P+$ greatest of $D_P-2*P$ and 2 | $(P-1)*P+C_P$ |
| Mixed power-of primes | $2*(P-1)*(Q-1)$ $+Q*A_P+P*A_Q$ | $4*(P-1)*(Q-1)$ $+Q*M_P+P*M_Q$ | $2*P*Q+$ greatest of $D_Q-2*Q$ and $D_P-2*P$ and 2 | $(P-1)*(2*Q-P)$ $+C_P+C_Q$ |
| Singleton | $2*(P-1)*(Q-1)$ $+Q*A_P+P*A_Q$ | $4*(P-1)*(Q-1)$ $+Q*M_P+P*M_Q$ | $2*P*Q+$ greatest of $D_Q-2*Q$ and $D_P-2*P$ and 2 | $(P-1)*(2*Q-P)$ $+C_P+C_Q$ |

*Key to Variables*

$N$ = number of points in an FFT

$M$ = number of FFT and IFFT points used to implement an $N$-point Bluestein algorithm

$A_{M/2}$ = number of adds in $M/2$-point FFT used for $N$-point Bluestein algorithm

$M_{M/2}$ = number of multiplies in $M/2$-point FFT used for $N$-point Bluestein algorithm

$D_{M/2}$ = number of memory locations used for data in $M/2$-point FFT used for $N$-point Bluestein algorithm

$C_{M/2}$ = number of memory locations used for constants in $M/2$-point FFT used for $N$-point Bluestein algorithm

$P$ = number of points in the first building block of an $N = P * Q$-point FFT

$M_P$ = number of multiplies required for $P$-point building block of $N = P * Q$-point FFT

$A_P$ = number of adds required for $P$-point building block of $N = P * Q$-point FFT

$D_P$ = number of memory locations used for data in $P$-point building block of $N = P * Q$-point FFT

$C_P$ = number of memory locations used for constants in $P$-point building block of $N$-point Bluestein algorithm

$Q$ = number of points in the second building block of an $N = P * Q$-point FFT

$M_Q$ = number of multiplies required for $Q$-point building block of $N = P * Q$-point FFT

$A_Q$ = number of adds required for $Q$-point building block of $N = P * Q$-point FFT

$D_Q$ = number of memory locations used for data in $Q$-point building block of $N = P * Q$-point FFT

$C_Q$ = number of memory locations used for constants in $Q$-point building block of $N$-point Bluestein algorithm

**Table 9-8**    FFT Algorithm Examples Comparison Matrix

| Algorithm | # of adds | # of multiplies | # of data locations | # of const. locations |
|---|---|---|---|---|
| **Convolution** | | | | |
| 15-point Bluestein | 790 | 464 | 72 | 162 |
| 15-point Winograd | 162 | 34 | 36 | 17 |
| **Prime Factor** | | | | |
| 15-point Kolba-Parks | 156 | 68 | 32 | 6 |
| 15-point SWIFT | 156 | 68 | 32 | 6 |
| **Mixed-Radix** | | | | |
| 16-point radix 4 | 144 | 24 | 40* | 6 |
| 16-point radix 8 and 2 | 148 | 28 | 34 | 6 |
| 15-point Singleton | 172 | 100 | 32 | 14** |

\* See Section 9.7.5 for why this does not match the formula in the Comparison Matrix in Table 9-7.

\*\* See Section 9.7.7 for why this does not match the formula in the Comparison Matrix in Table 9-7.

## 9.9 CONCLUSIONS

The algorithms detailed here have memory map relabeling instructions that will work for every algorithm building block in Chapter 8. Seven examples give detailed memory maps, with the relabeling incorporated, for each algorithm step. They have accompanying block diagrams to illustrate the data reorganization needed to combine small-point transforms in the examples and four general algorithms. These block diagrams help to see how to distribute data and algorithms on multiprocessor architectures that are explained in Chapter 12.

The next three chapters can be skipped if it is clear that a single processor will adequately compute the algorithm. However, if multiple processors are required, the next three chapters provide the information needed to learn how to map algorithms on multiprocessor architectures.

## REFERENCES

[1] L. I. Bluestein, "A Linear Filtering Approach to the Computation of Discrete Fourier Transform," *IEEE Transactions on Audio and Electroacoustics*, Vol. AU-18, pp. 451–455 (1970).

[2] S. Winograd, "On Computing the Discrete Fourier Transform," *Mathematics of Computation*, Vol. 32, No. 141, pp. 175–199 (1978).

[3] D. P. Kolba and T. W. Parks, "A Prime Factor FFT Algorithm Using High-Speed Convolution", *IEEE Transactions Acoustics, Speech, and Signal Processing*, Vol. ASSP-25, No. 4, pp. 281–294 (1977).

[4] Patent number 4,293,921, October 6, 1981, *Method and Signal Processor for Frequency Analysis of Time Domain Signals*, Winthrop W. Smith, Jr.

[5] R. C. Singleton, "An Algorithm for Computing the Mixed Radix Fast Fourier Transform," *IEEE Transactions on Audio and Electroacoustics*, Vol. AU-17, pp. 93–103 (1969).

[6] J. W. Cooley and J. W. Tukey, "An Algorithm for the Machine Calculation of Complex Fourier Series," *Mathematics of Computation*, Vol. 19, p. 297 (1965).

[7] J. W. Cooley, "The Structure of FFT Algorithms," *IEEE International Conference on Acoustics, Speech and Signal Processing Tutorial Session*, pp. 12–14 (1990).

# 10

# Arithmetic Building Blocks
# for Architectures

## 10.0 INTRODUCTION

Arithmetic building blocks are adders and multipliers combined in different ways that affect their cost and speed. This chapter does not contain a Comparison Matrix because these building blocks will already be imbedded in the processors by their vendors. Their memory and bus configurations are explained in Chapter 11. Arithmetic building blocks fall into three categories:

- Bit slice
- Integrated arithmetic
- Special purpose

The first two categories are known as general-purpose building blocks. Because most applications require more than just the computation of FFTs, general-purpose arithmetic architectures are typically used to allow the non-FFT functions to be computed on the same processor.

As a rule-of-thumb, if a DSP application requires more than four programmable DSP chips, and the FFT portion of the computations can be separated onto a dedicated processor, then a special-purpose arithmetic architecture, such as a hardware implementation of a 2-point FFT, is used for the dedicated processing. Once the special-purpose FFT architecture is part of an application, two things often happen. First, the number of programmable DSP chips can be reduced. Second, other functions being done on the programmable DSP chip, such as linear filtering and pattern matching, are often performed in the frequency domain (Chapter 6) using the special-purpose hardware, further reducing the number of programmable DSP chips needed.

## 10.1 FIVE PERFORMANCE MEASURES

All FFT algorithms have addition and multiplication steps. Sections 10.1.1 through 10.1.5 define five performance measures that can be used to characterize the following:

- How the data enters and leaves the arithmetic building block
- How the adder and multiplier are connected inside the building block
- How long it takes to perform adds and multiplies once the data is inside the building block

### 10.1.1 Input Data Organization

Since adders and multipliers each have two inputs, it is also vital to know whether two pieces of data to be added or multiplied can be entered into the building block simultaneously. If entry must be done sequentially, knowing the order of the sequence is important. Input data organization is described for each of the arithmetic building-block architectures and explained for each DSP chip in Chapter 14.

### 10.1.2 Output Data Organization

When a building block has both an adder and a multiplier, there are two potential outputs. It is important to know whether the building block has separate outputs for the adder and multiplier, a single output for both, or a single output that can be multiplexed between the adder and multiplier. This performance measure has a significant affect on how flexible the building block is for computing FFT algorithms. Output data organization is described for each of the arithmetic building-block architectures and explained for each DSP chip in Chapter 14.

### 10.1.3 Internal Data Bus Loading

How the adder and multiplier are connected by a bus, within an arithmetic building block, affects how much an algorithm loads the bus. The most common internal data bus configuration is a multiplier-accumulator (Figure 10-4). In that configuration the input data goes to the multiplier and the output comes from the adder. The output of the multiplier and the delayed adder output are the two inputs to the adder. Internal data bus loading is described for each arithmetic building-block architecture and explained for each DSP chip in Chapter 14.

### 10.1.4 Throughput from Computations

Throughput is the number of adds and multiplies per second that the arithmetic building block can perform if input data is supplied as fast as the building block can process it. Since the number of required adds and multiplies is a key performance measure of FFT algorithms, the ability to execute those arithmetic computations is an important performance measure. Throughput is described for each of the arithmetic building blocks and explained in more detail in Chapter 12 for algorithm mappings.

### 10.1.5  Latency from Computations

Latency is entirely different from throughput. Latency is the delay between when data enters the arithmetic building block and when answers are ready to be output. Latency becomes important in applications where the time it takes a system to respond to input data is critical. In a radar altimeter, if the plane is flying close to the ground, short latency is important in order to know rapidly any substantial loss of altitude. Latency is described for each of the arithmetic building-block architectures and explained in Chapter 12 for algorithm mappings.

## 10.2  BIT-SLICE ARITHMETIC

Addition and multiplication are linear operations. Just as linear operations allow multiple signals to be processed at one time (Section 2.3.3), a single signal can be decomposed into multiple signals, processed separately, and then recombined. One way of decomposing a single signal into two is to make the least significant digits one signal and the most significant digits another. For example, $21 = 20 + 01$ in decimal representation. Since $213 = (128+64+16+4+1) = 11010101$ in fixed-point binary arithmetic format (Section 13.2.1), it can be decomposed into $(208 + 5) = 11010000 + 00000101$ by separating the 4 least significant bits from the 4 most significant bits.

Addition is then performed by adding the corresponding 4 bit numbers and then recombining the results. For example:

$$(213 + 113) = (208 + 5) + (112 + 1)$$
$$= (208 + 112) + (5 + 1) = 320 + 6 = 326 \tag{10-1}$$

where:   $213 = 128 + 64 + 16 + 4 + 1 = 11010101 = 11010000 + 00000101$
$113 = 64 + 32 + 16 + 1 = 01110001 = 01110000 + 00000001$

A similar effect occurs with multiplication. Equation 10-2 shows the operations required for multiplication in a bit-slice arithmetic architecture.

$$A * B = (a_u * 2^M + a_l) * (b_u * 2^M + b_l)$$
$$= a_u * b_u * 2^{2M} + (a_l * b_u + a_u * b_l) * 2^M + a_l * b_l \tag{10-2}$$

where:   $a_u =$ upper bits of $A$
$a_l =$ lower $M$ bits of $A$
$b_u =$ upper bits of $B$
$b_l =$ lower $M$ bits of $B$

Multiplying rather than adding the numbers in Equation 10-1 gives:

$$(213) * (113) = (208 + 5) * (112 + 1)$$
$$= (208 * 112) + (208 * 1 + 5 * 112) + (5 * 1) \tag{10-3}$$
$$= 23{,}296 + 208 + 560 + 5$$

where:   $23{,}296 = 1011011000000000$
$768 = 0000011000000000$
$5 = 0000000000000101$

The results of the second and third multiplies and their sum have nonzero digits that are in the same locations as nonzero digits from the result of the first multiply. This approach requires four 4-bit multiplies and three 8-bit adds to obtain the results. This replaces doing one 16-bit multiply in order to reduce hardware. However, it increases computation time because of the sequence of operations that replace one 16-bit multiply.

The advantage of this architecture is that the multipliers and adders do not handle as many bits simultaneously. This was very important in the past, but is less important now because low-power full multipliers are commonly available. However, the technique can still be used to provide ultrafast arithmetic computations.

### 10.2.1 Multiplier

Equation 10-2 describes the functions that must be performed by the simplest bit-slice multiplier. For example, an 8-bit multiply can be performed by this equation using two 4-bit ($M = 4$), bit-slice multipliers. Similarly, a 16-bit multiply requires two 8-bit ($M = 8$) bit-slice multipliers using Equation 10-2.

Clearly, the technique can be extended to combining any number of bit-slice multipliers to form a larger multiplier. The algorithm is defined by writing the individual data words as their bit-slice components and then performing all of the required multiplies and adds. Equation 10-4 is an example for combining four 4-bit ($M = 4$) bit-slice multipliers into one large 16-bit multiply.

$$
\begin{aligned}
(a_0 + a_1 * 2^4 &+ a_2 * 2^8 + a_3 * 2^{12}) * (b_0 + b_1 * 2^4 + b_2 * 2^8 + b_3 * 2^{12} \\
&= a_0b_0 + a_0b_1 * 2^4 + a_0b_2 * 2^8 + a_0b_3 * 2^{12} \\
&+ a_1b_0 * 2^4 + a_1b_1 * 2^8 + a_1b_2 * 2^{12} + a_1b_3 * 2^{16} \\
&+ a_2b_0 * 2^8 + a_2b_1 * 2^{12} + a_2b_2 * 2^{16} + a_2b_3 * 2^{20} \\
&+ a_3b_0 * 2^{12} + a_3b_1 * 2^{16} + a_3b_2 * 2^{20} + a_3b_3 * 2^{24}
\end{aligned}
\tag{10-4}
$$

This set of equations can be implemented in several ways. At one extreme, 16 multipliers and 15 adders can be connected (Figure 10-1). At the other extreme, one bit-slice multiplier can be connected to an accumulator. In this case, control logic is required to sequentially feed the 16 pairs of $a_i$'s and $b_j$'s to the multiplier and properly shift the multiplier outputs into the adder by the number of bits equal to the exponent on the corresponding factor of 2 (Figure 10-2). For example, the $a_3b_1$ term must be shifted by 16 bits to properly contribute to the answer.

Between these two extremes are several choices. For example, Figure 10-3 shows the case of two multipliers and two adders. In this configuration, eight arithmetic cycles are required to accumulate all of the terms in Equation 10-4. During each of those eight cycles, two multiplies from Equation 10-4 are performed. The results of each pair of multiplies are shifted, added together, then sent to the accumulator. When all eight have been accumulated, the total is the hybrid bit-slice multiplier output. The design trade-off is speed versus hardware. If speed is more important than hardware, Figure 10-1 provides the best solution. If hardware is of paramount importance, Figure 10-2 provides the best solution. Figure 10-3 is a compromise between the speed of the implementation in Figure 10-1 and the minimal amount of hardware required in Figure 10-2.

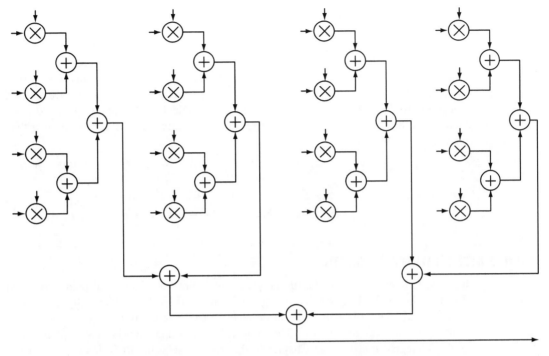

**Figure 10-1**    Full parallel 16-bit bit-slice multiplier.

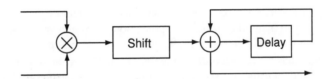

**Figure 10-2**    Sequential 16-bit bit-slice multiplication.

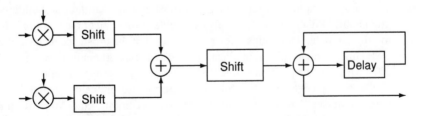

**Figure 10-3**    Hybrid (parallel/sequential) bit-slice multiplier.

### 10.2.2 Multiplier-Accumulator

There are two types of bit-slice multiplier-accumulators. The first was shown in Figure 10-2 as a way of implementing a bit-slice multiply algorithm sequentially. The second type is used to compute the sums of products of numbers. The core of this second type of architectural building block is the bit-slice multiplier. To it is added a bit-slice adder. Equations 10-5 and 10-6 are the bit-slice adder equivalents of Equations 10-2 and 10-4. Notice that the algorithm for implementing bit-slice addition is considerably simpler than bit-slice multiplication.

$$(a_u * 2^M + a_l) + (b_u * 2^M + b_l) = (a_u + b_u) * 2^M + (a_l + b_l) \qquad (10\text{-}5)$$

$$(a_0 + a_1 * 2^4 + a_2 * 2^8 + a_3 * 2^{12}) + (b_0 + b_1 * 2^4 + b_2 * 2^8 + b_3 * 2^{12}) =$$
$$(a_0 + b_0) + (a_1 + b_1) * 2^4 + (a_2 + b_2) * 2^8 + (a_3 + b_3) * 2^{12} \qquad (10\text{-}6)$$

## 10.3 INTEGRATED ARITHMETIC

Integrated circuit technology has progressed to the point that 16-bit fixed-point and 32-bit floating-point multipliers are commonly available on DSP chips. Generally, the output of these multipliers feeds one side of an adder because so many DSP functions involve multiply-accumulate operations. The drawback to this approach is in algorithms, such as the Winograd transform in Chapter 9, that require sequences of adds and sequences of multiplies, as well as multiply-accumulates. Then, during the addition sequences, the multiplier cannot be used, and during the multiply sequences the adder cannot be used.

### 10.3.1 Multiplier

At one point in the development of DSP technology, integrated 16-bit multiplier chips played a significant role in application development. However, with the advent of programmable DSP chips, multiplier chips have lost their popularity because so much of the computations in DSP algorithms involves multiplier-accumulator computations. However, for applications that just require multiplication, such as the weighting function multiplication prior to FFT algorithms, a multiplier provides the most computationally efficient use of hardware real estate.

### 10.3.2 Multiplier-Accumulator

The multiplier-accumulator is the most common arithmetic building block in programmable DSP chips. They are also available without all of the additional features built in to programmable DSP chips. However, because of the broad acceptance of programmable DSP chips in high-volume applications such as telecommunications, it is often more cost effective to buy the programmable DSP chip and only use its multiplier-accumulator feature.

The key advantage over bit-slice multiplier-accumulators is that the whole function is in one device. There is no added hardware to combine chips to perform the algorithms in Equation 10-6. The disadvantage is that the hardware cannot be tailored for specific applications. For example, a low-cost application that does not require high-speed multiplication but does require low power can use an adder to perform the multiplications and additions to save power and cost.

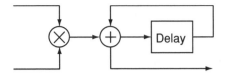

**Figure 10-4**    Multiplier-accumulator.

Figure 10-4 shows the most common multiplier-accumulator block diagram. All of the programmable DSP chips in Chapter 14 use this basic architecture with varying degrees of bells and whistles to enhance performance for a particular manufacturer's perceived market. One example is the number of bits in the accumulator, depending on the anticipated number of multiply-accumulates required to compute results for particular algorithms. To ensure that a fixed-point accumulator does not overflow, it needs to have at least $\log_2 N$ bits more than the multiplier output that feeds it, if $N$ multiplies must be accumulated prior to storing results.

## 10.4 SPECIAL PURPOSE

In applications that require more than four programmable DSP chips to perform the power-of-two FFT computations, hardware that has an architecture dedicated to FFT computations, special-purpose chips, should be used. The special-purpose FFT chips in Section 14.7 do power-of-two FFTs much faster than programmable DSP chips, because the common building blocks of FFT algorithms are imbedded in the hardware. For the power-of-two FFT algorithms in Section 9.7, the common arithmetic building block is the 2-point-building-block algorithm. Building blocks for non-power-of-two algorithms have not become popular because these algorithms are not common and because they require several building blocks, not a single one. Section 14.7 describes chips that have been built to implement the 2-, 4-, and 8-point building blocks from Chapter 8.

Since FFT equations assume complex inputs, the 2-point building block assumes complex input data. The 2-point building block can be implemented in full parallel form with two complex input signals entering the hardware simultaneously, or it can be implemented in half-complex form, where the real portion of the two input signals enters the arithmetic building block first, followed by the imaginary part. The linearity of FFTs allows this sequential computation, followed by a recombination of the results (Section 2.3.3).

Two forms of the 2-point FFT building block have been developed to implement the two approaches to decomposing the DFT to form the power-of-two FFT. The data separation pattern for each of these approaches is presented in Section 10.4.1. Then the 2-point building-block hardware for each approach is presented in Sections 10.4.2 and 10.4.3.

### 10.4.1 FFT Data Separation Patterns

The first FFT data separation approach is called decimation in time (DIT). In the DIT algorithm, which is used in Chapters 8 and 9, the input samples are first reordered into two subsets of input samples, one containing the odd-numbered samples and the other the even-numbered ones, shown in Figure 10-5 as the 1st decimation in time. Then each of these subsets is further reordered by taking every other one of its members and putting it into a new subset, shown in Figure 10-5 as the 2nd decimation in time. Once the data reordering is complete, the paired input data samples are used as the inputs to the 2-point FFT building block from Section 8.3. Since the input data sequences are usually thought of as sequences in time, they are being decimated in time by this reordering process.

The second approach, decimation in frequency (DIF), also starts by segmenting the input sequence into two subsets of data. The difference is that this algorithm puts the first half of the samples in the first subset and the second half in the second subset, shown in Figure 10-6 as the 1st decimation in frequency. The next step in the algorithm segments each of these subsets into new subsets, again by putting the first half of its members in the first subset and the rest in the other subset. This process is shown in Figure 10-6 as the 2nd decimation in frequency. These four subsets are the inputs to the first set of 2-point FFTs from Section 8.3. The outputs of the first set of 2-point FFTs are reordered following this same strategy. This process continues until the output frequencies are reached. At the output, the output frequency components are in subsets of even- and odd-numbered frequencies. Therefore, the output frequencies have been decimated, which led to calling this approach decimation in frequency.

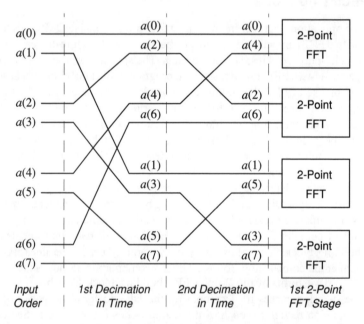

**Figure 10-5**   Eight-point FFT decimation-in-time input data organization.

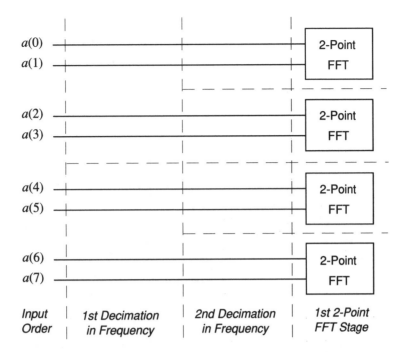

**Figure 10-6**   Eight-point FFT decimation-in-frequency input data organization.

### 10.4.2 Decimation-in-Time Building Block

The flow graph for the DIT 2-point hardware building block is shown in Figure 10-7 (on page 254). One advantage of this algorithm over the decimation-in-frequency algorithm is that it is organized to work easily with multiplier-accumulator arithmetic building blocks.

### 10.4.3 Decimation-in-Frequency Building Block

The flow graph for the DIF 2-point hardware building block is shown in Figure 10-8. The primary difference between this and the DIT flow graph is the multiplier on the output rather than the input. While this appears to cause problems with using multiplier-accumulator building blocks, it does not. The reason is that most FFT applications require a weighting function prior to the FFT. This weighting function multiplier is then added to the front end of the flow graph in Figure 10-8 for the first stage and then the back-end multiplier is moved to the front end of the next 2-point building block of the FFT algorithm.

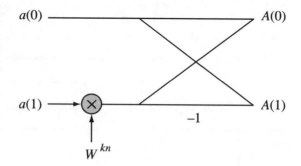

**Figure 10-7** Decimation-in-time 2-point FFT flow graph.

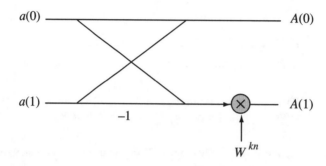

**Figure 10-8** Decimation-in-frequency 2-point FFT flow graph.

## 10.5 CONCLUSIONS

Prior to the introduction of programmable DSP chips, a detailed understanding of arithmetic building blocks was crucial in the creation of DSP processors on boards. This was because the number of processor clock cycles required to perform multiplies was significantly higher than for additions. Arithmetic building blocks are now imbedded in DSP chips. Understanding the nuances of how chip manufacturers connect the multipliers and accumulators helps in the selection of an algorithm from Chapters 8 and 9.

# 11

# Multiprocessor Architectures

## 11.0 INTRODUCTION

A single-processor architecture is the interconnection of arithmetic building blocks with memory, data I/O, and control logic. A multiprocessor architecture is an interconnection of two or more single processors. Several single and multiprocessor architectures are used to perform FFTs. This chapter explains how a single-processor architecture is created and then shows nine ways in which they are combined into multiprocessor architectures.

DSP architectures are composed of:

- Memory for storing data
- Memory for storing constants
- Memory for storing algorithm code
- Arithmetic units for doing adds and multiplies on the data
- Arithmetic units for generating data addressing sequences
- Bus for moving program instructions
- Bus for moving instruction addresses
- Bus or buses for moving data and control information
- Bus or buses for moving data addresses
- Bus or buses for moving data I/O

## 11.1 TWO SINGLE PROCESSORS

There are two popular single-processor architectures. The first, called Von Neumann [1], has only one bus and uses it to interconnect the arithmetic unit to the rest of the processor. The arithmetic unit is used for all algorithm computations and data address generation. The single bus and arithmetic unit are shared at each step for FFT arithmetic computations and data addressing. This "Von Neumann bottleneck" stimulated development of the second type of single processor, called Harvard. This architecture has separate arithmetic and addressing hardware and buses to alleviate the bottleneck. All the chips in Chapter 14 are Harvard architectures. Section 11.1.1 presents the Von Neumann architecture to illustrate specifically the inefficiencies associated with using it for signal processing applications.

### 11.1.1 Von Neumann Architecture

The Von Neumann architecture (Figure 11-1), has been the most popular approach to standard computers for many years because of its simplicity. This architecture has:

- One arithmetic unit shared between address generation and arithmetic computations
- One memory shared between data, constants, and program instructions
- One bus used for moving data addresses and instructions

The arithmetic unit includes not only the adder and multiplier for data computations but the "next instruction address," "present instruction," and "present data address" registers, as well as the logic for executing instructions.

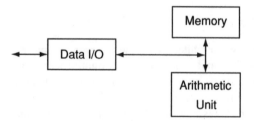

**Figure 11-1**   Von Neumann architecture block diagram.

The simplicity of this architecture allows it to run at high clock speeds and to be used for a general class of applications. For example, applications that access data sequentially do not require address generation algorithms, and applications that perform large numbers of computations on each new data sample use the arithmetic unit for data addressing infrequently. A simple example that illustrates both of these is converting an input data sequence into the logarithm of that sequence using the Taylor series expansion. In this algorithm, a data value is accessed from memory, followed by a long sequence of adds and multiplies on that data, to form the logarithm. The result is then stored in the same memory location. The processor then steps to the next memory location and repeats the process.

The two major disadvantages of this architecture for FFT algorithms are that it has a single bus for handling data I/O, data movement, and instruction movement, and it needs the arithmetic unit to perform the data reordering between algorithm steps as well as to perform the algorithm computations. A simple example is a single multiply accumulation of data values stored in nonsequential locations of memory. The arithmetic unit steps are as follows:

1. Use the next instruction address in the arithmetic unit register to access the next instruction from memory and store it in the present instruction register.

2. Decode the present instruction register to determine the computation to perform and the data memory address offset to the next piece of input data for the multiply-accumulate function.

3. Add the data memory address offset to the present address and store the result in the present data address register.

4. Use the present data address to access the next piece of data from memory.

5. Decode the present instruction register to determine the multiplier constant memory address offset.

6. Add the multiplier constant memory address offset to the present multiplier constant memory address and store in the present multiplier constant memory address register.

7. Use the present multiplier constant address to access the next multiplier constant from memory.

8. Perform the multiply function.

9. Store the result in the present data address.

10. Decode the present instruction register to determine the program memory offset to the next instruction, add that value to the next instruction address register and store the result in the next instruction address register.

Steps 3, 6, 8, and 10 use the arithmetic unit, steps 1, 4, 7, and 9 make use of the bus between arithmetic unit and memory, and steps 2, 5, and 10 use the instruction decoding logic. Steps 4 and 5 can be performed in parallel by the Von Neumann architecture. The result is a sequence of nine steps to perform the multiply-and-store function that is common to FFT algorithms. Note that step 10 uses the arithmetic unit as well as the instruction decoding logic. This is the most obvious example of reduced computation time that is obtained if the instruction and computational functions of the processor are separated. This separation is the basis of the Harvard architecture described in the next section.

### 11.1.2 Harvard Architecture

The Harvard [2] architecture (Figure 11-2) is the most popular single arithmetic unit processor for DSP applications. All of the programmable DSP chips in Chapter 14 use a variant of this architecture. Its main feature is that it physically separates the algorithm computations from the data and instruction memory addressing (control) functions. It also uses separate buses to interconnect the building blocks associated with the computational and control functions. This provides significant improvements in throughput and latency for FFT algorithms because it removes the Von Neumann bus bottleneck and allows the arithmetic unit to be used only for algorithm computations.

The multiply-accumulate steps in Section 11.1.1 are identical to those used by the Harvard architecture. However, they can be overlapped in the Harvard architecture to speed up the computations. The most recent generations of programmable DSP chips have two data memory to arithmetic unit buses, two data memories, and two address generators. This allows the data and multiplier constant address generation and memory accesses to be accomplished in parallel. For those chips, steps 2, 3, and 4 can be performed in parallel

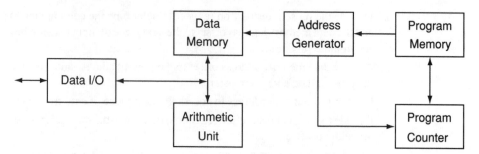

**Figure 11-2** Harvard architecture block diagram.

with steps 5, 6, and 7. Similarly, steps 8 and 9 can be performed in parallel with steps 10 and 1. The result is that the 10 steps can be performed as if they were 5, rather than having to do the 9 required by the Von Neumann architecture. Thus, the Harvard architecture can compute FFTs nearly twice as fast as the Von Neumann. That is why all the commercial DSP chips are based on this more efficient architecture.

## 11.2 THREE LINEAR ARRAYS

Linear array architectures, the simplest form of multiprocessor systems, fall into three classes:

- Pipeline, where the output of each processor provides the input for the next
- Linear bus, where all processors are connected to a common communication bus
- Ring bus, an extension of the linear bus with the ends of the common communication bus connected

Any of the arithmetic building blocks from Chapter 10 can be used as the processors in these three bus architectures. Further, either of the single processors described in Section 11.1 can be used. Because of this, the key differences between the linear array architectures are how their interconnections affect their ability to perform FFT algorithms. This section describes those three architectures, and Section 12.4 shows how they are used to compute the FFT algorithms from Chapter 9.

### 11.2.1 Pipeline

The pipeline [1, 3] architecture interconnects processors such that the output of one becomes the input to the next. The three-block version of the pipeline in Figure 11-3 can be used to illustrate the key features of this architecture. The most important design consideration is matching the data output rate from one processor to the input data rate of the next so that it keeps the next processor busy without overloading. If each processor is kept busy, then the performance of the overall architecture is the sum of the performances of each processor.

A multiplier-accumulator is a common example of a two-processor pipeline that is found in nearly all modern programmable DSP chips and is explained in more detail in Chapter 14. Processor 0 would be the multiplier and Processor 1 the accumulator, as

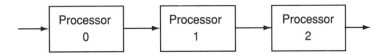

**Figure 11-3**   A pipeline architecture block diagram.

shown in Figure 10-4. The input to Processor 0 is the next data sample to be multiplied and its multiplier constant. Each time Processor 0 produces a multiplication result, it sends that result to Processor 1 to add to the accumulator. Processor 1 then performs the addition and stores the result in its accumulator register while Processor 0 is performing the next multiplication. At some point, the multiply-accumulation process is complete, and Processor 1 outputs its result to data memory.

Therefore, if the input data rate to Processor 0 is $R$ samples per second, the overall input rate to Processor 0 is $2 * R$ per second because it must also receive the multiplier constants. The output data rate from Processor 0 is $R$ per second, which then becomes the input data rate to Processor 1. If Processor 1 can perform $R$ adds and accumulator register stores per second, then the data rate between the two processors is ideal. Finally, notice that the output data rate from Processor 1 is lower than its input rate. If $M$ multiply-accumulates are performed before an output is produced, then Processor 1's output data rate is $R/M$ per second.

If further computations are needed on these results, then Processor 2 should be chosen to perform its portion of those computations at an input data rate of $R/M$ per second. A well-designed pipeline architecture uses processors at each stage that match the required data rates of the previous processor outputs.

### 11.2.2 Linear Bus

A linear bus [1] (Figure 11-4) is an architecture where a single bus is used to provide the path for all of the data communications among two or more processors. Overloading of the bus can occur because it handles all the interprocessor data transfers as well as the data I/O. If the bus can handle enough data so that each processor is kept busy, then the performance of the overall architecture is the sum of the performances of each processor.

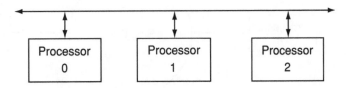

**Figure 11-4**   Linear bus architecture block diagram.

Some programmable DSP chips use this bus architecture when they have multiple arithmetic processors. These are described in more detail in Chapter 14. Again, the multiply-accumulate example can be used to illustrate the issues associated with using this architecture. Assume Processor 0 is the multiplier, Processor 1 is the accumulator, and Processor 2 is the data and multiplier constant memory. To keep the multiplier busy, it must have a new data word and multiplier constant each computation cycle. Since both of these come across the bus from Processor 2, this forces Processor 2 to handle two data accesses per computation cycle and puts a two-word-per-computation cycle load on the bus.

The multiplier also produces a new result each computation cycle, and this answer must be passed to the accumulator (Processor 1) to allow Processor 0 to continue performing multiplications and to allow Processor 1 to remain busy performing accumulations. This adds another word per computation cycle to the bus requirements. Finally, after $M$ accumulations the accumulator has an output that it must pass back to the data memory (Processor 2). This adds load on the bus of $1/M$ words per computations cycle.

In addition to these computational loads, data must be coming into the processor and be stored in the data memory so that data is available for multiply-accumulation. Assuming the new data must enter at the multiplier computation rate, this adds another data word per computation cycle to the bus requirements. Eventually, results must also exit the processor to be used elsewhere. If this is assumed to occur at the $1/M$ rate of the accumulator outputs, then the output function increases the total bus loading to $(4+2/M)$ words per computation cycle. If the computation rate is $R$ multiplies per second, then the data rate that must be sustained on the bus is at least $[R*(4+2/M)]$ words per second. A well-designed linear bus architecture uses processors and buses that match the required performance of the chosen algorithm.

### 11.2.3 Ring Bus

The ring bus [3] (Figure 11-5) is a special case of the linear bus, in which the ends of the linear bus are connected. Generally, algorithms are implemented on this type of bus using a combination of pipeline and linear bus techniques. Any arithmetic building block from Chapter 10 or processor from Section 11.1 can be one of the processor blocks in this architecture, and the number of processors can be as small as two or rather large.

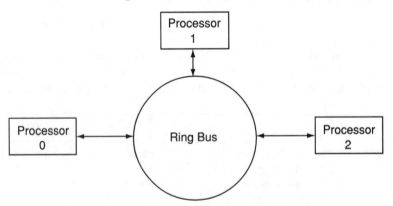

**Figure 11-5**   Ring bus architecture block diagram.

At first glance, this architecture does not appear to differ from the linear bus. In fact, it can be used in that manner. In this case it has the same properties as the linear bus. However, this architecture allows another type of processing, namely the input data can be thought of as being sequentially passed from one building block to the next along with a codeword that tells whether that processor is supposed to perform a function on that piece of data. The codeword also can tell the processor what function to perform if the processor is programmable. This allows multiple words to be on the bus at one time because each is stored in a data register at the input to one of the processors. This makes the architecture look like a series of linear buses between processors.

For example, consider the multiply-accumulation example again. However, this time consider Processor 0 to be one of the bit-slice multiplier building blocks described in Chapter 10. Chapter 10 showed that a complete multiplication can be performed with bit-slice building blocks by passing the various "slices" of the input data word and multiplier constant through the bit-slice multiplier, properly scaling the output and adding it to the accumulator.

Further, assume Processor 1 is a bit-slice adder, Processor 2 is a data memory, and the data words are bit-sliced into two pieces. From Chapter 10 the multiply process requires four bit-slice multiplies and three bit-slice adds, as shown in Equation 11-1. The accumulation portion of the multiply-accumulate can now be integrated with the addition portion of the bit-slice multiply.

$$A * B = (a_u * 2^M + a_l) * (b_u * 2^M + b_l)$$
$$= a_u * b_u * 2^{2M} + (a_l * b_u + a_u * b_l) * 2^M + a_l * b_l * 2^0 \tag{11-1}$$

The first step is to load the data ($A$) and multiplier constant ($B$) words from data memory (Processor 2) onto the bus along with a control code that tells the bit-slice multiplier (Processor 0) to multiply the two lower halves of the word. When $A$ and $B$ reach the bit-slice multiplier, it loads the lower portion of both words, performs the multiplication, and changes the codeword to indicate it has performed that portion of the task. While the multiplication is being performed, the two data words move along to Processor 1. However, the codeword accompanying these words tells that processor not to perform any computations. The same thing happens on the next clock when the data words are at the input register to Processor 2. Another clock later, the two data words are back at Processor 0, and this time the codeword, altered by Processor 0, tells Processor 0 to take the lower half of the multiplier constant and the upper half of the data constant and perform the multiplication. The two input data words make two more cycles around the ring bus to allow all four bit-slice multiplications to be performed.

Meanwhile, once the first bit-slice multiplication is complete, the result ($a_l * b_l$) is moved from Processor 0 to Processor 1 to perform the addition part of the multiplication and accumulation processes. Again, this partial result is accompanied with a codeword generated by Processor 0 that tells the bit-slice adder the scale factor of the word (in this case the factor is $2^0 = 1$). The codeword that accompanies this partial result also tells Processor 1 to remove the word from the ring bus. In other applications the word might stay on the ring bus and be used in a different way by one of the other processors. This feature is used in FFT algorithms because they generally use each computational result in two or more places.

The other three intermediate results, along with their codewords, are also put on the bus by Processor 0 to go to Processor 1 to be accumulated. After the input data has passed by Processor 0 four times and Processor 0's results fed to Processor 1, the multiplication and accumulation is complete and new data and multiplier words must be accessed to continue the multiply-accumulation process. Finally, the $M$ multiply-accumulations are complete, and the result is put on the bus by Processor 1 to return to data memory in Processor 2. The data memory processor not only stores the result but removes it from the bus.

The key concern with this architecture is bus contention, just as for the linear bus. Only this architecture has a more demanding requirement because data passes around the ring several times before the algorithm computations are complete. When bus contention occurs, the transmission of processor outputs must be delayed. This results in a reduction in throughput and an increase in latency.

One solution to bus contention is to allocate specific time slots to each processor connected to the ring. This completely removes the contention problem. However, the contention problem is then replaced with the need to design algorithms so that the processors finish their computations close to their ring bus time slot. Otherwise, the processors have the overhead of waiting for their turn to output results and input the next set of data. For FFT algorithms this approach can be efficient because the algorithms are highly modular. Section 14.11 shows a product family that uses this time-slot technique to remove bus contention.

## 11.3 THREE PARALLEL ARRAYS

Parallel arrays have two-dimensional interconnectivity that fit the following three classes:

- Crossbar, which is the most general and allows processors to be directly connected as needed to a large number of others in the array.

- Massively parallel, where the processors are generally connected to just their nearest neighbors and communications beyond the nearest neighbor requires passing information through other processors.

- Star, which has all processors connected to a central one. The central processor may use the connected processors as coprocessors, or it may be a central memory that is used by the surrounding processors. When the central processor is replaced with memory, this is called a shared-memory architecture.

### 11.3.1 Crossbar

A crossbar [1, 3] switch is a device that allows each of its inputs to be directly interconnected to any other one. For example, consider a crossbar switch to interconnect four processors that each have one I/O port. Table 11-1 shows the number of simultaneous interconnections available. If the number of processors is larger, or the processors have additional I/O ports, the number of different interconnection combinations grows exponentially.

Figure 11-6 is a block diagram of a crossbar architecture where the individual crossbar elements control the routing of four processors in an overall array of 16. Each crossbar switch can arbitrarily connect any of its four processors to any other one. The crossbar switch used in Figure 11-6 has an additional output that can be connected to any of the four inputs. This increases the number of combinations shown in Table 11-1 from 3 to 12 because for each combination any of the four processors can also be connected to the additional output to feed the larger network. Further, the central crossbar switch in Figure 11-6 can connect any of the four crossbar switches to another. The result is that with

**Table 11-1**    Four-Way Crossbar Interconnection Options

| Interconnect option | Set 1 | Set 2 |
|:---:|:---:|:---:|
| 1 | Processors 0 and 1 | Processors 2 and 3 |
| 2 | Processors 0 and 2 | Processors 1 and 3 |
| 3 | Processors 0 and 3 | Processors 1 and 2 |

these two levels of crossbar switching, any of the 16 processors can be directly connected to one of the others without going through another processor.

There are numerous variations to this architecture, depending on the vendor. For example, the crossbar switch described in Table 11-1 can also be designed to allow a processor's I/O to connect to more than one of the other processors. Table 11-2 shows the combinations available under these design constraints. Note that for this set of design rules (each processor only having one I/O port), if three processors are connected the fourth has nowhere to be connected. This architecture's interprocessor data I/O rate is not limited by the buses themselves, but by scheduling the processing tasks so that two or more processors do not have to feed data to the same one simultaneously. This is more accurately characterized as processor I/O contention, rather than bus contention.

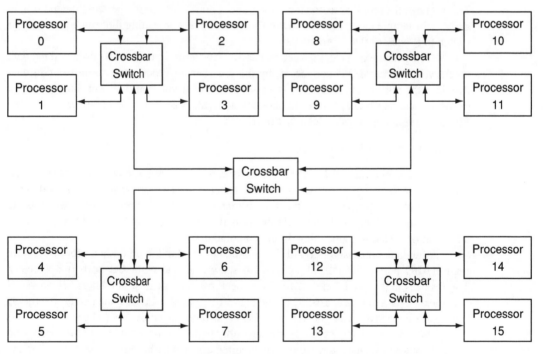

**Figure 11-6**    Crossbar switch architecture block diagram.

The multiply-accumulation example is again used to illustrate the processor I/O contention issues. For example, assume that the upper-left-hand crossbar switch in Figure 11-6 has Processor 0 containing the data memory and multiplier constants, Processor 1 contain-

**Table 11-2** Four-Way + Broadcast Crossbar Switch Options

| Interconnect option | Set 1 | Set 2 |
|---|---|---|
| 1 | Processors 0 and 1 | Processors 2 and 3 |
| 2 | Processors 0 and 2 | Processors 1 and 3 |
| 3 | Processors 0 and 3 | Processors 1 and 2 |
| 4 | Processors 0, 1, and 2 | N/A |
| 5 | Processors 0, 1, and 3 | N/A |
| 6 | Processors 0, 2, and 3 | N/A |
| 7 | Processors 1, 2, and 3 | N/A |
| 8 | Processors 0, 1, 2, and 3 | N/A |

ing the multiplier, Processor 2 containing the accumulator, and Processor 3 being the data I/O. Since data must be input as fast as it is being operated on by the multiply-accumulator, a single multiply-accumulate cycle will be assumed to also include receiving a new input data sample.

The first step is to connect Processor 0 to Processor 1 for two cycles to move a data word and multiplier constant from memory into the multiplier. During the next cycle the multiplier performs its computation and sends the result to the accumulator in Processor 2. This requires the crossbar to connect Processors 1 and 2. This is the perfect time to bring in a new data sample using the data I/O in Processor 3 and connecting it through the crossbar switch to Processor 0 to store the data.

During the next cycle, the accumulator in Processor 2 performs its task, and the data memory in Processor 0 is connected, by the crossbar, to Processor 1 to move additional data into the multiplier. This is a rather simplistic example that does not illustrate all of the power and flexibility of the crossbar network. This is addressed in conjunction with the FFT algorithm mappings in Section 12.5.1.

### 11.3.2 Massively Parallel

A massively parallel [1, 3] processor is defined as having more than 1000 smaller processors. Most often, the processors are connected in a two-dimensional array with only nearest-neighbor connections. If the array is rectangular, then the processors are connected either to four or all eight of their neighbors, as shown in Figures 11-7 and 11-8. There are a number of variations depending on the manufacturer.

A fundamental assumption of this architecture is that the individual processors have multiple I/O ports. Figures 11-7 and 11-8 show four and eight I/O ports, respectively. The result is that there is no data I/O bottleneck between nearest neighbors. However, if data must be passed to processors beyond nearest-neighbor locations, the nearest neighbors must participate in the data transfer. This I/O requirement occupies the I/O ports of multiple processors, thus reducing a processor's capability to pass its own data to another processor.

Another key characteristic of this architecture is whether all of the processors are controlled by one program or whether each one can implement its own. If all the processors must execute the same program, the architecture is called single-instruction, multiple-data (SIMD). If each processor can have its own program to execute, then it is called multiple-instruction, multiple-data (MIMD).

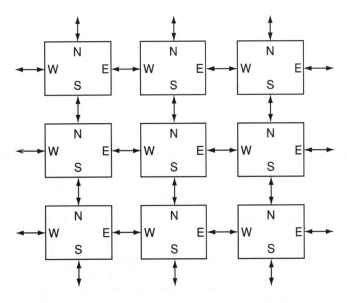

**Figure 11-7**    North-east-west-south connected massively parallel array
architecture block diagram.

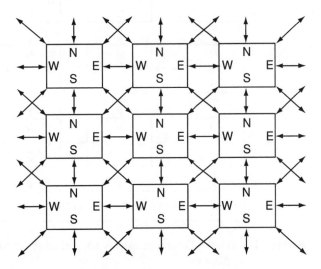

**Figure 11-8**    Completely connected nearest-neighbor array architecture
block diagram.

Most massively parallel processors have been SIMD architectures. There are two primary reasons for this and one significant drawback. The first reason is that technology has not allowed it to be cost efficient to implement a control processor for each of the 1000 or more processors. Second, it is much more difficult to think through how to control 1000 programs working at the same time. The drawback is that it is very difficult to map individual algorithms onto an array of 1000 or more processors and have them execute it efficiently.

More recently, programmable signal processor chips have been designed to be interconnected in larger arrays. Since each of these has its own program control, they are likely to be used in an MIMD configuration. While thousands of these devices are not likely to be connected in the near future, a trend is developing in that direction. Examples of this are shown in Section 14.11.

Massively parallel array architectures generally have their own special-purpose I/O subsystem that converts the input data from a sequential stream into data vectors that can be passed into the processing array along one of its edges. Figure 11-9 shows a specific example of this I/O strategy for the north-east-west-south (NEWS) connected massively parallel array in Figure 11-7. When the computations are complete, the results can be shifted down to the output data reorganizer and converted back to a sequential stream of data.

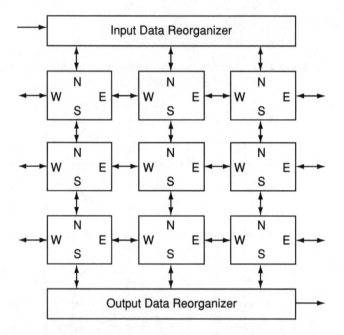

**Figure 11-9**   Data I/O for a massively parallel array architecture block diagram.

These more sophisticated architectures provide more opportunity for variation in the way an algorithm is implemented. The simple multiply-accumulation algorithm is no exception. A 2 × 2 NEWS array of processors is used to illustrate the two extremes of using a massively parallel processor for multiply-accumulate functions.

In the first approach assume that each processor is a single Harvard architecture processor and store the multiplier constants in each of these processors. Then as data arrives to the processors, store the data associated with particular multiplier constants in that processor's data memory. Every time $M$ data samples have been stored in each processor, all the processors can be told to perform the $M$-step multiply-accumulate process on its set of data. All the processors then execute the same instruction set and finish at the same time. When they are finished, multiply-accumulates have been performed on four sets of data. If during that computation period, $M$ new data samples can be loaded into each of the four processor's data memory, then the four processors can begin the multiply-accumulation process on the next set of data as soon as they have finished the present set and have output the results.

In the second approach each set of $M$ inputs is divided equally among the four processors. Then each of the four processors computes $M/4$ of the multiply-accumulates, and these four partial results are combined by adding. In more detail, one-quarter of the multiplier constants are stored in each of the four processors. Then the input data interface separates the input data words so that one-quarter of them go to each processor. Then each processor performs multiply-accumulation on its $M/4$ data words, using its $M/4$ multiplier constants.

Once these partial results are obtained, they must be added to form the final $M$ sample multiply-accumulation. One way to do this is to send the partial answers from the left two processors to memory locations in the right two processors, using the "E" output of the left-hand processors and the "W" input of the right-hand processors. Then the right two processors can add their partial results to those computed by the processor to their left. Finally, the top right processor can send its partial result to the bottom right processor for the final addition needed to produce the desired output.

The second approach takes longer to compute because of the data passing required and because all of the processors are not active during the final additions used to combine the partial results. However, the computation has less latency to produce its result. Namely, a new multiply accumulation starts every $M$ samples with the second approach, and therefore answers are output every $M$ samples. In the first approach the processor only starts a new multiply-accumulate computation every $4*M$ data samples. Therefore, it can only produce results every $4*M$ data samples. Hence, even though the individual multiply-accumulate is produced faster, it takes longer for the answers to be available for further computations.

### 11.3.3 Star

The star [1] architecture is most often used when one function or process dominates the application. It consists of one central processor with interconnections to numerous others, as shown in Figure 11-10. The star architecture does not have to have four processors surrounding the central one. It can have more or less, depending on the application.

The interprocessor communications in this architecture all occur via the central unit. This requires it to have the capability to handle multiple data streams simultaneously or the architecture will not be efficient. The most likely uses for this architecture are for applications where either:

1. The central block does the general computations and the surrounding ones are used as coprocessors to perform specific functions, such as nonlinear operations or database searching, or

2. The central processor is data memory (shared memory) that needs to be accessed by multiple processors at the same time, like a simultaneous database search from multiple remote locations.

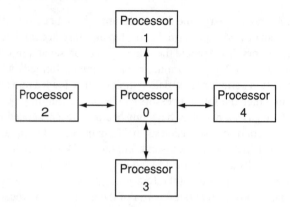

**Figure 11-10**   The star architecture block diagram.

Just like the massively parallel architecture, there are many ways to use a star architecture to implement a set of algorithms. Using the multiply-accumulate as an example, assume five processors connected to the central processor. In this case let four of the outlying processors be 8-bit bit-slice multipliers, and let the central processor be the data memory and an accumulator. Let the fifth outlying processor handle the data I/O functions.

The first step is to move 16-bit input data through the data I/O processor and store it in the data memory in the central processor. The next step is to have the central processor slice the 16-bit input words into 8-bit slices and pass the slices to each of the four bit-slice multipliers. The next step is for each of the bit-slice multipliers to perform one of the multiplications shown in Equation 11-1. Once the computations are complete, each bit-slice multiplier passes its result back to the central processor. The central processor is then responsible for performing the scaled additions shown in Equation 11-1. The final result for the first multiplication now resides in the central processor, and it can be added to the other multiplied data to form the $M$-step multiply-accumulation.

## 11.4 THREE MULTIDIMENSIONAL ARRAYS

Multidimensional arrays are one step beyond parallel arrays because they exhibit interconnectivity that has three or more dimensions. The three presented in this section are:

- Hypercube, which is the most common and is configured to minimize interprocessor communications distances.

- Three-dimensional massively parallel arrays, which have been built for special problems, such as fluid dynamics calculations, but are very difficult to program for problems that are not easily described in the same number of dimensions as the architecture.

- Hybrid, where each element in the array is itself at least a two-dimensional architecture of a different type than the high-level architecture. Again, these architectures are most useful for solving specific types of problems.

This type of architecture has been included because there are multidimensional FFT applications and because even one-dimensional applications can be conveniently written as a multidimensional FFT computation.

### 11.4.1 Hypercube

In mathematics a cube is a three-dimensional object with equal sides. The mathematical generalization of this equal-sided object to more than three dimensions is called a hypercube. A hypercube [1, 3] processing architecture is an organization of connections between processing elements that form cubes. Joining two hypercubes of the same dimension forms a hypercube of the next higher dimension. A single processor is a zero-dimensional hypercube. Connecting two of those forms a one-dimensional hypercube. Connecting two of these forms a square, which is a two-dimensional hypercube. Connecting two squares forms a cube, called a three-dimensional hypercube. It becomes difficult to envision higher-dimensional hypercubes. Figure 11-11 shows the four-dimensional hypercube. Note that it is composed of two interconnected (one inside the other), three-dimensional hypercubes.

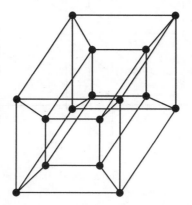

**Figure 11-11**    Four-dimensional hypercube architecture.

An $N$-dimensional hypercube has $2^N$ processing elements. For example, the four-dimensional hypercube in Figure 11-11 has $2^4 = 16$ processing elements. The most unique feature of the hypercube architecture is the efficiency of its interconnectivity. Namely, in an $N$-dimensional hypercube, data can be passed from one processor to any other in the architecture by passing through no more than $N$-1 other processing elements. In Figure 11-11 data can be passed from a processor to any other by passing through no more than three processors. This contrasts with a 16-processor NEWS connected architecture where passing data from one corner to the opposite one requires passing data through five other processors, $(N - 1) + (N - 2)$ in general. For larger arrays, such as 1024 elements, the difference is even more dramatic. This makes the hypercube architecture attractive for high-performance problems that require large amounts of data passing between arbitrary pairs of processing elements.

The biggest drawback to the hypercube is that, in order to obtain the data-passing efficiency with numerous processing elements, the array is very difficult to visualize. In fact, going beyond the four dimensions shown in Figure 11-11 (16 processor elements) is difficult to visualize. Processor arrays with large numbers of processing elements are also difficult to program efficiently. Once the visualization of the processor architecture is removed, it becomes even more difficult to program.

### 11.4.2 Massively Parallel

The simplest form of three-dimensional massively parallel [1] processing is multiple two-dimensional arrays (Figure 11-12) that lay on top of each other and are interconnected by giving each processor an "up" and "down" connection in addition to its NEWS connections.

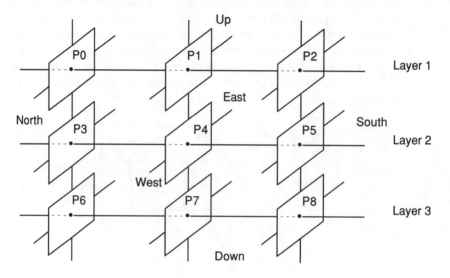

**Figure 11-12** Three-dimensional massively parallel-array block diagram.

Figure 11-12 is a simplified block diagram of such an interconnection. The top three processors (P0, P1, and P2) represent one row of the two-dimensional array in Figure 11-7. The middle (P3, P4, and P5) and bottom (P6, P7, and P8) sets of processors also represent a row of another two-dimensional array. The vertical interconnections are the up and down connections between these two-dimensional arrays. The six basic interconnections, north, east, west, south, up, and down, are labeled in Figure 11-12.

### 11.4.3 Hybrids

By definition a hybrid architecture is a combination of two or more of the architectures described in previous sections. The example is a high-level crossbar [1, 3] architecture (Figure 11-13) where half of the processors (2, 3, 6, 7, 10, 11, 14, and 15) are 3 × 3 arrays of Harvard [2] architecture processing elements connected in a massively parallel [1, 3] NEWS architecture for a total of 72 processors. The other half of the high-level crossbar processors is split between data memory (1, 5, 9, and 13) and data input/output (0, 4, 8, and 12). Therefore, this is a combination of Harvard, massively parallel, and crossbar architectures.

Figure 11-14 shows the 3 × 3 parallel processor array that exists at each of the processors 2, 3, 6, 7, 10, 11, 14, and 15 in Figure 11-13, and Figure 11-15 shows the

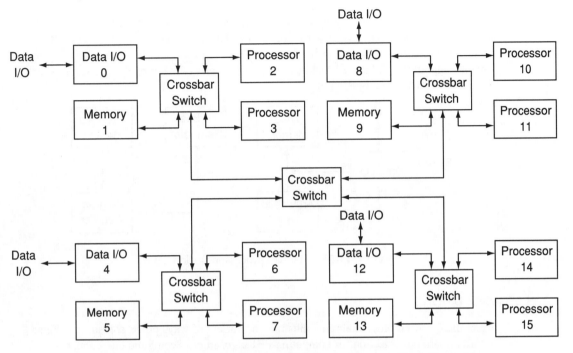

**Figure 11-13**    High-level crossbar architecture block diagram.

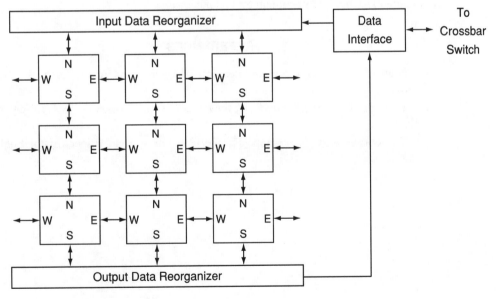

**Figure 11-14**    3 × 3 parallel processor array block diagram.

Harvard processor at each node of each of these $3 \times 3$ parallel processor arrays. Multiply-accumulate functions would be performed with the 72 Harvard processors. This means that 72 multiply-accumulations can be done at the same time and the answers combined at whatever level is necessary by using the NEWS and crossbar interconnections. The strength of this architecture is its processing power. However, the drawback, like all MIMD architectures, is the difficulty in programming the 72 processors to work efficiently on complex algorithms. Chapter 12 addresses the complexity of mapping the algorithms from Chapter 9 onto these architectures.

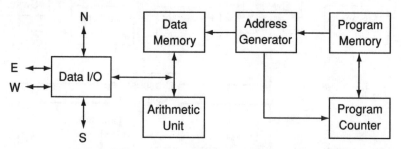

**Figure 11-15**   Harvard processor block diagram.

## 11.5 CONCLUSIONS

More than a dozen block diagrams illustrate the variety of ways processors are combined to offer enormous selection for computing FFT algorithms. Seeing the interconnection of the processors allows data movement overhead to be estimated. This helps to narrow the choices of how to map an algorithm onto an architecture, which is shown in the next chapter for minimum latency and maximum throughput examples.

## REFERENCES

[1]  T. Fountain, *Processor Arrays Architecture and Applications*, Academic Press, London, 1987.

[2]  S. K. Mitra, J. F. Kaiser, *Handbook for Digital Signal Processing*, Wiley, New York, 1993.

[3]  R. W. Hockney and C. R. Jesshope, *Parallel Computers*, Adam Hilger, Bristol, England, 1981.

# 12

# Algorithm and Data Mappings

## 12.0 INTRODUCTION

The method used to distribute and redistribute data and an algorithm in a single or multi-processor hardware architecture is called algorithm mapping. The process of choosing an algorithm mapping for a particular application is often complex. The data I/O requirements, processor interconnections and building-block algorithms must all be considered to reach an optimal approach for a particular application.

This chapter uses minimum latency and maximum throughput examples to illustrate how to map the algorithms from Chapter 9 onto the hardware architectures from Chapter 11. It is assumed that each processor takes one instruction cycle for each add, multiply, or data move. The measures of how well an architecture performs an FFT algorithm are:

- How much delay does the architecture introduce while obtaining the results (latency)?
- How many FFTs per second can be computed (throughput)?

## 12.1 FIVE PERFORMANCE MEASURES

The two major issues of an algorithm mapping's efficiency are the:

- Time to move data into, out of, and between processors
- Computational efficiency of the algorithm (latency and throughput combined) on the processor or processors

The first three performance measures apply to the first issue and the last two to the second.

### 12.1.1 Input Data Overhead

Input data overhead is the number of clock cycles to move the data into the hardware architecture and store it in the processor that will use it first.

### 12.1.2 Intermediate Results Reorganization Overhead

Intermediate results reorganization overhead is the number of clock cycles needed to reorganize intermediate results among processors prior to performing the next stage of algorithm computations.

### 12.1.3 Output Data Overhead

Output data overhead is a count of the number of clock cycles to organize and move the FFT algorithm results out of the hardware architecture.

### 12.1.4 Computational Throughput

Computational throughput is the average number of clock cycles per FFT for the hardware architecture to perform the arithmetic.

### 12.1.5 Processing Latency

Processing latency is the number of clock cycles from the time an input data sequence starts going into the hardware architecture until the results are output from that hardware architecture.

## 12.2 MAPPINGS

Algorithms and architectures are interesting to study. However, it is the efficiency with which an architecture can execute an FFT algorithm that is of paramount importance in making choices in the development of an application. The following sections use the performance measures to characterize how each algorithm from Chapter 9 will work on each architecture from Chapter 11.

In general, the best mapping of an algorithm onto processors is to allocate a processor to each algorithm building block. If a transform length is factored into $P$ smaller numbers, then:

1. The Bluestein algorithm needs $2P + 3$ hardware blocks. Three are used for the complex multiplies at the beginning, middle, and end of the algorithm. The other $2P$ are needed to implement the forward and inverse transforms, where $P$ is the number of building blocks needed to implement the FFT.

2. The Winograd algorithm needs three hardware building blocks to implement the two sets of adds and one set of multiplies.

3. The prime factor algorithms need $P$ hardware building blocks to compute the $P$ building-block algorithms.

4. The mixed-radix algorithms need $P$ hardware building blocks to compute the $P$ building-block algorithms and $P - 1$ more to implement the complex multiplications between the stages.

To allow the mapping comparisons to be as close to apples to apples as possible, the Harvard architecture described in Section 12.3 is used as the processor at all of the nodes of the multiprocessing architectures. The pipeline linear array architecture is used to illustrate how the various algorithms from Chapter 9 can be mapped onto a multiprocessor architecture. Then, for each architecture, mapping the 16-point radix-4 FFT algorithm example from Chapter 9 is described in detail, by providing the data movement steps and using the computational algorithm steps in Chapter 9. This provides a means for each of the architectures within a class to be compared as well as the same algorithm across architecture classes. Similar results would be obtained if any one of the other FFT examples from Chapter 9 were used.

## 12.3 SINGLE PROCESSOR

Single processors are the simplest form of hardware architecture used for computing FFTs. The memory holds the FFT algorithm steps, the multiplier constants, and the data being processed. For real-time processing the memory must include space for three sets. While the present set of complex samples is being operated on by the FFT algorithm, a new set of complex samples is entering for the FFT computations, and the results of the last FFT computations must be output. Table 12-1 shows how sets of complex samples are distributed among these three portions of the memory, starting with the present set flowing into the processor through the data I/O (input set) until it flows out of the processor via the data I/O (output results).

**Table 12-1**    Single-Processor Real-Time Data Mapping

| Time slot | Input set | Data RAM section 1 | Data RAM section 2 | Data RAM section 3 | Output results |
|-----------|-----------|--------------------|--------------------|--------------------|----------------|
| 1 | 1 | 1 | N/A | N/A | N/A |
| 2 | 2 | 1 | 2 | N/A | N/A |
| 3 | 3 | 1 | 2 | 3 | 1 |
| 4 | 4 | 4 | 2 | 3 | 2 |
| 5 | 5 | 4 | 5 | 3 | 3 |
| 6 | 6 | 4 | 5 | 6 | 4 |
| 7 | 7 | 7 | 5 | 6 | 5 |
| 8 | 8 | 7 | 8 | 6 | 6 |

Table 12-1 shows input set 1 flowing through the data I/O portion of the processor during time slot 1 and being stored in data RAM section 1. After one time slot for computation, the FFT outputs from input set 1 are passed out of the processor during time slot 3. This process is repeated for each set of complex samples. The only difference is the section of memory used for each set. Therefore, the processor's real-time computational requirement is to perform the entire FFT algorithm during the time slot for inputting one set of complex samples. This includes algorithm arithmetic and memory address calculations. If the processor is fast enough to perform all of these functions in real-time, a single processor is sufficient for the application and the throughput is an FFT per time slot. If it is not, multiple processors are needed, leading to one of the other architectures from Chapter 11.

The latency of this processing architecture is two time slots because the data goes into the processor during time slot 1 and the results exit the processor during time slot 3. This performance must also be adequate for the application in order for a single processor to be sufficient. If the latency must be less than two sets of complex samples, multiple processors must be used.

### 12.3.1 Data I/O Requirements

For a given transform length the data I/O rates are the same for all of the algorithms because all $N$-point FFTs use $N$ input complex samples and produce $N$ output frequency components. However, if data I/O is marginal, it is important to find the smallest transform length that meets the performance goals of the application. Generally, the smallest transform length is not a power-of-two.

The other factor affecting data I/O is the data sequence reordering needed to compute the algorithm. On the input, the data is almost always in time sequence order because it came from an A/D converter or out of some linear filtering function. However, all of the algorithms in Chapter 9 needed the data to be reorganized to be ready for the first building-block algorithm computations. This can be performed as the data enters the processor by the way it is stored in memory. Or it can be performed at the beginning of the first building-block computations by the way data is initially accessed from memory.

The FFT results are not in sequential order either. Since the next computational stage generally needs the frequency components in sequential order, another data reorganization is required. Since the addresses used for the last-stage computational outputs are based on the building-block addressing, this data reorganization is performed as the data moves from the data memory through the data I/O hardware.

The algorithms for performing these two reorganizations are given in Chapter 9, and all use multiplies, adds, and modulo arithmetic. Therefore, there is no significant advantage of one algorithm over another for this portion of the computations.

### 12.3.2 Memory Requirements

Memory requirements are the sum of the data memory, multiplier constant memory, and program memory. The Comparison Matrix in Table 9-8 shows that the amount of data memory needed for the different algorithms is nearly equal. Further, the number of multiplier constants is small compared to data memory, except for the mixed-radix algorithms.

The largest program memory requirement occurs when every required instruction is explicitly written out for an algorithm, rather than using the algorithm building-block code as subroutines that get called by the main program. This is called straight-line or in-line code and is the fastest possible code because no subroutine calls must be made and no data memory addresses computed during the execution of the code. However, the program memory is significantly larger than if subroutines are used and addresses are computed as needed. For the 15-point examples in Chapter 9, the building-block subroutine approach requires memory for the 3- and 5-point transforms and for memory addressing algorithms. Since the 15-point algorithm uses the 3-point transform five times and the 5-point transform three times, all with different input and output data addresses, program memory must store five copies of the 3-point algorithm and three copies of the 5-point algorithm in the straight-line approach. For the 16-point radix-4 FFT example in Chapter 9, eight copies of the 4-point building block are used in the straight-line approach, rather than the one copy for the building-block subroutine code approach.

### 12.3.3 Arithmetic Unit Requirements

The arithmetic unit is responsible for algorithm and data addressing computations. The algorithm computations are different for each algorithm. The I/O addressing is explained in Section 12.3.1. The other data addressing computations are to reorganize the data between each building-block algorithm stage. Each algorithm from Chapter 9 requires this data reorganization and uses multiplies, adds, and modulo arithmetic. Therefore, there is no significant advantage of one algorithm over another for this portion of the computations.

The arithmetic unit must be capable of computing all of these tasks in the time slot allotted by the real-time requirements of the application. Millions of instructions per second (MIPS) and millions of operations per second (MOPS) are only crude measures of a processor's ability to execute the needed FFT algorithm in real time, because no hardware architecture is 100% efficient at computing FFTs.

The chip Comparison Matrices in Chapter 14 show 1024-point complex FFT timings for most DSP chips on the market. Section 14.1.1 describes how to estimate timings for other FFT lengths, based on the 1024-point benchmark. This is a better measure of chip performance than MIPS and MOPS because it incorporates internal overhead of the chip. When processors are connected into larger arrays, additional overhead is incurred when data must be passed between processors. That additional overhead is explained for each algorithm mapped in this chapter.

### 12.3.4 Von Neumann Architecture

The straightforward approach to implementing all of the algorithms from Chapter 9 on the Von Neumann [1] architecture is to have a subroutine for each building-block algorithm and its data addressing. Then input and output data addressing algorithms can be programmed for each stage in the Chapter 9 algorithm. To perform these algorithms for sets of complex samples that are in the three different sections of memory, an address "offset" is used to move each starting address to the necessary location. Then the FFT algorithm is performed by sequencing through the various subroutines for computations, data addressing, and address offsets.

If the algorithm can be performed in real-time using one arithmetic unit, then the Von Neumann architecture (Figure 12-1) provides the simplest solution. If not, there are four options. The first is to change to a different algorithm that may have less arithmetic or address computations. The second is to change to the Harvard architecture where the addressing is performed by different hardware. The third is to change from a subroutine-based program to straight-line code that has all of the addresses precalculated and built into the code. Finally, multiple-processor architectures can be used. Options 2 and 4 are described in other sections of this chapter. Option 3 is explained next. This chapter's performance measures, in conjunction with those from Chapters 9 and 14, can be used to assess the difference in performance of the various algorithms.

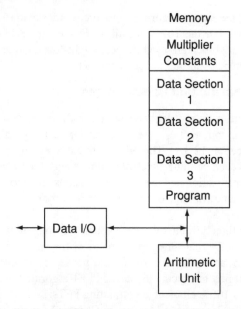

**Figure 12-1**   Von Neumann architecture block diagram.

### 12.3.5 Harvard Architecture

The Harvard [2] architecture is the most popular for programmable DSP chips and DSP applications in general, because DSP functions generally have numerous computations as well as complex data addressing. The FFT algorithms in Chapter 9 are no exception, and the programmable DSP chips in Chapter 14 all use the Harvard architecture. Figure 12-2 shows the basic Harvard architecture with the data memory separated into three sections for real-time operation.

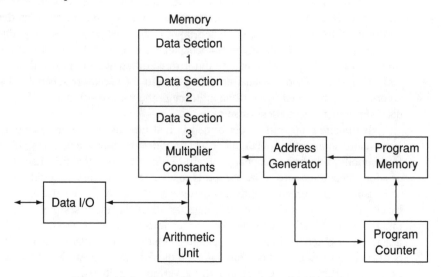

**Figure 12-2**   Basic Harvard architecture block diagram.

Since additional hardware is used to compute memory addresses and sequence through the program, this architecture coupled with building-block subroutine code generally has better performance than a Von Neumann architecture using straight-line code. Additionally, the larger memory needed for straight-line code is replaced with a small amount of control logic in the Harvard architecture.

The extent of the performance improvement over the Von Neumann architecture depends on the sophistication of the address generators. In the more recent generations of DSP chips, the address generators, often multiple ones, allow the complex memory address sequences to be generated at the same speed as the arithmetic computations are performed. In the early generations of DSP chips, the address generator was nothing more than a counter. For these less sophisticated address generators, straight-line coding provided additional performance gain over using building-block subroutines. All of the other data I/O, memory, and arithmetic unit considerations are virtually the same for the Harvard and Von Neumann architectures.

### 12.3.6  Harvard 16-Point Radix-4 FFT Example

Because only one processor is being used, any of the FFT examples from Chapter 9 can be used to illustrate the mapping process. If the 16-point, radix-4 FFT is used and it is assumed that (1) the data addressing is all accomplished by an address generator, in parallel with the computations, and (2) the arithmetic unit performs either an add or a multiply in a clock cycle, then 232 clock cycles are required because there are 144 real adds, 24 real multiplies, and 64 data I/O operations (32 to input 16 complex data samples and 32 to output 16 complex frequency components) to execute. Therefore, the throughput is one 16-point radix-4 FFT every 232 clock cycles with a processing latency that is also 232 clock cycles. If the arithmetic unit allows multiplies and adds on the same clock cycle, the clock cycle total is reduced as a function of how many places in the algorithm adds and multiplies can be done in parallel.

## 12.4  THREE LINEAR ARRAYS

Linear arrays were early architectures for increasing the performance of an FFT algorithm beyond the capability of a single processor. The primary difference between the various algorithms on this architecture is the number of processors that are efficient for decomposing the algorithm into smaller pieces. Table 12-2 shows how each of the FFT examples from Chapter 9 can be mapped onto a three-processor linear-array architecture. These mappings are then described in more detail for each linear-array architecture from Chapter 11. Finally, the 16-point radix-4 FFT example is described in more detail. Throughout this section, when the $k$-th input data sample is written as $a(k)$, it means both the real and imaginary parts of the sample. Specifically, $a(k) = a_R(k) + j * a_I(k)$. This same shorthand notation is also used for intermediate results and output frequency components.

### 12.4.1  Pipeline

The pipeline [1, 3] architecture was one of the first real-time architectures used to implement the power-of-two FFT. It interconnects processors such that the output of each one becomes the input to the next. Then an FFT algorithm is implemented by segmenting

**Table 12-2** Chapter 9 Example Algorithms Mapped onto a Three-Processor Linear Array

| Chapter 9 FFT examples | Processor 0 | Processor 1 | Processor 2 |
|---|---|---|---|
| 15-point Bluestein | 16-point FFT | Complex multiplier | 16-point IFFT |
| 15-point Winograd | 15-point input adds | Multiplier | 15-point output adds |
| 15-point prime factor | 3-point FFT | Not used | 5-point FFT |
| 16-point radix-4 | 4-point FFT | Complex multiplier | 4-point FFT |
| 16-point radix-8 and -2 | 8-point FFT | Complex multiplier | 2-point FFT |
| 15-point Singleton | 3-point FFT | Complex multiplier | 5-point FFT |

it into a sequence of smaller building-block algorithms and performing each algorithm on one of the processors. Figure 12-3 is a pipeline architecture with three processors.

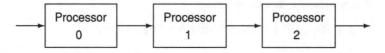

**Figure 12-3** Pipeline architecture block diagram.

Each FFT algorithm in Chapter 9 requires the input samples, intermediate results, and output results to be reorganized. These reorganizations are implemented by the sequence in which data is read into each processor in Figure 12-3 or by the address pattern used to store the data in the processor. Therefore, the time for data reorganization is similar for all algorithms.

In terms of algorithm computational efficiency, the key is to provide enough computational capability in each processor so that it can process the outputs from the previous processor as fast as provided and can provide inputs for the next processor as fast as needed. If each processor meets these criteria, the $P$-stage pipeline processor can process $P$ times as much data as a single processor. The pipeline approach allows each processor to be tailored to execute the computations in that portion of the algorithm.

There are three contributors to processing latency in a pipeline architecture. The first is the individual latencies of each of the processors, once they have received the necessary data to perform the computations. The second is added latency due to one processor not working fast enough to feed results to the next one. Then the next processor must wait for data prior to performing its computations.

The final contributor to pipeline processor latency is whether a processor waits until it has an entire set of complex samples before it begins processing. If it does, the processing latency of each processor is as described in Section 12.3. However, it is possible to start processing data prior to the entire set of complex samples being present. This can be observed by looking at the algorithm steps in Chapter 9 for the 15- and 16-point examples. In all of the 15-point examples the first computations can be performed once the complex $a(0)$, $a(5)$, and $a(10)$ samples are received. For the 16-point radix-4 example, computations can start once complex samples $a(0)$, $a(4)$, $a(8)$, and $a(12)$ are received. The 16-point mixed power-of-primes example must wait until sample $a(14)$ is received.

For algorithms where a 2-point transform is computed first, computations can start after receiving the first sample in the second half of the data. This technique was used extensively in early pipeline implementations of power-of-two (power-of-primes algorithm with the "prime" being 2) FFT algorithms to reduce processing latency.

Figures 12-4 through 12-9 are examples of how each of the example algorithms from Chapter 9 can be implemented with the pipeline architecture. At the inputs to each processor the data addressing portion of the algorithms must also be implemented.

**Figure 12-4**    Pipeline architecture block diagram for the 15-point Bluestein algorithm.

The Bluestein algorithm requires much more processing power for the first and third blocks than for the second block. This can be accommodated by using blocks with different processing power or by subdividing the computations for the 16-point algorithm into smaller blocks. For example, the first and/or third blocks in Figure 12-4 can be replaced with the three blocks in Figure 12-7, resulting in a pipeline with five or seven blocks with more comparable amounts of computations. The advantage of this is the possibility of having all the computational blocks be the same hardware architecture, or at least fill the same amount of board space. The disadvantage of this approach is that it adds processing latency to the algorithm, even though it does not decrease the system input data rate.

The Winograd algorithm provides the best chance for optimizing the hardware to the algorithm because it segregates adds and multiplies. This allows the first and third processors to be constructed using only adders. Only the center processor needs the multiplication capability. For the 15-point FFT this algorithm also allows the first and third processors to be decomposed into a sequence of 3- and 5-point add processors. However, with the cost of programmable DSP chips decreasing rapidly, it may still be most cost effective to use those chips for each of the three blocks needed for the 3- and 5-point FFTs.

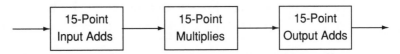

**Figure 12-5**    Pipeline architecture block diagram for the 15-point Winograd algorithm.

The prime factor algorithm (Figure 12-6) has two potentially attractive features because multipliers are not needed between the stages. The first is that a two-stage algorithm can be implemented with processors that are much closer to having the same computational requirements than if the multiply stage were in the middle. The second is the potential for a smaller processing latency because of the lack of the multiplier processor.

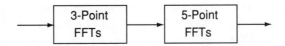

**Figure 12-6**    Pipeline architecture block diagram for the 15-point prime factor algorithm.

Additionally, these two blocks can be further decomposed into smaller building blocks to meet the computational requirements. For example, the Winograd building blocks from Chapter 8 allow each block in Figure 12-6 to be divided into three blocks. In that case, each processor can be optimized as described for the adds and multiplies required by the Winograd algorithm.

The power-of-primes algorithm in Figure 12-7 has the special feature that the first and third blocks are the same. Further, when they are 4-point FFTs, they do not have multiplications. Therefore, they can be implemented by using only adder blocks for the arithmetic unit. Again, the 4-point FFT requires more computations than the complex multiplies. This means more processing power is needed in the first and third blocks than in the second block. If the processor latency requirements allow, the 4-point algorithm can be computed with a pair of 2-point FFTs. This increases processor latency by turning a three-block process into a five-block process. However, it makes the processing requirements of each block similar.

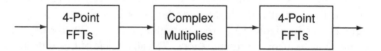

**Figure 12-7** Pipeline architecture block diagram for the 16-point powers-of-primes algorithm.

The mixed powers-of-primes algorithm in Figure 12-8 has the worst mismatch of computational tasks of any of the examples because all three blocks have different require-ments. Again, this can be improved by decomposing the 8-point FFT into three 2-point or 4- and 2-point mixed-radix FFT algorithms. The three 2-point FFT algorithms offer the best computational match because the 2-point FFT requires four adds and each complex multiply consists of four multiplies and two adds.

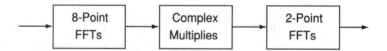

**Figure 12-8** Pipeline architecture block diagram for the 16-point mixed powers-of-primes algorithm.

A third option for decomposing the 8-point FFT is to use the Winograd 8-point algorithm. Then it can be decomposed into a sequence of adds, then multiplies, and then adds again. Since the 2-point FFT is also just adds, it can be implemented with the same hardware architecture as the Winograd input and output adds. Further, the Winograd multiplies and the complex multiplies can be implemented with the same hardware architecture.

The block diagram in Figure 12-9 is very similar to the prime factor algorithm in Figure 12-6. The two drawbacks to this algorithm, over the prime factor algorithm, are that the processing latency is one more set of complex samples because of the complex multiplies, and the complex multiplies need a simpler computational architecture than the 3- and 5-point FFTs. The second issue can be resolved by decomposing the 3- and 5-point FFTs into smaller building blocks. However, this decomposition results in added processing latency.

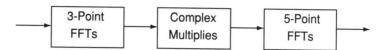

**Figure 12-9**   Pipeline architecture block diagram for the 15-point Singleton mixed-radix algorithm.

## 12.4.2 Linear Bus

A linear [1] bus is an architecture where a single bus is used to provide the path for all of the data communications among the arithmetic processors. Figure 12-10 is a block diagram of the linear bus architecture. There are numerous ways each of the examples from Chapter 9 can be executed on this architecture. One is to allocate functions to each processor in the same way as allocated in the pipeline architecture (Table 12-2). Then the only difference between this architecture and the pipeline is that only one set of data can move on the bus at one time. In the pipeline architecture the input and output of all processors can work simultaneously.

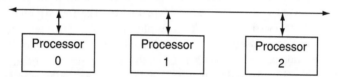

**Figure 12-10**   Linear bus architecture block diagram.

## 12.4.3 Ring Bus

The ring [3] bus is a special case of the linear bus where the ends of the linear bus are connected. Figure 12-11 shows a three-hardware-processor ring bus architecture. Table 12-2 shows how each of the example FFTs from Chapter 9 can be implemented on this architecture. The key issue with this architecture is bus contention, just as for the linear bus. However, this architecture has a more demanding requirement because data may pass around the ring several times before the algorithm computations are complete. When bus contention occurs, the transmission of processor outputs must be delayed. This results in both a reduction in throughput and an increase in latency.

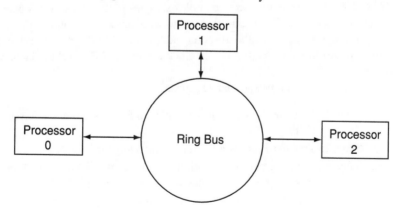

**Figure 12-11**   Ring bus block diagram.

As explained in Chapter 11, data in this architecture flow along the bus from one processor to the next, accompanied by a codeword. The codeword tells the next processor if it has computations to perform on the next set of data and what those computations are. Additionally, just as in the pipeline section, each processor can be further decomposed so that there are more smaller processors connected to the ring.

### 12.4.4 Pipeline 16-Point Radix-4 Example

There are two extremes for processing in this class of architectures. One extreme distributes the algorithm across all of the processors (Option 1), and the other uses each processor to compute an entire transform (Option 2). For these architectures and this FFT length, Option 1 provides maximum throughput and minimum latency. This is not usually the case, as is seen for the parallel array and multidimensional array architectures.

### *Option 1: All Processors Used to Compute One 16-Point Radix-4 FFT*

Assuming one of the Harvard processors is used at each processor location in Figure 12-7, the 4-point computations will need more time than the complex multiplies. From Chapter 8 each 4-point FFT takes 16 real adds. The four 4-point FFTs are computed by Processor 0 in $4*16 = 64$ clock cycles plus the 32 for data input (clock cycles 0–95). Then 32 clock cycles are used to move these partial results to Processor 1 to perform the complex multiplies (clock cycles 96–127). Once this has occurred, another 96 clock cycles (clock cycles 128–223) are used to move the next set of data into Processor 0 and perform the four 4-point input FFTs. Then the second set of results is ready for input to Processor 1 at clock 224.

Even though the 12 complex multiplications use 24 real multiplies, 16 real adds, and 32 data output clock cycles (72 clock cycles), 96 clock cycles are allotted because no new data is available until then. Therefore, the first set of complex multiply results is output from clock cycles 192 to 223 in preparation for receiving the next set of data. Therefore, at clock 224, Processor 2 has data for computing the four 4-point output FFTs. Since this takes 64 clock cycles to compute and 32 to output the results (the same time as Processor 0), the results are completely output from Processor 2 at clock 320. Therefore, the processing latency for the pipeline architecture and 16-point radix-4 algorithm is 320 clock cycles. Meanwhile, the second set of complex samples moves to Processor 1 from clock cycles 224 to 255. Therefore, this set of complex samples is 128 clock cycles behind the first set of complex samples. This means that a new set of answers is output from this architecture every 128 clock cycles for the 16-point radix-4 algorithm. Therefore, the computational throughput of this architecture is 128 clock cycles per FFT, and the latency is 320 clock cycles.

This process can be summarized in stages:

Stage 1: Input set 1 of complex samples to Processor 0 and compute input 4-point FFTs.

Stage 2: Transfer Processor 0's set 1 results to Processor 1.

Stage 3: Compute complex multiplications on set 1 in Processor 1 and input set 2 to Processor 0 and compute input 4-point FFTs.

Stage 4: Transfer Processor 0 set 2 results to Processor 1; transfer Processor 1 set 1 results to Processor 2.

Stage 5:  Compute complex multiplications on set 2 in Processor 1, compute the set 1 output 4-point FFTs in Processor 2, and input set 3 to Processor 0 and compute input 4-point FFTs.

This process is repeated for multiple sets of complex samples. Table 12-3 summarizes these events as a function of clock cycles from the beginning of the process.

**Table 12-3**    Timing for 16-Point Radix-4 FFT on a Three-Processor Pipeline

| Clock cycle | Task |
|---|---|
| 0–95 | Input 1st set into Processor 0 and compute four input 4-point FFTs. |
| 96–127 | Move Processor 0 results from 1st set to Processor 1. |
| 128–223 | Input 2nd set into Processor 0 and compute four input 4-point FFTs. |
| 128–191 | Compute complex multiplies on 1st set in Processor 1. |
| 192–223 | Move Processor 1 results from the 1st set into Processor 2. |
| 224–319 | Compute four output 4-point FFTs on 1st set and output results. |
| 224–255 | Move Processor 0 results from 2nd set to Processor 1. |
| 256–341 | Input 3rd set into Processor 0 and compute four input 4-point FFTs. |

## Option 2: Each Processor Computes One 16-Point Radix-4 FFT

### Stage 1: Distribute One Set of Complex Samples to Each Processor

In the pipeline architecture the input data samples that are to be processed by the second processor are passed through the first processor. Similarly, the input data samples to be processed by the third processor are passed through the first and second processors. Assuming this step takes one clock cycle for each input data word, the first set is moved into the first processor in 32 clock cycles. As 32 clock cycles are used to move the second set of complex samples into the first processor, the first processor passes the first set into the second processor. As 32 more clock cycles are used to move the third set of data into the first processor, the first set is moved from the second to the third processor, and the second set of data samples is moved from the first to the second processor. Therefore, these three sets of 16 complex input data samples take 96 clock cycles to input to the pipeline.

### Stage 2: Compute Three 16-Point Radix-4 FFTs

It takes 168 clock cycles to compute the 16-point radix-4 FFT using the Harvard architecture assumptions from Section 12.3.5. Since all three processors are computing the algorithm, it takes only 168 clock cycles to compute all three 16-point radix-4 FFTs.

### Stage 3: Collect the Results of the Three 16-Point Radix-4 FFT Computations

Assuming this step takes two clock cycles to output each complex frequency component, the first set of output frequency components is moved out of the third processor in 32 clock cycles. At the same time, the second set of complex frequency components is moved from the second processor to the third processor. Also, these same 32 additional clock cycles are used to move the third set of output frequency components from the first processor to the second processor. During the next set of 32 clock cycles, the second set of output frequency components is moved out of the third processor and the third set of output frequency components is moved from the second to the third processors. Finally, during the last set of 32 clock cycles, the third set of output frequency components is moved out of the third processor. Therefore, the three sets of 16 complex output frequencies are output in 96 clock cycles. Therefore, this option takes a total of 360 clock cycles, which is the latency and defines the throughput rate of $360/3 = 120$ clock cycles per FFT.

### 12.4.5 Linear and Ring Bus 16-Point Radix-4 FFT Examples

There are two extremes for processing in linear and ring bus architectures. One extreme distributes the algorithm across all of the processors (Option 1), and the other uses each processor to compute an entire transform (Option 2). For these architectures and this FFT length, Option 1 provides maximum throughput and minimum latency. This is not usually so, as is seen for the parallel-array and multidimensional-array architectures.

#### *Option 1: All Processors Used to Compute One 16-Point Radix-4 FFT*

Assuming one of the Harvard processors is used at each processor location, the 4-point computations will need more time than the complex multiplies. From Chapter 8 each 4-point FFT takes 16 real adds. The four 4-point FFTs are computed by Processor 0 in $4*16 = 64$ clock cycles plus the 32 for data input (clock cycles 0–95). Then 32 clock cycles are used to move these partial results to Processor 1 to perform the complex multiplies (clock cycles 96–127). Once this has occurred, another 96 clock cycles (clock cycles 128–223) are used to move the next set of data into Processor 0 and perform the four 4-point input FFTs. Then the second set of results is ready for input to Processor 1 at clock 224.

Even though the 12 complex multiplications use 24 multiplies, 16 adds, and 32 data output clock cycles (72 clock cycles), 96 clock cycles are used because no new data is available until then. Therefore, the first set of complex multiply results is output from clock cycles 192 to 223 in preparation for receiving the next set of data. Therefore, at clock 224, Processor 2 has data for computing the four 4-point output FFTs. Since this takes 64 clock cycles to compute and 32 to output the results (the same time as Processor 0), the results are completely output at clock 320. Therefore, the processing latency for the linear array architecture and 16-point radix-4 algorithm is 320 clock cycles. Meanwhile, the second set of complex samples moves to Processor 1 from clock cycles 224 to 255. Therefore, this set of complex samples is 128 clock cycles behind the first set of complex samples. This means that a new set of answers is output from this architecture every 128 clock cycles for the 16-point radix-4 algorithm. Therefore, the computational throughput of this architecture is 128 clock cycles per FFT and the latency is 320 clock cycles.

To summarize this process in stages:

Stage 1: Input set 1 of complex samples to Processor 0 and compute input 4-point FFTs.

Stage 2: Transfer Processor 0's set 1 results to Processor 1.

Stage 3: Compute complex multiplications on set 1 in Processor 1, and input set 2 to Processor 0 and compute input 4-point FFTs.

Stage 4: Transfer Processor 0 set 2 results to Processor 1; transfer Processor 1 set 1 results to Processor 2.

Stage 5:  Compute complex multiplications on set 2 in Processor 1, compute the set 1 output 4-point FFTs in Processor 2, and input set 3 to Processor 0 and compute input 4-point FFTs.

This process is repeated for multiple sets of complex samples. Table 12-4 summarizes these events as a function of clock cycles from the beginning of the process.

**Table 12-4**    Timing for 16-Point Radix-4 FFT on a Linear Array

| Clock cycle | Task |
|---|---|
| 0–95 | Input 1st set into Processor 0 and compute four input 4-point FFTs. |
| 96–127 | Move Processor 0 results from 1st set to Processor 1. |
| 128–223 | Input 2nd set into Processor 0 and compute four input 4-point FFTs. |
| 128–191 | Compute complex multiplies on 1st set in Processor 1. |
| 192–223 | Move Processor 1 results from the 1st set into Processor 2. |
| 224–319 | Compute four output 4-point FFTs on 1st set and output results. |
| 224–255 | Move Processor 0 results from 2nd set to Processor 1. |
| 256–341 | Input 3rd set into Processor 0 and compute four input 4-point FFTs. |

### *Option 2: Each Processor Computes One 16-Point Radix-4 FFT*

### Stage 1: Distribute One Set of Complex Samples to Each Processor

Assuming this step takes one clock cycle for each input data word, the three sets of 16 complex input data points take 96 clock cycles to be distributed to the three processors.

### Stage 2: Compute Three 16-Point Radix-4 FFTs

It takes 168 clock cycles to compute the 16-point radix-4 FFT by using the Harvard architecture assumptions from Section 12.3.5. Since all three processors are computing the algorithm, it takes only 168 clock cycles to compute all three 16-point radix-4 FFTs.

### Stage 3: Collect the Results of the Three 16-Point Radix-4 FFT Computations

Assuming this step takes one clock cycle for each output result, the three sets of 16 complex output frequencies take 96 clock cycles. Therefore, this option takes a total of 360 clock cycles, which is the latency and defines the throughput rate of $360/3 = 120$ clock cycles per FFT.

## 12.5 THREE PARALLEL ARRAYS

Processors can be combined into parallel arrays in numerous ways, and there are many ways to use the array to compute each of the algorithms in Chapter 9. At the two data mapping extremes are:

1. One set of complex samples is distributed among all of the processors in the array and then computed in one FFT. This approach usually results in minimum latency processing.

2. A set of complex samples is distributed to each of the processors and then a number of FFTs are performed in parallel. This usually results in maximum throughput but has more latency than the first approach.

Each extreme is described by mapping the 16-point radix-4 FFT onto each of the three parallel arrays from Chapter 11. Throughout this section, when the $k$-th input data sample is written as $a(k)$, it means both the real and imaginary parts of the sample. Specifically, $a(k) = a_R(k) + j * a_I(k)$. This same shorthand notation is also used for intermediate results and output frequency components.

### 12.5.1 Crossbar 16-Point Radix-4 FFT Examples

Fast Fourier transforms can be computed on the crossbar [1, 3] architecture in many ways. At one extreme all processors are used to compute one transform (Option 1); at the other each processor is used to compute an entire transform (Option 2). In each case a common way to handle the data I/O is to have one of the processors, say Processor 0, receive the input data and output the FFT results. Options 1 and 2 are described with the 16-point radix-4 FFT from Chapter 9 mapped onto the crossbar architecture in Figure 12-12.

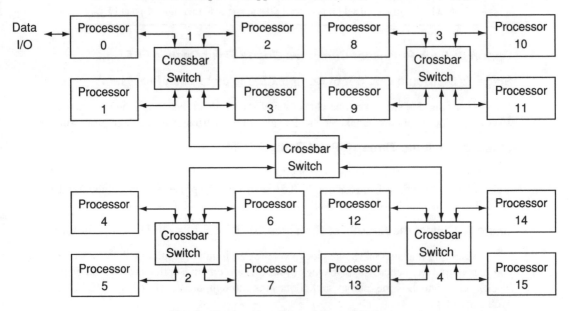

**Figure 12-12**    Crossbar switch architecture.

### *Option 1: All Processors Used to Compute One 16-Point Radix-4 FFT*

The strategy is to use the four-processor clusters in Figure 12-12 (Processors 0–3, Processors 4–7, Processors 8–11, Processors 12–15) to compute the 4-point building blocks. Therefore, the input data is mapped so that the sets of four complex samples needed for each of the four 4-point building blocks from Chapter 8 are each located in a processor cluster. Once the input 4-point building-block algorithms are computed, the complex multiplies can be performed. Then the data can be reorganized so that the intermediate results, needed as inputs to each of the second set of 4-point building blocks, are in a processor cluster. The output 4-point building blocks are then computed and the final results sent out of the architecture. The data mapping in each of the processors is the same as used in Section 8.5, because the computations in an individual processor are only 4-point building blocks.

### Stage 1: Distribute the Input Data onto the Processors

Use Processor 0 to load the 16 complex samples and use the crossbar network to distribute one of the data points to each of the other 16 processors. Group the data points such that $a(0)$, $a(4)$, $a(8)$, and $a(12)$ are in Processors 4, 5, 6, and 7, respectively. Similarly, group $a(1)$, $a(5)$, $a(9)$, and $a(13)$ in Processors 8, 9, 10, and 11, respectively; $a(2)$, $a(6)$, $a(10)$, and $a(14)$ in Processors 12, 13, 14, and 15, respectively; and $a(3)$, $a(7)$, $a(11)$, and $a(15)$ in Processors 0, 1, 2, and 3, respectively. It takes two clock cycles to input and store each complex data sample in the processor, if no additional clock cycles are assumed for passing data through the crossbar switches. This is a total of 32 clock cycles. Figure 12-13 shows which of the 16 processors has each of the 16 complex samples, intermediate results, and output results after each stage of the 16-point radix-4 algorithm by listing them in their processor on the same line as the label on the left side of the figure that defines the stage of the algorithm.

### Stage 2: Compute Input 4-Point FFTs

Compute 4-point FFTs in each processor cluster. Use Stage 1 of the 16-point radix-4 FFT example in Chapter 9 as the guideline, along with the memory mapping scheme in Chapter 8, and each processor cluster's crossbar switch to move data between processors. Specifically, processor cluster 0–3 is used to compute the fourth of four input 4-point building blocks in Stage 1 of Section 9.7.5. Similarly, processor cluster 4–7 is used to compute the first of four input 4-point building blocks. Processor cluster 8–11 is used to compute the third of the four input 4-point building blocks. Finally, processor cluster 12–15 is used to compute the second of the four input 4-point building blocks in Stage 1 of Section 9.7.5.

To illustrate how a processor cluster can be used to compute these input 4-point building blocks, consider processor cluster 4–7. One approach for this cluster is to use the crossbar switch to connect Processor 4 to Processor 6 and to connect Processor 5 to Processor 7. Then:

Step 1: Copy $a(0)$ from Processor 4 into Processor 6 and copy $a(4)$ from Processor 5 into Processor 7, simultaneously.

Step 2: Copy $a(8)$ from Processor 6 into Processor 4 and copy $a(12)$ from Processor 7 into Processor 5, simultaneously.

Step 3: Use the equations from the 16-point radix-4 example in Section 9.7.5 to compute

$$b(0) = a(0) + a(8) \text{ in Processor 4}$$
$$b(2) = a(4) + a(12) \text{ in Processor 5}$$
$$b(1) = a(0) - a(8) \text{ in Processor 6}$$
$$b(3) = a(4) - a(12) \text{ in Processor 7 simultaneously}$$

Use the crossbar switch to connect Processor 4 to Processor 5 and to connect Processor 6 to Processor 7. Then:

| | | | |
|---|---|---|---|
| Results of Stage 1 | $a(3)$ | $a(11)$ | $a(1)$ | $a(9)$ |
| Results of Stage 2 | $c(12)$ | $c(13)$ | $c(8)$ | $c(9)$ |
| Results of Stage 3 | $c(12)$ | $e(13)$ | $c(8)$ | $e(9)$ |
| Results of Stage 4 | $c(3)$ | $e(11)$ | $c(1)$ | $e(9)$ |
| Results of Stage 5 | $A(3)$ | $A(7)$ | $A(1)$ | $A(5)$ |
| | Processor 0 | Processor 2 | Processor 8 | Processor 10 |

| | | | |
|---|---|---|---|
| Results of Stage 1 | $a(7)$ | $a(15)$ | $a(5)$ | $a(13)$ |
| Results of Stage 2 | $c(14)$ | $c(15)$ | $c(10)$ | $c(11)$ |
| Results of Stage 3 | $e(14)$ | $e(15)$ | $e(10)$ | $e(11)$ |
| Results of Stage 4 | $e(7)$ | $e(15)$ | $e(5)$ | $e(13)$ |
| Results of Stage 5 | $A(11)$ | $A(15)$ | $A(9)$ | $A(13)$ |
| | Processor 1 | Processor 3 | Processor 9 | Processor 11 |

| | | | |
|---|---|---|---|
| Results of Stage 1 | $a(0)$ | $a(8)$ | $a(2)$ | $a(10)$ |
| Results of Stage 2 | $c(0)$ | $c(1)$ | $c(4)$ | $c(5)$ |
| Results of Stage 3 | $c(0)$ | $c(1)$ | $c(4)$ | $e(5)$ |
| Results of Stage 4 | $c(0)$ | $c(8)$ | $c(2)$ | $e(10)$ |
| Results of Stage 5 | $A(0)$ | $A(8)$ | $A(2)$ | $A(6)$ |
| | Processor 4 | Processor 6 | Processor 12 | Processor 14 |

| | | | |
|---|---|---|---|
| Results of Stage 1 | $a(4)$ | $a(12)$ | $a(6)$ | $a(14)$ |
| Results of Stage 2 | $c(2)$ | $c(3)$ | $c(6)$ | $c(7)$ |
| Results of Stage 3 | $c(2)$ | $c(3)$ | $e(6)$ | $e(7)$ |
| Results of Stage 4 | $c(4)$ | $c(12)$ | $e(6)$ | $e(14)$ |
| Results of Stage 5 | $A(4)$ | $A(12)$ | $A(10)$ | $A(14)$ |
| | Processor 5 | Processor 7 | Processor 13 | Processor 15 |

**Figure 12-13**    Data map for crossbar implementation of 16-point radix-4 FFT.

Step 4:  Copy $b(0)$ from Processor 4 into Processor 5 and copy $b(1)$ from Processor 6 into Processor 7 simultaneously.

Step 5:  Copy $b(2)$ from Processor 5 into Processor 4 and copy $b(3)$ from Processor 7 into Processor 6 simultaneously.

Step 6:  Use the equations from the 16-point, radix-4 example in Section 9.7.5 to compute

$$c(0) = b(0) + b(2) \text{ in Processor 4}$$
$$c(1) = b(1) - jb(3) \text{ in Processor 6}$$
$$c(2) = b(0) - b(2) \text{ in Processor 5}$$
$$c(3) = b(1) + jb(3) \text{ in Processor 7 simultaneously.}$$

At the same time these computations and data movements are taking place, perform the equivalent functions in the other three processor clusters, using the data in their processors and the equations from Section 9.7.5. The data movements and adds each take a clock cycle, for a total of 12 clock cycles. Figure 12-13 shows the locations of the results of these computations as the second entry in each of the 16 processor blocks.

### Stage 3: Compute Complex Multiplications

Compute the complex multiplications in Stage 2 of the 16-point radix-4 FFT example in Section 9.7.5. Since one complex pair, $c(k)$, is in each processor, these can be performed, in parallel, in each of the 16 processors and will take a maximum of six clock cycles (four multiplications and two additions). The maximum occurs in Processors 2, 3, 10, and 11. At this point, $c(0)$, $c(1)$, $c(2)$, and $c(3)$ are in Processors 4, 6, 5, and 7, respectively; $c(4)$, $e(6)$, $e(5)$, and $e(7)$ are in Processors 12, 13, 14, and 15, respectively; $c(8)$, $e(10)$, $e(9)$, and $e(11)$ are in Processors 8, 9, 10, and 11, respectively; and, $c(12)$, $e(14)$, $e(13)$, and $e(15)$ are in Processors 0, 1, 2, and 3, respectively. Figure 12-13 shows the locations of the results of these computations as the third entry in each of the 16 processor blocks.

### Stage 4: Reorganize Intermediate Results

Use the crossbar switches to move data among processors so that $c(0)$, $c(4)$, $c(8)$, and $c(12)$ are in Processors 4, 5, 6, and 7, respectively; $c(1)$, $e(5)$, $e(9)$, and $e(13)$ are in Processors 8, 9, 10, and 11, respectively; $c(2)$, $e(6)$, $e(10)$, and $e(14)$ are in Processors 12, 13, 14, and 15, respectively; and $c(3)$, $e(7)$, $e(11)$, and $e(15)$ are in Processors 0, 1, 2, and 3, respectively. Twelve of the 16 intermediate results must be moved. They can be moved in pairs by using the following steps.

Step 1: First use crossbar switches 1, 2, and 5 to connect Processor 0 to Processor 7 and crossbar switches 3, 4, and 5 to connect Processor 9 to Processor 14. Then move
  **(a)** $c(12)$ from Processor 0 to Processor 7 and
  **(b)** $c(10)$ from Processor 9 to Processor 14, simultaneously
Step 2: Using the same crossbar interconnections, move
  **(a)** $c(3)$ from Processor 7 to Processor 0 and
  **(b)** $e(5)$ from Processor 14 to Processor 9, simultaneously
Step 3: Use crossbar switches 1, 3, and 5 to connect Processor 2 to Processor 11 and crossbar switches 2, 4, and 5 to connect Processor 5 to Processor 12. Then move
  **(a)** $e(3)$ from Processor 2 to Processor 11 and
  **(b)** $c(2)$ from Processor 5 to Processor 12, simultaneously
Step 4: Using the same crossbar interconnections, move
  **(a)** $e(11)$ from Processor 11 to Processor 2 and
  **(b)** $c(4)$ from Processor 12 to Processor 5, simultaneously
Step 5: Use crossbar switches 1, 4, and 5 to connect Processor 1 to Processor 15 and crossbar switches 2, 3, and 5 to connect Processor 6 to Processor 8. Then move
  **(a)** $e(14)$ from Processor 1 to Processor 15 and
  **(b)** $c(1)$ from Processor 6 to Processor 8, simultaneously
Step 6: Using the same crossbar interconnections, move
  **(a)** $e(7)$ from Processor 15 to Processor 1 and
  **(b)** $c(8)$ from Processor 8 to Processor 6, simultaneously

Since only the real or imaginary part of one sample can move on a crossbar connection during any clock cycle, these data moves take 12 clock cycles. Figure 12-13 shows the locations of the results of this reorganization of intermediate results as the fourth entry in each of the 16 processor blocks.

### Stage 5: Compute the Output 4-Point FFTs

Compute 4-point FFTs in each processor cluster, using Stage 3 of the radix-4 16-point example as the guideline. This uses each processor cluster's crossbar switch to move data between processors. Specifically, processor cluster 0–3 is used to compute the fourth of four output 4-point building blocks in Stage 3 of Section 9.7.5. Similarly, processor cluster 4–7 is used to compute the first of four output 4-point building blocks. Processor cluster 8–11 is used to compute the second of the four output 4-point building blocks. Finally, processor cluster 12–15 is used to compute the third of the four 4-point output building blocks in Stage 3 of Section 9.7.5.

To illustrate how a processor cluster can be used to compute these output 4-point building blocks, consider processor cluster 4–7. One approach for this processor cluster uses crossbar switch 2 to connect Processor 4 and Processor 5 and to connect Processor 6 to Processor 7. Then:

Step 1: Copy $c(0)$ from Processor 4 into Processor 5 and copy $c(8)$ from Processor 6 into Processor 7 simultaneously.

Step 2: Copy $c(4)$ from Processor 5 into Processor 4 and copy $c(12)$ from Processor 7 into Processor 6 simultaneously.

Step 3: Use the equations from the radix-4 16-point example to compute

$$f(0) = c(0) + c(4) \text{ in Processor 4}$$
$$f(2) = c(8) + c(12) \text{ in Processor 6}$$
$$f(1) = c(0) - c(4) \text{ in Processor 5}$$
$$f(3) = c(8) - c(12) \text{ in Processor 7 simultaneously}$$

Use crossbar switch 2 to connect Processor 4 to Processor 6 and to connect Processor 5 to Processor 7. Then:

Step 4: Copy $f(0)$ from Processor 4 into Processor 6 and copy $f(1)$ from Processor 5 into Processor 7 simultaneously.

Step 5: Copy $f(2)$ from Processor 6 into Processor 4 and copy $f(3)$ from Processor 7 into Processor 5 simultaneously.

Step 6: Use the equations from the radix-4 16-point example to compute

$$A(0) = f(0) + f(2) \text{ in Processor 4}$$
$$A(4) = f(1) - jf(3) \text{ in Processor 5}$$
$$A(8) = f(0) - f(2) \text{ in Processor 6}$$
$$A(12) = f(1) + jf(3) \text{ in Processor 7 simultaneously}$$

At the same time these computations and data movements are taking place, perform the equivalent functions in the other three-processor clusters, using the data in their processors and the algorithm steps in Section 9.7.5. This stage also takes 12 clock cycles.

## Stage 6: Output the Results Using 32 Clock Cycles

The total is 104 clock cycles for throughput and 104 clock cycles for processing latency.

### Option 2: Each Processor Computes One 16-Point Radix-4 FFT

If computational throughput is the most important criterion, then the 16 processors should all be used to compute complete algorithms. This provides the best throughput because no interprocessor communications are used during the algorithm. However, it has the worst processing latency because 16 sets of complex samples are needed to fill up the array for processing. The stages are as follows.

## Stage 1: Distribute One of the 16 Sets of Complex Samples to Each of the 16 Processors

In this case the crossbar switches are used to distribute sets of complex samples from the input processor to the other 15 processors. It takes 32 clock cycles for data input (2 clock cycles for each complex data sample) for each of the 16 sets of complex samples, for a total of $32 * 16 = 512$ clock cycles.

## Stage 2: Compute the Sixteen 16-Point Radix-4 FFTs

It takes 168 clock cycles to compute the 16-point radix-4 FFT using the Harvard architecture assumptions from Section 12.3. Since all 16 processors are computing the algorithm, it takes only 168 clock cycles to compute all sixteen 16-point radix-4 FFTs. During these computations, the crossbar interconnections are not used. This simplifies the routing of data but makes poor use of the crossbar interconnection capability.

## Stage 3: Collect the Results of the Sixteen 16-Point Radix-4 FFT Computations

This stage collects the results for output to the next portion of the application via Processor 0. It takes 32 clock cycles for data output (2 clock cycles for each complex data sample) for each of the 16 sets of complex samples, for a total of 512 clock cycles.

The total number of clock cycles for this approach is the number of clock cycles to perform the 16-point radix-4 FFT plus the data I/O time for 16 sets of complex samples. The result is a total of 1192 clock cycles for the 16 FFTs. This is an average of 74.5 clock cycles per FFT in data throughput load and 1192 latency clock cycles.

## 12.5.2 Massively Parallel 16-Point Radix-4 FFT Examples

The FFT algorithms from Chapter 9 can be implemented in multiple ways on a massively parallel [1, 3] architecture. In fact, the two extremes are the same as for the crossbar architecture. A single set of complex samples can be distributed across all of the processors (Option 1), or each processor can be provided a full set of data to compute (Option 2). The stages for each option are presented below.

Because of the restricted interconnection structure in a massively parallel array, FFT data I/O is generally performed differently than on a crossbar array. Specifically, using one processor for data input imposes a severe restriction on latency and throughput because of the long time needed to pass data across the entire array via intermediate processors. As a result, these architectures generally have their own special-purpose I/O subsystem that converts the input data from a sequential stream into data vectors that can be passed into the processing array along one of its edges. Additionally, the outputs are passed out of the array along another, usually opposite, edge and converted back to a sequential set of passband filter outputs for further processing. Figure 12-14 shows a specific example of this I/O strategy for the $4 \times 4$ NEWS connected massively parallel array described in Section 11.3.2 and used later in the implementation example.

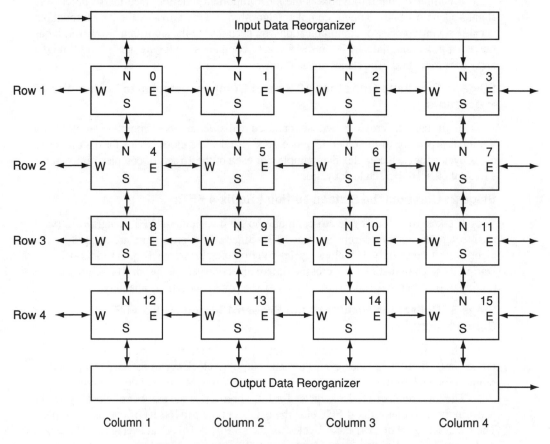

**Figure 12-14** $4 \times 4$ massively parallel array.

The details of the I/O data reorganizers depend on whether the computational portion of the FFT algorithm is distributed across all of the processors (minimal latency) or whether each processor computes an entire FFT (maximum throughput).

### Option 1: All Processors Used to Compute One 16-Point Radix-4 FFT

For minimal latency the input data reorganizer is just a shift register, and one set of complex samples is processed at a time. For the 16-point radix-4 example and the hardware architecture in Figure 12-14, a data processing sequence has the following stages.

### Stage 1: Distribute the Input Data onto the Processors

Load a set of complex input data using the following steps:

Step 1: Load complex samples $a(0)$, $a(1)$, $a(2)$, and $a(3)$ into the input shift register so that sample $a(0)$ is above Processor 3 (8 clock cycles because the samples are complex). Then shift this set of four complex samples into the top four processors. This takes 2 clock cycles because the data is complex, for a total of 10 clock cycles.

Step 2: Load complex samples $a(4)$, $a(5)$, $a(6)$, and $a(7)$ into the input shift register so that sample $a(4)$ is above Processor 3. This takes 8 clock cycles. Then shift this set of four complex samples into the top four processors. At the same time, shift the first four complex samples from the top row of processors to the second row of processors. This takes 2 clock cycles, for a total of 10 clock cycles. Figure 12-15 shows which of the 16 processors has each of the 16 complex samples, intermediate and output results at the end of each stage of this algorithm by listing them in their processor on the same line as the label on the left side of the figure that defines the stage of the algorithm.

| | | | |
|---|---|---|---|
| Results of Stage 1 | $a(15)$ | $a(14)$ | $a(13)$ | $a(12)$ |
| Results of Stage 2 | $c(15)$ | $c(7)$ | $c(11)$ | $c(3)$ |
| Results of Stage 3 | $e(15)$ | $e(7)$ | $e(11)$ | $e(3)$ |
| Results of Stage 4 | $A(15)$ | $A(7)$ | $A(2)$ | $A(3)$ |
| | Processor 0 | Processor 2 | Processor 8 | Processor 10 |

| | | | |
|---|---|---|---|
| Results of Stage 1 | $a(11)$ | $a(10)$ | $a(9)$ | $a(8)$ |
| Results of Stage 2 | $c(13)$ | $c(5)$ | $c(9)$ | $c(1)$ |
| Results of Stage 3 | $e(13)$ | $e(5)$ | $e(9)$ | $e(1)$ |
| Results of Stage 4 | $A(13)$ | $A(5)$ | $A(9)$ | $A(1)$ |
| | Processor 1 | Processor 3 | Processor 9 | Processor 11 |

| | | | |
|---|---|---|---|
| Results of Stage 1 | $a(7)$ | $a(6)$ | $a(5)$ | $a(4)$ |
| Results of Stage 2 | $c(14)$ | $c(6)$ | $c(10)$ | $c(2)$ |
| Results of Stage 3 | $e(14)$ | $e(6)$ | $e(10)$ | $c(2)$ |
| Results of Stage 4 | $A(14)$ | $A(6)$ | $A(10)$ | $A(2)$ |
| | Processor 4 | Processor 6 | Processor 12 | Processor 14 |

| | | | |
|---|---|---|---|
| Results of Stage 1 | $a(3)$ | $a(2)$ | $a(1)$ | $a(0)$ |
| Results of Stage 2 | $c(12)$ | $c(4)$ | $c(8)$ | $c(0)$ |
| Results of Stage 3 | $c(12)$ | $c(4)$ | $c(8)$ | $c(0)$ |
| Results of Stage 4 | $A(12)$ | $A(4)$ | $A(8)$ | $A(0)$ |
| | Processor 5 | Processor 7 | Processor 13 | Processor 15 |

**Figure 12-15**    Data map for massively parallel implementation of 16-point radix-4 FFT.

Step 3: Load complex samples $a(8)$, $a(9)$, $a(10)$, and $a(11)$ into the input shift register so that sample $a(8)$ is above Processor 3. This takes 8 clock cycles. Then shift this set of four complex samples into the top four processors. At the same time, shift the second four complex samples from the top row of processors to the second row and the first set of complex samples from the second row to the third. This takes 2 clock cycles, for a total of 10 clock cycles.

Step 4: Load complex samples $a(12)$, $a(13)$, $a(14)$, and $a(15)$ into the input shift register so that sample $a(12)$ is above Processor 3. This takes 8 clock cycles. Then shift this set of four complex samples into the top four processors. At the same time, shift the first four complex samples from the third to fourth rows, the second set from the second row to the third, and the third set from the first row to the second. This takes 2 clock cycles, for a total of 10 clock cycles.

Stage 1 takes a total of 40 clock cycles. The results are:

(i) Complex samples $a(0)$, $a(1)$, $a(2)$, and $a(3)$ in Processors 15, 14, 13, and 12 (row 4)

(ii) Complex samples $a(4)$, $a(5)$, $a(6)$, and $a(7)$ in Processors 11, 10, 9, and 8 (row 3)

(iii) Complex samples $a(8)$, $a(9)$, $a(10)$, and $a(11)$ in Processors 7, 6, 5, and 4 (row 2)

(iv) Complex samples $a(12)$, $a(13)$, $a(14)$, and $a(15)$ in Processors 3, 2, 1, and 0 (row 1)

Figure 12-15 shows the locations of the input data samples in the first row of each processor block.

## Stage 2: Compute the Input 4-Point FFTs

To do this, notice that the complex samples in the columns are the ones that must be combined. Therefore, whatever processing steps are used for one column can be performed on all four columns at once to compute the four 4-point input FFTs. The steps are as follows:

Step 1: Move the complex samples $a(4)$, $a(5)$, $a(6)$, and $a(7)$ in row 3 to row 2 and the complex samples $a(8)$, $a(9)$, $a(10)$, and $a(11)$ in row 2 to row 3. This step takes 4 clock cycles because each data point is complex.

Step 2: Copy the complex samples $a(4)$, $a(5)$, $a(6)$, and $a(7)$ in row 2 into row 1 and copy the complex samples $a(12)$, $a(13)$, $a(14)$, and $a(15)$ from row 1 into row 2 so that rows 1 and 2 both have the same complex samples. At the same time do the same copy function in rows 3 and 4. This step takes 4 clock cycles.

Step 3: In rows 2 and 4 add the two sets of complex samples. At the same time subtract the complex samples in rows 1 and 3, following the equations in Section 9.7.5. This step takes 2 clock cycles. At the end of this step:

(i) Intermediate results $b(0)$, $b(8)$, $b(4)$, and $b(12)$ are in Processors 15, 14, 13, and 12 (row 4).

(ii) Intermediate results $b(1)$, $b(9)$, $b(5)$, and $b(13)$ are in Processors 11, 10, 9, and 8 (row 3).

(iii) Intermediate results $b(2)$, $b(10)$, $b(6)$, and $b(14)$ are in Processors 7, 6, 5, and 4 (row 2).

(iv) Intermediate results $b(3)$, $b(11)$, $b(7)$, and $b(15)$ are in Processors 3, 2, 1, and 0 (row 1).

Step 4: Move the intermediate results $b(2)$, $b(10)$, $b(6)$, and $b(14)$ in row 2 to row 3 and the intermediate results $b(1)$, $b(9)$, $b(5)$, and $b(13)$ in row 3 to row 2. This step takes 4 clock cycles.

Step 5: Copy the intermediate results $b(1)$, $b(9)$, $b(5)$, and $b(13)$, in row 2 into row 1 and copy the intermediate results $b(3)$, $b(11)$, $b(7)$, and $b(15)$ from row 1 into row 2 so that rows 1 and 2 both have the same intermediate results. At the same time do the same copy function in rows 3 and 4. This step takes 4 clock cycles.

Step 6: In rows 2 and 4 add the two sets of intermediate results. In rows 1 and 3 subtract the intermediate results, using the equations in Section 9.7.5. This step takes 2 clock cycles.

This stage takes a total of 20 clock cycles and:

(i) Intermediate results $c(0)$, $c(8)$, $c(4)$, and $c(12)$ are in Processors 15, 14, 13, and 12 (row 4).

(ii) Intermediate results $c(1)$, $c(9)$, $c(5)$, and $c(13)$ are in Processors 7, 6, 5, and 4 (row 3).

(iii) Intermediate results $c(2)$, $c(10)$, $c(6)$, and $c(14)$ are in Processors 11, 10, 9, and 8 (row 2).

(iv) Intermediate results $c(3)$, $c(11)$, $c(7)$, and $c(15)$ are in Processors 3, 2, 1, and 0 (row 1).

Figure 12-15 shows the locations of these intermediate results in the second row of each processor block.

## Stage 3: Compute Complex Multiplications

Perform the complex multiplications in each individual processor. Since a complex multiply uses four real multiplies and two real adds, the Harvard architecture defined in Section 12.3.5 takes 6 clock cycles for this computation. At the end of this stage:

(i) Intermediate results $c(0)$, $c(8)$, $c(4)$, and $c(12)$ are in Processors 15, 14, 13, and 12 (row 4).

(ii) Intermediate results $c(1)$, $e(9)$, $e(5)$, and $e(13)$ are in Processors 7, 6, 5, and 4 (row 3).

(iii) Intermediate results $c(2)$, $e(10)$, $e(6)$, and $e(14)$ are in Processors 11, 10, 9, and 8 (row 2).

(iv) Intermediate results $c(3)$, $e(11)$, $e(7)$, and $e(15)$ are in Processors 3, 2, 1, and 0 (row 1).

Figure 12-15 shows the locations of these intermediate results in the third row of each processor block.

### Stage 4: Compute the Output 4-Point FFTs

Compute the four 4-point output FFTs by using the intermediate results that are now located in the rows of the array. The steps are similar to those used in the columns to compute the 4-point input FFTs. The columns are defined as numbered from left to right. The steps are:

Step 1: Move the intermediate results in column 2 to column 3 and the intermediate results in column 3 to column 2. This step takes 4 clock cycles.

Step 2: Copy the intermediate results in column 2 into column 1 and the intermediate results from column 1 into column 2 so that columns 1 and 2 both have the same intermediate results. At the same time do the same function in columns 3 and 4. This step takes 4 clock cycles.

Step 3: In columns 2 and 4 add the two sets of intermediate results. At the same time subtract the intermediate results in columns 1 and 3, following the Algorithm Steps in Section 9.7.5. This step takes 2 clock cycles. At the end of this step

  (i) Intermediate results $f(0)$, $f(8)$, $f(4)$, and $f(12)$ are in Processors 15, 11, 7, and 3 (column 4).

  (ii) Intermediate results $f(1)$, $f(9)$, $f(5)$, and $f(13)$ are in Processors 14, 10, 6, and 2 (column 3).

  (iii) Intermediate results $f(2)$, $f(10)$, $f(6)$, and $f(14)$ are in Processors 13, 9, 5, and 1 (column 2).

  (iv) Intermediate results $f(3)$, $f(11)$, $f(7)$, and $f(15)$ are in Processors 12, 8, 4, and 0 (column 1).

Step 4: Move the intermediate results in column 2 to column 3 and the intermediate results in column 3 to column 2. This step takes 4 clock cycles.

Step 5: Copy the intermediate results in column 2 into column 1 and the intermediate results from column 1 into column 2 so that columns 1 and 2 both have the same intermediate results. At the same time do the same function in columns 3 and 4. This step takes 4 clock cycles.

Step 6: Follow the 16-point radix-4 equations to add or subtract the pairs of intermediate results in columns 2 and 4 and in columns 1 and 3. This step takes 2 clock cycles and the output frequency components:

  (i) $A(0)$, $A(2)$, $A(1)$, and $A(3)$ are in Processors 15, 11, 7, and 3 (column 4).

  (ii) $A(8)$, $A(10)$, $A(9)$, and $A(11)$ are in Processors 14, 10, 6, and 2 (column 3).

  (iii) $A(4)$, $A(6)$, $A(5)$, and $A(7)$ are in Processors 13, 9, 5, and 1 (column 2).

  (iv) $A(12)$, $A(14)$, $A(13)$, and $A(15)$ are in Processors 12, 8, 4, and 0 (column 1).

This stage takes 20 clock cycles, and Figure 12-15 shows the locations of these output frequency components in the fourth row of each processor block.

## Stage 5: Output the Results

When the computations are complete, the results can be shifted down to the output data reorganizer and converted back to a sequential stream of data. Again, this can be accomplished by using the shift register I/O concept for converting the input data, and it also takes 40 clock cycles to move the data to the output data reorganizer.

The total number of clock cycles for this algorithm mapping is 122, and the processing latency is 122 clock cycles.

### Option 2: Each Processor Computes One 16-Point Radix-4 FFT

At the other extreme, 16 sets of complex samples can be loaded into the processor array and then each processor can compute a 16-point radix-4 FFT and output the results.

## Stage 1: Distribute One of the 16 Sets of Complex Samples to Each of the 16 Processors

Option 1 showed that it takes 40 clock cycles for the data input for one set of complex samples, so it takes $16 * 40 = 640$ clock cycles for 16 sets of complex samples. However, the input data reorganizer must be configured differently because the goal is to have all of the data from one set of complex samples in one processor. The simplest way to implement the input data reorganizer for this option is to have memory for four sets of 16 complex words at the top of each column, rather than the single pair of memory locations needed in this architecture's Option 1.

If all 16 sets of complex samples are lined up in sequence to input to the input data reorganizer, it will take $16 * 32 = 512$ clock cycles to move it all in as described. Now, the shifting process into the array takes one-fourth as long (128 clock cycles) because four words move into the array at once, one into each column of processors. Therefore, moving 16 sets of 16 complex samples into the array takes 640 clock cycles. When the input process is complete, all of the data is in the proper processor for performing the 16 FFTs.

## Stage 2: Compute the Sixteen 16-Point Radix-4 FFTs

It takes 168 clock cycles to compute the 16-point radix-4 FFT using the Harvard architecture assumptions from Section 12.3. Since all 16 processors are computing the algorithm, it takes only 168 clock cycles to compute all sixteen 16-point radix-4 FFTs.

## Stage 3: Collect the Results of the Sixteen 16-Point Radix-4 FFT Computations

Option 1 showed that it takes 40 clock cycles for the data input for one set of complex samples. Therefore, it takes $16 * 40 = 640$ clock cycles for 16 sets of complex samples.

The total number of clock cycles for this approach is the number of clock cycles to perform the 16-point radix-4 FFT plus the data I/O time for 16 sets of complex samples.

The result is a total of 1448 clock cycles, which is the processing latency. The processing throughput is an average of $1448/16 = 90.5$ clock cycles per FFT. Notice that the data I/O clock cycle total is much larger (1280) than the computational clock cycles (168). This time can be improved to 1280 clock cycles by requiring each processor to perform data I/O and computations simultaneously.

### 12.5.3 Star 16-Point Radix-4 FFT Examples

The star [1] architecture is most often used when one function or process dominates the application. It consists of one central processor with interconnections to numerous others as shown in Figure 12-16. The number of processing elements depends on the FFT algorithm to be computed. Figure 12-16 is a natural configuration for the 16-point radix-4 FFT because of the four 4-point FFTs computed on the input and output. For this example, Processor 0 is the data I/O processor and global memory. The other four processors have the Harvard architecture from Section 12.3.5.

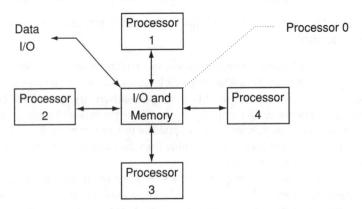

**Figure 12-16** Star architecture for 16-point radix-4 FFT example.

This architecture can also be used in the two extremes of minimum processing latency (Option 1) and maximum processing throughput (Option 2) described for the crossbar and massively parallel architectures. Both are described.

### Option 1: All Processors Used to Compute One 16-Point Radix-4 FFT

The strategy for this option is to use Processor 0 as the data I/O processor and to use Processors 1–4 to perform all the computations. Between the input and output 4-point building blocks, the data must be reorganized. In this example this is accomplished by moving all the intermediate results from Processors 1–4 back to Processor 0 and then redistributing the intermediate results to Processors 1–4 based on which ones are grouped for an output 4-point building-block computation in the algorithm steps in Section 9.7.5.

Since Processors 1–4 only perform 4-point building-block computations, the memory mapping in Chapter 8 for the 4-point building block can be used for all four processors. Figure 12-17 shows which of the four processors has each of the 16 complex samples, intermediate results, and output frequency components at the end of each stage by listing them in their processor on the same line as the label to the left of the figure that defines the stage of the algorithm.

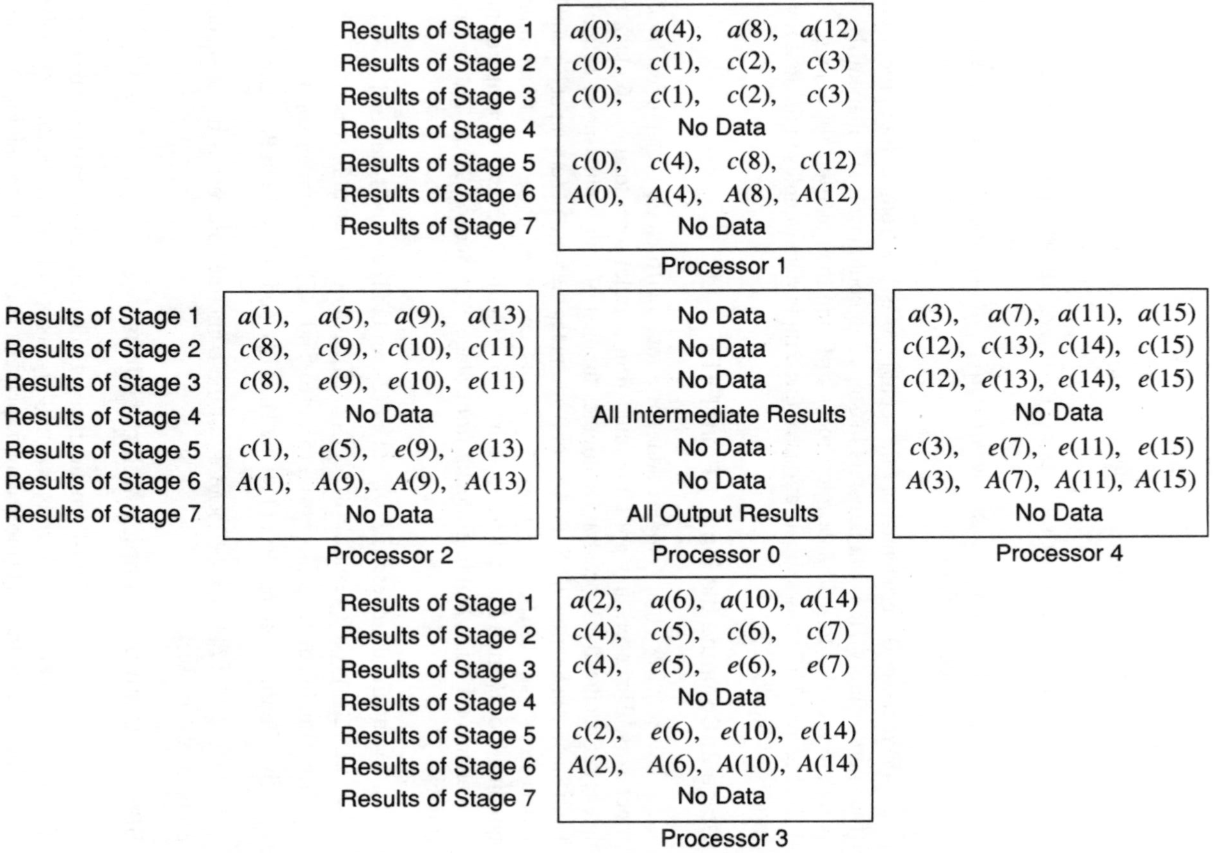

**Processor 1**

| | | | |
|---|---|---|---|
| Results of Stage 1 | $a(0)$, | $a(4)$, | $a(8)$, | $a(12)$ |
| Results of Stage 2 | $c(0)$, | $c(1)$, | $c(2)$, | $c(3)$ |
| Results of Stage 3 | $c(0)$, | $c(1)$, | $c(2)$, | $c(3)$ |
| Results of Stage 4 | No Data | | | |
| Results of Stage 5 | $c(0)$, | $c(4)$, | $c(8)$, | $c(12)$ |
| Results of Stage 6 | $A(0)$, | $A(4)$, | $A(8)$, | $A(12)$ |
| Results of Stage 7 | No Data | | | |

**Processor 2**

| | | | |
|---|---|---|---|
| Results of Stage 1 | $a(1)$, | $a(5)$, | $a(9)$, | $a(13)$ |
| Results of Stage 2 | $c(8)$, | $c(9)$, | $c(10)$, | $c(11)$ |
| Results of Stage 3 | $c(8)$, | $e(9)$, | $e(10)$, | $e(11)$ |
| Results of Stage 4 | No Data | | | |
| Results of Stage 5 | $c(1)$, | $e(5)$, | $e(9)$, | $e(13)$ |
| Results of Stage 6 | $A(1)$, | $A(9)$, | $A(9)$, | $A(13)$ |
| Results of Stage 7 | No Data | | | |

**Processor 0**

| |
|---|
| No Data |
| No Data |
| No Data |
| All Intermediate Results |
| No Data |
| No Data |
| All Output Results |

**Processor 3**

| | | | |
|---|---|---|---|
| Results of Stage 1 | $a(2)$, | $a(6)$, | $a(10)$, | $a(14)$ |
| Results of Stage 2 | $c(4)$, | $c(5)$, | $c(6)$, | $c(7)$ |
| Results of Stage 3 | $c(4)$, | $e(5)$, | $e(6)$, | $e(7)$ |
| Results of Stage 4 | No Data | | | |
| Results of Stage 5 | $c(2)$, | $e(6)$, | $e(10)$, | $e(14)$ |
| Results of Stage 6 | $A(2)$, | $A(6)$, | $A(10)$, | $A(14)$ |
| Results of Stage 7 | No Data | | | |

**Processor 4**

| | | | |
|---|---|---|---|
| $a(3)$, | $a(7)$, | $a(11)$, | $a(15)$ |
| $c(12)$, | $c(13)$, | $c(14)$, | $c(15)$ |
| $c(12)$, | $e(13)$, | $e(14)$, | $e(15)$ |
| No Data | | | |
| $c(3)$, | $e(7)$, | $e(11)$, | $e(15)$ |
| $A(3)$, | $A(7)$, | $A(11)$, | $A(15)$ |
| No Data | | | |

**Figure 12-17**  Data map for star implementation of 16-point radix-4 FFT.

The stages are as follows:

### Stage 1: Distribute the Input Data onto the Processors

Step 1: Load the input data into Processor 0. This step takes 32 clock cycles.

Step 2: Move complex samples $a(0)$, $a(4)$, $a(8)$, and $a(12)$ to Processor 1 using 8 clock cycles.

Step 3: Move complex samples $a(1)$, $a(5)$, $a(9)$, and $a(13)$ to Processor 2 using 8 clock cycles.

Step 4: Move complex samples $a(2)$, $a(6)$, $a(10)$, and $a(14)$ to Processor 3 using 8 clock cycles.

Step 5: Move complex samples $a(3)$, $a(7)$, $a(11)$, and $a(15)$ to Processor 4 using 8 clock cycles.

If Processor 0 were a memory that could move data from all four processors at once (four-port memory), the data transfers in Steps 2–5 could occur simultaneously. The total is 64 clock cycles to load data into Processor 0 and then distribute it among Processors 1–4. Figure 12-17 shows the locations of these input data samples in the first row of each processor block.

### Stage 2: Compute the Input 4-Point FFTs

This requires eight complex adds for a total of 16 clock cycles in each of the four processors. However, they are all computed in parallel for a total of 16 clock cycles of latency. Specifically, Processor 1 computes the first of four input 4-point building blocks from Stage 1 in Section 9.7.5. Processor 2 computes the third of four input 4-point building blocks from Stage 1 in Section 9.7.5. Processor 3 computes the second of four input 4-point building blocks from Stage 1 in Section 9.7.5. Finally, Processor 4 computes the fourth of four input 4-point building blocks from Stage 1 in Section 9.7.5. At the end of this stage:

(i) Intermediate results $c(0)$, $c(1)$, $c(2)$, and $c(3)$ are in Processor 1.

(ii) Intermediate results $c(8)$, $c(9)$, $c(10)$, and $c(11)$ are in Processor 2.

(iii) Intermediate results $c(4)$, $c(5)$, $c(6)$, and $c(7)$ are in Processor 3.

(iv) Intermediate results $c(12)$, $c(13)$, $c(14)$, and $c(15)$ are in Processor 4.

Figure 12-17 shows the locations of these intermediate results in the second row of each processor block.

### Stage 3: Compute Complex Multiplications

Since Processors 2 and 4 contain three intermediate results that must be multiplied by a complex constant, and each complex multiply takes 6 clock cycles, this stage takes a total of 18 clock cycles and is performed in the processors prior to reorganizing the intermediate results, using equations in Section 9.7.5. At the end of this stage:

(i) Intermediate results $c(0)$, $c(1)$, $c(2)$, and $c(3)$ are in Processor 1.

(ii) Intermediate results $c(8)$, $e(9)$, $e(10)$, and $e(11)$ are in Processor 2.

    (iii)  Intermediate results $c(4)$, $e(5)$, $e(6)$, and $e(7)$ are in Processor 3.

    (iv)  Intermediate results $c(12)$, $e(13)$, $e(14)$, and $e(15)$ are in Processor 4.

Figure 12-17 shows the locations of these intermediate results in the third row of each processor block.

### Stage 4: Move Intermediate Results Back to Processor 0

Move the results of these calculations from Processors 1 and 4 back to Processor 0. This step takes 32 clock cycles (2 clock cycles for each of the 16 complex results) unless Processor 0 is a four-port memory. If Processor 0 can send and receive data from all four processors at once (work as a four-port memory), this stage only requires 8 clock cycles. Figure 12-17 shows the locations of all the data to be in Processor 0 in the fourth row of each processor block.

### Stage 5: Redistribute Intermediate Results for Output 4-Point FFT Computations

This process takes 32 clock cycles using the following steps to move intermediate results from Processor 0 to the appropriate processor for computing the output 4-point FFTs.

    Step 1:  Move intermediate results $c(0)$, $c(4)$, $c(8)$, and $c(12)$ to Processor 1, using 8 clock cycles.

    Step 2:  Move intermediate results $c(1)$, $e(5)$, $e(9)$, and $e(13)$ to Processor 2, using 8 clock cycles.

    Step 3:  Move intermediate results $c(2)$, $e(6)$, $e(10)$, and $e(14)$ to Processor 3, using 8 clock cycles.

    Step 4:  Move intermediate results $c(3)$, $e(7)$, $e(11)$, and $e(15)$ to Processor 4, using 8 clock cycles.

Stages 4 and 5 can be done with 16 fewer clock cycles because one of the four results from each processor output of Stage 3 ends up back in the same processor for the output 4-point FFT computations. This means it does not have to be moved from its location at the end of Stage 3 into Processor 0 and then back out to the same location in the same processor. Moving these four complex intermediate results twice takes 16 clock cycles. Therefore, Stages 4 and 5 can be performed with 48, not 64, clock cycles. Figure 12-17 shows the locations of the intermediate results in the fifth row of each processor block.

### Stage 6: Compute the Output 4-Point FFTs

This requires eight complex adds for a total of 16 clock cycles in each of the four processors. Each processor computes one set of the output 4-point building-block algorithm steps from Section 9.7.5. However, they are all computed in parallel, for a total of 16 clock cycles of latency. Specifically, Processor 1 computes the first of four output 4-point building blocks from Stage 3 in Section 9.7.5. Processor 2 computes the second of four output 4-point building blocks from Stage 3 in Section 9.7.5. Processor 3 computes the third of four output 4-point building blocks from Stage 3 in Section 9.7.5. Finally, Processor 4 computes the fourth of four output 4-point building blocks from Stage 3 in Section 9.7.5. At the end of this stage the output frequency components:

(i) $A(0)$, $A(4)$, $A(8)$, and $A(12)$ are in Processor 1.

(ii) $A(1)$, $A(5)$, $A(9)$, and $A(13)$ are in Processor 2.

(iii) $A(2)$, $A(6)$, $A(10)$, and $A(14)$ are in Processor 3.

(iv) $A(3)$, $A(7)$, $A(11)$, and $A(15)$ are in Processor 4.

Figure 12-17 shows the locations of these output frequency components in the sixth row of each processor block.

### Stage 7: Move the Output Results Back to Processor 0

Move the results of these calculations from the processors back to Processor 0. This step takes 32 clock cycles unless Processor 0 is a four-port memory. In that case it would only take 8 clock cycles. Figure 12-17 shows in row 7 of each processor that all of the output frequency components are in Processor 0 awaiting output through the data I/O path.

### Stage 8: Output the Results from Processor 0

This step takes 32 clock cycles. The total for this algorithm mapping is 226 clock cycles, and the processing latency is also 226 clock cycles. The largest contributor to these clock cycles is the data I/O and movement to the computational processors.

### *Option 2: Each Processor Computes One 16-Point Radix-4 FFT*

### Stage 1: Distribute One of the 16 Sets of Complex Samples to Each of the 16 Processors

The data input is for four sets of 16 complex samples, which takes 128 clock cycles to move to Processor 0 and another 128 clock cycles to move out to the four processors.

### Stage 2: Compute the Four 16-Point Radix-4 FFTs

It takes 168 clock cycles to compute the 16-point radix-4 FFT using the Harvard architecture from Section 12.3.5. Since all four processors are computing the algorithm at the same time, it takes only 168 clock cycles to compute all four 16-point radix-4 FFTs.

### Stage 3: Collect the Results of the Four 16-Point Radix-4 FFT Computations

It takes 128 clock cycles to move the four sets of 16-point complex results from the four processors to Processor 0 and another 128 clock cycles to move it out of the processor array.

The total number of clock cycles for this option is the number of clock cycles to perform the 16-point radix-4 FFT plus the data I/O time for four sets of complex samples. This is a total of 680 clock cycles, which is the processing latency. The processing throughput is an average of $680/4 = 170$ clock cycles per FFT.

## 12.6 THREE MULTIDIMENSIONAL ARRAYS

Processors can be combined into multidimensional arrays in numerous ways, and there are many ways to use the array to compute each of the algorithms in Chapter 9. At the two data mapping extremes are:

1. One set of complex samples is mapped onto all of the processors in the array and then one FFT is computed. This option usually results in minimum latency processing.

2. A set of complex samples is mapped onto each of the processors and then a number of FFTs are performed in parallel. This usually results in the maximum throughput but has more latency than the first option.

Each extreme is described for mapping the 16-point radix-4 FFT onto the four-dimensional hypercube architecture from Section 11.4.1. Mapping onto the massively parallel and hybrid arrays from Chapter 11 is described in general terms, but a detailed example is not presented because these complex architectures are not suited to implementing the 16-point radix-4 FFT efficiently. Throughout this section, when the $k$-th input data sample is written as $a(k)$, it means both the real and imaginary parts of the sample. Specifically, $a(k) = a_R(k) + j * a_I(k)$. This same shorthand notation is also used for intermediate results and output frequency components.

### 12.6.1 Hypercube 16-Point Radix-4 FFT Examples

The four-dimensional hypercube [1, 3] in Figure 12-18 has 16 processing nodes. For the 16-point radix-4 FFT, the two extremes for algorithm mapping are to distribute one set of complex samples among all the processors (Option 1) or to load a set of complex samples into each processor (Option 2). Option 1 requires more computational power than Option 2 to meet a fixed throughput requirement, but has the lowest processing latency. Option 2 reduces the computational costs because no data interchanges are required to perform the FFT algorithm. However, the processing latency is large because 16 sets of complex samples are loaded before any computations are performed.

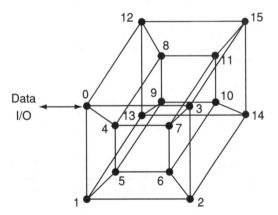

**Figure 12-18**    Four-dimensional hypercube.

### *Option 1: All Processors Used to Compute One 16-Point Radix-4 FFT*

One logical data distribution for this option is based on noting that there are four square arrays of four processors in this four-dimensional hypercube. The processors in each of these squares are 0–3, 4–7, 8–11, and 12–15. Each square array of processors can be used to compute one of the input 4-point FFTs, followed by complex multiplications within the processors and a reordering of the data so that the same squares or another set of squares can be used to compute the output 4-point FFTs. Figure 12-19 shows which of the 16 processors has each of the 16 complex samples, intermediate results, and output results at the end of each stage by listing them in their processor on the same line as the label to the left of the figure that defines the stage of the algorithm.

| | Processor 0 | Processor 1 | Processor 2 | Processor 3 |
|---|---|---|---|---|
| Results of Stage 1 | $a(0)$ | $a(8)$ | $a(4)$ | $a(12)$ |
| Results of Stage 2 | $c(0)$ | $c(1)$ | $c(3)$ | $c(2)$ |
| Results of Stage 3 | $c(0)$ | $c(1)$ | $e(3)$ | $c(2)$ |
| Results of Stage 4 | $A(0)$ | $A(1)$ | $A(3)$ | $A(2)$ |

| | Processor 4 | Processor 5 | Processor 6 | Processor 7 |
|---|---|---|---|---|
| Results of Stage 1 | $a(1)$ | $a(9)$ | $a(5)$ | $a(13)$ |
| Results of Stage 2 | $c(8)$ | $c(9)$ | $c(11)$ | $c(10)$ |
| Results of Stage 3 | $c(8)$ | $e(9)$ | $e(11)$ | $e(10)$ |
| Results of Stage 4 | $A(9)$ | $A(9)$ | $A(11)$ | $A(10)$ |

| | Processor 8 | Processor 9 | Processor 10 | Processor 11 |
|---|---|---|---|---|
| Results of Stage 1 | $a(2)$ | $a(10)$ | $a(6)$ | $a(14)$ |
| Results of Stage 2 | $c(4)$ | $c(5)$ | $c(7)$ | $c(6)$ |
| Results of Stage 3 | $c(4)$ | $e(5)$ | $e(7)$ | $e(6)$ |
| Results of Stage 4 | $A(12)$ | $A(13)$ | $A(15)$ | $A(14)$ |

| | Processor 12 | Processor 13 | Processor 14 | Processor 15 |
|---|---|---|---|---|
| Results of Stage 1 | $a(3)$ | $a(11)$ | $a(7)$ | $a(15)$ |
| Results of Stage 2 | $c(12)$ | $c(13)$ | $c(15)$ | $c(14)$ |
| Results of Stage 3 | $c(12)$ | $e(13)$ | $e(15)$ | $e(14)$ |
| Results of Stage 4 | $A(4)$ | $A(5)$ | $A(7)$ | $A(6)$ |

**Figure 12-19**  Data map for four-dimensional hypercube implementation of 16-point radix-4 FFT.

The stages for implementing this option are as follows.

### Stage 1: Distribute the Input Data onto the Processors

Any of the processors can be used as the data I/O path. For this example, Processor 0 is used. Data is moved from Processor 0 to another processor by stepping it through the hypercube architecture. Table 12-5 shows the number of clock cycles required to move a data word from Processor 0 to one of the other processors in the architecture, assuming one clock cycle to move a data word between any two processors. Notice that, as mentioned in Chapter 11, the longest path length for a four-dimensional hypercube is 4. In this example the path from Processor 0 to Processor 10 is longest. Since each of the input samples is complex, the numbers in Table 12-5 must be doubled to determine the actual number of clock cycles used for each complex data input. This stage takes 42 clock cycles.

**Table 12-5**    Data I/O Transfer Clock Costs

| Processor # | # Clock cycles |
|:-----------:|:--------------:|
| 0 | 0 |
| 1 | 1 |
| 2 | 2 |
| 3 | 1 |
| 4 | 1 |
| 5 | 2 |
| 6 | 3 |
| 7 | 2 |
| 8 | 2 |
| 9 | 3 |
| 10 | 4 |
| 11 | 3 |
| 12 | 1 |
| 13 | 2 |
| 14 | 3 |
| 15 | 2 |

The specific steps for this stage are:

Step 1: Load complex samples $a(0)$, $a(4)$, $a(8)$, and $a(12)$ into Processors 0, 2, 1, and 3, using 8 clock cycles.

Step 2: Load complex samples $a(1)$, $a(5)$, $a(9)$, and $a(13)$ into Processors 4, 6, 5, and 7 by first loading them into Processors 0, 2, 1, and 3 and then moving them to Processors 4, 6, 5, and 7 in parallel in 2 additional clock cycles. This step takes 10 clock cycles.

Step 3: Load complex samples $a(2)$, $a(6)$, $a(10)$, and $a(14)$ into Processors 8, 10, 9, and 11 by first loading them into Processors 0, 2, 1, and 3 and then moving them through Processors 4, 6, 5, and 7 in parallel to Processors 8, 10, 9, and 11 in 4 additional clock cycles. This step takes 14 clock cycles.

Step 4: Load complex samples $a(3)$, $a(7)$, $a(11)$, and $a(15)$ into Processors 12, 14, 13, and 15 by first loading them into Processors 0, 2, 1, and 3 and then moving them to Processors 12, 14, 13, and 15 in parallel in 2 additional clock cycles. This step takes 10 clock cycles.

Figure 12-19 shows the locations of the complex input samples in the first row of each processor block.

### Stage 2: Compute the Input 4-Point FFTs

The steps are:

Step 1: Copy the complex sample $a(0)$ from Processor 0 into Processor 1 and copy the complex sample $a(8)$ from Processor 1 into Processor 0. At the same time, perform this same copy of complex samples operation between Processors 2 and 3, 4 and 5, 6 and 7, 8 and 9, 10 and 11, 12 and 13, and 14 and 15. This step takes 4 clock cycles because all of the pairs of complex sample moves can be done in parallel.

Step 2: In Processors 0, 3, 4, 7, 8, 11, 12, and 15 add the two complex samples, using two clock cycles. For example, Processor 0 computes $b(0) = a(0) + a(8)$, which is part of the first of four input 4-point building blocks in Stage 1 of the algorithm in Section 9.7.5.

Step 3: In Processors 1, 2, 5, 6, 9, 10, 13, and 14 subtract the two complex samples at the same time as Step 2. For example, Processor 1 computes $b(1) = a(0) - a(8)$, which is part of the first of four input 4-point building blocks in Stage 1 of the algorithm in Section 9.7.5. At the end of these computations:

   (i) intermediate results $b(0)$, $b(1)$, $b(2)$, and $b(3)$ are in Processors 0, 1, 3, and 2.
   (ii) intermediate results $b(8)$, $b(9)$, $b(10)$, and $b(11)$ are in Processors 4, 5, 7, and 6.
   (iii) intermediate results $b(4)$, $b(5)$, $b(6)$, and $b(7)$ are in Processors 8, 9, 11, and 10.
   (iv) intermediate results $b(12)$, $b(13)$, $b(14)$, and $b(15)$ are in Processors 12, 13, 15, and 14.

Step 4: Copy the intermediate results from Processor 0, $b(0)$, into Processor 3 and copy the intermediate results from Processor 3, $b(2)$, into Processor 0. At the same time, perform this same copying of the intermediate results between Processors 1 and 2, 4 and 7, 5 and 6, 8 and 11, 9 and 10, 12 and 15, and 13 and 14. This step takes 4 clock cycles because all of the pairs of complex sample moves can be done in parallel.

Step 5: In Processors 0, 1, 4, 5, 8, 9, 12, and 13, add the two complex intermediate results. This step takes two clock cycles. For example, Processor 0 computes $c(0) = b(0) + b(2)$, which is part of the first of four input 4-point building blocks in Stage 1 of the algorithm in Section 9.7.5.

Step 6: In Processors 3, 2, 7, 6, 11, 10, 15, and 14, subtract the two complex numbers at the same time as Step 5 because these processors are not performing other functions during the time Step 5 is being performed. For example, Processor 3 computes $c(2) = b(0) - b(2)$, which is part of the first of four input 4-point building blocks in Stage 1 of the algorithm in Section 9.7.5. At the end of these computations:

   (i) intermediate results $c(0)$, $c(3)$, $c(1)$, and $c(2)$ are in Processors 0, 2, 1, and 3.
   (ii) intermediate results $c(8)$, $c(11)$, $c(9)$, and $c(10)$ are in Processors 4, 6, 5, and 7.
   (iii) intermediate results $c(4)$, $c(7)$, $c(5)$, and $c(6)$ are in Processors 8, 10, 9, and 11.

(iv) intermediate results $c(12)$, $c(15)$, $c(13)$, and $c(14)$ are in Processors 12, 14, 13, and 15.

Figure 12-19 shows the locations of these intermediate results in the second row of each processor block.

## Stage 3: Compute Complex Multiplications

These can be computed within the individual processors. Since each takes four multiplies and two adds, the complex multiplies use 6 clock cycles. At this point:

(i) Intermediate results $c(0)$, $c(3)$, $c(1)$, and $c(2)$ are in Processors 0, 2, 1, and 3.

(ii) Intermediate results $c(8)$, $e(11)$, $e(9)$, and $e(10)$ are in Processors 4, 6, 5, and 7.

(iii) Intermediate results $c(4)$, $e(7)$, $e(5)$, and $e(6)$ are in Processors 8, 10, 9, and 11,

(iv) Intermediate results $c(12)$, $e(15)$, $e(13)$, and $e(14)$ are in Processors 12, 14, 13, and 15.

Figure 12-19 shows the locations of these intermediate results in the third row of each processor block.

## Stage 4: Compute the Output 4-Point FFTs

Following the algorithm steps in Section 9.7.5, the steps are:

Step 1: Reorganize the intermediate results in Processors 8 through 15 of the hypercube in preparation for computing the output 4-point FFTs. To do this:

    (i) Move intermediate result $c(4)$ from Processor 8 to Processor 12, using 2 clock cycles.

    (ii) Move intermediate result $c(12)$ from Processor 12 to Processor 8, using 2 clock cycles.

    (iii) Move intermediate result $e(5)$ from Processor 9 to Processor 13, using 2 clock cycles.

    (iv) Move intermediate result $e(13)$ from Processor 13 to Processor 9, using 2 clock cycles.

    (v) Move intermediate result $e(6)$ from Processor 11 to Processor 15, using 2 clock cycles.

    (vi) Move intermediate result $e(14)$ from Processor 15 to Processor 11, using 2 clock cycles.

    (vii) Move intermediate result $e(7)$ from Processor 10 to Processor 14, using 2 clock cycles.

    (viii) Move intermediate result $e(15)$ from Processor 14 to Processor 10, using 2 clock cycles.

Step 2: Copy the intermediate results, $c(0)$, from Processor 0 into Processor 12 and copy the intermediate results, $c(4)$, from Processor 12 into Processor 0. At the same time, perform this operation between Processors 4 and 8, 5 and 9, 1 and 13, 7 and 11, 6 and 10, 3 and 15, and 2 and 14. This step takes 4 clock cycles because all of the pairs of complex sample moves can be done in parallel.

Step 3: In Processors 0, 1, 2, 3, 4, 5, 6, and 7, add the two complex intermediate results. This step takes two clock cycles. For example, Processor 0 computes $f(0) = c(0) + c(4)$, which is part of the first of four output 4-point building blocks in Stage 3 of the algorithm in Section 9.7.5.

Step 4: In Processors 8, 9, 10, 11, 12, 13, 14, and 15, subtract the two intermediate results, following the algorithm steps in Section 9.7.5 and using 2 clock cycles. For example, Processor 8 computes $f(2) = c(8) + c(12)$, which is part of the first of four output 4-point building blocks in Stage 3 of the algorithm in Section 9.7.5. At the end of these computations:

  (i) Intermediate results $f(0)$, $f(1)$, $f(2)$, and $f(3)$ are in Processors 0, 12, 4, and 8.

 (ii) Intermediate results $f(8)$, $f(9)$, $f(10)$, and $f(11)$ are in Processors 3, 15, 7, and 11.

(iii) Intermediate results $f(4)$, $f(5)$, $f(6)$, and $f(7)$ are in Processors 1, 13, 5, and 9.

(iv) Intermediate results $f(12)$, $f(13)$, $f(14)$, and $f(15)$ are in Processors 2, 14, 6, and 10.

Step 5: Load the intermediate results, $f(0)$, from Processor 0 into Processor 4 and load the intermediate results, $f(2)$, from Processor 4 into Processor 0. At the same time, perform this operation between Processors 3 and 7, 1 and 5, 2 and 6, 8 and 12, 9 and 13, 10 and 14, and 11 and 15. This only takes 4 clock cycles because all of these operations can be done in parallel.

Step 6: In Processors 0, 1, 2, 3, 12, 13, 14, and 15, add the two intermediate results, using two clock cycles. For example, Processor 0 computes $A(0) = f(0) + f(2)$, which is part of the first of four output 4-point building blocks in Stage 3 of the algorithm in Section 9.7.5.

Step 7: In Processors 4, 5, 6, 7, 8, 9, 10, and 11, subtract the two intermediate results, using two clock cycles, using the equations in Section 9.7.5. For example, Processor 4 computes $A(8) = f(0) - f(2)$, which is part of the first of four output 4-point building blocks in Stage 3 of the algorithm in Section 9.7.5. Then the output frequency components:

  (i) $A(0)$, $A(1)$, $A(2)$, and $A(3)$ are in Processors 0, 1, 3, and 2,

 (ii) $A(8)$, $A(9)$, $A(10)$, and $A(11)$ are in Processors 4, 5, 7, and 6.

(iii) $A(4)$, $A(5)$, $A(6)$, and $A(7)$ are in Processors 12, 13, 15, and 14.

(iv) $A(12)$, $A(13)$, $A(14)$, and $A(15)$ are in Processors 8, 9, 11, and 10.

Figure 12-19 shows the locations of the output frequency components in the fourth row of each processor block.

### Stage 5: Output the Results Using Processor 0

Since all of the outputs are available at one time, the steps are based on the same logic for inputting data and Table 12-4.

Step 1: Move output frequency components $A(0)$, $A(1)$, $A(2)$, and $A(3)$ out of the hypercube first. This step takes 8 clock cycles based on adding the number of clock cycles in Processors 0, 1, 2, and 3 in Table 12-4 and multiplying by 2 to account for complex data.

Step 2: Move the answers in Processors 12, 13, 14, and 15 ($A(4)$, $A(5)$, $A(7)$, and $A(6)$, respectively) into Processors 0, 1, 2, and 3, respectively. This step takes 2 clock cycles because all four moves can be done at once.

Step 3: Move $A(4)$, $A(5)$, $A(6)$, and $A(7)$ out of the hypercube. Since $A(4)$, $A(5)$, $A(6)$, and $A(7)$ are now in Processors 0, 1, 3, and 2, this step takes 8 clock cycles. As in Step 1 of this stage, this is based on adding the number of clock cycles in Processors 0, 1, 2, and 3 in Table 12-4 and multiplying by 2 to account for complex data.

Step 4: Move the answers in Processors 4, 5, 6, and 7 ($A(8)$, $A(9)$, $A(11)$, and $A(10)$, respectively) into Processors 0, 1, 2, and 3, respectively. At the same time, the answers in Processors 8, 9, 10, and 11 ($A(12)$, $A(13)$, $A(15)$, and $A(14)$, respectively) can be moved into Processors 4, 5, 6, and 7. This step takes 2 clock cycles.

Step 5: Move $A(8)$, $A(9)$, $A(10)$, and $A(11)$ out. Since they are now in Processors 0, 1, 3, and 2, this step takes 8 clock cycles. As in Step 1 of this stage, this is based on adding the number of clock cycles in Processors 0, 1, 2, and 3 in Table 12-4 and multiplying by 2 to account for complex data.

Step 6: Move the answers in Processors 4, 5, 6, and 7 (now $A(12)$, $A(13)$, $A(15)$, and $A(14)$ from Step 3 of this stage) into Processors 0, 1, 2, and 3, respectively. This step takes 2 clock cycles because all four moves can be done at once, each by one pair of processors.

Step 7: Move $A(12)$, $A(13)$, $A(14)$, and $A(15)$ out. Since they are now in Processors 0, 1, 3, and 2, this step takes 8 clock cycles. As in Step 1 of this stage, this is based on adding the number of clock cycles in Processors 0, 1, 2, and 3 in Table 12-4 and multiplying by 2 to account for complex data.

The total is 134 clock cycles of processing load and processing latency.

### *Option 2: Each Processor Computes One 16-Point Radix-4 FFT*

The four-dimensional hypercube is used to compute sixteen, 16-point radix-4 FFTs in parallel. The stages for doing that are as follows.

### Stage 1: Distribute the 16 Sets of Complex Samples onto the Processors

These complex sample moves take 16 times as many clock cycles as used to move one set of complex samples into the 16 processors in Stage 1 of Option 1, in this section. This is a total of $42 * 16 = 672$ clock cycles.

### Stage 2: Compute the Sixteen, 16-Point Radix-4 FFTs

Using a Harvard architecture processor at each node, this takes 168 clock cycles, based on the assumptions in Section 12.3.5.

### Stage 3: Output the Results of the Sixteen, 16-Point Radix-4 FFTs

This takes 16 times as long as it takes to move the answers from one set of data out through Processor 0. Based on the results in Stage 5 of Option 1, in this section, this is a total of $38 * 16 = 608$ clock cycles.

The total for this approach is 1448 clock cycles. Therefore, the processing latency is 1448 clock cycles, and the average processing load per FFT is $1448/16 = 90.5$ clock cycles.

### 12.6.2 Massively Parallel 16-Point Radix-4 FFT Examples

The simplest form of three-dimensional massively parallel [1] processing is multiple two-dimensional arrays, as shown in Figure 12-20, that lay atop each other and are interconnected using "up" and "down" links in addition to the standard, two-dimensional NEWS connections.

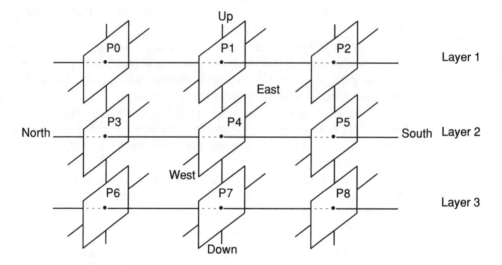

**Figure 12-20** Three-dimensional massively parallel processor.

The top three processors represent one row of the massively parallel processor array in Section 11.4.2. The middle and bottom sets of processors each represent a row of an additional two-dimensional array. The vertical interconnections are the "up" and "down" connections between these two-dimensional arrays that makes the resulting array three dimensional. This is a very complex architecture to efficiently use to compute the small FFT examples from Chapter 9. In all likelihood, if this architecture had to compute the 16-point radix-4 FFT, it would use one of the two approaches described for the two-dimensional massively parallel processor in Section 12.5.2. The two additional layers of two-dimensional processors would process more sets of data, but the interconnections between vertical layers would not be used. The result is that the computational throughput and latency would be multiplied by how many layers of two-dimensional processors were in the array.

### 12.6.3 Hybrid 16-Point Radix-4 FFT Examples

A hybrid architecture is a combination of two or more of the architectures described in previous sections. This example is an array of 16 programmable DSP chips interconnected as a NEWS parallel processing architecture (Figure 12-14). Each processor is then a programmable DSP chip using a Harvard architecture (Figure 12-2). Inside the DSP chip are multiple arithmetic processing units interconnected on a linear bus (Figure 12-10). Finally, the multiplier-accumulator arithmetic processing unit is a pipeline combination of the multiplier and accumulator (Figure 10-4). Figure 12-21 shows the additional interconnects needed to interface the conventional Harvard architecture in Figure 12-2 into a NEWS architecture. Note that this hybrid example is exactly the same as the two-dimensional massively parallel processing architecture described in Section 12.5.2. Therefore, its computational performance and processing latency are also the same.

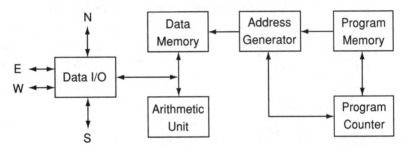

**Figure 12-21**   Harvard architecture block from parallel array.

## 12.7 ALGORITHM MAPPING EXAMPLES COMPARISON MATRIX

All entries are in clock cycles per FFT (see Table 12-6 on page 314).

## 12.8 CONCLUSIONS

Algorithms and data are distributed and redistributed among the processors in the course of computing the entire algorithm. The data map figures for four parallel and multidimensional arrays depict where the data resides at the end of each stage of computing an algorithm. This awareness makes it easier to understand how the reorganization of the data among the processors was done in the examples. This chapter concludes the portion of the book on architectures and algorithms. The next four chapters deal with selecting hardware and testing it.

**Table 12-6** Algorithm Mapping Examples Comparison Matrix

| Architecture examples mappings | Input overhead | Reorgan. overhead | Output overhead | Comp. thruput | Process. latency |
|---|---|---|---|---|---|
| **Single Processors** | | | | | |
| Harvard | 32 | 0 | 32 | 232 | 232 |
| **Linear Arrays (Option 1)** | | | | | |
| 3-processor pipeline | 32 | 64 | 32 | 128 | 320 |
| 3-processor linear bus | 32 | 64 | 32 | 128 | 320 |
| 3-processor ring bus | 32 | 64 | 32 | 128 | 320 |
| **Linear Arrays (Option 2)** | | | | | |
| 3-processor pipeline | 96 | 0 | 96 | 120 | 360 |
| 3-processor linear bus | 96 | 0 | 96 | 120 | 360 |
| 3-processor ring bus | 96 | 0 | 96 | 120 | 360 |
| **Parallel Arrays (Option 1)** | | | | | |
| 16-processor crossbar | 32 | 12 | 32 | 106 | 106 |
| 16-processor 2-D massively par. | 40 | 0 | 40 | 122 | 122 |
| 5-processor star | 64 | 48 | 64 | 226 | 226 |
| **Parallel Arrays (Option 2)** | | | | | |
| 16-processor crossbar | 512 | 0 | 512 | 74.5 | 1192 |
| 16-processor 2-D massively par. | 640 | 0 | 640 | 90.5 | 1448 |
| 5-processor star | 128 | 0 | 128 | 170 | 680 |
| **Multidimensional Arrays (Option 1)** | | | | | |
| 16-processor 4-D hypercube | 64 | 0 | 38 | 134 | 134 |
| 3-D massively parallel array | 64 | 0 | 38 | 134 | 134 |
| Hybrid | 64 | 0 | 38 | 134 | 134 |
| **Multidimensional Arrays (Option 2)** | | | | | |
| 16-processor 4-D hypercube | 672 | 0 | 608 | 90.5 | 1448 |
| 3-D massively parallel array | 672 | 0 | 608 | 90.5 | 1448 |
| Hybrid | 672 | 0 | 608 | 90.5 | 1448 |

## REFERENCES

[1] T. Fountain, *Processor Arrays Architecture and Applications*, Academic Press, London, 1987.

[2] S. K. Mitra and J.F. Kaiser, *Handbook for Digital Signal Processing*, Wiley, New York, 1993.

[3] R. W. Hockney and C. R. Jesshope, *Parallel Computers*, Adam Hilger, Bristol, England, 1981.

# 13

# Arithmetic Formats

## 13.0 INTRODUCTION

After the hardware architecture selection is made, the exact chip can only be chosen by deciding what arithmetic format will best meet the specification. The primary effect of the format choice is in the accuracy of the results. Three arithmetic formats are used for computing FFTs:

- Fixed-point, which uses integer arithmetic
- Floating-point, which has the binary point in a fixed place and an exponent for each number
- Block-floating-point, which is fixed-point arithmetic with one exponent for all the data

Prior to the development of DSP chips, the choice of fixed-point arithmetic resulted in faster and smaller hardware architectures than floating-point or block-floating-point arithmetic. However, the opposite is generally true today, as can be seen in the Comparison Matrices of Chapter 14.

## 13.1 THREE PERFORMANCE MEASURES

Since the primary effect of choosing the arithmetic format is the accuracy of the results, the performance measures here are those that quantify the computational accuracy of FFT algorithms.

### 13.1.1 Dynamic Range

Dynamic range is the ratio of the largest-magnitude number to the smallest-magnitude number that can be represented in an arithmetic format. Arithmetic formats where the smallest-magnitude number is $10^{-m}$ and the largest-magnitude number is $10^{+p}$ have a dynamic range of $10^{+p} \div 10^{-m} = 10^{p+m}$, regardless of what $p$ and $m$ are. For example, if $m = 0$ and $p = 16$, the dynamic range is $10^{16}$. If $m = 8$ and $p = 8$, the dynamic range is still $10^{16}$, even though the numbers that can be represented are quite different.

### 13.1.2 Arithmetic Accuracy

Arithmetic accuracy is the precision with which an arithmetic format can represent numbers. In the example in Section 13.1.1 if $m = 0$, the smallest numbers that can be represented are integers, because $10^0 = 1$. The arithmetic accuracy is then 0.5 because it is the largest error that can occur by rounding off a number to the nearest integer. If $m = 8$, then the smallest numbers that can be represented are $10^{-8}$, much smaller than an integer. The arithmetic accuracy is then $0.5 * 10^{-8}$ because it is the largest error that can occur by rounding off a number to the nearest $10^{-8}$. The arithmetic accuracies of these two examples are very different, but their dynamic ranges are the same.

### 13.1.3 Quantization Noise Escalation

Quantization noise is the error in digital computations caused by the need for digital computers to round off numbers. These errors are caused by two effects. First, when analog data is digitized, to allow it to enter a digital computer, digital numbers are assigned to the continuous analog voltages. Since there are only a finite number of possible digital numbers, each analog data sample is represented by the closest digital number to its analog voltage. The result is that the digital signal is the real analog signal minus an error signal. This error signal is called quantization noise. Since the FFT is a linear function (Section 2.2.3), its output is the FFT of the actual analog signal minus the FFT of the error signal.

The second type of quantization error is round-off in the digital computer to control the number of bits used to represent a number. For example, when two 16-bit numbers are multiplied, the result has 32 bits. To represent this output as a 16-bit result, the bottom 16 bits must be removed. Usually this is accomplished by rounding the 32-bit number to the closest 16-bit number. The result is a quantization noise error. Each of these errors is processed by the remaining step in the FFT algorithm and appears as errors in the amplitude of the output frequency components.

Quantization noise is difficult to describe theoretically because it is a nonlinear process. Simulation studies have shown rules-of-thumb for how quantization noise increases (escalates) as the FFT algorithm increases in transform length by factors of two. This escalation factor is presented for each arithmetic format.

## 13.2 THREE ARITHMETIC FORMATS

This section explains each of the three arithmetic formats in terms of the three performance measures. While the Comparison Matrix provides the first level of decision between arithmetic formats, there are often several choices within the format. The explanation here can be used to further refine the arithmetic format decision to specific bit lengths. For example, 16-, 20-, and 24-bit fixed-point programmable DSP chips are commercially available and described in Chapter 14.

### 13.2.1 Fixed-Point

Fixed-point [1] numbers are like working with integers. The format has a specific number of bits, say 16, to represent the numbers, and the binary point (comparable to the decimal point for base 10 numbers) is located at a fixed position among the bits. It might be to the right of all the bits. In this case all of the numbers are represented as integers. It might be to the left of all the bits. In this case all the numbers are less than 1, (i.e., fractions).

The other feature of fixed-point arithmetic formats is that one of the bits is used to represent the sign of the numerical value. Generally, the sign bit is the most significant bit with 0 representing positive numbers and 1 representing negative numbers. For an $n$-bit format where all of the numbers are represented as fractions, the binary point is between the sign bit and the other $n - 1$ bits. All of the fixed-point DSP chips in Chapter 14 have a multiplier-accumulator block diagram similar to that in Figure 13-1 to implement fixed-point arithmetic.

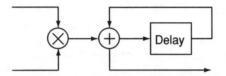

**Figure 13-1**    Fixed-point arithmetic multiplier-accumulator block diagram.

**Dynamic Range.**    The dynamic range of a fixed-point format is independent of the location of the binary point. It is controlled completely by the number of bits. For an $n$-bit fixed-point format, $(n - 1)$ bits are used to provide dynamic range. With the binary point to the right of all $n$ bits, the smallest number is $2^0 = 1$ and the largest is $2^{(n-1)} - 1$ (1's in all $(n - 1)$ bits). The dynamic range is the ratio of these two numbers, which is $2^{(n-1)} - 1$. Moving the binary point anywhere from the right to the sign bit only changes the numbers that can be represented. The ratio of the largest to smallest number does not change. Therefore, once the dynamic range of the input data and the FFT computations is determined to be $D$, it is easy to compute the number of bits required of a fixed-point format as:

$$n = 1 + \log_2[D + 1] \qquad (13-1)$$

**Arithmetic Accuracy.**    The binary point in a fixed-point format controls its arithmetic accuracy. If the binary point is all the way to the right, numbers are all represented as integers. Therefore, the numbers are only accurate to $1/2$. If the binary point in an $n$-bit format is just to the right of the sign bit, then there are $(n - 1)$ fractional bits. This makes the largest fractional bit $2^{-1}$ and the smallest fractional bit $2^{-(n-1)}$, which translates into numbers being represented to an accuracy of $2^{-n}$. For example, in a 16-bit format with the binary point just to the right of the sign bit, the least significant bit is $2^{-15}$, which means numbers are accurate to $2^{-16}$. Therefore, the location of the binary point depends on the required accuracy of the computations.

**Quantization Noise Escalation.** Fixed-point quantization noise is a nonlinear phenomena that depends on the data and the sequence of computations. Analysis of quantization noise for power-of-two FFTs has determined a rule-of-thumb for growth of the noise relative to the signal as a function of the transform length of roughly 1/2 bit per power-of-two [1]. For example, a 1024-point FFT has twice the quantization noise, relative to the signal level, as a 256-point FFT has. The actual levels depend on the signal being analyzed.

The drawback of the fixed-point format is that this quantization noise is relatively independent of the size of the frequency component. Therefore, the signal-to-noise level for strong frequency components is large and for small-frequency components is small. This sometimes causes small-frequency components to be masked by the quantization noise.

Quantization noise for fixed-point FFTs has also been analyzed for the Winograd [2] algorithm. The growth trend is roughly the same as for power-of-two algorithms, and the actual amount of quantization noise is slightly larger than for power-of-two algorithms.

## 13.2.2 Floating-Point

Floating-point [3] numbers are like performing computations in scientific notation. The allotted digits that represent each number are divided between the exponent and the mantissa of the number. In a decimal floating-point format, numbers such as 536 are represented as $5.36 * 10^2$. In a binary floating-point format, 536 would be represented based on decomposing it by powers-of-two. Namely, $536 = 512 + 16 + 8$. Just as for decimal scientific notation, this number can be written as 1000011000, or normalized as $1.000011000 \times 2^8$. Therefore, a binary floating-point number has a certain number of digits to represent the mantissa (1.000011000 in the example) and to represent the exponent ($8 = 01000$ in the example). Notice that to represent numbers with magnitudes less than 1, the exponent is negative. In those cases one of the bits in the exponent must be used as a sign bit. Figure 13-2 is a functional block diagram for floating-point addition, and Figure 13-3 is a functional block diagram for floating-point multiplication, as they are typically implemented by the floating-point DSP chips in Chapter 14.

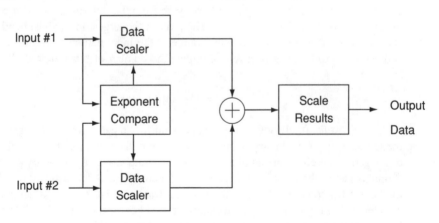

**Figure 13-2** Floating-point addition block diagram.

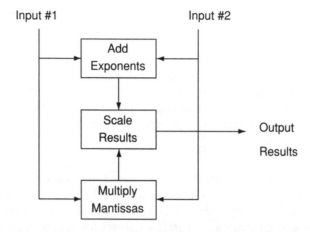

**Figure 13-3**   Floating-point multiplication block diagram.

   If the sign bit is inserted to the left of the mantissa, then the bits to the left of the binary point are always 01 for positive numbers because the binary point is always set after the first nonzero digit. Similarly for negative numbers, the digits to the left of the binary point are always 10. Therefore, there is no need to have two digits to the left of the binary point. The sign bit implies the next bit. This allows an extra bit in the mantissa to be used for representing fractional numbers. For 32-bit floating-point numbers, the IEEE has defined a standard, called IEEE-754, that allocates the lowest 23 bits to mantissa, the next 8 bits to exponent, and the most significant bit to the sign of the number.

   **Dynamic Range.**   The dynamic range of a floating-point arithmetic format is controlled by the number of bits allocated for the exponent. Suppose that "$e$ bits" are allocated for the exponent and one of these is a sign bit. If $e = 8$, then the exponent covers numbers from roughly $2^{127}$ to $2^{-128}$. This is a dynamic range of roughly $2^{255} = 5.79 * 10^{76}$. Therefore, a very small number of bits allocated to the exponent provides huge amounts of dynamic range.

   **Arithmetic Accuracy.**   Arithmetic accuracy is variable for floating-point numbers since the mantissa bits are multiplied by the exponent. This becomes important when analyzing signals where there is a significant difference between the signal strengths of the various frequencies. As the FFT algorithm progresses from stage to stage, it is collecting the information associated with each frequency into smaller and smaller numbers of intermediate data values. Since floating-point arithmetic adjusts numbers at each step to keep the most significant bit of the data in the most significant bit of the mantissa, the small numbers associated with noise and small-frequency components continue to have the accuracy of the full set of mantissa bits. The result is that each frequency component has the accuracy of the mantissa, regardless of the size of the signal. This is in contrast to fixed-point arithmetic, where the largest frequency component controls the most significant bit and does not allow the smaller frequency components the full advantage of all the fixed-point bits.

   **Quantization Noise Escalation.**   Floating-point quantization noise is a nonlinear phenomenon that depends on the data and the sequence of computations. Analysis of quantization noise for power-of-two FFTs [3] has determined a rule-of-thumb for growth of the noise relative to the signal as a function of the transform length of roughly $\log_2(N)$ bits for an $N$-point power-of-two FFT. For example, a 1024-point FFT has $10/8 = 1.25$ the amount of quantization noise, relative to the signal level, than does a 256-point FFT. The actual levels depend on the signal being analyzed and are controlled by the number of bits in the mantissa: the larger the number of mantissa bits, the smaller the quantization noise level.

Quantization noise for floating-point FFTs has also been analyzed for the prime factor [4] algorithm. The growth trend is roughly the same, and the actual amount of quantization noise is slightly larger than for power-of-two algorithms.

### 13.2.3 Block-Floating-Point

Block-floating-point [5] numbers were developed to provide a compromise between the accuracy of fixed-point numbers and the dynamic range of floating-point numbers, without the full complexity or speed penalty associated with full complex floating-point arithmetic computations. Figure 13-4 shows the generic functions required for block-floating-point arithmetic. The only current DSP chips using block-floating-point arithmetic are dedicated to computing FFTs or linear filtering and pattern matching in the frequency domain (Chapter 6). Current block-floating-point DSP chips are 5–10 times faster for FFTs than fixed or floating-point chips because they are dedicated to computing FFTs (see Section 14.7).

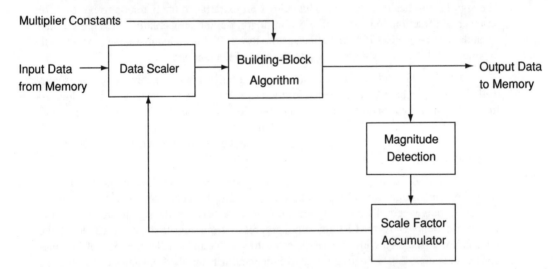

**Figure 13-4** Block-floating-point arithmetic block diagram.

The arithmetic in each building block of the FFT algorithm is performed as fixed-point arithmetic. However, from stage to stage, the intermediate answers are evaluated to ensure that the full dynamic range of the fixed-point numbers is being utilized. If not, all of the intermediate values are scaled enough so that the largest value uses roughly half of the full dynamic range. Then the next stage of computations is performed and the results reevaluated. The processor keeps track of the net scaling that has occurred from stage to stage as an exponent that effectively increases the dynamic range of the processor. The scaling only uses half the dynamic range because the next stage of a power-of-two FFT algorithm will have a gain of 2 for sine-wave inputs. This keeps the fixed-point computation from overflowing.

**Dynamic Range.**    The dynamic range of a block-floating-point FFT processor is controlled by the number of bits allocated to keeping track of the shifts between stages. Suppose that $e$ bits are allocated for the exponent and one of these is a sign bit. If $e = 8$, then the exponent covers numbers from roughly $2^{127}$ to $2^{-128}$. This is a dynamic range of roughly $2^{255} = 5.79 * 10^{76}$. Therefore, a very small number of bits allocated to the exponent provides huge amounts of dynamic range.

**Arithmetic Accuracy.**    The arithmetic accuracy of a block-floating-point format is between that for fixed- and floating-point formats. It has an advantage over fixed-point formats because the scaling between stages keeps more of the bits active in the computations for any input signal. However, the exponent is not changed for each intermediate value, only on the block of values out of each computational stage. Therefore, for small-frequency components in noise it does not keep as many of the bits active as a floating-point format does.

A comparison between block-floating-point and floating-point is data dependent. The only consistent feature is that block-floating-point arithmetic accuracy degrades at roughly the same rate as floating-point arithmetic accuracy as the length of the FFT increases.

**Quantization Noise Escalation.**    Block-floating-point quantization noise effects have better characteristics than fixed-point and worse ones than floating-point for the same reasons as arithmetic accuracy does. A direct comparison between block-floating-point and floating-point is data dependent. The only consistent feature is that block-floating-point arithmetic accuracy degrades at roughly the same rate as floating-point arithmetic accuracy as the length of the FFT increases.

## 13.3  ARITHMETIC FORMAT COMPARISON MATRIX

**Table 13-1**    Arithmetic Format Comparison Matrix

| Arithmetic format | Dynamic range | Arithmetic accuracy | Quantization noise escalation |
|---|---|---|---|
| Fixed-point | $2^{n-1} - 1$ | $0.5*(\text{LSB})$ | Add 0.5 bit |
| Floating-point | $2^p$ | $0.5*(\text{mantissa LSB})$ | Multiply by $\log_2(2 * N) / \log_2(N)$ |
| Block-floating-point | $2^p$ | $0.5*(\text{mantissa LSB})$ | Between fixed and floating point |

*Key to Variables*

$n$ = number of bits in a fixed-point arithmetic format

LSB = numerical value of least significant bit of fixed-point arithmetic format

$p = 2^e$, where $e$ is number of bits used to represent the exponent

Mantissa LSB = numerical value of least significant bit of floating-point mantissa

$N$ = number of points in FFT

## 13.4 CONCLUSIONS

An application usually has a specification for dynamic range and/or arithmetic accuracy. This chapter shows how to determine which arithmetic format best meets the product specification. If a format cannot meet the specifications, the chips in the next chapter that use that format are automatically eliminated from consideration. This is usually the first decision in selecting a chip.

## REFERENCES

[1] P. D. Welch, "A Fixed-Point Fast Fourier Transform Error Analysis," *IEEE Transactions on Audio and Electroacoustics*, Vol. AU-17, pp. 151–157 (1969).

[2] R. W. Patterson and J. H. McClellan, "Fixed-Point Error Analysis Winograd Fourier Transform Algorithms," *IEEE Transactions on Acoustics, Speech, and Signal Processing*, Vol. ASSP-26, No. 4, pp. 447–455 (1978).

[3] C. J. Weinstein, "Roundoff Noise in Floating Point Fast Fourier Transform Computation," *IEEE Transactions on Audio and Electroacoustics*, Vol. AU-17, pp. 209–215 (1969).

[4] D. C. Munson, Jr. and B. Liu, "Floating Point Roundoff Error in the Prime Factor FFT," *IEEE Transactions on Acoustics, Speech, and Signal Processing*, Vol. ASSP-29, No. 4, pp. 877–882 (1981).

[5] A. V. Oppenheim and C. J. Weinstein, "Effects of Finite Register Length in Digital Filtering and the Fast Fourier Transform," *Proceedings of IEEE*, Vol. 60, No. 8, pp. 957–976 (1972).

# 14

# Chips

## 14.0 INTRODUCTION

This chapter gives an objective description of commonly available DSP chips for executing FFT algorithms. A unique feature is the "generic" DSP chip block diagram, to which all the commercial DSP chips are standardized and compared, to simplify understanding their differences. Making the decision about which chip to use depends on the arithmetic format, algorithm and data mapping process (Chapter 12), and the architecture's efficiency at performing that algorithm. FFT code can be written for any programmable processor chip; however, Harvard architectures are specifically designed to execute FFTs efficiently and thus are the only type used in this chapter.

Programmable DSP chips fall into four categories:

- General purpose, both fixed-point and floating-point
- Special purpose
- Application-specific integrated circuits (ASICs)
- Multiple processors on a single chip

The most popular category is general-purpose programmable chips. These chips are designed to efficiently execute FFT and FIR filter algorithms. However, they also have enough general-purpose instructions to be used in a variety of non-DSP functions, particularly when the functions can utilize the on-chip multipliers. Motor controllers, modems, and matrix arithmetic are good examples of these more general-purpose applications. The earliest of these chips used fixed-point arithmetic because the more complex floating-point computations and buses required too much integrated circuit area to be practical. More recent generations are available in fixed- and floating-point arithmetic formats (Chapter 13).

The second category is special-purpose programmable chips, designed to implement just FFT algorithms. Their programmability is limited to choosing the transform length or to configuring the chip to perform linear filtering or pattern matching in the frequency domain (Chapter 6). These chips only implement standard power-of-two FFT algorithms. Their advantage is that they perform power-of-two FFT algorithms 5–10 times faster than general-purpose programmable DSP chips. The disadvantage is that they are limited to FFT computations. Block-floating-point arithmetic has been adopted by the manufacturers of these chips because FFT algorithms are particularly well suited for that arithmetic format, and it provides considerably more dynamic range than fixed point without the complexity of floating-point (Chapter 13).

A recent addition to the DSP chip marketplace is application-specific integrated circuits (ASICs), with DSP processors as building blocks. Once a programmable DSP processor is provided as an ASIC building block, the data I/O, control, and synchronization functions can be added to develop efficient DSP applications on a single chip. The front-end design of these chips generally costs more than designing a board with the equivalent functions. However, the resulting product will require less power and board area and often run faster because the I/O from the DSP building blocks to peripheral devices is often inside the chip.

Another new trend in programmable DSP chips is to have multiple processors on a single chip. Choosing one of these chips implies not only understanding the performance of the individual processors but also their interconnection architecture. Each of the two presented in this chapter uses a fixed-point processor. One uses a ring bus with 24-bit fixed-point processors, and the other a crossbar switch to interconnect 16-bit fixed-point processors.

Each chip manufacturer has its own programming languages, development systems, and support libraries. Those development tools can be found in the referenced vendor material. The algorithms in Chapters 8 and 9 have been given in a form that is easily converted into either chip-specific assembly language or high-level languages.

## 14.1 FIVE FFT PERFORMANCE MEASURES

The following performance measures are the keys to characterizing the ability of a programmable DSP chip to efficiently compute FFT algorithms.

### 14.1.1 1024-Point Complex FFT

The 1024-point complex FFT performance measure is the time, in milliseconds, it takes a chip to perform a 1024-point complex FFT. Chip manufacturers often quote this time as a measure of FFT performance.

### 14.1.2 Data I/O Ports

The data I/O ports performance measure is the number of serial and parallel ports that can be used to move data and program instructions in and out of the chip. Serial ports are often used to initially move data into the chip and to move results off of the chip. Parallel ports may also be used for these data I/O functions. For complex input data it takes $2 * N$

input cycles and $2 * N$ output cycles to move data on and off the chip. The parallel ports are also used to move data and program instructions into the chip from off-chip memory. If the data and program fit in the on-chip memory, these parallel port functions are not needed.

### 14.1.3 On-Chip Data Memory Words

The on-chip data memory words performance measure is the total number of words of RAM available on a DSP chip for storing the FFT input, output, and intermediate data values. This is important because it defines how large an FFT can be computed, with all of the data in the on-chip memory. An $N$-point complex FFT requires at least $2 * N$ data memory locations on the chip for the entire algorithm to be performed on-chip. The Comparison Matrices in Chapters 8 and 9 show the data memory required to compute each algorithm, and the Comparison Matrices in this chapter show the data memory available in each chip. All chips in this chapter have temporary registers. If these registers are not being used when they are needed by the algorithms in Chapters 8 and 9, they may be used to reduce the data memory required for intermediate computational results.

### 14.1.4 On-Chip Program Memory Words

The on-chip program memory words performance measure is the total number of words of memory available on a DSP chip for the FFT program. This is important because it defines how large the FFT program can be without using off-chip program memory. When off-chip program memory is required, it reduces the efficiency of the chip because accessing instructions from off-chip memory is usually slower than accessing them from on-chip memory.

### 14.1.5 Number of Address Generators

Address generators are used to compute where to get the data for the next computation and where to store the results of the present computation (the Memory Map), so that the arithmetic units can spend all of their time computing the Algorithm Steps. There is usually one address generator for each on-chip data memory block. The address generators that are capable of stepping multiple, as well as single, address locations can be used by all of the FFT algorithms given in Chapters 8 and 9.

## 14.2 GENERIC PROGRAMMABLE DSP CHIP

This section describes the function that each block in Figure 14-1 performs in computing FFTs. This "generic" block diagram of a programmable DSP chip is a unique feature of the book. All the vendor block diagrams have been standardized to this generic one to make it easy to compare them and to see where and how they differ. The following methods are used to identify how a specific chip varies from the generic diagram: bold lines indicate where a new connection exists; double bold lines indicate where one or more buses are added to an existing one; dotted lines show where a connection does not exist; shaded blocks are modified functions; diagonal shaded blocks are new functions; and dotted line blocks are ones that do not exist. Differences that do not affect FFT performance are not covered.

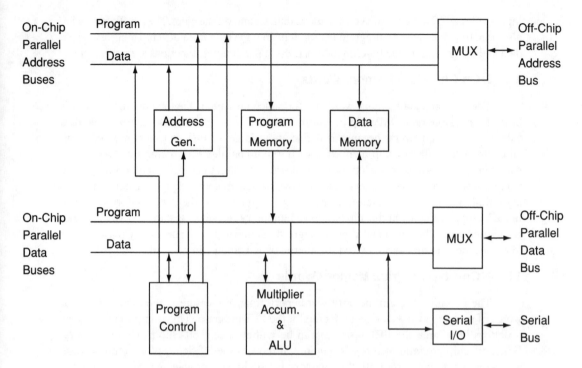

**Figure 14-1** Generic programmable DSP chip block diagram.

### 14.2.1 Block Diagram

Figure 14-1 is a generic block diagram of the Harvard architecture used for programmable DSP chips. These chips are complex devices designed to accomplish a variety of computationally intensive tasks. All of the chips in this chapter have temporary registers. If these registers are not being used when they are needed by the algorithms in Chapters 8 and 9, they may be used to reduce the data memory required for intermediate computational results.

### 14.2.2 On-Chip Data Memory

The role of on-chip data memory was explained in Section 14.1.3. The only amplification to that description is that weighting function coefficients and FFT multiplier coefficients may also be stored in data memory. Since weighting function coefficients are symmetric about the center data sample as described in Chapter 4, $N/2$ (for $N$ even) and $(N + 1)/2$ (for $N$ odd) data memory locations are required to store them. The number of FFT multiplier coefficients varies widely with the FFT algorithm. The largest number of coefficients is for radix-2 mixed-radix algorithms, and the smallest number is for the Winograd algorithm. The Comparison Matrices at the end of Chapters 8 and 9 list the number of memory locations for each algorithm's constants (coefficients).

Some DSP chips have one bank of data memory, and others have two. The advantage of two banks is that one is used for multiplier constants and weighting functions and the other for data. Then, at each multiplication step in the algorithm, both inputs to the multiplier (data value and multiplier constant) can be accessed from memory in one clock cycle rather than sequentially addressing them in one data memory bank.

### 14.2.3 On-Chip Program Memory

The role of on-chip program memory was explained in Section 14.1.4. The algorithms that require the least amount of program memory are the ones with simple computational building blocks and the simplest memory maps. The power-of-primes algorithms from Chapter 9 fit this description if the multiplier coefficients are stored in data memory. If these coefficients are stored in program memory, then the prime factor algorithms can result in the smallest program memory because they only require a few multiplier coefficients and are also computed with simple building blocks. The exact length of program memory can only be determined by writing the code.

### 14.2.4 On-Chip Data Buses

All of the DSP chips in this chapter have at least one on-chip bus dedicated to data movement. Some chips have two data buses, each connected to a data memory. For FFT algorithms these dual buses make it convenient to store FFT or weighting function constants in one memory and data in the second. FFT algorithms that are structured for the maximum use of the multiply-accumulate function have an advantage on the multiple-data-bus architectures because both multiplier and multiplicand can be pulled from data memory in one instruction cycle. The SWIFT, Singleton, and PTL algorithms from Chapters 8 and 9 are the best examples of multiply-accumulate-intensive FFT algorithms.

### 14.2.5 Off-Chip Data Bus

The purpose of the off-chip data bus is to access data blocks that are too large to store on-chip. Because of pin limitations, there is generally only one off-chip data bus. There are exceptions, and they are explained under the appropriate chip family. Ideally, the time required to access off-chip data memory should be the same as for on-chip memory. However, DSP chip I/O limitations, off-chip data memory speed, or cost factors often result in the off-chip data access time being larger than the access time for internal data. This causes FFT performance to degrade when off-chip data memory is required.

Even if off-chip data memory accesses are at the same speed as internal ones, the chip will be slower executing from off-chip data memory if there are two internal data buses. The reason is that the external data inputs to the multiplier or adder must be accessed one at a time rather than in parallel. This adds clock cycles to the computation, which results in longer FFT execution times.

If off-chip program memory is used, this bus is also used to carry program memory instructions to the chip. This reduces the data I/O rate that can be supported. Accessing externally stored program instructions is generally implemented by moving substantial chunks of program code to the chip's internal program RAM and then executing that code until another set of code is required. The building-block formulation of the FFT algorithms in Chapters 8 and 9 is ideal for this approach because each building block's code can be moved into the chip and executed on the entire data set. Then code for the next building block is moved into the chip and the process repeated. This implies that mixed-radix algorithms with identical small building blocks, power-of-primes, are ideal in this situation. Of these, the power-of-two algorithms are the best because they require the smallest amount of code to be transferred into the chip.

### 14.2.6 On-Chip Address Buses

On-chip address buses have two functions. The first is to provide the address needed to point to the next program memory location. Second, they are used for providing the addresses to data memory to access input and intermediate data values and multiplier constants. Figure 14-1 shows a program address bus and a data address bus. DSP chips have the same number of data buses as they have data memories and the same number of address buses as they have program and data memory. This makes the address buses extensions of data and program memory in terms of their affect on FFT algorithms.

### 14.2.7 Off-Chip Address Bus

For most DSP chips, the off-chip address bus plays a dual role. If data must be stored off-chip, this bus provides the addresses to access the off-chip data for processing and for returning answers to the off-chip data memory. If the FFT program is too large to store in the DSP chip, this bus supplies the address sequence to the off-chip program memory. DSP chip I/O limitations, off-chip data memory speed, or cost factors often result in the off-chip access time being larger than the access time for internal memory. This causes FFT performance to degrade.

However, FFT performance can also degrade when the off-chip memory accesses work at the full internal rates. This happens when there are independent address buses inside the chip for program and data memory. Outside the chip, pin limitations usually result in those buses being multiplexed (MUX) as shown in Figure 14-1. Additionally, if there are multiple internal data address buses, the off-chip address bus is further shared, resulting in additional performance decreases.

### 14.2.8 Address Generators

The building-block architecture of FFT algorithms allows FFT code to be written with building-block subroutines and nested loops. Chapters 8 and 9 show that the input data to these building-block algorithms is not sequential and therefore requires addressing with non-unit-step sizes. Likewise, the required data mapping relabeling explained in Section 9.4 also needs nonsequential addressing of data memory. All of the DSP chips in this book have dedicated hardware to perform some form of these types of addressing. In earlier generations it was preprogrammed to provide the reverse binary sequence required for power-of-two FFTs. In more recent generations the address generators are capable of arbitrary step size addressing as well as reverse binary operations for power-of-two FFTs.

Figure 14-2 is a generic block diagram of an address generator. Specific chip families have added bells and whistles to enhance their address generators for specific applications. The most important feature for FFT computations is the ability to change the data memory address in arbitrary step sizes. This is controlled by the register connected to the address increment control in Figure 14-2. If the address generator can perform that function, any of the algorithms in Chapters 8 and 9 can be implemented efficiently. For example, for the 16-point radix-4 FFT in Section 9.7.5, the sequence of data input addresses is shown next to the algorithm steps for the first stage of additions and subtractions. In the second and third columns are the initial address and address increment to accomplish this addressing. The fourth column lists the data memory addressing sequence for each group of input data values that resulted from the inputs to the address generator.

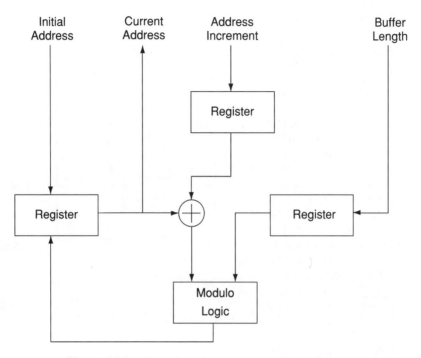

**Figure 14-2**    Generic address generator block diagram.

## 14.2.9  Serial I/O Ports

The role of the serial I/O ports was explained in Section 14.1.2. Figure 14-3 is a typical block diagram for a serial I/O interface in a programmable DSP chip. Some chips have one serial port and some have as many as six. These appear to have been originally provided to allow a convenient data interface with inexpensive voice bandwidth A/D and D/A converters for modem applications. However, more recent generations of DSP chips also use them for interchip communications in multiprocessor architectures. The value of this interface is that it requires few pins and reduces the interrupt overhead to the main processing circuitry to one clock cycle per input or output word.

In Figure 14-3, data is input to the receive shift register one bit at a time. Once an entire word is loaded, it is shifted in parallel to the receive buffer used to load it into the main processor. The main processor then uses one instruction cycle to move the data from the receive buffer to its data memory. The receive buffer allows the main processor to load the new data word asynchronously with the reception of the word through the serial port. The reverse sequence of operations is used to output parallel data words through the serial port. For FFT applications the reduction of interrupt overhead to one instruction cycle makes it less likely for the data I/O rate to become the system bottleneck.

**Table 14-1** Address Generator Sequences for the 16-Point Radix-4 FFT Example

| Algorithm steps | Initial address | Address increment | Address sequence |
|---|---|---|---|
| $b_R(0) = a_R(0) + a_R(8)$<br>$b_I(0) = a_I(0) + a_I(8)$<br>$b_R(1) = a_R(0) - a_R(8)$<br>$b_I(1) = a_I(0) - a_I(8)$ | 0 | 8 | 0,8,16,24 |
| $b_R(2) = a_R(4) + a_R(12)$<br>$b_I(2) = a_I(4) + a_I(12)$<br>$b_R(3) = a_R(4) - a_R(12)$<br>$b_I(3) = a_I(4) - a_I(12)$ | 4 | 8 | 4,12,20,28 |
| $b_R(4) = a_R(2) + a_R(10)$<br>$b_I(4) = a_I(2) + a_I(10)$<br>$b_R(5) = a_R(2) - a_R(10)$<br>$b_I(5) = a_I(2) - a_I(10)$ | 2 | 8 | 2,10,18,26 |
| $b_R(6) = a_R(6) + a_R(14)$<br>$b_I(6) = a_I(6) + a_I(14)$<br>$b_R(7) = a_R(6) - a_R(14)$<br>$b_I(7) = a_I(6) - a_I(14)$ | 6 | 8 | 6,14,22,30 |
| $b_R(8) = a_R(1) + a_R(9)$<br>$b_I(8) = a_I(1) + a_I(9)$<br>$b_R(9) = a_R(1) - a_R(9)$<br>$b_I(9) = a_I(1) - a_I(9)$ | 1 | 8 | 1,9,17,25 |
| $b_R(10) = a_R(5) + a_R(13)$<br>$b_I(10) = a_I(5) + a_I(13)$<br>$b_R(11) = a_R(5) - a_R(13)$<br>$b_I(11) = a_I(5) - a_I(13)$ | 5 | 8 | 5,13,21,29 |
| $b_R(12) = a_R(3) + a_R(11)$<br>$b_I(12) = a_I(3) + a_I(11)$<br>$b_R(13) = a_R(3) - a_R(11)$<br>$b_I(13) = a_I(3) - a_I(11)$ | 3 | 8 | 3,11,19,27 |
| $b_R(14) = a_R(7) + a_R(15)$<br>$b_I(14) = a_I(7) + a_I(15)$<br>$b_R(15) = a_R(7) - a_R(15)$<br>$b_I(15) = a_I(7) - a_I(15)$ | 7 | 8 | 7,15,23,31 |

Multiple serial ports also provide a way to interconnect multiple DSP chips into the architectures defined in Chapter 11, without significant overhead. The programmable DSP chips described in this chapter have one, two, four, or six serial ports. Figure 14-4 is an example of how to form a pipeline multiprocessor architecture using two serial ports. Figure 14-5 shows how to form a 2-D array massively parallel architecture using four serial ports. Figure 14-6 shows how to form a 3-D massively parallel multiprocessor architecture using six serial ports. The ports that go to the adjacent layers are labeled. Refer to Chapter 12 for details on the features of each of these architectures for the various FFT algorithms in Chapters 8 and 9.

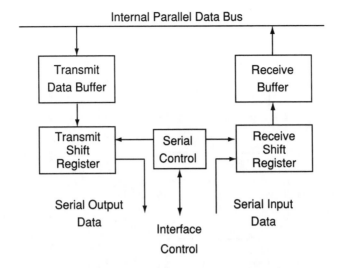

**Figure 14-3**   Generic serial interface block diagram.

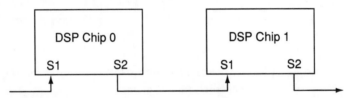

**Figure 14-4**   Two serial ports to form a bus/pipeline architecture.

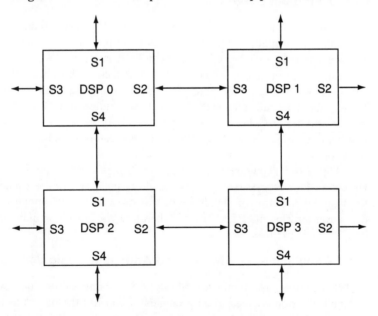

**Figure 14-5**   Four serial ports used to form a two-dimensional massively parallel architecture.

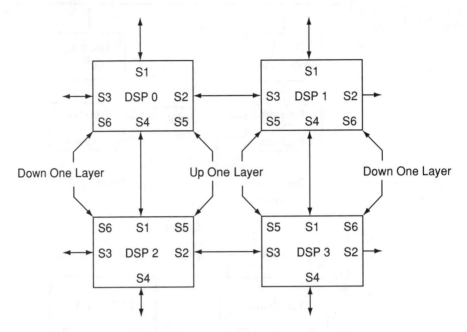

**Figure 14-6**   Six serial ports used to form a three-dimensional massively parallel architecture.

### 14.2.10 Program Control

Zero-overhead looping is a powerful tool for reusing building-block code, written as a subroutine, for multiple input data sets without paying the price to test for the end of a loop. For example, for a radix-2, 1024-point FFT, each time the 2-point FFT is called, only four adds are performed. In a dual-data-bus architecture, this only requires four instruction cycles. However, if the loop counter logic adds as much as one extra instruction cycle per 2-point subroutine call, it has added 25% to the execution time. Therefore, for chips without the zero-overhead looping feature, larger building-block algorithms provide more efficient algorithm performance because the looping overhead is a smaller portion of the total code execution time.

Figure 14-7 shows the overhead looping process for the radix-2, 1024-point example. The end-of-loop process (Line Y) at the end of each access of the 2-point subroutine (Line X+1) can be performed in hardware or software. For a 1024-point FFT each of the 10 stages uses the 2-point FFT 512 times. Therefore, the inner loop in Figure 14-7 is executed $10 * 512 = 5120$ times.

### 14.2.11 Multiplier-Accumulator and Arithmetic Logic Unit

The multiplier-accumulator (MAC) provides single-instruction-cycle multiplication and multiplication coupled to an accumulator and has the basic functional form shown in Figure 14-8. In $n$-bit fixed-point chips, the multiplier inputs are $n$ bits wide and the multiplier output is $2 * n$ bits wide. Multiplier results can be rounded off to $N$ bits and returned to data memory or fed into an accumulator that is at least $2 * n$ bits wide. The

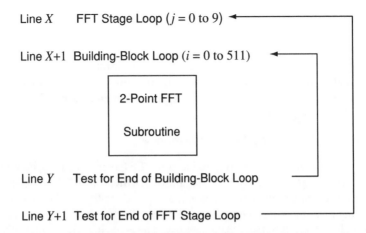

Line $X$    FFT Stage Loop ($j = 0$ to $9$)

Line $X+1$  Building-Block Loop ($i = 0$ to $511$)

2-Point FFT

Subroutine

Line $Y$    Test for End of Building-Block Loop

Line $Y+1$  Test for End of FFT Stage Loop

**Figure 14-7**    End of loop testing process.

accumulator output can also be rounded off to $n$ bits and the results returned to data memory. Several bells and whistles have been added by the individual vendors to optimize the MAC for specific tasks. The most visible one is shifting logic that aligns the binary point for the add and multiply processes. This function is not included in Figure 14-8 because it occurs in different places for different chip families and its location has little effect on the overall computation time for an FFT algorithm.

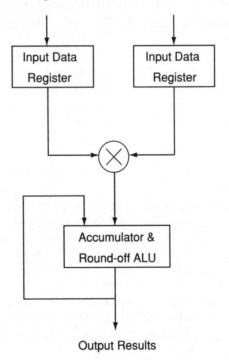

Input Data
Register

Input Data
Register

Accumulator &
Round-off ALU

Output Results

**Figure 14-8**    Generic multiplier-accumulator block diagram.

In $n$-bit floating-point chips, multiplication and addition require additional functions over fixed-point arithmetic. Block diagrams for these functions are presented in Chapters 10 and 13. The fundamental difference is that the multiplier requires an adder for the floating-point exponents, and a shifter is needed to align the mantissa of the floating-point words prior to addition. However, Figure 14-8 still represents the generic functions performed by the floating-point MAC. The details of the implementation have little effect on the performance of the FFT algorithms.

### 14.2.12 Estimating FFT Performance

Chip vendors usually provide some FFT benchmark for how long it takes its chip to perform some power-of-two-length FFT. Often the 1024-point FFT is used. From the given benchmark the performance of any power-of-two FFT length $N$ can be estimated by using one of two techniques, depending on whether the chip can perform the FFT entirely on-chip or needs external data memory. The estimated 1024-point FFT benchmarks in the Comparison Matrices of this chapter are based on the techniques described below.

*Case* 1: *Benchmark and Desired FFT Both Use On-Chip or Off-Chip Data Memory*

In this case, the following equation can be used:

$$N\text{-point FFT time} = (1024\text{-point FFT time})$$
$$*5 * N * \log(N)/[5 * 1024 * \log(1024)] \tag{14-1}$$

For example, to estimate the time it takes to perform a 256-point complex FFT, compute $5 * 256 * \log(256)/[5 * 1024 * \log(1024)] = 0.2$ times the 1024-point FFT time.

*Case* 2: *Benchmark Uses On-Chip Data Memory and the Desired FFT Uses Off-Chip Memory*

The only place Equation 14-1 fails to provide accurate estimates is when the FFT length gets too long for the FFTs to be computed with on-chip data memory. When off-chip data memory is required, the efficiency of the chip is reduced because accessing off-chip memory is slower than accessing on-chip memory. When this occurs, understanding the building-block approach to the FFT algorithm becomes the key to estimating the performance of the chip for the needed FFT length. The steps to estimating the chip's performance are as follows:

*Step* 1: *Divide the FFT Length into Building-block Lengths with Known FFT Performance*

Chapter 9 presents three categories of FFT algorithms. All three use the building-block approach. In each case, if the $N$-point FFT can be factored into $P$-point and $Q$-point building blocks ($N = P * Q$), then the FFT algorithm requires $P$ $Q$-point building-block computations, followed by $Q$ $P$-point building-block computations. For those computations, some algorithms need some complex multiplications. Factor $N$ such that the chips can perform the $P$- and $Q$-point FFTs using only on-chip memory. Further, choose $P$ and $Q$ such that their on-chip performance is known. If it is not known, choose $P$ and $Q$ so that their performance can be calculated by using Equation 14-1.

*Step* 2: *Compute the Time Required to Compute All the $P$- and $Q$-point FFTs*

This is done by computing:

$$\text{FFT Time} = P * (Q\text{-point FFT's time}) + Q * (P\text{-point FFT's time}) \tag{14-2}$$

*Step* 3: *Compute the Time for Moving Data On and Off the Chip*

Assume all data is stored in off-chip data memory. To compute a $P$-point FFT, move $P$ data samples onto the chip, perform the $P$-point FFT, and return the answers to off-chip memory. Since this is done $Q$ times, all of the data is moved onto the chip and the answers back off again once for the $P$-point FFTs and once for the $Q$-point FFTs. Therefore, the data transfer time is:

Data transfer time = (Data word transfer time) $*$ (2 words) $*$ (2 for on and off) $*N$ (14-3)

*Step* 4: *Compute the Time for Complex Multiplies*

DSP chips usually specify the time required to perform a multiply. Determine the number, $X$, of complex multiplies required for the desired algorithm and FFT length. Then compute

$$\text{Complex multiply time} = X* \text{(complex multiply time)} \qquad (14\text{-}4)$$

*Step* 5: *Add All Times that Contribute*

The total FFT performance time estimate is:

Total time estimate = FFT time + data transfer time + complex multiply time   (14-5)

If all of the data can be stored on-chip, the data transfer time is not part of the total time estimate. The effect of this on the chip's FFT performance depends on the data I/O speed of the chip and the speed of the off-chip memory. Table 14-2 illustrates that Equation 14-1 works and also illustrates the performance degradation suffered by using off-chip memory, with two generations of fixed-point DSP chips from Texas Instruments. In moving from 64 to 256 points, the computation time is expected to increase by roughly a factor of $5 * 256 * \log(256)/[5 * 64 * \log(64)] = 5.333$. Similarly, moving from 256 to 1024 points should increase the computation time by roughly a factor of $5 * 1024 * \log(1024)/[5 * 256 * \log(256)] = 5$. The TMS320C5x series follows these ratios closely because this generation of chips has enough on-chip RAM to compute any of these three FFT lengths. The TMS320C2x series follows closely for the transition from 64 to 256 points because it has enough RAM for the 256-point FFT. However, the ratio for moving from 256 to 1024 points is larger than expected because off-chip data memory is required.

**Table 14-2**   On- versus Off-Chip FFT Performance Comparison

| TI chip family | 64-pt clock cycles | 256-pt clock cycles | 1024-pt clock cycles |
|---|---|---|---|
| TMS320C2x | 3088 | 17,602 (5.7 : 1) | 109,755 (6.2 : 1) |
| TMS320C5x | 1515 | 8131 (5.36 : 1) | 41,665 (5.12 : 1) |

## 14.3 PROGRAMMABLE FIXED-POINT CHIP FAMILIES

The first programmable DSP chip to become popular was introduced by Texas Instruments in 1982. This chip, the TMS32010, was a 16-bit fixed-point chip designed primarily for speech processing and data communications applications. Since that time others, such as Analog Devices, Motorola, AT&T, NEC, DSP Semiconductor, SGS-Thomson, Star Semiconductor, Zilog, and Zoran have introduced production fixed-point DSP chips. Traditionally, the biggest market for these chips has been telecommunications applications such as modems and fax. However, today these chips are used for a broad range of applications that require high-speed arithmetic computations and can tolerate the dynamic range constraints of fixed-point arithmetic explained in Chapter 13.

### 14.3.1 Analog Devices ADSP-21xx Family

The ADSP-21xx family is a series of 16-point DSP chips that offers a variety of bells and whistles to meet specific application needs. However, few of these have a dramatic impact on FFT performance. The primary impact is in the data I/O capability for an application. The members of this family are ADSP-2100A, ADSP-2101, ADSP-2103, ADSP-2105, ADSP-2111, ADSP-2115, ADSP-216x, ADSP-2171, ADSP-2175, and ADSP-21msp5xx, where the "x" means that there are several subfamily members of that family member (see Figure 14.9) [1–4].

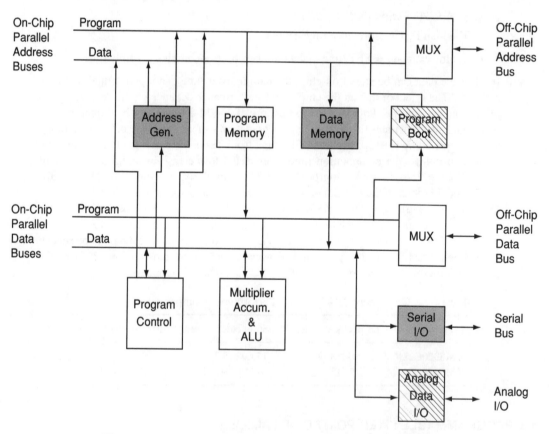

**Figure 14-9**    Analog Devices ADSP-21xx family block diagram.

Serial I/O.    All of this family, except the ADSP-2105, have dual serial ports with hardware companding circuitry. This additional serial port provides the capability to interface these devices into linear bus, pipeline, and ring bus architectures for multiprocessor applications (Section 14.2.9) without having to use the parallel bus that may be addressing off-chip data or program memory.

The companding hardware is an advantage in applications where the FFT is obtaining its data from an A/D converter or sending its results to a D/A converter. If the A/D and D/A converters are connected to networks such as the telephone system, the voltages they convert may be logarithmically compressed by using either the A-law (European standard) or $\mu$-law (U.S. standard). Since the FFT is assuming linear data, the input data must be converted to linear form. This function is called companding. If companding is performed in software, it takes several instruction cycles. If the process takes 10 instruction cycles, the total data I/O time for an $N$-point complex FFT increases from $4 * N$ to at least $10 * 4 * N$ instruction cycles. Since the FFT takes roughly $5 * N * \log_2(N)$ instructions, an FFT becomes I/O limited when $10 * 4 * N > 5 * N * \log_2(N)$. This occurs for $N < 256$ points. The companding hardware removes the need for these 10 cycles and allows the data I/O overhead to return to one cycle per word so that I/O limiting only occurs for 2-point FFTs, based on the inequality.

**Other Data I/O.**    The ADSP-21msp50 and ADSP-21msp51 provide a full voice band analog interface which includes 16-bit Sigma-Delta A/D and D/A converters, antialiasing and antiimaging filters, and automatic gain control (AGC). Voice applications, such as speech recognition, that use FFTs (see the example in Chapter 17) can use this feature to reduce the cost of development and production.

**Data Memory.**    Only the ADSP-2171 and ADSP-2175 have enough on-chip data RAM to perform a 1024-point FFT, and the ADSP-2171 is marginal since it has just 2048 data memory words. It would require all of the weighting function and multiplier constants to be in program memory. Therefore, the 1024-point FFT benchmarks for the other chips in this family already reflect the slowdown incurred by having to store data off-chip. This means that Equation 14-1, the FFT performance estimator, will work for FFT performance above 1024 points but gives answers that are too large for smaller transform lengths. The Programmable Fixed-Point Chips Comparison Matrix (Section 14.4) shows that the ADSP-2171 and ADSP-2175 have significantly better 1024-point FFT computation times than the other devices in this family because of the additional on-chip data memory.

**Address Generators.**    All of the members of this family have dual address generators. This maximizes the ability to address both data and multiplier constants to feed to the MAC unit on each instruction cycle. The flexibility of the address step sizes for these generators also allows them to be easily used to execute non-power-of-two algorithms as well as standard FFTs. Address generator 1 also has bit-reverse logic to accommodate standard power-of-two algorithms.

**Program Boot.**    This is additional logic to allow the on-chip program RAM to be loaded during the power-up phase of the application's operation from a low-speed 24-bit-wide EPROM to lower the cost of the overall application. It also allows multiple programs to be swapped in and out of the chip's on-chip program memory without having to store them in high-speed off-chip program RAM.

### 14.3.2 AT&T DSP16 Family

Unlike other DSP chip manufacturers, AT&T introduced the DSP16 line of fixed-point chips after having a floating-point chip (DSP32) in the market. The most characteristically different feature of this fixed-point family is the instruction cache provided to run inner-loop computations rapidly. The members of this family are DSP16 and DSP16A (see Figure 14-10) [5, 6].

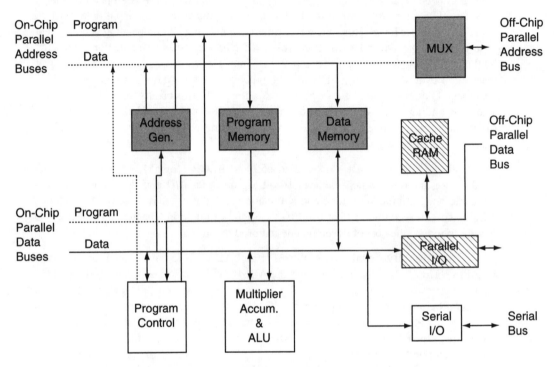

**Figure 14-10**    The AT&T DSP16 family block diagram.

Cache RAM.    The 15 instructions of on-chip cache RAM can execute a set of repetitive operations up to 127 times to increase the throughput and coding efficiency. This is particularly valuable for power-of-prime FFT algorithms where the same building block is used throughout the computations. In particular, the 2-point building block would easily fit into this RAM. The 4-point building block is a series of four 2-point building-block computations, and the 3-point building block uses two complete 2-point building blocks and two partial ones (just the add). Therefore, it may also be possible to efficiently implement 3- and 4-point building blocks with this cache memory.

MUX/Parallel I/O.    The MUX/parallel I/O chip does not use multiplexers (MUX) for interfacing the on-chip address bus to outside the chip because there is only one on-chip address bus. Even though there are two on-chip data buses, they are not interfaced to a single bus outside the chip because there are two off-chip parallel bus interfaces. This additional off-chip bus allows additional freedom in the internal organization of the chip and a way for data to be input to the on-chip data memory while off-chip data memory is being used to provide data to the MAC and ALU to perform computations.

If the FFT is small enough to execute entirely on-chip, then this architecture works best if all data is in the data RAM and all multiplier coefficients are in on-chip program memory ROM. If the FFT must be executed with off-chip memory, storing the data in off-chip memory and the multiplier coefficients in on-chip data RAM is the easiest way to program the algorithm. However, if the off-chip memory is slow, it may be more efficient to load portions of the data from off-chip to on-chip memory through the parallel I/O port and execute the FFT internally, in steps, using multiplier coefficients stored in on-chip program ROM. The manufacturer provides detailed data books to help make those decisions.

**Address Generators.**    Both members of this family have dual address generators. This maximizes the ability to address both data and multiplier constants to feed to the MAC unit on each instruction cycle. The flexibility of the address step sizes for these generators allows them to be easily used to execute non-power-of-two algorithms as well as standard FFTs.

**Program Memory.**    All on-chip program memory in this family is in ROM, and the programming strategy is to use this memory for programs and multiplier coefficients. The architecture does allow off-chip program RAM up to 64K words.

**Data Memory.**    The DSP16 has 512 words and the DSP16A has 2048 words of on-chip RAM. Therefore, the maximum on-chip complex FFT that can be performed by the DSP16 is 256 points and by the DSP16A is 1024 points. This assumes all of the multiplier constants and weighting function constants are stored in program memory. This means that the FFT performance formula will work for FFT performance above 1024 points (256 points for the DSP16A) but gives answers that are too large for smaller transform lengths. The Programmable Fixed-Point Chip Comparison Matrix (Section 14.4) shows that the DSP-16A has significantly better 1024-point FFT computation times than the DSP16 because of this additional internal data RAM.

### 14.3.3  AT&T DSP161x Family

This series of 16-bit fixed-point chips is focused on the digital cellular marketplace. However, they are general-purpose programmable DSP chips that can be used to execute FFT algorithms. In addition to the specific market focus, the primary difference between this family and the DSP16 family is on-chip RAM for programs. The members of this family are DSP1610, DSP1616, DSP1617, and DSP1618 (see Figure 14-11) [7–10].

**Cache RAM.**    The 15 instructions of on-chip cache memory can execute a set of repetitive operations up to 127 times to increase the throughput and coding efficiency. This is particularly valuable for power-of-prime FFT algorithms where the same building block is used throughout the computations. In particular, the 2-point building block would easily fit into this RAM. The 4-point building block is a series of four 2-point building-block computations and the 3-point building block uses two complete 2-point building blocks and two partial ones (just the add). Therefore, it may also be possible to efficiently implement 3- and 4-point building blocks using this cache memory.

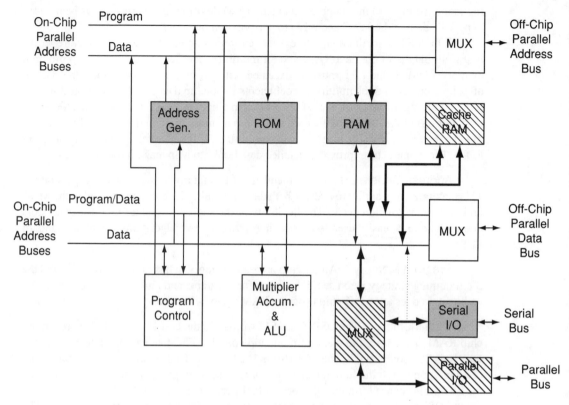

**Figure 14-11**   The AT&T DSP161x family block diagram.

Serial Ports.   All members of this family have dual serial ports. This additional serial port provides the capability to interface these devices into linear bus, pipeline, and ring bus architectures for multiprocessor applications (Section 14.2.9) without having to use the parallel bus that may be addressing off-chip data or program memory.

Parallel I/O/Interface Bus.   In addition to the two on-chip data buses that are interfaced off-chip by using multiplexers, there is an additional parallel interface, just like the one in the DSP16 family. The difference is that it is multiplexed onto a bus that is then interfaced with one of the on-chip data buses.

Data Memory.   All of the devices in this family, except the DSP1618, have at least 2048 words of data RAM with two access ports. Therefore, the 1024-point FFT can be performed on-chip if the weighting function and multiplier coefficients are stored in program memory. The DSP1617 and DSP1618 have 4096 words of dual-ported data RAM, so they can compute up to 2048-point complex FFTs without going off the chip. The DSP1610 has 8192 words of data RAM. It can compute up to 4096-point complex FFTs without going off the chip.

Read-Only Memory (ROM).   All of the devices in this family have on-chip program ROM. The DSP1610 has 512 words, the DSP1616 has 12K words, the DSP1617 has

24K words, and the DSP1618 has 16K words. For high-volume applications, this ROM can be used to store FFT algorithms. Otherwise, the on-chip RAM can be used to store the program. However, storing the program in data RAM reduces the location available for data, which results in a smaller FFT length that is computable with only on-chip memory.

### 14.3.4 Motorola DSP56001 Family

The DSP56001 was the first programmable DSP chip family from Motorola. Its most characteristically different feature is that it is a 24-bit fixed-point processor. The members of this family are DSP56001, DSP56002, DSP56L002, and DSP56004 (see Figure 14-12) [11–13].

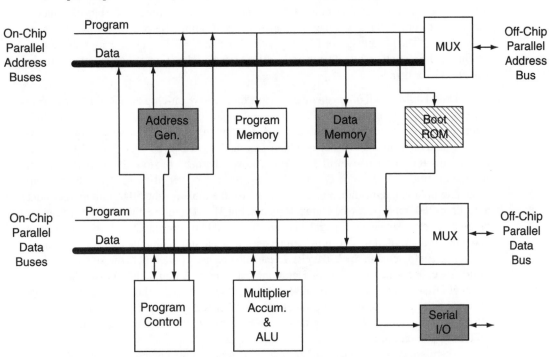

**Figure 14-12**    Motorola DSP56001 family block diagram.

**Serial Ports.**    All members of this family have dual serial ports. This additional serial port provides the capability to interface these devices into linear bus, pipeline, and ring bus architectures for multiprocessor applications (Section 14.2.9) without having to use the parallel bus that may be addressing off-chip data or program memory.

In conjunction with these ports, the X-data memory has a built-in table of A-law and $\mu$-law companding coefficients to simplify the interface with companded data sources. Since the FFT is assuming linear data, the companded input data must be converted to linear form. If companding is performed in software, it takes several instruction cycles. If the process takes 10 instruction cycles, the total data I/O time becomes at least $10 * 4 * N$ instruction cycles. Since the FFT takes roughly $5 * N * \log_2(N)$ instructions, an FFT will be I/O limited when $10 * 4 * N > 5 * N * \log_2(N)$. This occurs for $N < 256$ points. The companding table removes the need for these 10 cycles and allows the data I/O overhead to return to one cycle per word. At one cycle per data I/O word, the device is only I/O limited for 2-point FFTs.

**Data Memory.**    All members of this family have 512 words of data RAM on-chip. Therefore, the largest FFT that can be computed with only on-chip memory is 256 points. Therefore, the performance numbers in the Programmable Fixed-Point Chips Comparison Matrix (Section 14.4) already reflect the penalty paid for having to access off-chip data memory. Further, the data RAM is divided into two 256-word memories called X-data memory and Y-data memory.

The other nonstandard fact about this family is that it is 24-bit fixed point. This allows it to be used for digital compact disc (CD) products that require roughly 20 bits of dynamic range and accuracy. This was the first family of fixed-point DSP processors to offer more than 16 bits. The advantage for FFT algorithms is that it has less quantization noise than 16-bit fixed-point chips by a factor of 24 dB. See the explanation of quantization error in Chapter 13 for details.

**Data ROM.**    All of the members in this family have on-chip data ROM. The X-data memory ROM is programmed with A-law and $\mu$-law companding functions to simplify interfaces with companded data sources such as telephone lines. The Y-data memory ROM is programmed with a full, four-quadrant sine table that can be used for the multiplier coefficients for power-of-two FFTs. This removes the need to store these coefficients in program memory. This table can also be used for non-power-of-two FFTs with the help of an interpolation algorithm. For example, to use the table for the 504-point mixed-radix algorithm, 360° must be divided into 504 pieces, not 512. Therefore, the table entries cannot be used directly. However, for each needed value, the two surrounding phase angle values and a linear interpolation algorithm can be used to accurately compute the correct value.

The coefficients in the Y-data ROM can also be used to compute the sine lobe, Hanning, sine cubed, sine to the fourth, Hamming, Blackman, 3-sample Blackman-Harris, and 4-sample Blackman-Harris weighting functions in Sections 4.2.3 through 4.2.10. This removes the need to store weighting function coefficients if the chip's computational power allows the weighting function coefficients to be computed as needed within the required FFT computation time.

There are two drawbacks to the Y-data memory ROM having the sine table. This table is specifically designed for power-of-two algorithms. Therefore, it does not contain the multiplier constants needed for non-power-of-two algorithms. Further, the table is fixed in the Y-data memory ROM. Therefore, to pull a multiplier coefficient and data value during the same instruction cycle, the data must be in the X-data memory. For radix-2 algorithms this is not a problem because the data can always be partitioned so that the values that require the multiplications are in the X-memory, because only half of the data in the radix-2 building block ever gets multiplied by other than 1. In general, mixed-radix algorithms require $N - 1$ of the $N$-point building-block inputs to be multiplied by a complex number. For full-speed operation this requires that the data must be modified prior to being input to the $N$-point building block to be stored in the X-memory. If that data is stored in the Y-memory, two memory access clock cycles are required to get the data and multiplier constant. This slows FFT performance.

**Address Generators.**    All of the members of this family have dual address generators. This maximizes the ability to address both data and multiplier constants to feed to the MAC unit on each instruction cycle. The flexibility of the address step sizes for these generators also allows them to be easily used to generate non-power-of-two algorithms as well as standard FFTs. Both address generators also have bit-reverse logic to accommodate standard power-of-two algorithms.

**Data Address and Data Buses.**    To accommodate the extra data memories, there is an extra data memory bus and an extra data memory address bus. This provides a simpler way of thinking about programming the devices, because the natural thought process of pulling two data values from data memory can be programmed.

**Boot ROM.**    Boot ROM is additional memory to allow the on-chip program RAM to be loaded during the power-up phase of the application's operation from a low-speed 24-bit-wide EPROM to lower the cost of the overall application. It also allows multiple programs to be swapped in and out of the chip's on-chip program memory without having to store them in high-speed off-chip program RAM.

### 14.3.5 Motorola DSP561xx Family

The DSP561xx family of 16-bit fixed-point chips is based on the 24-bit fixed-point DSP560xx series from Motorola.    The members of this family are DSP56156, DSP56156ROM, DSP56166, and DSP56166ROM (see Figure 14-13) [14–17].

**Serial I/O and A/D-D/A I/O.**    All members of this family have dual serial ports. This additional serial port provides the capability to interface these devices into linear bus, pipeline, and ring bus architectures for multiprocessor applications (Section 14.2.9) without having to use the parallel bus that may be addressing off-chip data or program memory.

All members of the family also provide 14-bit Sigma Delta A/D and D/A conversion to simplify the application of these devices to telecommunications and digital cellular applications. Example 3 of Chapter 17 uses these on-chip A/D and D/A converters to simplify doing the pitch detection portion of speech recognition algorithms.

**Data Memory.**    Both DSP56156 devices have 2048 words of data RAM, and both DSP56166 devices have 4096 words of data RAM. Therefore, the 1024-point FFT can be performed on-chip if the weighting function and multiplier coefficients are stored off-chip or in program memory for the DSP56166 devices and even without that constraint for the DSP56166 devices.

**Buses and Multiplexers.**    This family has dual data address buses and an additional data bus for moving the serial and analog I/O port data on and off the chip. The result is that the multiplexers for combining on-chip buses to one off-chip bus are both 3:1 rather than the more standard 2:1 found in other chip families. The additional data bus enhances the chip's capability to input data in parallel while performing computations. This improves its FFT performance.

**Address Generators.**    Unlike the DSP5600x family, this family only has one address generator. However, its logic is fast enough to compute two addresses per instruction

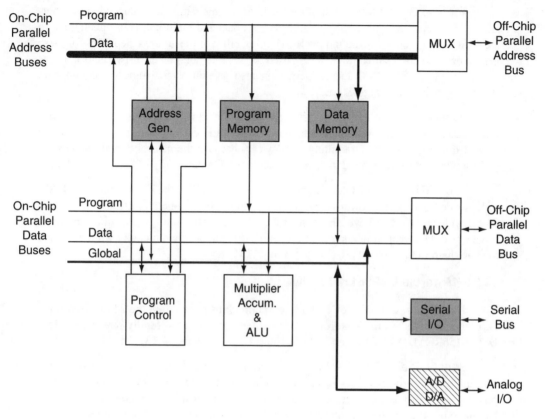

**Figure 14-13** Motorola DSP56156/166 family block diagram.

cycle. Thus, it functions like two address generators and still provides the FFT performance advantages described for dual-generator architectures.

**Program Memory.** A set of addresses in program memory is used to allow the on-chip program RAM to be loaded during the power-up phase of the application's operation from a low-speed 24-bit-wide EPROM to lower the cost of the overall application. It also allows multiple programs to be swapped in and out of the chip's on-chip program memory without having to store them in high-speed off-chip program RAM. Both the DSP56156 and DSP56166 have 2048 additional words of on-chip program ROM. The DSP56156ROM and DSP56166ROM devices have 12K and 8K of on-chip program ROM, respectively.

### 14.3.6 NEC μPD77xxx Family

The distinguishing feature of this family is that it only has one on-chip bus. However, the on-chip circuitry runs fast enough to move two data words to the MAC and the next instruction cycle to its register during an instruction cycle. The members of this family are μPD77C20A, μPD7720A, μPD77P20, μPD77C25, and μPD77P25 (see Figure 14-14) [18, 19].

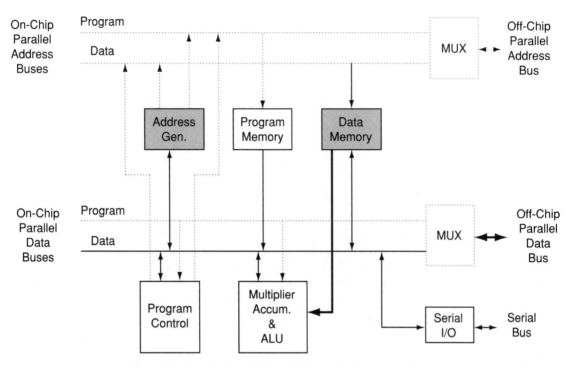

**Figure 14-14**    NEC $\mu$PD77xxx family block diagram.

**Buses and Multiplexers.**    The distinguishing feature of the $\mu$PD77xxx is its single bus to carry data and program words.  However, this does not slow down the processor's ability to perform single-cycle multiply-accumulate operations because it can access two data words for the multiplier and the next instruction, all during one instruction cycle. Notice in the block diagram that there is an independent path from data memory to the multiplier that is used for one of the data words per instruction cycle.

**Address Generator.**    The simplicity of this device's address generator is perhaps the biggest drawback for FFT computations.  The address generator is a program counter with four registers to hold return addresses for up to four levels of nested looping.  This makes this architecture inefficient for data addressing that has non-unit-step sizes.  The offset addressing required for FFT algorithms is accommodated using values programmed as part of the instruction ROM.

**Data Memory.**    This device has two on-chip data memories.  One is a ROM ($1024 \times 16$ for the $\mu$PD77C25 and $512 \times 23$ for the $\mu$PD77C20) for storing multiplier and weighting function coefficients.  The other is a RAM ($256 \times 16$ for the $\mu$PD77C25 and $128 \times 16$ for the $\mu$PD77C20).  This means that the best case is being able to compute 128-point FFTs ($\mu$PD77C25) and 64-point FFTs ($\mu$PD77C20) using only on-chip memory.  Therefore, the 1024-point performance numbers in the Programmable Fixed-Point Chips Comparison Matrix (Section 14.4) assume off-chip data memory.

For FFTs larger than 128 points, FFT performance will lose efficiency because the off-chip data interface is only 8 bits wide. Therefore, two accesses are required to move one 16-bit word into and out of the chip. However, since the 16-bit word is stored in a buffer register prior to becoming two 8-bit words, it only takes one instruction cycle away from the processor to move data onto and off the chip. Furthermore, the buffer register is controlled from off-chip timing signals. Therefore, if the off-chip logic can operate at twice the on-chip instruction speed, the 8-bit I/O inefficiency is removed. Read the detailed timing information in the manufacturer's data book to determine the effect of the 8-bit interface.

The 8-bit interface is used because the family was designed to interface to 8-bit microprocessor hosts. The 8-bit interface also slows the data I/O before and after the FFT algorithm. However, the degree to which this affects overall FFT performance depends on the speed of the off-chip data transfer, just as it was for off-chip data memory accesses during the FFT computations.

### 14.3.7 NEC μPD7701x Family

The 16-bit fixed-point NEC μPD7701x family was developed for the digital cellular and modem/fax telecommunications markets. However, the Programmable Fixed-Point Chips Comparison Matrix (Section 14.4) shows it has good performance for FFT computations. The members of this family are the μPD77016 and μPD77017 (see Figure 14-15) [20].

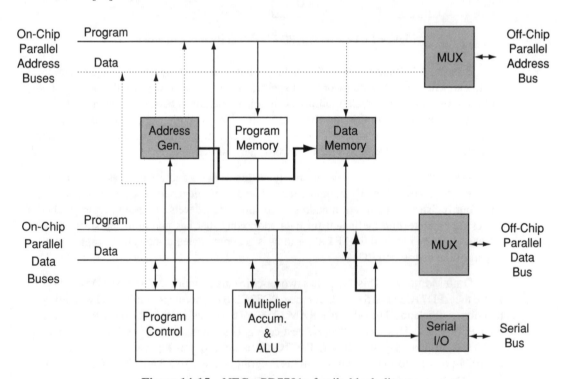

**Figure 14-15** NEC μPD7701x family block diagram.

**Serial Ports.**    Both the $\mu$PD77016 and $\mu$PD77017 have dual serial ports. This additional serial port provides the capability to interface these devices into linear bus, pipeline, and ring bus architectures for multiprocessor applications (Section 14.2.9) without having to use the parallel bus that may be addressing off-chip data or program memory.

**Address Generators.**    Both devices have dual address generators that are very similar to Figure 14-2. However, they are directly connected to the two data RAM blocks rather than to dual address buses because there is only one bus used for carrying address information, and it carries program memory addresses and other control data. The flexibility of these address generators makes them useful for computing all of the algorithms in Chapters 8 and 9. For the standard power-of-two algorithms, both address generators have hardware for performing bit-reversed addressing arithmetic.

**Buses and Multiplexers.**    Both of the on-chip data memory buses are also available outside the chip. This eliminates the need for the multiplexers shown in the block diagram. Furthermore, the reduced number of on-chip buses (two for data and one for program addressing) and multiple-address generators results in the address generators providing their output directly to their respective memories.

**Data Memory.**    Both devices have two data memories. Each data memory has 2048 sixteen-bit words of RAM. The $\mu$PD77017 also has 4096 words of data ROM in each data memory. Therefore, both devices can compute a 1024-point FFT on-chip if all of the multiplier and weighting function coefficients are stored in program memory. Even though the on-chip data buses are not multiplexed to the outside of the chip, going off-chip for data does slow down the computations. This is because the two off-chip data buses must be used for both data and addressing. Therefore, only one data memory value can be accessed during an instruction cycle, not two as can happen when the data is internal to the chip.

### 14.3.8 NEC $\mu$PD77220 Family

The key distinguishing characteristics of this family are that it uses 24-bit fixed-point arithmetic rather than the 16 bits used by most fixed-point DSP chip families, and it has a single main bus. The members of this family are $\mu$PD77220 and $\mu$PD77P220 (see Figure 14-16) [18, 21]. This family is very similar to the $\mu$PD77230 family of 32-bit floating-point devices described in Section 14.5.6.

**Buses and Multiplexers.**    The reduction to one main bus removes the need for multiplexers on the data and address buses to go off-chip. Further, this bus reduction forces several direct connections between functional blocks. Each of these is described below. These connections offset the degradation in FFT performance associated with only having one main bus.

**Data Memory.**    This device has two 256-word data RAM blocks and one 1024-word data ROM for storing multiplier constants and weighting function coefficients. Externally, the device supports a 12-bit address word which corresponds to addressing 4096 data words. This limits this device to performing 2048-point FFTs, even using off-chip memory. Using on-chip memory with real and imaginary components in respective 256-word blocks of data memory provides the capability to perform 256-point complex FFTs.

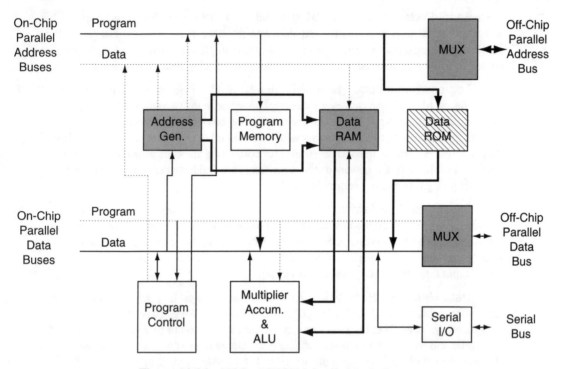

**Figure 14-16** NEC $\mu$PD77220 family block diagram.

Data memory does not use the main bus to transfer data to the multiplier. Each data RAM has its own direct path to the multiplier. However, the results from the multiplier or accumulator are stored in data RAM using the main bus.

**Address Generators.** This device has an address generator for each of the data RAMs to avoid having to use the main bus. These generators are simple base address plus offset calculators that require the offset to be programmed into the instructions for nonunit values. Therefore, they are not ideally suited for computing non-power-of-two FFT algorithms.

### 14.3.9 Texas Instruments TMS320C1x Family

The TMS320C1x is TI's first family of CMOS programmable DSP chips and is still used for low-cost applications. It is a follow-on to the NMOS TMS32010 series introduced in 1982. The members of this family are TMS320C10, TMS320C14, TMS320P14, TMS320E14, TMS320C15, TMS320P15, TMS320E15, TMS320C16, TMS320C17, TMS320P17, and TMS320E17 (see Figure 14-17) [22]. The "E" indicates the presence of on-chip EPROM for program memory, and the "P" indicates 3.3-V versions of the chip.

**Serial I/O.** The TMS320C14, TMS320P14, TMS320E14, TMS320C17, TMS320P17, and TMS320E17 have one serial port, but the other members of this family do not have serial ports. This means that the only input path for data and output path for results are through the parallel port. This is not a problem for applications where the

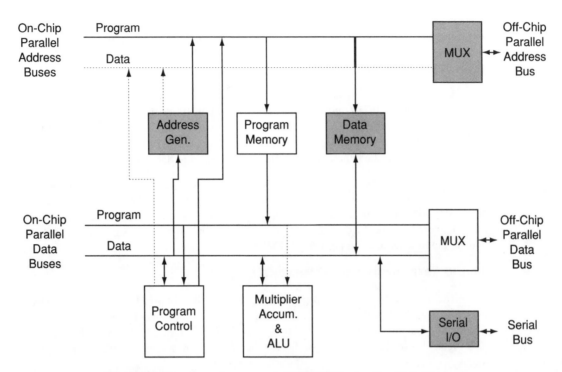

**Figure 14-17**    Texas Instruments TMS320C1x family block diagram.

input comes from a data buffer and the outputs go to a data buffer. For applications where the data I/O is asynchronous, overhead cycles are required to synchronize these DSP chips with the source of data or destination of results. These overhead cycles reduce the effective throughput rate of the chip.

The conversion of data to a linear form (frequency analysis with FFTs requires the data to be in linear form) is called companding. The TMS320C17 and TMS320E17 have companding hardware, which is an advantage in applications where the FFT is obtaining its data from an A/D converter or sending its results to a D/A converter. If the A/D and D/A converters are connected to networks such as the telephone system, the voltages they convert may be logarithmically compressed by using either the A-law (European standard) or $\mu$-law (U.S. standard).

If companding is performed in software, it takes several instruction cycles. If the process takes 10 instruction cycles, the total data I/O time for an $N$-point complex FFT increases from $4*N$ to at least $10*4*N$ instruction cycles. Since the FFT takes roughly $5*N*\log_2(N)$ instructions, an FFT will be I/O limited when $10*4*N > 5*N*\log_2(N)$. This occurs for $N < 256$ points. The companding hardware removes the need for these 10 cycles and allows the data I/O overhead to return to 1 cycle per word so that I/O limiting only occurs for 2-point FFTs, based on the inequality.

**Buses and Multiplexers.**    The data address bus is highlighted because it does not exist in this family. This eliminates the need for the I/O multiplexer for on-chip address buses. Additionally, the MAC is only connected to the data bus. To multiply numbers, one cycle is used to load one number, the second cycle to load the other and perform the multiplication. This two-cycle process, as opposed to one cycle for multiple-bus architectures, results in the significantly higher 1024-point FFT times shown in the Programmable Fixed-Point Chips Comparison Matrix in Section 14.4.

**Data Memory.** There are only 256 words of data RAM in this family of devices. Actually, the TMS320C10 only has 144 data words. This limits the complex FFTs that can be performed on-chip to 128 and 64 points, respectively. Therefore, the 1024-point FFT performance numbers in the Programmable Fixed-Point Chips Comparison Matrix (Section 14.4) already reflect the penalty paid for addressing off-chip data memory.

**Address Generators.** There are no special address generators for data memory in this family. Nonsequential addressing is done by coding the instructions to perform indirect addressing. This includes loading auxiliary registers with address offsets and loading data page pointers because the data memory is partitioned into 128-word pages. Each of these adds to the time required to perform an FFT.

### 14.3.10 Texas Instruments TMS320C2x Family

The TMS320C2x, a second generation of 16-bit fixed-point DSP chips, was introduced by TI in 1986 with the TMS32020. This device has subsequently been discontinued. The members of this family are TMS320C25, TMS320E25, TMS320C26, and TMS320C28 (see Figure 14-18) [23]. The "E" indicates the presence of on-chip EPROM for program memory.

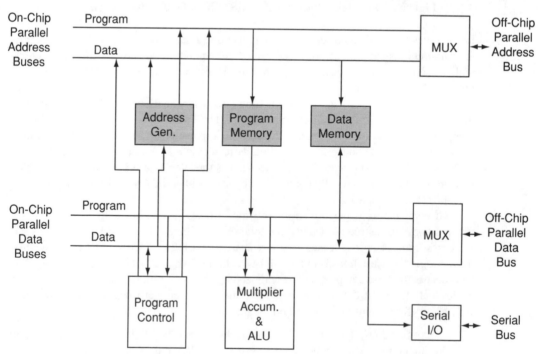

**Figure 14-18** Texas Instruments TMS320C2x family block diagram.

**Address Generator.**   Like the TMS320C10 family, this family has an increment-
ing counter for program memory addressing and auxiliary registers to offset data memory
addresses. Data memory address generation operates by loading an offset into an auxiliary register and moving the auxiliary register pointer to the correct register. Then indirect
address instructions address the offset data location. For power-of-two FFTs there is reverse binary addressing supported in hardware to alleviate the problems associated with
nonsequential memory addressing. However, this support does not help the nonsequential
addressing needed for non-power-of-two algorithms. Therefore, they are less efficient on
this chip family than comparable power-of-two algorithms.

**Data Memory.**   The TMS320C25/E25 and TMS320C28 members of this family
have 544 words of on-chip RAM that can be used for data. This means that the maximum
complex FFT that can be implemented on-chip is 256 points, assuming the multiplier coefficients and weighting function coefficients are stored in ROM/EPROM program memory.

The TMS320C26 has 1568 words of RAM. Of that, 32 words are dedicated to data
and the other 1536 words are in three 512-word blocks that can be used for either data or
program memory. This allows a 512-point complex power-of-two algorithm and roughly
a 768-point complex FFT if all weighting function and multiplier coefficients are stored in
program memory. Since $768 = 256 * 3$, this FFT can be computed with existing mixed-
radix 256-point code with the 3-point building block from Chapter 8 added to the front end
or back end of the algorithm.

In all cases, the 1024-point FFT performance numbers in the Programmable Fixed-
Point Chips Comparison Matrix (Section 14.4) reflect the data being in off-chip memory.
If multiplier and/or weighting function coefficients are stored in data memory, this further
reduces the maximum FFT length, depending on the required number of multiplier coefficients. In this case, larger FFTs can be implemented using the Winograd and prime factor
algorithms from Chapters 8 and 9 because they require fewer multiplier coefficients and
have FFT lengths between 128 and the maximum on-chip FFT length of 256 points.

**Program Memory.**   The TMS320C25/E25 family members have 4096 words of
ROM/EPROM dedicated to programs. Additionally, a 256-word block of RAM can be
used for either data or program memory. If it is used for program memory, the maximum
allowable on-chip FFT length is reduced. This leads to a complex trade because the Winograd and prime factor algorithms from Chapters 8 and 9 require fewer multiplier coefficients
but more program memory. Only detailed implementation can be used to determine the
maximum length in this situation. In the TMS320C26, the program ROM is a 256-word
boot program, and in the TMS320C28 the program memory is 8192 words.

### 14.3.11  Texas Instruments TMS320C5x Family

The TMS320C5x is the fifth family of programmable DSP chips introduced by TI
and the third 16-bit fixed-point family. The members of this family are TMS320C50,
TMS320C51, TMS320C52, and TMS320C53 (see Figure 14-19) [24]. For FFT computations the primary differences between this family and the TMS320C2x family are instruction
cycle speed and more on-chip data and program memory to avoid off-chip accesses.

**Address Generator.**   Like the TMS320C10 family, this family has an increment-
ing counter for program memory addressing and auxiliary registers to offset data memory

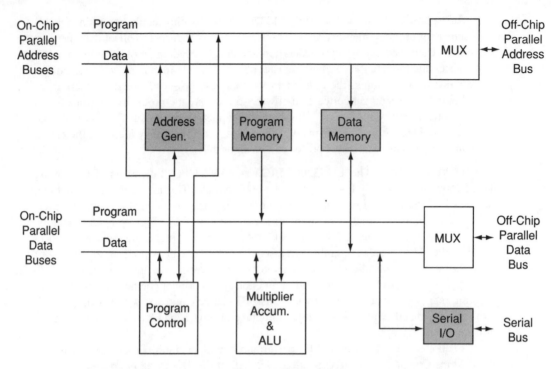

**Figure 14-19** Texas Instruments TMS320C5x family block diagram.

addresses. Data memory address generation operates by loading an offset into an auxiliary register and moving the auxiliary register pointer to the correct register. Then indirect
address instructions address the offset data location. For power-of-two FFTs there is reverse binary addressing supported in hardware to alleviate the problems associated with
nonsequential memory addressing. However, this support does not help the nonsequential
addressing needed for non-power-of-two algorithms. Therefore, they are less efficient on
this chip family than comparable power-of-two algorithms.

Data Memory. All members of this family have 1056 words of on-chip RAM
dedicated to data. Additionally, the TMS320C50/51/52/53 have 9K/1K/1K/3K of on-chip
RAM, respectively, that can be used for either data or programs. As a result, all members of this family have the ability to compute 1024-point complex FFTs on-chip. The
TMS320C51 and TMS320C52 require the complex multiplier coefficients to be stored in
program memory to allow enough room for all 2048 data words. This, combined with the
faster instruction cycle times (35 and 50 ns versus 80 and 100 ns for the TMS320C2x family), are the reasons for the improved 1024-point FFT performance in the Programmable
Fixed-Point Comparison Matrix (Section 14.4).

Program Memory. The TMS320C50/51/52/53 have 2K/8K/4K/16K of on-chip
program ROM as well as 9K/1K/1K/3K of on-chip RAM that can be used for either data
or programs. If some of the RAM is used for program memory, the maximum allowable

on-chip FFT is reduced. This results in a complex trade because the Winograd and prime factor algorithms from Chapters 8 and 9 require fewer multiplier coefficients but more program memory. Only detailed implementation can used to determine the maximum length in this situation.

**Serial Ports.**   The TMS320C50/51/53 have dual serial ports. This additional serial port provides the capability to interface these devices into linear bus, pipeline, and ring bus architectures for multiprocessor applications (Section 14.2.9) without having to use the parallel bus that may be addressing off-chip data or program memory. The TMS320C52 only has one serial port.

### 14.3.12  Zilog Z89Cxx Family

The Zilog Z89Cxx is a family of bare-bones 16-bit fixed-point processors. The most distinguishing feature of this processor is that the accumulator holds only 24 bits out of the $16 \times 16$ multiplier. This means that multiplier outputs are rounded from 32 bits to 24 bits prior to entering the accumulator. This introduces more quantization noise in the FFT outputs than accumulators that hold 32 bits or more. The only general-purpose member of this family is the Z89C00 (see Figure 14-20) [25]. Other members are customized to audio and multimedia applications.

**Multiplexers and Serial I/O.**   This processor does not have a serial I/O function. Additionally, the device has an off-chip program memory port and off-chip I/O port. Data

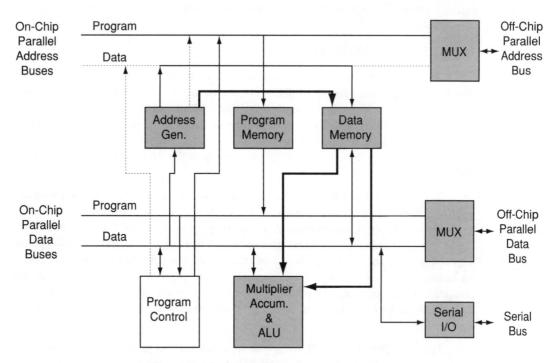

**Figure 14-20**   Zilog Z89Cxx family block diagram.

is input through the I/O port and no multiplexer exists because the program data bus is only used to connect the program memory with the program control function. Likewise, there is no multiplexer needed for the address buses because there is only one external address bus. Data memory addresses are generated and directly connected to each of the two data memories as shown in Figure 14-20.

**Data Memory.**   The Z89C00 has two 256-word data memories. Assuming all the multiplier coefficients and weighting function coefficients can be stored in program memory, this device can execute up to a 128-point FFT on-chip. Moving data from data memory to the multiplier is simplified by having it directly connected to the two data memory blocks as shown in Figure 14-20. This eliminates the need for two data buses in order to feed two data words to the multiplier during one instruction.

**Program Memory.**   This device has a 4K ROM internal program memory, but no internal RAM for program memory.

**Address Generators.**   Each data RAM has its own dedicated address generator that is based on programming offset address pointers rather than having an ALU to compute the offset address. This makes this device's address generation scheme similar to the first two generations of TI chips, the TMS320C1x and TMS320C2x.

**Multiplier-Accumulator.**   The $16 \times 16$ multiplier output is 24 bits and is fed to an ALU before going to the 24-bit accumulator. The output of the multiplier can also return to data memory. The multiplier and ALU outputs are returned to data memory through the chips' bus.

### 14.3.13 Zoran ZR38000 Family

This is the first family of fixed-point DSP chips to compute the 1024-point complex FFT in less than 1 ms. A second distinguishing feature for FFT computations is that it performs 20-bit, not 16-bit, integer arithmetic. These additional 4 bits reduce the algorithm-generated quantization noise by 12 dB and increase the dynamic range by 24 dB. Another distinguishing feature for these fixed-point processors is the six half-duplex (three two-way) serial ports. The only member of this family is the ZR38000 (see Figure 14-21) [26].

**Data/Program Memory.**   This chip has 2048 twenty-bit words of data memory and 8192 thirty-two-bit words of program/data ROM. Assuming all multiplier coefficients and weighting function coefficients are stored in program/data ROM, a 1024-point FFT can be computed on-chip. Therefore, Equation 14-1 works for FFTs less than 1024 points but not for those above 1024 points. However, the standard product only uses the ROM for bootstrapping the loading of the main operating program. Therefore, the standard product can only perform 512-point complex FFTs with on-chip data memory because it needs the rest of the data memory to store multiplier and weighting function coefficients.

**Address Generator.**   This chip has only one address generator, and its output is connected to the data memory address bus. However, this generator and the data memory

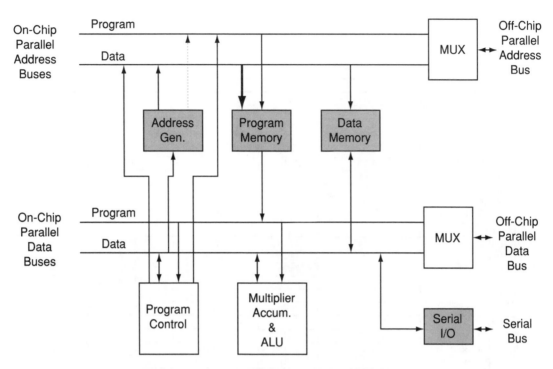

**Figure 14-21**    Zoran ZR38000 family block diagram.

are able to support the update of two data memory address locations per instruction cycle and two accesses of data memory per instruction cycle. The address generator also has built-in hardware that supports bit-reversed addressing for the power-of-two FFT algorithms in Chapter 9. The generator also supports modulo addressing, which is useful in implementing the non-power-of-two FFT algorithms in Chapter 9.

Serial I/O.    This device has six half-duplex serial ports. Therefore, it has the capability of moving data in and out of the processor as if there were three full-duplex serial ports.

## 14.4 PROGRAMMABLE FIXED-POINT CHIPS COMPARISON MATRIX

The data in the Comparison Matrix in Table 14-3, on page 354, comes from the referenced vendor material. In the case of the 1024-point complex FFT performance, this is the fastest number available in the material. Different versions of a 1024-point FFT may produce slightly different performance numbers. Versions of the chips that run at slower speeds will have times that are slower. Conversely, newer versions of these chips, which run faster, will have faster times. Performance numbers with asterisks are estimated because times for the 1024-point FFT were not available from the vendor.

**Table 14-3** Programmable Fixed-Point Chips Comparison Matrix

| Fixed-point chip | 1024-point complex FFT (MS) | Data I/O ports | On-chip data memory words | On-chip prog. memory words | # of address generators |
|---|---|---|---|---|---|
| **Analog Devices** | | | | | |
| ADSP-2100A | 2.77 | 0s/1p | 0 | 16384 | 2 |
| ADSP-2101 | 1.73 | 2s/1p | 1024 | 2048 | 2 |
| ADSP-2103 | 3.40 | 2s/1p | 1024 | 2048 | 2 |
| ADSP-2105 | 2.49 | 1s/1p | 512 | 1024 | 2 |
| ADSP-2111 | 1.73 | 2s/1p | 1024 | 2048 | 2 |
| ADSP-2115 | 1.73 | 2s/1p | 512 | 1024 | 2 |
| ADSP-216x | 2.08 | 2s/1p | 512 | 0 | 2 |
| ADSP-2171 | 1.04 | 2s/1p | 2048 | 2048 | 2 |
| ADSP-2175 | 1.04 | 2s/1p | 16384 | 16384 | 2 |
| ADSP-21msp5xx | 2.67 | 2s/1p | 1024 | 2048 | 2 |
| **AT&T** | | | | | |
| DSP16 | 6.54* | 1s/2p | 512 | 2048 | 2 |
| DSP16A | 2.97 | 1s/2p | 2048 | 2048 | 2 |
| DSP1610 | 2.97 | 2s/2p | 8192 | 4096 | 2 |
| DSP1616 | 2.38 | 2s/2p | 2048 | 12288 | 2 |
| DSP1617 | 2.38 | 2s/2p | 4096 | 24576 | 2 |
| DSP1618 | 2.38 | 2s/2p | 4096 | 16384 | 2 |
| **Motorola** | | | | | |
| DSP56156 | 1.53 | 2s/1p | 2048 | 2048 | 2 |
| DSP56166 | 1.53 | 2s/1p | 4096 | 2048 | 2 |
| DSP56001 | 1.797 | 2s/1p | 512 | 512 | 2 |
| DSP56002 | 0.908 | 2s/1p | 512 | 512 | 2 |
| DSP56L002 | 1.497 | 2s/1p | 512 | 512 | 2 |
| DSP56004 | 1.497 | 2s/1p | 512 | 512 | 2 |
| **NEC** | | | | | |
| $\mu$PD77C20A | 48.5* | 1s/1p | 256 | 2048 | 1 |
| $\mu$PD7720A | 48.5* | 1s/1p | 256 | 2048 | 1 |
| $\mu$PD77P20 | 48.5* | 1s/1p | 256 | 2048 | 1 |
| $\mu$PD77C25 | 24.3* | 1s/1p | 256 | 2048 | 1 |
| $\mu$PD77P25 | 24.3* | 1s/1p | 256 | 2048 | 1 |
| $\mu$PD77016 | 0.95 | 2s/1p | 4096 | 1536 | 2 |
| $\mu$PD77220 | 8.5* | 1s/1p | 512 | 2048 | 2 |
| $\mu$PD77P220 | 8.5* | 1s/1p | 512 | 2048 | 2 |
| **TI** | | | | | |
| TMS320C10 | 66.2 | 0s/1p | 144 | 1536 | 1 |
| TMS320C14 | 53.0 | 1s/1p | 256 | 4096 | 1 |
| TMS320C15 | 66.2 | 0s/1p | 256 | 4096 | 1 |
| TMS320C16 | 37.7 | 0s/1p | 256 | 8192 | 1 |
| TMS320C17 | 66.2 | 1s/1p | 256 | 4096 | 1 |
| TMS320C25 | 4.54 | 1s/1p | 544 | 4096 | 1 |
| TMS320C26 | 4.54 | 1s/1p | 1568 | 256 | 1 |
| TMS320C28 | 5.67 | 1s/1p | 544 | 8192 | 1 |
| TMS320C50 | 2.40 | 2s/1p | 10240 | 2048 | 1 |
| TMS320C51 | 2.40 | 2s/1p | 2048 | 8192 | 1 |
| TMS320C52 | 2.60* | 1s/1p | 1024 | 4096 | 1 |
| TMS320C53 | 2.40 | 2s/1p | 4096 | 16384 | 1 |
| **Zilog** | | | | | |
| Z89C00 | 3.16* | 0s/1p | 512 | 4096 | 2 |
| **Zoran** | | | | | |
| ZR38000 | 0.88 | 6s/1p | 2048 | 8192 | 1 |

\* = estimated time; s = serial ports; p = parallel ports.

## 14.5  PROGRAMMABLE FLOATING-POINT CHIPS

All of the general-purpose floating-point DSP chips in this chapter use 32-bit arithmetic with 8 bits of exponent and 24 bits of mantissa. In addition to these chips, the Intel i860 has also been included. While this chip was initially developed for graphics applications, its FFT performance is so good that it has been used by many DSP board manufacturers. The i860 uses the same configuration of 32-bit floating-point numbers described above. The way the different vendors treat the smallest and largest number varies slightly but has no effect on the computational performance, except in rare instances when the top or bottom numbers in the dynamic range are reached.

### 14.5.1  Analog Devices 21020 Family

The 21020 is Analog Devices first family of 32-bit floating-point processors. Its most distinguishing feature is that it has no on-chip program or data memory. However, the on-chip buses are designed to work at full speed with off-chip memory to produce high-performance computing that does not depend on the inability to get large amounts of memory on-chip. The only member of this family is the ADSP-21020 (see Figure 14-22) [27].

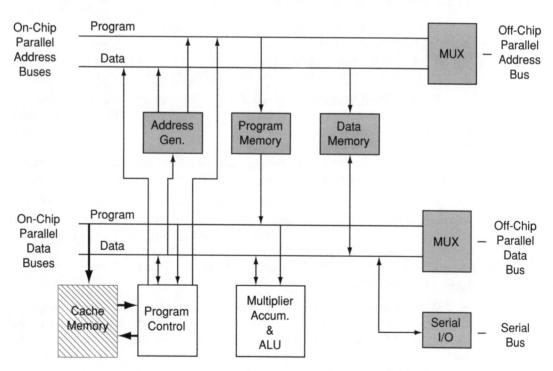

**Figure 14-22**    Analog Devices 21020 family block diagram.

Serial I/O.    This device does not have a serial I/O port.

Multiplexers.    This device does not use the MUX hardware because it provides I/O pins for all four on-chip data and address buses.

Data and Program Memory.    This device does not have any on-chip data or program memory. It is all accessed directly using off-chip memory. As a result, the FFT performance numbers in the Programmable Floating-Point Chips Comparison Matrix (Section 14.7) can be scaled to estimate larger or smaller FFT computation times using Equation 14-1.

Address Generators.    The ADSP-21020 has dual address generators. This maximizes the ability to address both data and multiplier constants to feed to the MAC unit on each instruction cycle. The flexibility of the address step sizes for these generators also allows them to be easily used to generate non-power-of-two algorithms as well as standard FFTs. Address generator 1 also has bit-reverse logic to accommodate standard power-of-two algorithms.

Cache Memory.    This device has a 48-word instruction cache memory to run frequently used instruction sequences without having to access off-chip program memory. Building-block FFT algorithms can be executed from this memory. Because of the small size, it is likely that only 2-, 3-, and possibly 4-point building blocks from Chapter 8 can be programmed to fit in the cache.

### 14.5.2 Analog Devices ADSP-21060 Family

The ADSP-21060 is the second generation of Analog Devices programmable floating-point DSP chips. Its most distinguishing feature is its FFT performance, large on-chip RAM, and six link ports for interfacing it into multiprocessor networks (see Figure 14-23). The members of this family are ADSP-21060 and ADSP-21062 [28].

Program/Data Memory.    The ADSP-21060 has 4 Mbits of dual-ported RAM, organized as two 2-Mbit blocks for different combinations of data and program instructions. Configured as 32-bit words, each block holds 65,536 words. This allows a 32,768-point FFT to be performed using on-chip memory if all the multiplier coefficients and weighting function coefficients are stored in one block and the data in the other. The multiple-bus architecture allows both memories to be accessed in a single cycle for FFT arithmetic.

Address Generators.    The ADSP-21060 has dual address generators. This maximizes the ability to address both data and multiplier constants to feed to the MAC unit on each instruction cycle. The flexibility of the address step sizes for these generators also allows them to be easily used to generate non-power-of-two algorithms as well as standard FFTs. Address generator 1 also has bit-reverse logic to accommodate standard power-of-two algorithms.

Cache Memory.    This device has a 48-word instruction cache memory to run frequently used instruction sequences without having to access off-chip program memory. Building-block FFT algorithms can be executed from this memory. Because of the small size, it is likely that only 2-, 3-, and possibly 4-point building blocks from Chapter 8 can be programmed to fit in the cache.

Link and Serial Ports.    The ADSP-21060 has two serial ports and six serial link ports designed for interfacing to other ADSP-21060s to form multiprocessor architectures. All eight of these inputs are interfaced to the main processor using I/O port (IOP) registers and a DMA controller. The DMA controller allows data to move in through the link ports

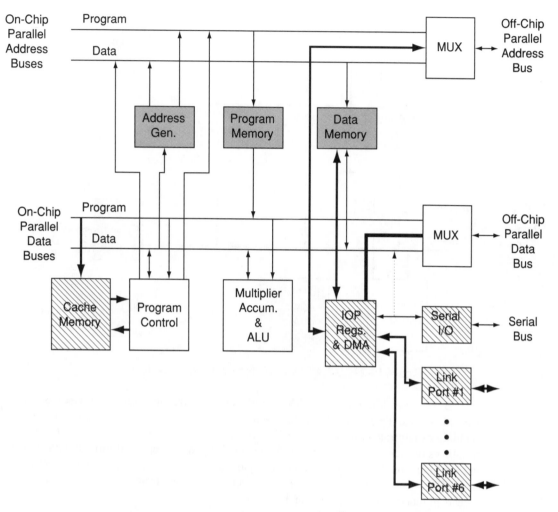

**Figure 14-23**    Analog Devices 21060 family block diagram.

and to be stored either in on-chip RAM or in off-chip RAM via the interface multiplexers. These six communications ports allow this device to be connected into a variety of one-, two-, and three-dimensional architectures. The three-dimensional massively parallel processor example in Figure 14-6 is one example. Others are described in Chapter 11.

### 14.5.3 AT&T DSP32C Family

The DSP32C is AT&T's first CMOS family of 32-bit floating-point processors and is a follow-on to their DSP32 introduced in 1984. The most distinguishing feature of this family is that it operates like a Harvard architecture even though it is actually a Von Neumann architecture. This is accomplished by allowing multiple uses of the data and program buses during one instruction cycle. The members of this family are DSP32C, DSP3210, and DSP3207 (see Figure 14-24) [29, 30].

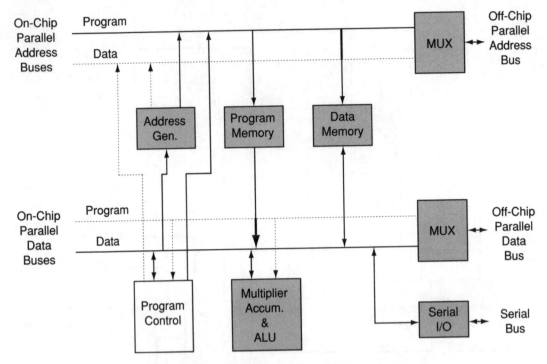

**Figure 14-24** AT&T DSP32C family block diagram.

**Buses and Multiplexers.** This family's architecture uses only one data bus and one address bus. Therefore, all functions must be connected to these, and there is no need to multiplex multiple buses to access off-chip data and program memory. This high-speed bus allows the device to access two 32-bit operands from memory, perform multiplication and accumulation operations on a previous pair of operands, and write a previous result to an I/O port or memory in one instruction cycle. Therefore, from the outside the device appears to function like a Harvard architecture.

**Address Generator.** With only one address bus, there is only need for one address generator if it can produce the multiple addresses supportable by the address bus during an instruction cycle. The address generator in this device family is capable of that. Additionally, the address generator has an ALU that can be used to perform addressing in nonunit increments. This makes it useful for implementing any of the FFT algorithms in Chapter 9. However, the devices are more efficient for power-of-two FFT algorithms because bit-reversed addressing is directly supported for reorganizing data for these FFTs.

**Data/Program Memory.** The DSP32C supports one of two on-chip memory configurations that can be used for data or program. The first is 1024 words of RAM and 4096 words of ROM. The second is 1536 words of RAM. Therefore, the largest power-of-two complex FFT that can be executed on-chip is 512 points. The limit on the largest non-power-of-two FFT is more difficult to calculate without getting an estimate on the complexity of the code that must be stored in on-chip memory. It is likely that code will need to be written to determine the largest allowable FFT. For the 4096-word ROM option, the answer is clearly 512 points, assuming all multiplier coefficients and weighting function coefficients are stored in ROM.

The primary difference between the DSP32C and the DSP3210 for executing FFT algorithms is the larger on-chip memory space. The DSP3210 has two banks of 1024 words of RAM and a small 256-word boot ROM. Program instructions and data can reside in any of the 2048 RAM locations, and the boot ROM is preprogrammed to load the on-chip RAM from off-chip EPROM for lower-cost operation. Again, the largest FFT depends on the size of the FFT algorithm code, but will not be larger than 512 points for power-of-two algorithms because the next largest size (1024 points) would not leave any room for the FFT program code. The largest non-power-of-two algorithm depends on the size of its code.

Serial I/O.    All members of the device family, except the DSP3207 have one serial I/O port. The DSP3207 has no serial ports.

Multiplier-Accumulator and ALU.    Because there is only one data bus in this chip family, all data must be moved sequentially. Since the data bus can support two of those data accesses per instruction cycle, the MAC and ALU function can also support two inputs during an instruction cycle. This makes the MAC/ALU unit appear as if it has two ports.

### 14.5.4 Intel i860 Family

This family of programmable 32-bit floating-point processors is not usually considered a DSP chip. The family was initially targeted for engineering and three-dimensional graphics workstations as well as numerical accelerators. However, DSP board manufacturers discovered that the devices had superior performance for FFT algorithms. The result has been the widespread use of this chip family in high-speed DSP applications. The most significant feature of this family for FFT algorithms is the multiple instruction and computational functions that are pipelined for speed. While this increases the speed of the i860, it makes it much more difficult to program in assembly language to take advantage of that speed. The members of this family are i860XR and i860XP (see Figure 14-25) [31].

On-Chip Buses/Off-Chip Buses.    The on-chip bus structure for the i860 family is different from standard DSP chips. There are three data buses to and from the floating-point multiplier and adder units, rather than the one or two for more standard DSP chips. Conversely, there is only one data bus from on-chip data memory to the floating-point control unit. The off-chip address bus is highlighted because the i860 family only has this as a unidirectional bus for addressing off-chip memory.

Bus Control Unit.    Intel calls its interface to off-chip data memory the bus control unit. The i860 family's single on-chip data bus architecture removes the need for the bus control unit to perform the data bus MUX function found in conventional DSP chips for off-chip data access.

Memory Management Unit/Address Generators.    The memory management unit performs the functions usually accomplished by the address generators in a conventional DSP chip. This includes the addressing of external memory which removes the need for the address bus MUX found in conventional chips.

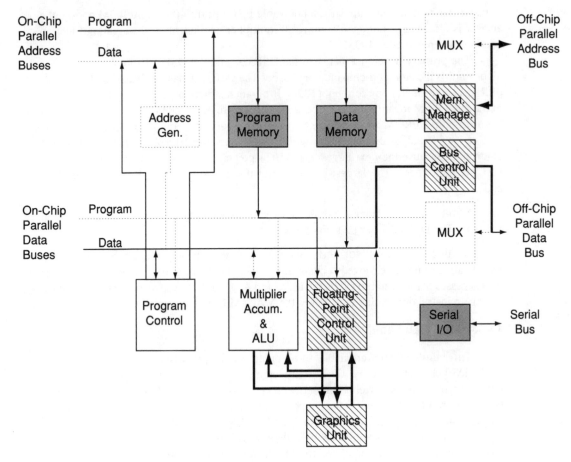

**Figure 14-25** Intel i860 family block diagram.

**Floating-Point Control Unit.** The floating-point control unit is also a different feature of the i860 family. It provides an interface between the instruction and data memories and the computational units. Conventional DSP chips directly connect the memories to the computational units, as is shown in Figure 14-1.

**Program/Data Memory.** Both members of this family have on-chip data and program memory, called cache memory by Intel. Stored as 32-bit floating-point words, the i860XR has 1024 words of data memory and the i860XP has 2048 words. Similarly, there are 512 sixty-four-bit instructions that can be stored in the i860XR's on-chip instruction cache and 1024 sixty-four-bit words in the i860XP. Assuming all multiplier and weighting function coefficients can be stored in program memory, the i860XR can perform up to a 512-point complex FFT on-chip, and the i860XP can execute a 1024-point complex FFT on-chip.

**Serial I/O.** This family does not have a serial I/O port.

**Multiply Accumulator and ALU.**    The i860 family has a separate multiplier and adder. Both are pipelined for maximum computation rate. This means that multiple cycles are used to perform each arithmetic computation. Conventional DSP chips perform these functions in one instruction cycle.

**Graphics Unit.**    The i860 chip family was designed with built-in support for high-speed graphics. While this feature does not modify its capability to compute FFT algorithms, it is a unique feature worth mentioning. Specifically, this hardware performs the integer operations necessary for shading and hidden line removal. The $4 \times 4$ transforms needed for orienting points are performed by the floating-point hardware.

### 14.5.5  Motorola DSP96002 Family

The DSP96002 is Motorola's first 32-bit floating-point family and is aimed at the multimedia market. It is basically a 32-bit floating-point extension of the 24-bit fixed-point DSP5600x family. Its most distinguishing features are the large number of on-chip buses, dual parallel interfaces off the chip, and an arithmetic unit that has Newton-Raphson-based square root and 1/(square root) functions. The only member of this family is the 96002 (see Figure 14-26) [32].

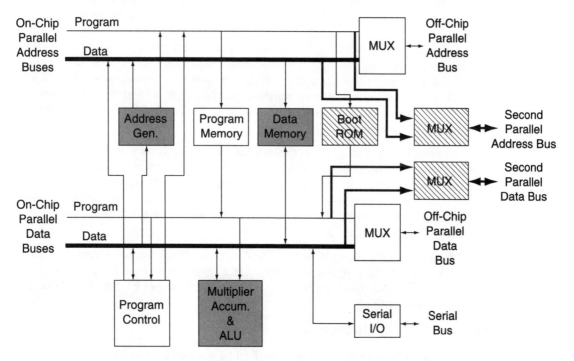

**Figure 14-26**    Motorola 96002 family block diagram.

**Buses and Multiplexers.**    In addition to the buses in the Motorola DSP5600x architecture (three address and four data), the DSP96002 provides a DMA data bus. Another feature of the DSP96002 is the dual parallel interfaces off the chip. This additional off-chip parallel interface allows these devices to be connected into linear bus, pipeline, and ring bus architectures for multiprocessor applications (Section 14.2.9) without having to use the parallel bus that may be addressing off-chip data or program memory.

**Data RAM and ROM.**   The DSP96002 has 1024 words of data RAM on-chip. Therefore, the largest FFT that can be computed with on-chip memory is 512 points. The performance numbers in the Programmable Floating-Point Chips Comparison Matrix (Section 14.7) already reflect the penalty paid for having to access off-chip data memory. Further, the data RAM is divided into two 512-word memories called X-data memory and Y-data memory. To accommodate these extra memories, there is an extra data memory bus and extra data memory address bus.

Grouped with each of these 512-word RAMs is a 512-word ROM. The X-data ROM contains a full cycle of the "cosine" function, and the Y-data ROM contains a full cycle of the "sine" function to be used by power-of-two FFT algorithms directly as the multiplier constants. Specifically, the 360° phase angle is divided into 512 pieces. These tables can also be used for non-power-of-two FFTs with the help of an interpolation algorithm. For example, to use the table for the 504-point mixed-radix algorithm, 360° must be divided into 504 pieces, not 512. Therefore, the table entries cannot be used directly. However, for each needed value, the two surrounding phase angle values and a linear interpolation algorithm can be used to accurately compute the correct value.

The coefficients in the X- and Y-data ROMs can also be used to compute the sine lobe, Hanning, sine cubed, sine to the fourth, Hamming, Blackman, three-sample Blackman-Harris, and four-sample Blackman-Harris weighting functions in Sections 4.2.3 through 4.2.10. This removes the need to store weighting function coefficients if the chip's computational power allows the weighting function coefficients to be computed as needed within the required FFT computation time.

**Address Generators.**   All of the members of this family have dual address generators. This maximizes the ability to address both data and multiplier constants to feed to the MAC unit on each instruction cycle. The flexibility of the address step sizes for these generators also allows them to be easily used to generate non-power-of-two algorithms as well as standard FFTs. Both address generators also have bit-reverse logic to accommodate standard power-of-two algorithms.

**Multiply Accumulator and ALU.**   The ALU has a "divide and square root" unit that uses the Newton-Raphson algorithm to compute the square root(x) and 1/(square root(x)) in 12 and 11 instruction cycles, respectively. This is not critical for FFT algorithms but can accelerate an overall application.

### 14.5.6  NEC $\mu$PD77240/230A Family

The $\mu$PD77240/230A family of 32-bit floating-point chips from NEC has nearly the same architecture as the $\mu$PD77220 24-bit fixed-point series. The members of this family are $\mu$PD77240 and $\mu$PD77230A (see Figure 14-27) [18].

**Buses and Multiplexers.**   The reduction to one main bus removes the need for multiplexers on the data and address buses in the standard DSP chip approach. Further, this bus reduction forces several direct connections between functional blocks. Each of these is

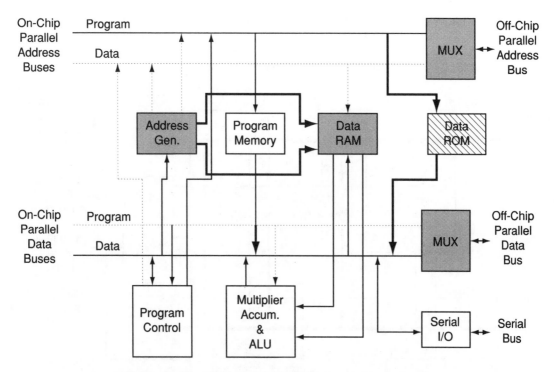

**Figure 14-27**    NEC $\mu$PD77240/230A family block diagram.

described below. These connections offset the degradation in FFT performance associated with having only one main bus.

**Data Memory.**    Both devices have two 512-word data RAM blocks and the $\mu$PD77230A has 1024- and 2048-word data ROMs for storing multiplier constants and weighting function coefficients. Externally, both devices support a 12-bit address word which corresponds to addressing 4096 data words. This limits them to performing 2048-point FFTs, even using off-chip memory. Using on-chip memory with real and imaginary components in respective 512-word blocks of data memory provides the capability to perform 512-point complex FFTs.

Data memory does not use the main bus to transfer data to the multiplier. Each data RAM has its own direct path to the multiplier. However, the results from the multiplier or accumulator are stored in data RAM using the main bus.

**Address Generators.**    Both devices have an address generator for each of the data RAMs to avoid having to use the main bus. These generators are simple base address plus offset calculators that require the offset to be programmed into the instructions for nonunit values. Therefore, they are not ideally suited for computing non-power-of-two FFT algorithms.

### 14.5.7 Texas Instruments TMS320C3x Family

The TMS320C3x is TI's first generation of programmable 32-bit floating-point DSP chips. The architecture of this chip family is more efficient for computing FFTs than the earlier fixed-point generations primarily because of the additional buses that allow multiple tasks to occur during the same instruction cycle. The primary distinguishing feature of this device family is the multiple data and address ports. The members of this family are TMS320C30 and TMS320C31 (see Figure 14-28) [33].

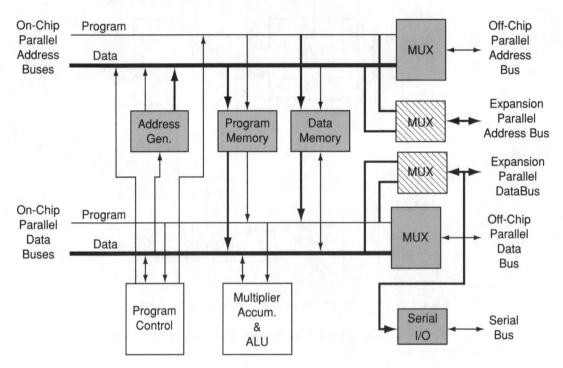

**Figure 14-28** Texas Instruments TMS320C3x family block diagram.

**Buses and Multiplexers.** The large number of on-chip buses is a primary characteristic of this family. There are four on-chip data buses and three on-chip address buses, which make it possible to access multiple pieces of data during one instruction cycle. This improves the performance of this TI family over the TMS320C1x and TMS320C2x fixed-point families, which only access one data word per instruction cycle. Additionally, the on-chip buses are multiplexed off the chip twice. The additional off-chip parallel interface allows these devices to be connected into linear bus, pipeline, and ring bus architectures for multiprocessor applications without having to use the parallel bus that may be addressing off-chip data or program memory.

**Data/Program Memory.** This family has two 1024-word RAMs and one 4096-word ROM. Each RAM and ROM can support two memory accesses each instruction cycle, and the multiple buses allow for parallel program fetches, data reads/writes, and DMA operations. Additionally, a 64-word instruction cache is provided to store often used pieces of code so that they need not be stored off-chip to slow down execution. If all multiplier constants and weighting function coefficients are stored in program ROM, this chip family can be used to compute up to a 1024-point complex FFT on-chip.

**Address Generators.**    This is the first generation of TI DSP chips to have a full-function address generator. This family has two that can do addressing in nonunit steps to support non-power-of-two FFT algorithms. They can compute two addresses per instruction cycle to address two pieces of data using two of the four data buses. The address generators also support bit-reversed addressing for power-of-two FFT algorithms.

**Serial I/O.**    The TMS320C30 has two serial I/O ports. This additional serial port provides the capability to interface these devices into linear bus, pipeline, and ring bus architectures for multiprocessor applications (Section 14.2.9) without having to use the parallel bus that may be addressing off-chip data or program memory. The TMS320C31 only has one serial port.

Another fundamental difference of this family architecture is that the serial ports interface to the expansion I/O buses rather than directly to the on-chip buses. The advantage of this is allowing the serial data port to interface directly to all of the on-chip data buses. The disadvantage of this is that the serial port data cannot be input to the on-chip data buses while the expansion I/O bus is active to some other peripheral. If the serial port were tied to one of the on-chip data buses, it could be active while the expansion I/O bus was connected to one of the other on-chip data buses.

## 14.5.8 Texas Instruments TMS320C40 Family

The TMS320C40 is the second generation of 32-bit floating-point chips from TI. The primary distinguishing feature of this family is the six serial ports designed to support using this device in large multiprocessor arrays without significant overhead penalties for the central processing unit. The first member of this family is the TMS320C40 (see Figure 14-29) [34].

**Buses and Multiplexers.**    The large number of on-chip buses is a primary characteristic of this device. There are four on-chip data buses and three on-chip address buses, which make it possible to access multiple pieces of data during one instruction cycle. This improves the performance of this TI family over the previous TI fixed-point families that could only access one data word per instruction cycle. Additionally, the on-chip buses are multiplexed off the chip twice. The additional off-chip parallel interface allows these devices to be connected into linear bus, pipeline, and ring bus architectures for multiprocessor applications without having to use the parallel bus that may be addressing off-chip data or program memory. However, the intent is to interface to additional peripheral devices and let the communication (Comm) ports interface into the larger array of similar processors.

The multiplexer that connects the six communications ports to the on-chip address buses also includes a DMA controller to move data directly into on-chip memory. The connection from that MUX to the data address bus provides the addressing information, and the connection to the data bus provides the data bus interface.

**Data/Program Memory.**    This family has two 1024-word RAMs and one 4096-word ROM. Each RAM and ROM can support two memory accesses each instruction cycle, and the multiple buses allow for parallel program fetches, data reads/writes, and DMA operations. Additionally, a 64-word instruction cache is provided to store often used pieces of code so that they need not be stored off-chip to slow down execution. If all multiplier

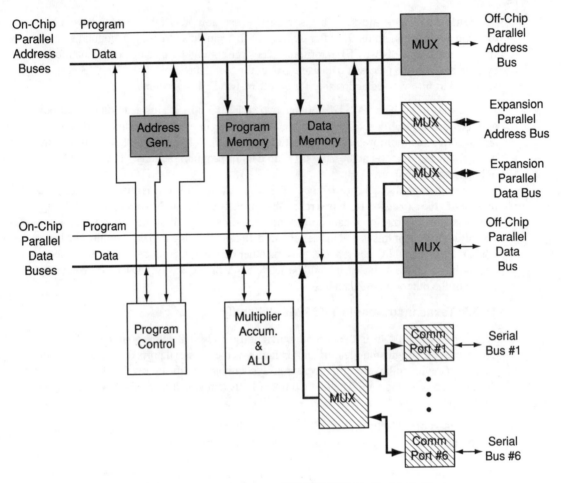

**Figure 14-29** Texas Instruments TMS320C40 family block diagram.

constants and weighting function coefficients are stored in program ROM, this chip family can be used to compute up to a 1024-point complex FFT on-chip.

**Address Generators.** This is the second generation of TI DSP chips to have a full-function address generator. This family has two that can do addressing in nonunit steps to support non-power-of-two FFT algorithms. They can compute two addresses per instruction cycle to address two pieces of data using two of the four data buses. The address generators also support bit-reversed addressing.

**Serial I/O (Comm Ports 1–6).** The TMS320C40 has six serial I/O ports, which are called communications ports. These ports are independently multiplexed into the on-chip buses to provide full bus utilization flexibility. These six communications ports allow this device to be connected into one-, two-, and three-dimensional architectures. The three-dimensional architecture in Section 12.6.2 shows one option.

## 14.6 PROGRAMMABLE FLOATING-POINT CHIPS COMPARISON MATRIX

The data in the Comparison Matrix in Table 14-4 comes from the referenced vendor material. For the 1024-point complex FFT performance, this is the fastest number available in the referenced material. Different versions of a 1024-point FFT may produce slightly different performance numbers. Versions of the chips that run at slower speeds will have times that are slower. Conversely, newer versions of these chips, which run faster, will have faster times. Finally, some of the entries in the on-chip memory columns have two numbers. This means there are two versions of the chip available.

**Table 14-4**    Programmable Floating-Point Chips Comparison Matrix

| Floating-point chip | 1024-point complex FFT (MS) | Data I/O ports | On-chip data memory words | On-chip prog. memory words | # of address generators |
|---|---|---|---|---|---|
| **Analog Devices** | | | | | |
| ADSP-21020 | 0.58 | 0s/2p | 0 | 0 | 2 |
| ADSP-21060 | 0.46 | 8s/1p | 65,536 | 65,536 | 2 |
| **AT&T** | | | | | |
| DSP32C | 3.2 | 1s/1p | 1024/1536 | 4096/0 | 1 |
| DSP3210 | 2.4 | 1s/1p | 1024/2048 | 1024/256 | 1 |
| DSP3207 | 1.9 | 0s/1p | 1024/2048 | 1024/256 | 1 |
| **Intel** | | | | | |
| i860XR | 0.74 | 0s/1p | 1024 | 256 | 1 |
| i860XP | 0.55 | 0s/1p | 2048 | 1024 | 1 |
| **Motorola** | | | | | |
| DSP96002 | 1.04 | 0s/2p | 1024 | 1024 | 2 |
| **NEC** | | | | | |
| $\mu$PD77240 | 7.07 | 1s/1p | 1024 | 0 | 2 |
| $\mu$PD77230A | 11.78 | 1s/1p | 1024 | 1024/2048 | 2 |
| **TI** | | | | | |
| TMS320C30 | 1.97 | 2s/2p | 2048 | 4096 | 2 |
| TMS320C31 | 1.97 | 1s/2p | 2048 | 4096 | 2 |
| TMS320C40 | 1.54 | 6s/2p | 2048 | 4096 | 2 |

s = serial ports; p = parallel ports

## 14.7 FFT-SPECIFIC CHIPS AND CHIP SETS

Several dedicated chips and chip sets have been developed to compute power-of-two FFTs. These chips also can be programmed to perform linear filtering and pattern matching in the frequency domain using the algorithms described in Chapter 6. Because these chips are dedicated to computing FFTs, they are 5 to 10 times faster at computing FFTs than are programmable DSP chips. Additionally, they can be combined, using the architectural approaches described in Chapter 11, to perform FFTs at even higher rates.

The primary features of these chip sets are their raw FFT computation performance, the building blocks they offer, and the largest FFT that can be performed by a single chip/chip set. Since these chips are designed to perform FFTs, it is more relevant to show block diagrams of how the chips are connected to off-chip memory and address controllers than to show the internal block diagram of the chip. These block diagrams can then be combined to form the multiprocessor architectures in Chapter 11. Refer to the manufacturer's data books and application notes for details on the limitations of each chip for multiprocessor operation.

The primary disadvantage of these chips is they are not designed to perform general-purpose functions, such as user interface and decision making, often required to complete an application. A second disadvantage is that these chips can only perform power-of-two FFTs. However, for the Bluestein algorithm in Section 9.5, these chip/chip sets can be used to perform non-power-of-two algorithms by customizing the complex multiplications to the transform length of interest by using the Bluestein approach. While this approach is less efficient than power-of-two algorithms with these chips, they do perform those algorithms 5 to 10 times faster than programmable DSP chips. Therefore, even a factor of 2 or 3 inefficiency still results in higher-speed computations than can be obtained from programmable DSP chips. For some applications this can be the difference in success or failure.

Because these chips are specifically designed to perform FFTs, their performance can be measured by using more FFT specific items. These are:

1. 1024-point complex FFT performance ($\mu$s)
   This is the same as the first performance measure for the programmable DSP chips.

2. Programmed FFT building blocks
   This performance measure is the list of FFT building blocks that have code built into the chip.

3. Largest complex FFT size
   This is the largest complex FFT length that can be programmed into the chip.

4. Number of block-floating-point mantissa bits
   This is the number of mantissa bits built-in to the arithmetic units of each chip. All of these chips use the block-floating-point arithmetic format (Chapter 13).

### 14.7.1 Array Microsystems a66110/66210 Chip Set

The array Microsystems a66110/66210 chip set [35] is designed to perform real and complex FFTs, IFFTs, as well as linear filtering and pattern matching in the time and frequency domains. The chip has radix-2 and -4 FFT building-block instructions that are connected using the mixed power-of-primes algorithm from Chapter 9 to implement up to a 65,536-point complex FFT. The chip uses both the Two-Signal Algorithm and Double-Length Algorithm from Chapter 2 to compute FFTs of real input data. It uses the Overlap-and-Add Algorithm from Chapter 6 for performing linear filtering and pattern matching in the frequency domain. All arithmetic is 16-bit mantissa block-floating-point.

Figure 14-30 is a block diagram of one of several ways to interface this chip set with data memory and algorithm control logic. In addition to the a66110 (269 pins), the address generator function is also provided as a chip and is called the a66210 (180 pins). Array Microsystems also provides a reduced pinout version of this chip set (a66111/a66211), each having 144 pins. The primary distinguishing feature of this chip set is that it performs FFTs up to 65,536 points.

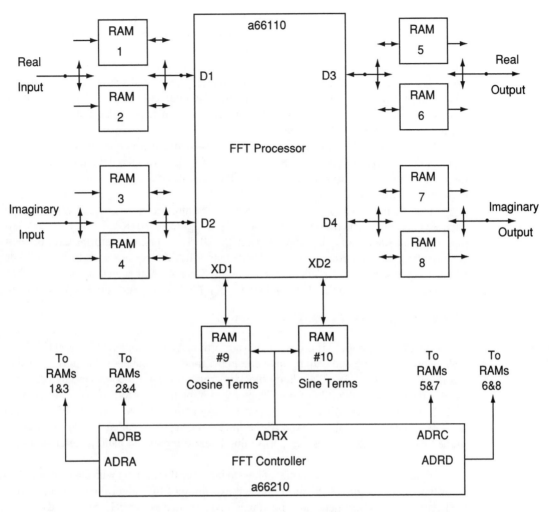

**Figure 14-30**  Array Microsystems a66110/66210 chip set block diagram.

The operational strategy for the configuration in Figure 14-30 is to start by loading a set of data into RAMs 1 and 3. Then, that set of data is moved through the processor to output RAMs 5 and 7 while the first stage of FFT computations is performed. Then, these intermediate results are passed back through the processor to RAMs 1 and 3 to perform the second stage of the algorithm. This process continues until the final computations result in the output frequency components being in RAMs 5 and 7.

During each pass, the appropriate complex multiplier coefficients are addressed from RAMs 9 and 10 to satisfy the mixed-radix algorithm. During the first stage, these coefficients can be the weighting function. This capability is also used during frequency-domain filtering/pattern matching to input the needed complex filter coefficients between the input FFT and output inverse FFT. The chip supports both 25% and 50% overlapped data sets, as explained in Chapter 6.

While the first FFT is being computed, the next set of data to be transformed is being loaded into RAMs 2 and 4. After the first set of data is transformed, RAMs 2 and 4 become the input, and RAMs 6 and 8 work with those RAMs to produce the next set of outputs. At the same time, the controller addresses RAMs 5 and 7 to output the results of the previous FFT. This architecture allows data to be continuously input and the results to be output while computations are performed. It also allows the input and output data clocks to work at a different rate than the processing clock, as long as the data is loaded and output before the end of the present FFT computation.

For computing FFTs of real data, the processor has instructions that support both types of data reorganization described in Chapter 6. However, the data must be input in the proper form for the transform to work. Once that has occurred, an output instruction performs the necessary unraveling of the data.

A subtle point with this chip set is that an odd number of FFT stages is required to have the output in the memories on the right side of Figure 14-30 (RAMs 5–8). This means that if 2-point stages are being used, 128-, 512-, 2048- ... point transforms have the best performance. To get a 1024-point FFT to the output RAMs requires an extra pass of data through the processor if 2-point stages are used. Since 4-point stages are also available, they should be used for 64-, 1024-, and 4096-point FFTs to have an odd number of stages.

### 14.7.2 Sharp LH9124/LH9320 Chip Set

The Sharp chip set [36] is designed to perform real and complex FFTs, and IFFTs, as well as linear filtering and pattern matching in the time and frequency domains. The chip has radix-2, -4, and -16 FFT building-block instructions that are connected by using the mixed power-of-two algorithm from Chapter 9 to implement up to a 4096-point complex FFT. The chip uses the Two-Signal Algorithm from Chapter 2 to compute FFTs of real input data and the Overlap-and-Add Algorithm from Chapter 6 (called overlap and discard in the Sharp application notes) for performing linear filtering and pattern matching in the frequency domain.

Figure 14-31 is a block diagram of how to interface this chip set with data memory and algorithm control logic for the most efficient execution of FFT algorithms. In addition to the LH9124, the address generator function is also provided as a chip by Sharp and is called their LH9320. The primary distinguishing feature of this chip set is that it performs FFTs using 24-bit block-floating-point arithmetic. This makes the random quantization noise at the output of the FFT computation 8 bits less than using a 16-bit block-floating-point processor. This allows frequency components that are 24 dB lower to become visible above quantization noise.

In Figure 14-31, the Q-port is used to input data and to output results from the processor. The C-port is used to provide weighting function coefficients, complex multiplier coefficients, and frequency-domain linear filter/pattern matching coefficients. This allows any weighting function or filter coefficients to be used by the processor.

The A- and B-ports are used to store intermediate results during the various stages of the computations. If data is stored in the RAM connected to data port A, then the next step is to pass that data into the processor to execute the next stage of the FFT algorithm and store the results in the data RAM connected to port B. The opposite process occurs at the next stage of computations.

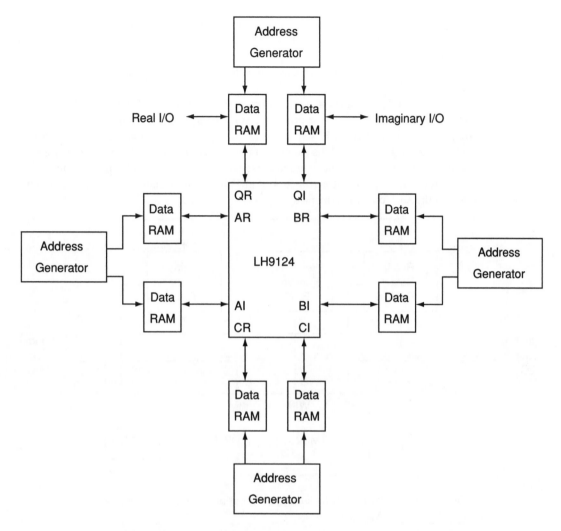

**Figure 14-31**    Full-speed single LH9124 FFT implementation block diagram.

Unlike the array Microsystems chip set, either intermediate RAM can feed data to the output. However, the same data RAM is used for both input and output data, as shown in Figure 14-31. This requires more coordination between the input of data and the output of results than is required by the array Microsystems chip set.

### 14.7.3 Raytheon TMC2310 Chip

The Raytheon TMC2310 chip [37] is designed to perform real and complex FFTs, and IFFTs, and linear filtering and pattern matching in the time domain. The chip has radix-2 FFT building-block instructions that are connected using the primes-to-a-power algorithm from Chapter 9 to implement 16-, 32-, 64-, 128-, 256-, 512-, and 1024-point real or complex FFTs. The chip does not support sequencing for executing real FFTs or linear filtering in the frequency domain. However, both real FFT algorithms from Chapter 2 and frequency-domain filtering/pattern matching algorithms from Chapter 6 can be implemented with off-chip logic because the chip does support complex and real multiplication.

Figure 14-32 is a block diagram of how to interface this FFT chip with data memory and algorithm control logic. The primary distinguishing features of this chip is that it can compute all power-of-two FFTs from 16 to 1024 points and has the complex multiplier coefficients for these algorithms stored in an on-chip ROM. Its 16-bit block-floating-point arithmetic provides better quantization noise performance than 16-bit fixed-point processors, and its off-chip weighting function RAM allows any weighting function or complex filter coefficients to be implemented.

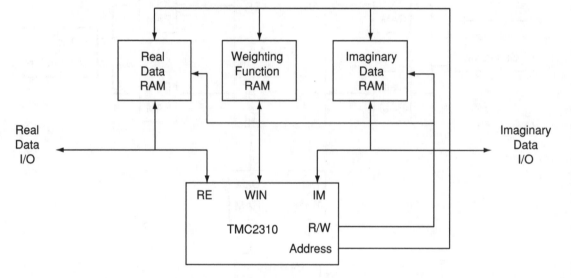

**Figure 14-32**   Hardware block diagram for computing FFTs using the TMC2310.

### 14.7.4 Plessey Semiconductor PDSP16510 Chip

The Plessey PDSP16510 [38] performs the radix-4 mixed-radix FFT and IFFT algorithms on real or complex data of 256 or 1024 points. The device can also compute sixteen 16-point or four 64-point FFTs. All of the computations are performed with block-floating-point arithmetic with 16-bit mantissas. The internal organization of the chip allows it to simultaneously input new data, transform the previous input data set, and output the results from the data set prior to the one being transformed.

Figure 14-33 is a block diagram of how to interface this FFT chip with data memory and algorithm control logic. The primary distinguishing features of this chip are that it has the complex multipliers for up to a 256-point FFT stored in on-chip ROM and either Hamming or Blackman-Harris (67-dB version) weighting functions (Sections 4.2.7 and 4.2.9b) can be applied to the input data by the chip because they are also stored inside.

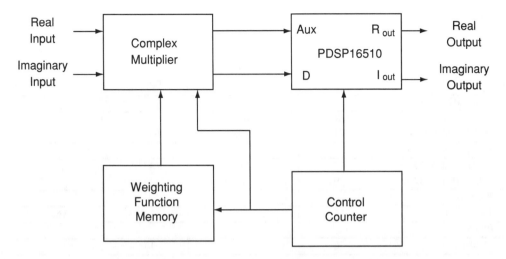

**Figure 14-33**   Arbitrary weighting or frequency-domain filtering/pattern matching block diagram.

If another weighting function is required, it must be applied before inputting the data to the chip. Similarly, if the device is to be used to perform linear filtering or pattern matching in the frequency domain, an off-chip complex multiplier must be connected as shown in Figure 14-33. No off-chip data memory is needed up to 256-point FFTs. Figure 14-34 shows the configuration required for 1024-point FFTs. Plessey makes a companion chip (PDSP16540) to perform the needed data memory addressing function, including the address and clock timing interfaces.

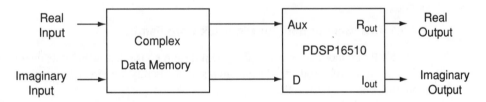

**Figure 14-34**   Off-chip buffer configuration for 1024-point FFTs.

## 14.8 FFT-SPECIFIC CHIP AND CHIP SET COMPARISON MATRIX

The data in the Comparison Matrix in Table 14-5 comes from the referenced vendor material. For the 1024-point complex FFT performance, this is the fastest number available in the referenced material. Different versions of a 1024-point FFT may produce slightly different performance numbers. Versions of the chips that run at slower speeds will have times that are slower. Conversely, newer versions of these chips, which run faster, will have faster times.

**Table 14-5** FFT-Specific Chip and Chip Set Comparison Matrix

| FFT-specific chip/set | 1024-point complex FFT $\mu s$ | Programmed FFT building blocks | Largest complex FFT | # of block floating-point mantissa bits |
|---|---|---|---|---|
| **array Microsystems** | | | | |
| a66110/a66210 | 131 | 2 and 4 points | 65,536 | 16 |
| a66111/a66211 | 131 | 2 and 4 points | 65,536 | 16 |
| **Sharp Electronics** | | | | |
| LH9124/LH9320 | 87 | 2, 4, and 16 points | 4,096 | 24 |
| LH9124L/LH9320 | 129 | 2, 4 and 16 points | 4,096 | 24 |
| **Raytheon** | | | | |
| TMC2310 | 514 | 2 point | 1,024 | 16 |
| **Plessey** | | | | |
| PDSP16510 | 96 | 4 point | 1,024 | 16 |

## 14.9 APPLICATION-SPECIFIC INTEGRATED CIRCUITS

Application-specific integrated circuits (ASICs), with programmable DSP processors as building blocks, are a recent addition to the DSP market. Once these processors are provided as an ASIC building block, the data I/O, control, and synchronization functions can be added to develop efficient DSP applications on a single chip. The front-end design of these chips generally costs more than designing a board with the equivalent functions. However, the resulting product will require less power and board area and often run faster because the I/O from the DSP building block to peripheral devices is inside the chip.

### 14.9.1 DSP Semiconductor Pine/Oak Core Family

DSP Semiconductor is a DSP system design house that licenses its own fixed-point DSP core for ASIC products. The members of this family are Pine DSP core and Oak DSP core (see Figure 14-35) [39].

Serial Ports.  This family contains the basic DSP core without serial ports because it is a core for an ASIC chip.

Multiplexer.  This family does not multiplex its on-chip data and address buses off the chip because these devices are DSP core designs to be integrated into a larger device on a single chip.

Address Generators.  All of the members of this family have dual address generators. This maximizes the ability to address both data and multiplier constants to feed to the MAC unit on each instruction cycle. The flexibility of the address step sizes for these generators also allows them to be easily used to generate non-power-of-two algorithms as well as standard FFTs.

Data Memory.  Both of the members of this family have from a minimum of 144 words up to 2048 words of data RAM. This allows them to compute up to a 1024-point

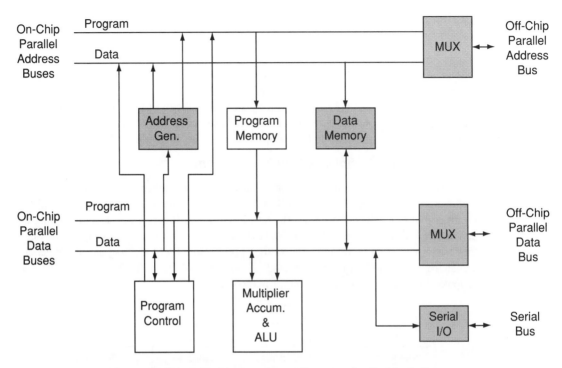

**Figure 14-35**    DSP Semiconductor pine core family block diagram.

complex FFT without adding data memory to the ASIC design. Program memory must be added to store the algorithm code and the multiplier constants.

## 14.10 ASIC PROGRAMMABLE DSP CHIP CORES COMPARISON MATRIX

The data in the Comparison Matrix in Table 14-6 comes from the referenced vendor material. For the 1024-point complex FFT performance, this is the fastest number available in the referenced material. Different versions of a 1024-point FFT may produce slightly different performance numbers. Versions of the chips that run at slower speeds will have times that are slower. Conversely, newer versions of these chips, which run faster, will have faster times.

**Table 14-6**    ASIC Programmable DSP Chip Cores Comparison Matrix

| ASIC programmable DSP chip core | 1024-point complex FFT (MS) | Data I/O ports | On-chip data memory words | On-chip prog. memory words | # of address generators |
|---|---|---|---|---|---|
| **DSP Semiconductor** | | | | | |
| Pine core | 2.2 | 0s/0p | 2048 | 0 | 2 |
| Oak core | 2.2 | 0s/0p | 2048 | 0 | 2 |

s = serial port; p = parallel port.

## 14.11 MULTIPLE PROCESSORS ON A SINGLE CHIP

Another new trend in programmable DSP chips is to have multiple processors on a single chip. Choosing one of these chips implies understanding not only the performance of the individual processors but also their interconnection architecture. For this reason this section first presents the top-level processor interconnection architecture for each chip family and describes its operation. This is followed by a block diagram of the individual processors that are integrated onto the chip. In each case these processors are Harvard architectures that work much like the generic DSP chip block diagram in Figure 14-1.

### 14.11.1 Star Semiconductor SPROC-1000 Family

The SPROC-1000 family [40] of 24-bit fixed-point DSP chips has a multiprocessor architecture fed by a single program RAM and a single data RAM. The members of this family are SPROC1400, SPROC1200, and SPROC1210. Figure 14-36 is a block diagram of the SPROC1400. The SPROC1200/1210 chips have the same block diagram except they have two, rather than four, general signal processors. A block diagram of the general signal processors is shown in Figure 14-37. The overall chip architecture is described first, followed by a description of the general-purpose DSP.

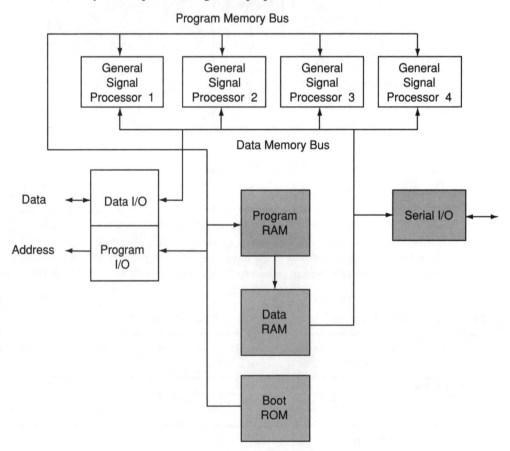

**Figure 14-36** Star Semiconductor SPROC-1400 family block diagram.

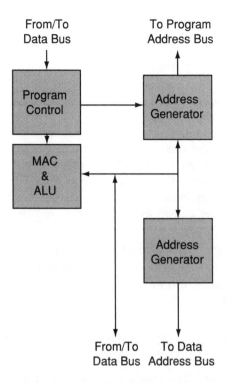

**Figure 14-37**    Star Semiconductor general signal processor block diagram.

The multiprocessor architecture is similar to the linear bus described in Section 11.2.2 with multiple processors and data memory on the bus. Star Semiconductor has devised a unique time-division-multiplexing scheme to remove the complexity of the four (two for the SPROC1200/1210) processors trying to access the data memory from the same bus. For example, the program memory bus has a five-cycle sequence. Each of the four processors is assigned to use the bus during one of the five cycles, and the fifth cycle is for data I/O. The same is true of the data memory bus.

Each of the four general-purpose DSPs has a five-stage pipeline processing cycle to match the five-cycle bus multiplexing scheme. By time-multiplexing the program and data accesses of each processor, all five can be kept busy without causing bus contention. Each processor has its own 24-bit fixed-point MAC (multiply-accumulator; Figure 14-37).

The building-block form of FFT algorithms matches well with this architecture. At a top level, consider the implementation of a 256-point radix-4 FFT algorithm. The algorithm has four stages, and at each stage it requires 64 four-point FFT computations. One strategy for performing this algorithm on the SPROC1400 is to allocate 16 of the 64 four-point FFTs at each stage to one of the four processors. Since each 4-point building block is identical, each processor has the exact same code to execute and therefore finishes its portion of each stage at the same time.

This approach also makes this architecture good for computing the Winograd, prime factor, or mixed-radix algorithms from Chapter 9. For example, consider the $3*5*8 = 120$-point prime factor algorithm. The 3-point stage requires computing $120/3 = 40$ three-point building blocks. For the SPROC1400 this means each processor performs 10 three-point FFTs. The 5-point stage requires computing $120/5 = 24$ five-point building blocks. For the SPROC1400 this means each processor performs 6 five-point FFTs. Finally, the eight-point stage requires $120/8 = 15$ eight-point building-block computations. For this stage, three of the four processors compute 4 eight-point FFTs and one only computes three. The single central data RAM makes accessing the proper inputs for each of these building-block computations straightforward.

At first glance, having all the processors repeat the same algorithm causes lost cycles, while each processor waits for its turn to obtain input data and output results. In reality, the solution is simple. At the end of the first time the processors finish a block of algorithm code, the processors send results out in sequence and receive new data in sequence. From that point on, the processors are out of synchronization by one, two, three, and four clocks and therefore have outputs available, in time sequence, so that processor cycles are not lost.

**Serial I/O.**   All members of this family have two serial input ports and two serial output ports. This additional serial port provides capability to interface these devices into linear bus, pipeline, and ring bus architectures for multiprocessor applications (Section 14.2.9) without having to use the parallel bus that may be addressing off-chip data or program memory.

**Program RAM.**   The SPROC1200 and SPROC1210 have 512 words of program RAM, and the SPROC1400 has 1024 words of program RAM.

**Data RAM.**   The SPROC1200 and SPROC1210 have 512 twenty-four-bit words of data RAM, and the SPROC1400 has 1024 twenty-four-bit words of data RAM. This limits the complex FFTs that can be performed on-chip to 256 and 512 points, respectively. Therefore, the 1024-point FFT performance numbers in the Multiple Processor Programmable DSP Chips Comparison Matrix (Section 14.12) already reflect the penalty paid for addressing off-chip data memory.

**Boot ROM.**   Boot ROM is additional on-chip memory to allow the on-chip program RAM to be loaded during the power-up phase of the application's operation from a low-speed 24-bit-wide EPROM to lower the cost of the overall application. It also allows multiple programs to be swapped in and out of the on-chip program memory without having to store them in high-speed off-chip program RAM.

**Multiply-Accumulator (MAC) and Arithmetic Logic Unit (ALU).**   Unlike the generic programmable DSP chip block diagram (Figure 14-1), the MAC and ALU in this architecture have only one bus to input data and output results. This is not a problem for computing FFTs because the multiply-accumulate function takes three clock cycles to implement, not one cycle like the generic programmable DSP chip, and a data interface with the main chip architecture can only occur every five cycles.

**Address Generators.**   Each general signal processor has two address generators. One handles program memory addressing and one handles data memory addressing. These generators are capable of direct and indexed addressing needed to implement the FFT algorithms in Chapters 8 and 9.

Program Control.   Program control logic controls the sequencing of the various functions in the general signal processor, such as address generation and the three steps in each multiply computation.

### 14.11.2  Texas Instruments TMS320C8x Family

The TMS320C8x is the first programmable DSP chip to have four DSP blocks connected by a crossbar switch and controlled by a RISC floating-point processor.  The first block diagram, Figure 14-38, shows how the four processors are interconnected with each other and on-chip memory.  The second block diagram, Figure 14-39, shows the internal architecture of the programmable DSP blocks.  The only member of this family is the TMS320C80 [41].

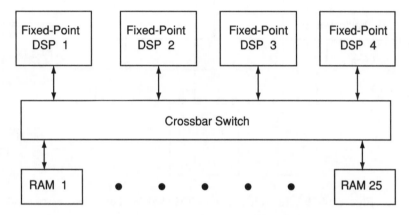

**Figure 14-38**   High-level block diagram of TMS320C80 family.

Each fixed-point DSP has a $16 \times 16$ fixed-point multiplier, so it is a 16-bit fixed-point processor, and the processor to memory buses are configured as 16 bits.  Section 12.5.1 provides a detailed look at the pros and cons of implementing FFT algorithms on a crossbar architecture.  The building-block form of FFTs matches well with this architecture.  At a top level consider the implementation of a 256-point radix-4 FFT algorithm.  The algorithm has four stages, and at each stage it requires 64 four-point FFT computations.  One strategy for performing this algorithm on the TMS320C80 is to allocate 16 of the 64 four-point FFTs at each stage to one of the four processors.  Since each four-point building block is identical, each processor has the exact same code to execute and therefore finishes its portion of each stage at the same time.

This approach also makes this architecture good for computing the Winograd, prime factor, or mixed-radix algorithms from Chapter 9.  For example, consider the $3*5*8 = 120$-point prime factor algorithm.  The 3-point stage requires computing $120/3 = 40$ three-point building blocks.  For the TMS320C80 this means each processor performs 10 three-point FFTs.  The five-point stage requires computing $120/5 = 24$ five-point building blocks.  For the TMS320C80 this means each processor performs 6 five-point FFTs.  Finally, the eight-point stage requires $120/8 = 15$ eight-point building-block computations.  For this stage, three of the four processors compute 4 eight-point FFTs and one only com-

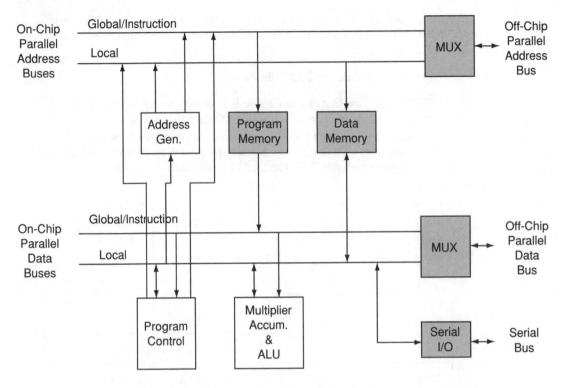

**Figure 14-39** Texas Instruments TMS320C8x family processor block diagram.

putes three. The crossbar switch interface to data RAM makes accessing the proper inputs for each of these building-block computations straightforward.

The architecture of the individual fixed-point DSPs is shown in Figure 14-39. Each has two address generators and no data or program memory or multiplexers to combine the data and program buses. Additionally, there is a third address and data bus pair, called the global bus. The serial I/O is also missing from the DSPs because it is not needed in this highly integrated internal chip architecture.

## 14.12 MULTIPLE-PROCESSOR PROGRAMMABLE DSP CHIPS COMPARISON MATRIX

The data in the Comparison Matrix in Table 14-7 comes from the referenced vendor material. In the case of the 1024-point complex FFT performance, this is the fastest number available in the referenced material. Different versions of a 1024-point FFT may produce slightly different performance numbers. Versions of the chips that run at slower speeds will have times that are slower. Conversely, newer versions of these chips, which run faster, will have faster times. Performance numbers with an asterisk behind them are estimated because times for the 1024-point FFT were not available from the vendor.

**Table 14-7**    Multiple-Processor Programmable DSP Chips Comparison Matrix

| Multiple-processor programmable chip | 1024-point complex FFT (MS) | Data I/O ports | On-chip data memory words | On-chip prog. memory words | # of address generators |
|---|---|---|---|---|---|
| **Star Semiconductor** | | | | | |
| SPROC1400 | 2.4 | 2s/1p | 1024 | 1024 | 1 |
| SPROC1200 | 4.8* | 2s/1p | 512 | 512 | 1 |
| SPROC1210 | 4.8* | 2s/1p | 512 | 1024 | 1 |
| **TI** | | | | | |
| TMS320C80 | 0.163* | 0s/1p | 50K total | 50K total | 8 |

\* = estimate; s = serial port, p = parallel port.

## 14.13 CONCLUSIONS

Choices, choices, and more choices! Few engineers have the time to keep abreast of the rapid changes and hundreds of options available for creating DSP products in general and FFT products in particular. This comprehensive inventory would be hard to choose from without the guidelines given with a "standardized" approach to block diagrams for each chip family. At this stage of the book, the reader is ready to select a chip or multiples of it for processing the algorithm chosen from the information in Chapters 8, 9, and 12.

The number of board-level companies and products for FFT applications is many times higher than at the chip level. Therefore, only guidelines for selecting off-the-shelf boards are provided in the next chapter.

## REFERENCES

[1] *ADSP-2101 and ADSP-2102 User's Manual—Architecture*, Analog Devices, Inc., Norwood, MA, 1990.

[2] *ADSP-2111 User's Manual—Architecture*, Analog Devices, Inc., Norwood, MA, 1990.

[3] *Mixed-Signal Processor with Host Interface Port - ADSP-21msp50A/55A/56A*, Analog Devices, Inc., Norwood, MA.

[4] *ADSP-2171 DSP Microcomputer*, Analog Devices, Inc., Norwood, MA, 1993.

[5] *WE DSP16 and DSP16A Digital Signal Processors Information Manual*, AT&T Microelectronics, Allentown, PA, 1989.

[6] *WE DSP16C Digital Signal Processor/Codec*, AT&T Microelectronics, Allentown, PA, 1991.

[7] *DSP1610 Signal Coding Processor*, AT&T Microelectronics, Allentown, PA, 1993.

[8] *DSP1616-x11 Digital Signal Processor*, AT&T Microelectronics, Allentown, PA, 1993.

[9] *Piranha Digital Signal Processor, DSP1616-x30*, AT&T Microelectronics, Allentown, PA, 1993.

[10] *DSP1617 Digital Signal Processor*, AT&T Microelectronics, Allentown, PA, 1993.

[11] *DSP56000/DSP56001 Digital Signal Processor User's Manual*, Motorola, Inc., Phoenix, AZ, 1990.

[12] *DSP56002 Digital Signal Processor User's Manual*, Motorola, Inc., Phoenix, AZ, 1993.

[13] Motorola Semiconductor Technical Data, *DSP560004 Rev 1, 24-Bit General Purpose Digital Signal Processor*, Motorola, Inc., Phoenix, AZ, 1993.

[14] *DSP56116 Digital Signal Processor User's Manual*, Motorola, Inc., Phoenix, AZ, 1990.

[15] Motorola Semiconductor Product Information, *DSP56156 and DSP56156ROM, 16-bit Digital Signal Processor*, Motorola, Inc., Phoenix, AZ, 1994.

[16] Motorola Semiconductor Product Information, *DSP56156 and DSP56156ROM, 16-bit Digital Signal Processor*, Motorola, Inc., Phoenix, AZ, 1994.

[17] Motorola Semiconductor Product Information, *DSP56166 and DSP56166ROM, 16-bit Digital Signal Processor*, Motorola, Inc., Phoenix, AZ, 1994.

[18] *Digital Signal Processor (DSP) and Speech Processor Products Data Book*, NEC Electronics, Inc., Mountain View, CA, 1992.

[19] *µPD77C25/P25 16-Bit Fixed Point CMOS Digital Signal Processor User's Manual*, NEC Electronics, Inc., Mountain View, CA, 1991.

[20] *µPD77016 (SPRX), 16-Bit Fixed-Point Digital Signal Processor*, NEC Electronics, Inc., Mountain View, CA, 1993.

[21] *µPD77220 Digital Signal Processor User's Manual*, NEC Electronics, Inc., Mountain View, CA, 1991.

[22] *First-Generation TMS320 User's Guide, Digital Signal Processing Products*, Texas Instruments, Inc., Dallas, TX, 1989.

[23] *TMS320C2x User's Guide, Digital Signal Processing Products*, Texas Instruments, Inc., Dallas, TX, 1993.

[24] *TMS320C5x User's Guide, Digital Signal Processing Products*, Texas Instruments, Inc., Dallas, TX, 1993.

[25] *Z89C00 Digital Signal Processor User's Manual*, Zilog, Inc. Campbell, CA, 1993.

[26] *ZR38000 Programmable Digital Signal Processor*, ZORAN Corporation, Santa Clara, CA, 1994.

[27] *ADSP-21020 and ADSP-21010 User's Manual*, Analog Devices, Inc., Norwood, MA, 1993.

[28] *ADSP-21060 SHARC Super Harvard Architecture Computer*, Analog Devices, Inc., Norwood, MA, 1993.

[29] *WE DSP32C Digital Signal Processor*, AT&T Microelectronics, Allentown, PA, 1990.

[30] *DSP3210 Digital Signal Processor, The Multimedia Solution*, AT&T Microelectronics, Allentown, PA, 1991.

[31] *Intel, i860 Microprocessor Architecture*, Osborne McGraw-Hill, Berkeley, CA, 1994.

[32] *DSP96002 IEEE Floating-Point Dual-Port Processor User's Manual*, Motorola, Inc., Phoenix, AZ, 1989.

[33] *TMS320C3x User's Guide, Digital Signal Processor Products*, Texas Instruments, Inc., Dallas, TX, 1990.

[34] *TMS320C4x Technical Brief, Digital Signal Processing Products*, Texas Instruments, Inc., Dallas, TX, 1991.

[35] *Digital Signal Processing a66540 FDaP User's Guide, Revision a66540IG/2.0*, array Microsystems, Inc., Colorado Springs, CO, 1992.

[36] *Application Notes, Integrated Circuits, Liquid Crystal Displays, RF Components, Optoelectronics*, Sharp Electronics Corporation, Portland, OR, 1993.

[37] *1994 Data Book, ASSP, Standard Products, ASIC Arrays & Standard Cells*, Raytheon Semiconductor, Mountain View, CA, 1993.

[38] *Digital Video & Digital Signal Processing IC Handbook*, GEC Plessy Semiconductors, Scotts Valley, CA, 1993.

[39] S. Berger, "An Application Specific DSP for Personal Communications Applications," *Proceedings of the 1994 DSPx Exposition & Symposium*, pp. 63–69 (June 1994).

[40] *SPROC-1400 Programmable Signal Processor Data Sheet*, STAR Semiconductor Corp., San Jose, CA, 1993.

[41] TMS320C80, "TI's First Multiprocessor DSP, Product Overview," Arrow Electronics, Inc., Carrollton, TX, 1994.

# 15

# Board Decisions and Selection

## 15.0 INTRODUCTION

Getting to market with an FFT product is usually less expensive and faster if commercial-off-the-shelf (COTS) hardware is available to run the algorithm efficiently. Even if the end product will not be at the board level, a commercial board can be an inexpensive way to develop and demonstrate the proof of concept. With several dozen manufacturers selling a wide variety of DSP boards for PC, VME, SBus, and embedded applications, it is unrealistic to describe and evaluate them in this chapter. That endeavor is surely an entire book by itself. This chapter provides guidelines that engineers, managers, and students can use to make their own decisions about appropriate COTS boards or the need to design one.

The key board specifications are:

- Processor
- Off-chip memory
- Analog I/O ports
- Instruction cycle time
- Parallel and serial I/O ports (buses)
- Host interface

## 15.1 FIVE BOARD SELECTION CATEGORIES

Though each application has its own specifications that affect board selection, issues can be grouped in five categories that are used to narrow board choices after the chip has been selected.

### 15.1.1 Algorithm Performance

Besides the FFT algorithm that will be computed with the DSP chip or chips on a board, data I/O, data reorganization, and additional signal processing algorithms are often part of the total processing. Knowing the FFT performance of the DSP chip does not mean that it will perform at that speed on a given board. Two factors that slow chip performance are the clock rate of the board being slower than the maximum instruction cycle time of the chip, and the on-board memory not being fast enough to send data or program instructions to the chip at the maximum rate it can receive them.

### 15.1.2 I/O Performance

The DSP chip or chips on a board may be capable of computing FFTs faster than data can move on and off the board. This makes it important to compare the board's data I/O rate with the chip's FFT benchmark. When the chip can perform FFTs faster than the I/O rate, it will be limited to that rate. The preferable situation is when the I/O rate is faster than the chip performs the FFT.

### 15.1.3 Software Support

Software support tools include assemblers, linkers, and compilers for writing code; simulators and debuggers to remove programming errors; and algorithm libraries to reduce the amount of code that must be written. The caliber of these tools affects the time required to develop a product.

### 15.1.4 Expansion Capability

Since boards are marketed to a broad customer base, a board may not meet all of the needs of an application. Daughter-card connectors and/or prototyping area are sometimes provided to allow user modifications to boards. A daughter card is a small board that connects to a main board. A prototyping area is space left empty on a board to allow a designer to add components to the board to enhance its capabilities. Both options are less expensive than designing a board from scratch. Sometimes board manufacturers offer daughter boards that provide the most common extra features, such as memory and I/O interface. For low-volume and custom designs, these options offer the ability to upgrade the product to meet changing customer requirements.

### 15.1.5 Multiprocessing

In a multiprocessor application, a COTS solution can be a single board with more than one chip connected in the selected architecture, or multiple boards, with one or more chips, that can be connected in the selected architecture. Chapters 11 and 12 provide extensive information on how to select multiprocessor architectures. When boards are connected in one of those architectures, performance is reduced if data I/O between the processors is slower than the processor's I/O instruction rate.

## 15.2 BOARD SELECTION QUESTIONS AND ANSWERS

This section deals with issues designers face when selecting or designing a board. If a single-chip solution meets the specifications, the last three questions do not apply.

**Question**

1. Which boards have the selected DSP chip?

**Answer**

The fastest way to narrow the number of board candidates is by eliminating those that do not have the chip already chosen. If two or more chips would meet product specifications, all of the boards without those are eliminated.

**Question**

2. Does the board slow the FFT performance of the chip?

**Answer**

The timing on the chip does not always translate to the same timing on the board because of slower board instruction cycle time and/or memory speed. Board vendors list instruction cycle time or clock rate (which can be the same or a multiple of the instruction cycle time) in the board specifications. Memory speed is listed by vendors in terms of the number of ws (wait state). If the off-chip memory runs at the same speed as the chip can access it, this is called 0 ws. If it runs at half the speed the chip can access it, the ws is 1, because the chip must wait one instruction cycle after it requests data.

**Question**

3. What digital I/O ports does the board have?

**Answer**

There are three types of digital interfaces found on COTS boards. The first is the standard bus interface such as PC, VME, or SBus. These are always parallel and generally slower than a DSP chip is capable of transferring data, which slows the chip's performance. The second is a serial interface, such as RS-232C. Most of the general-purpose DSP chips in Sections 14.3 and 14.5 have serial interfaces that work with an RS-232C.

The third and most preferable type of interface is a dedicated parallel interface, designed to run at the DSP chip's parallel I/O instruction rate. Not all boards have this feature because it requires adding a special-purpose connector and interface logic to the board. However, when this is available, the board's DSP chip is able to function at its maximum rate. This is a key element of a multiprocessor hardware architecture's ability to perform at peak efficiency.

**Question**

4. Does the board have analog I/O ports?

**Answer**

Not every board has analog I/O ports because some are designed to only receive and send digital data. The analog I/O port or ports use A/D and D/A functions in the DSP chip or on the board to convert analog signals to digital ones that the chip can process. The performance measures for A/D and D/A are the number of bits per sample and the number of samples per second that they convert.

## Question

5. Does the board have enough off-chip data and program memory?

## Answer

The amount of memory an application needs is determined by the FFT algorithm and transform length. The portion of that memory that will be off-chip is a function of the chip selected. Some may even be off-board, depending on which board is used. The on-chip memory is subtracted from the total memory to see how much the board needs to have. If there is too much remaining for a board to handle, an external source such as host processor RAM or hard disk, or a separate memory board, must be available.

## Question

6. Which boards work with the selected high-level language?

## Answer

Various versions of C and FORTRAN are common programming languages for engineers and scientists. In recent years, graphical user interface (GUI) software has become a popular way to go from block diagram design to C code. If the manufacturer of the board, or the DSP chip on it, supports application software, including library routine calls, in one of these languages, development time is reduced. The price paid for faster software development is the inefficiency of cross compilers when converting C and FORTRAN code to DSP chip code. Code converted from high-level languages can take two to five times longer to execute than DSP chip assembly language.

## Question

7. Does the algorithm library provide the needed FFT length?

## Answer

If the chip's algorithm library does not have the needed FFT length, maybe the board's library will. The more code an algorithm library provides, the less must be written in high-level or assembly languages. This reduces development time and speeds up processing because the algorithm library routines are usually written in assembly language. Even if entire algorithms are not available in the algorithm library, decomposing the needed algorithms into building blocks that are available speeds execution of the algorithm and shortens development time. If code is not available in a chip or board algorithm library, it may be available from a third-party supplier.

## Question

8. Do the algorithm library routines have a common I/O format?

## Answer

Ideally, an application can be constructed by using a sequence of routines from the algorithm library. However, if the data I/O formats for these routines are not the same, additional algorithms must be executed between the algorithm library routines to allow the data to flow from one routine to the next.

For example, suppose the application requires an FIR filter followed by an FFT. The input to and output from the FIR filter library routine is likely to be in sequential order, simply because that is how FIR filters are implemented. Then the filter routine will perform all the multiplies and adds to produce a new output each time a new input data value enters the routine.

On the other hand, the $N$-point FFT routine needs a set of $N$ samples at one time. Therefore, a buffer must be set up between the FIR filter routine and the $N$-point FFT routine to accumulate $N$ FIR outputs to use for the next $N$-point FFT input set (Figure 15-1). The output of the FFT library routine provides $N$ answers at one time. To convert this block of data back to a sequence of results requires another data buffer routine. All of this adds to the application execution time and to the development time and cost.

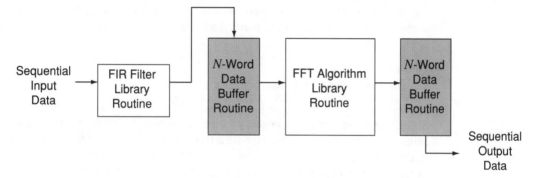

**Figure 15-1**    Connecting algorithm library routines.

### Question

9. Does the board support real-time operating systems (RTOS)?

### Answer

In real-time applications, a common but complex portion of the design is the code that controls the interface between the DSP chip and the data I/O interface hardware. Real-time operating systems (RTOS) are software subroutines that reduce the programming necessary to accomplish this portion of the design.

### Question

10. What control, data I/O, and graphical display software are available?

### Answer

Board manufacturers provide algorithm library software to reduce the time required for the application developer to implement required functions. Most applications also require software to control the operation of the board, control the movement of data on and off the board once the RTOS has synchronized the data interface, and interface to graphical display software and hardware. If basic algorithms are also provided by the board manufacturer for these functions, the time to market is reduced. This is because not only are these functions usually required by the application, but they can also be used to enter data and view results as part of the algorithm debugging process. Therefore, it is important to identify which of these functions are relevant for the application and determine if they are available from the board manufacturer, chip manufacturer, or a third-party supplier.

**Question**

11. Can the board be expanded with a daughter card?

**Answer**

One way to expand the capability of a board is by connecting a smaller board (daughter card) to it. This has two advantages over adding more boards. The first is cost. The small boards are generally less expensive than large ones and add little space to the volume required by the application. The second is performance. The connections to the daughter cards are much shorter, and therefore faster, than those between full cards.

**Question**

12. Does the board have prototyping area?

**Answer**

Some boards may meet the majority of the needs of an application but be missing something vital. For example, suppose a board can perform all of the computations in the required time but does not have the A/D and D/A converters needed. If the board vendor provides a prototyping area, then the application developer can put these functions in the prototyping area. The resulting product only requires one board rather than an additional A/D and D/A interface board. This reduces the cost, size, and weight of the product.

**Question**

13. Does the board have the selected architecture?

**Answer**

The fastest way to narrow the number of board candidates is by eliminating those that do not have the chip and architecture which have already been selected. If more than one board meets those specifications, the issues dealt with in the preceding questions and answers are used to further narrow the choice. If no single board is suitable, the answer to Question 14 must be used.

**Question**

14. Can the board be connected to one or more copies of itself, using the selected architecture?

**Answer**

The digital I/O ports on the board determine what kinds of multiprocessor architectures can be implemented. The text and figures in Section 14.2.9 show how to use chip serial I/O ports to form multiprocessor architectures. These same concepts can be applied to board interconnections by replacing the DSP chips in those figures with DSP boards, whether the I/O ports are parallel or serial. If no board exists that can be configured into the selected architecture, a custom board must be designed or the architecture decision must be revised.

**Question**

15. Can the board move data at the processor's I/O instruction rate?

**Answer**

An architecture was chosen because of its throughput and/or latency performance with a particular algorithm. Chapters 11 and 12 dealt with how efficiently architectures compared, assuming each processor takes one instruction cycle for each add, multiply, or data move. If the data input, intermediate, or output results overhead (which comprise total I/O instruction time) take more than one cycle, that portion of the architectures's throughput or latency will be slowed. It is important to be aware of this possible slowdown and what causes it. This is most likely to occur when a board uses a standard bus, and is least likely to happen when a board has a dedicated parallel interface.

## 15.3 CONCLUSIONS

Many factors must be carefully evaluated to be certain that a COTS board will do the job that meets the specifications of a product. Designers should know how to answer these questions for their application before purchasing a board or when deciding on the specifications for a custom-designed board. The next chapter gives the test signals and methods needed to detect and isolate errors that occur during software development on the board chosen using these guidelines.

# 16

# Test

## 16.0 INTRODUCTION

The book would not be complete without explaining how to test the performance of the FFT algorithms it shows how to construct and implement. This chapter provides test signals and shows how to use them to detect and isolate the errors that occur during development of FFT algorithms, conversion of them to code, and operation of them in a product. Each area is explained separately. A recommended set of test signals is described, and its ability to detect and isolate errors is illustrated, using the 4-point FFT example from Section 8.5 and the 16-point radix-4 FFT example from Section 9.7.5.

## 16.1 EXAMPLE

This chapter uses the 16-point radix-4 FFT example to illustrate the test signals and methods explained here. This algorithm is a mixed-radix technique from Chapter 9 and uses the 4-point building block from Chapter 8. Figure 16-1 is a flow graph of the 4-point building block, and Figure 16-2 is a flow graph of the 16-point radix-4 FFT. Unlike Chapters 8 and 9, where Memory Maps are more useful than flow graphs, flow graphs are the most powerful way to understand the test process, because it is so easy to see the path from the error to the FFT outputs. This allows the output error patterns to be easily understood.

## 16.2 ERRORS DURING ALGORITHM DEVELOPMENT

Algorithm development includes the Algorithm Steps and Memory Maps for the needed building-block algorithms as well as for combining them into the complete $N$-point FFT. The building blocks from Chapter 8 and algorithms in Chapter 9 have been checked, using the techniques described in this section, to ensure there are no algorithm errors. If another

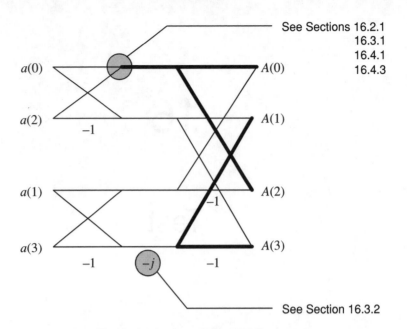

**Figure 16-1** Four-point FFT flow graph.

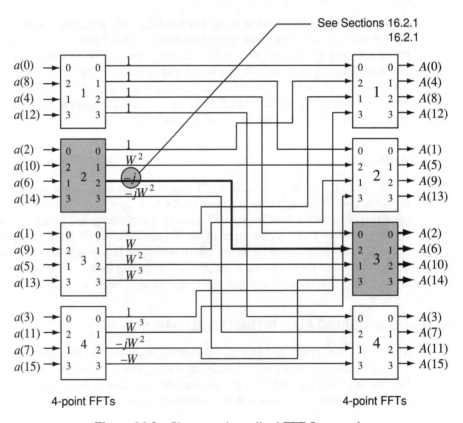

**Figure 16-2** Sixteen-point radix-4 FFT flow graph.

building block or algorithm is going to be used, it is recommended that test signals be used to verify the Algorithm Steps and Memory Maps prior to implementing the algorithm in code.

### 16.2.1 Arithmetic Check

Algorithm Step (arithmetic) errors can occur at the building-block level or in defining the complex multipliers between the stages. The most complete method for ensuring the correctness of the arithmetic is to start from each complex output frequency term, $A(i)$, and write the Algorithm Step for the terms with the Algorithm Step that is used to calculate it. Then continue to move back through the algorithm and replace each term that makes up those terms. This process continues until the equation is in terms of the complex input data, $a(i)$. Then compare that equation with the corresponding DFT equation to ensure they are the same.

The 4-point FFT, shown in Figure 16-1, provides a simple example that illustrates this approach. The Algorithm Steps for each of the output frequency terms (Equation 16-1) are listed first, followed by the corresponding 4-point DFT (Equation 16-2).

$$A_R(0) = b_R(0) + b_R(2) = [a_R(0) + a_R(2)] + [a_R(1) + a_R(3)]$$
$$A_I(0) = b_I(0) + b_I(2) = [a_I(0) + a_I(2)] + [a_I(1) + a_I(3)]$$
$$A_R(1) = b_R(1) + b_I(3) = [a_R(0) - a_R(2)] + [a_I(1) - a_I(3)]$$
$$A_I(1) = b_I(1) - b_R(3) = [a_I(0) - a_I(2)] - [a_R(1) - a_R(3)]$$
$$A_R(2) = b_R(0) - b_R(2) = [a_R(0) + a_R(2)] - [a_R(1) + a_R(3)] \qquad (16\text{-}1)$$
$$A_I(2) = b_I(0) - b_I(2) = [a_I(0) + a_I(2)] - [a_I(1) + a_I(3)]$$
$$A_R(3) = b_R(1) - b_I(3) = [a_R(0) - a_R(2)] - [a_I(1) - a_I(3)]$$
$$A_I(3) = b_I(1) + b_R(3) = [a_I(0) - a_I(2)] + [a_R(1) - a_R(3)]$$

$$A(0) = \sum_{n=0}^{3} a(n) * e^{-j2\pi 0n/4} = a(0) + a(1) + a(2) + a(3)$$

$$A(1) = \sum_{n=0}^{3} a(n) * e^{-j2\pi n/4} = a(0) - j*a(1) - a(2) + j*a(3)$$

$$\qquad\qquad\qquad\qquad\qquad\qquad\qquad\qquad\qquad\qquad\qquad (16\text{-}2)$$

$$A(2) = \sum_{n=0}^{3} a(n) * e^{-j\pi n} = a(0) - a(1) + a(2) - a(3)$$

$$A(3) = \sum_{n=0}^{3} a(n) * e^{-j3\pi n/2} = a(0) + j*a(1) - a(2) - j*a(3)$$

where

$$a(n) = a_R(n) + j*a_I(n)$$
$$j*a(n) = -a_I(n) + j*a_R(n)$$

If the real and imaginary parts of input data, $a(n)$, are substituted in Equation 16-2, the result is

$$A_R(0) = a_R(0) + a_R(1) + a_R(2) + a_R(3)$$
$$A_I(0) = a_I(0) + a_I(1) + a_I(2) + a_I(3)$$
$$A_R(1) = a_R(0) + a_I(1) - a_R(2) - a_I(3)$$
$$A_I(1) = a_I(0) - a_R(1) - a_I(2) + a_R(3)$$
$$A_R(2) = a_R(0) - a_R(1) + a_R(2) - a_R(3) \qquad (16\text{-}3)$$
$$A_I(2) = a_I(0) - a_I(1) + a_I(2) - a_I(3)$$
$$A_R(3) = a_R(0) - a_I(1) - a_R(2) + a_I(3)$$
$$A_I(3) = a_I(0) + a_R(1) - a_I(2) - a_R(3)$$

The final step is to compare Equations 16-1 and 16-3 to see that they are mathematically identical. Notice that the order of the $a(i)$ terms in the two sets of equations is different. This is caused by the sequence of Algorithm Steps used to reduce the total computations. However, the equations all have the same terms. Therefore, all of the building-block arithmetic is correct.

If there is an error, the flow graph in Figure 16-1 is invaluable in tracing the source of that error. For example, suppose the node in Figure 16-1 that adds $a(0)$ to $a(2)$ is a subtract instead of an add. Then, using Figure 16-1, that error affects $A(0)$ and $A(2)$ but not $A(1)$ and $A(3)$. Therefore, if $a(2)$ has the wrong sign in $A(0)$ and $A(2)$, it must have been subtracted from, not added to, $a(0)$. Each arithmetic error in the algorithm has its own pattern that can be easily discerned by looking at how the error propagates to the output of the flow graph.

This same process can be used at the complete algorithm level to verify the accuracy of the complex multiplications between the building blocks and that the output of the first-stage building blocks is input to the proper places in the second-stage building blocks. At first this looks like a very large set of computations to perform. Fortunately, the regularity of the building-block interconnection algorithms and the fact the building blocks have been checked can be used to simplify these checks significantly.

The 16-point radix-4 FFT, shown in Figure 16-2 and used later as an example, illustrates these features. The input to each of the four output 4-point FFTs is 4 of the 16 input building-block outputs, modified by the appropriate complex multipliers. Since the 4-point building-block arithmetic is known to be correct, checking any one of its four 4-point outputs verifies that the correct data has been sent to it. Therefore, only four output frequency terms must be checked to verify the algorithm, one from each of the four output 4-point FFTs.

For example, suppose the third output of the second input 4-point FFT is multiplied by $+j$, not $-j$. Then the error propagates into the third output 4-point FFT and affects frequency outputs $A(2)$, $A(6)$, $A(10)$, and $A(14)$. All of the other outputs will be correct. Since all four of the outputs of this 4-point FFT are affected by the error, it is immaterial which is chosen to check the algorithm arithmetic.

### 16.2.2 Memory Map Check

Memory mapping errors can occur at the building-block level or when combining the building blocks to form the complete FFT. The most complete method for avoiding these errors is to follow an approach similar to the steps used to detect arithmetic errors in Section 16.2.1. The Memory Map verification process is primarily looking for places where a memory location's data is modified before its present results have been used by all of the subsequent Algorithm Steps. The most efficient way to perform these checks is to start with the input Memory Map and work through to the Memory Map for the output frequency components. Because of the building-block nature of FFT algorithms, the memory mapping checks must be performed at two levels. First the memory mapping is checked at the building-block level. Then the building blocks are combined and the overall algorithm memory mapping is checked.

The 4-point FFT in Figure 16-1 is again used as an example. The Algorithm Steps and Memory Map in the first list below are from Chapter 8. The second list shows the sequence of values stored in each data memory location as the algorithm is executed. For the 4-point FFT all of the computations are performed by pulling two pieces of data from memory, doing the arithmetic, and storing the results in the same locations used by the two pieces of input data. For most of the building blocks, additional memory locations are needed to avoid writing over a data value needed later in the computations. The Comparison Matrix at the end of Chapter 8 shows the number of additional memory locations used by each of the building-block algorithms.

### Four-Point FFT Algorithm Steps and Memory Map

| Algorithm Steps | Memory Map |
|---|---|
| $b_R(0) = a_R(0) + a_R(2)$ | $b_R(0) \Rightarrow M(0)$ |
| $b_R(1) = a_R(0) - a_R(2)$ | $b_R(1) \Rightarrow M(2)$ |
| $b_I(0) = a_I(0) + a_I(2)$ | $b_I(0) \Rightarrow M(4)$ |
| $b_I(1) = a_I(0) - a_I(2)$ | $b_I(1) \Rightarrow M(6)$ |
| $b_R(2) = a_R(1) + a_R(3)$ | $b_R(2) \Rightarrow M(1)$ |
| $b_R(3) = a_R(1) - a_R(3)$ | $b_R(3) \Rightarrow M(3)$ |
| $b_I(2) = a_I(1) + a_I(3)$ | $b_I(2) \Rightarrow M(5)$ |
| $b_I(3) = a_I(1) - a_I(3)$ | $b_I(3) \Rightarrow M(7)$ |
| $A_R(0) = b_R(0) + b_R(2)$ | $A_R(0) \Rightarrow M(0)$ |
| $A_I(0) = b_I(0) + b_I(2)$ | $A_I(0) \Rightarrow M(4)$ |
| $A_R(2) = b_R(0) - b_R(2)$ | $A_R(2) \Rightarrow M(1)$ |
| $A_I(2) = b_I(0) - b_I(2)$ | $A_I(2) \Rightarrow M(5)$ |
| $A_R(1) = b_R(1) + b_I(3)$ | $A_R(1) \Rightarrow M(2)$ |
| $A_R(3) = b_R(1) - b_I(3)$ | $A_R(3) \Rightarrow M(7)$ |
| $A_I(1) = b_I(1) - b_R(3)$ | $A_I(1) \Rightarrow M(3)$ |
| $A_I(3) = b_I(1) + b_R(3)$ | $A_I(3) \Rightarrow M(6)$ |

**Four-Point FFT Memory Map History**

$$M(0) : a_R(0) \Rightarrow b_R(0) \Rightarrow A_R(0)$$
$$M(1) : a_R(1) \Rightarrow b_R(2) \Rightarrow A_R(2)$$
$$M(2) : a_R(2) \Rightarrow b_R(1) \Rightarrow A_R(1)$$
$$M(3) : a_R(3) \Rightarrow b_R(3) \Rightarrow A_I(1)$$
$$M(4) : a_I(0) \Rightarrow b_I(0) \Rightarrow A_I(0)$$
$$M(5) : a_I(1) \Rightarrow b_I(2) \Rightarrow A_I(2)$$
$$M(6) : a_I(2) \Rightarrow b_I(1) \Rightarrow A_I(3)$$
$$M(7) : a_I(3) \Rightarrow b_I(3) \Rightarrow A_R(3)$$

Once the individual building-block memory mapping schemes have been checked and used to form the complete FFT, it must also be checked. For a $P * Q = N$-point FFT, there are $Q$ $P$-point FFTs performed as the input computations and $P$ $Q$-point FFTs performed as the output computations. This leads to a two-stage memory mapping check of the complete algorithm. First the input $P$-point FFT memory mapping is checked. If the memory mapping strategy from Section 9.4 is used for the input building blocks, this check is simple. In that strategy, the Memory Map of the input data to each of the input FFT building blocks is different and follows the pattern of the building-block Memory Maps from Chapter 8.

The only exception to this is the additional data memory locations that most of the building blocks require in the center of their computations. The simplest answer to the additional memory location problem is to allocate those locations to a separate area of memory not used by any of the building blocks. As mentioned in Chapter 9, only one set of extra memory locations is required for most applications. This means that, since the building-block memory mapping is already checked before combining the building blocks into a larger transform, the only thing to check is that the data memory areas for each building block do not overlap. The algorithms in Chapter 9 were checked using this approach. A similar argument ensures that the output $Q$-point FFTs do not interfere with each other.

## 16.3 ERRORS DURING CODE DEVELOPMENT

Once the algorithms have been verified, the next step is to convert the Algorithm Steps and Memory Map into the code used by the chosen programmable DSP hardware. If the code is written in a high-level language, such as C or FORTRAN, the language will allocate the data memory locations when variables are chosen. Therefore, the only errors to be introduced are in coding the Algorithm Steps. However, for many product applications, the code must be written in assembly language to obtain optimized computational speed to minimize the cost of the processor used. In this case, Algorithm Step and Memory Map errors can be introduced by the code conversion process. These can occur in the building blocks, the complex multiplier constants, the data reorganization memory mapping, and the data relabeling required by the available data memory locations.

### 16.3.1 Coding the Building-Block Algorithm

Any error in coding the Algorithm Steps of a building block propagates to the output of the building block and to the output of the complete FFT when the code is combined by using the algorithms in Chapter 9. Debugging the FFT code during development is simplified by debugging the individual building blocks before they are combined into the complete FFT. For example, with the 4-point FFT building-block algorithm in Table 16-1, if the computation of $b_R(0) = a_R(0) + a_R(2)$ is incorrectly programmed, $A_R(0)$ and $A_R(2)$ will be incorrect because $b_R(0)$ is used to compute these two outputs. Figure 16-1 shows the same thing, where $b_R(0)$ is the real part of the node that combines $a(0)$ and $a(2)$. Other arithmetic errors can also cause the same two outputs to be incorrect.

These errors can be checked with the sequence of steps described in Section 16.2 for the algorithm development stage. However, because the code is in a computer at this point and has been verified at the algorithm level, test input signals provide the most efficient means for finding coding errors. The test signals described in Section 16.5 are specifically designed to isolate errors based on the patterns they exhibit at the building block and complete FFT outputs. In both cases, the flow graph of the building block makes it easier to trace and isolate errors.

### 16.3.2 Coding the Multiplier Constants

There are three ways that the multiplier constants, both in building blocks and complex constants between building blocks, can be incorrectly converted to code. In all three cases, the error propagates to the building block and complete FFT outputs to cause errors in the answers.

The first incorrect conversion is to use the wrong equation for computing the constant. The arguments of the sines and cosines or the way they are combined to form a constant can be wrong. This causes incorrect numerical values for the constants or a sign error. For example, in the 4-point FFT, the $-j$ multiplier in Figure 16-1 is $-j * \sin(90°)$. If the argument of the sine term were $-90°$, then the multiplier would have been $+j$ and an error would have occurred in $A(1)$ and $A(3)$.

The second incorrect conversion is to use the wrong round-off technique for the arithmetic format chosen for the application. For this reason all the multiplier constants for the algorithms in Chapters 8 and 9 are in equation form rather than just numerical values. Generally, standard round-off to the nearest least significant bit is the correct approach. If the constants are truncated instead, small errors are introduced into all of the outputs. The characteristics of these quantization errors are explained in Chapter 13.

The third incorrect conversion is the result of storing the multiplier constants in the wrong locations. Then, when the multiplier constants are accessed, completely uncontrolled numbers are used. These errors propagate to the output frequency components and have the same error patterns as incorrect arithmetic computations.

### 16.3.3 Coding the Memory Mapping

Data reorganization occurs at the input and between the building-block stages of an FFT. Additionally, the complete FFT requires the building blocks to memory-map blocks of data located in multiple locations in data memory. If either of these two memory mapping schemes is incorrectly converted to code, the FFT outputs will be dramatically altered.

If the equation for input data reorganization is incorrectly implemented, it reorders the input data sequence and causes the FFT to analyze a shuffled input signal. If the equation for data reorganization between the building-block stages is incorrect, the partial patterns computed by the input building-block FFTs are destroyed and the output is also drastically altered. Finally, if the incorrect memory map conversion results in using locations that do not contain data, then a portion of the input sequence is altered. The result is a substantial change in the output of the FFT. All three of these errors can be isolated by using the test sequences in Section 16.5.

### 16.3.4 Coding the Relabeled Memory Maps

Relabeling of the memory mapping scheme developed for each building block is required for most FFT algorithms because the data does not exit the first building-block algorithms in order. When a relabeling technique, like the one recommended in Section 9.4 is needed, it is possible to make a mistake in the relabeling process. When this occurs, the algorithm memory mapping uses incorrect data for some portion of the computations. Once the error is made, it generally propagates to several output frequencies. The error pattern that occurs when each of the test signals is applied can be used to isolate this error.

## 16.4 ERRORS DURING PRODUCT OPERATION

At some point in the life of all products, a portion of its hardware fails. At that time the product can be thrown away or fixed, depending on cost and other considerations. If the decision is to fix the product, a technique must be available for isolating the failed component. If the entire product is implemented on a single DSP chip, the decision is simple. Replace the DSP chip. However, in many cases the data I/O, program memory, and data memory are external to the DSP chip. When the product is implemented with discrete circuits, rather than DSP chips, each function in Figure 16-3 may be a different piece of hardware.

The following sections describe the kinds of errors that appear when each of the functional blocks in Figure 16-3 fails and the methods for using test signals to isolate the errors. Figure 16-3 is assumed to represent the entire hardware functional block diagram for the product, and the FFT algorithm is assumed to be stored in program memory.

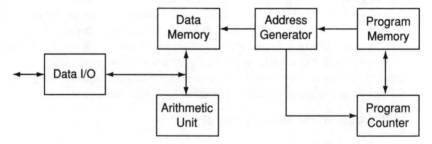

**Figure 16-3**   Harvard architecture product functional block diagram.

### 16.4.1 Arithmetic Unit

The arithmetic unit has a multiplier, adder, and accumulator register connected as shown in Figure 16-4. If one of these fails, the output of most of the arithmetic operations will be wrong. For the 16-point radix-4 FFT in Figure 16-2 and the 4-point building block

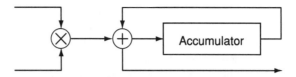

**Figure 16-4**   Multiply-accumulator.

in Figure 16-1, these arithmetic errors propagate to the output and generally cause all of the results to be wrong. Because this is a catastrophic arithmetic failure, any test signal is also likely to have all of its outputs wrong.

One exception is the zero test signal. In most cases a zero input sequence will result in zero outputs. The exception is if one of the bits of the multiplier, adder, or accumulator outputs is stuck high. However, these bits represent a very small portion of the total transistor count in the arithmetic unit. If this occurs, the zero input sequence is likely to produce the same nonzero outputs for all of the frequency components. The reason for this is that the only thing generating the nonzero numbers is the failed bit. Therefore, regardless of the arithmetic to be performed, the answer is likely to look the same.

### 16.4.2 Address Generator

The address generator is generally composed of an adder, a counter, and offset address register. If any of these fails, the address generator will produce incorrect memory maps to use to access data from memory and store results. This also causes catastrophic failure because the data to be operated on by the algorithm is not the actual input data or intermediate computational results. The failed address generator will access data in other portions of the data memory that have no relationship to the real data.

This catastrophic failure is also not able to be isolated using the test sequences described below. However, the zero test sequence can again be used to distinguish the failure. Since the output bears no relationship to the data, the output for the zero input sequence is likely to be a random sequence of numbers. This separates this failure from the arithmetic unit failure. The exception to this is when the address generator ends up accessing data from a portion of memory that has all zeros in it. However, in this case, the results of the computations will be all zeros, regardless of the input test signal used.

### 16.4.3 Data Memory

The likely failure in data memory is a bit in a memory location failing. If this occurs, one of the input data values or intermediate results changes value. With the building-block flow graph in Figure 16-1, the algorithm flow graph in Figure 16-2, and the memory map history in Table 16-2, an error in a data memory location can be propagated forward to the output frequency components. The result is failure of all of the outputs. However, this failure is detectable by using the right kind of input sequence.

For example, consider the 4-point FFT in Figure 16-1 and data memory location $M(0)$ failing by having one of its bits short to zero all the time. If the input test sequence had the $a(0)$ term equal to zero, the first set of computations would be correct because the short would not modify the input data value. However, when the answer for $b(0)$ was placed back in data memory location $M(0)$, it may or may not be in error depending on the specific value of $a(2)$. From Figure 16-1 this means that the error can propagate to $A(0)$ and $A(2)$ but not to $A(1)$ and $A(3)$. In fact, depending on the specific values of the other inputs, none of the outputs may be incorrect. One input sequence that can be used to catch this type of error in any of the memory locations is one that has a nonzero value for only one location. This is called the unit pulse when it is described in Section 16.5.1.

### 16.4.4 Program Memory

A failure in a program memory address results in a failure in one of the Algorithm Steps to be properly executed. If the error is in a memory address, the result will look much like the errors described for the address generator, except they will have a more localized pattern at the output. If the error is in a computational instruction, the errors will look much like those from the arithmetic unit, except they will not proliferate throughout the frequency outputs. They will produce a pattern of errors that can be traced back to the source using the test signals described in Section 16.5.

The most catastrophic error in program memory is in an instruction branching operation or program address offset. If this occurs, the program is likely to go off into another area of program memory and completely hang up the application.

### 16.4.5 Data I/O

There are three likely data I/O failures. The first is with the interrupt control logic that synchronizes the input of data to the processor and the output of results from the processor. When this occurs, the input data sequence is no longer correct, which results in incorrect FFT outputs.

The second and third likely failures are associated with the input and output connections for the data itself. If one of these fails, on either side of the data I/O circuitry in Figure 16-3, the signal is modified. Since the FFT is a linear computation, the resulting FFT provides answers as if there are two signals present, the actual signal and the signal which represents the data modification.  •

## 16.5  TEST SIGNAL FEATURES

This section describes the basic features of each of the four types of test signals recommended for debugging FFT algorithms. Many other combinations of signals can also be used. These recommendations are based on many years of FFT development experience coupled with a practical need to minimize the work required to ensure that FFT algorithms work. These same signals can be used during algorithm development, when the algorithm is being converted to DSP chip code, and to find failures after the product is operating. The columns in Table 16-1 show examples of each of these test signals for a 4-point complex test sequence. Table 16-2 shows the responses to those test signals as they go through the 4-point FFT in Figure 16-1.

### 16.5.1 Unit Pulse

The unit pulse is a digital signal where one of the complex values is nonzero and the others are all zero. In Table 16-1 the $a(0)$ term is chosen as the nonzero entry. However,

**Table 16-1**    Examples of Test Signals for the 4-Point FFT

|  |  |  | Sine wave 1 |
| Unit pulse | Constant | Sine wave 1 | + constant |
|---|---|---|---|
| $a_R(0) = 100$ | $a_R(0) = 100$ | $a_R(0) = 100$ | $a_R(0) = 200$ |
| $a_I(0) = 50$ | $a_I(0) = 50$ | $a_I(0) = 0$ | $a_I(0) = 50$ |
| $a_R(1) = 0$ | $a_R(1) = 100$ | $a_R(1) = 0$ | $a_R(1) = 100$ |
| $a_I(1) = 0$ | $a_I(1) = 50$ | $a_I(1) = 100$ | $a_I(1) = 150$ |
| $a_R(2) = 0$ | $a_R(2) = 100$ | $a_R(2) = -100$ | $a_R(2) = 0$ |
| $a_I(2) = 0$ | $a_I(2) = 50$ | $a_I(2) = 0$ | $a_I(2) = 50$ |
| $a_R(3) = 0$ | $a_R(3) = 100$ | $a_R(3) = 0$ | $a_R(3) = 100$ |
| $a_I(3) = 0$ | $a_I(3) = 50$ | $a_I(3) = -100$ | $a_I(3) = -50$ |

**Table 16-2**    Four-Point FFT Algorithm Responses to the Test Signals

| Responses to the unit pulse | Responses to the constant | Responses to sine wave 1 | Responses to sine wave 1 + constant |
|---|---|---|---|
| $b_R(0) = 100$ | $b_R(0) = 200$ | $b_R(0) = 0$ | $b_R(0) = 200$ |
| $b_I(0) = 50$ | $b_I(0) = 100$ | $b_I(0) = 0$ | $b_I(0) = 100$ |
| $b_R(1) = 100$ | $b_R(1) = 0$ | $b_R(1) = 200$ | $b_R(1) = 200$ |
| $b_I(1) = 50$ | $b_I(1) = 0$ | $b_I(1) = 0$ | $b_I(1) = 0$ |
| $b_R(2) = 0$ | $b_R(2) = 200$ | $b_R(2) = 0$ | $b_R(2) = 200$ |
| $b_I(2) = 0$ | $b_I(2) = 100$ | $b_I(2) = 0$ | $b_I(2) = 100$ |
| $b_R(3) = 0$ | $b_R(3) = 0$ | $b_R(3) = 0$ | $b_R(3) = 0$ |
| $b_I(3) = 0$ | $b_I(3) = 0$ | $b_I(3) = 200$ | $b_I(3) = 200$ |
| $A_R(0) = 100$ | $A_R(0) = 400$ | $A_R(0) = 0$ | $A_R(0) = 400$ |
| $A_I(0) = 50$ | $A_I(0) = 200$ | $A_I(0) = 0$ | $A_I(0) = 200$ |
| $A_R(1) = 100$ | $A_R(1) = 0$ | $A_R(1) = 400$ | $A_R(1) = 400$ |
| $A_I(1) = 50$ | $A_I(1) = 0$ | $A_I(1) = 0$ | $A_I(1) = 0$ |
| $A_R(2) = 100$ | $A_R(2) = 0$ | $A_R(2) = 0$ | $A_R(2) = 0$ |
| $A_I(2) = 50$ | $A_I(2) = 0$ | $A_I(2) = 0$ | $A_I(2) = 0$ |
| $A_R(3) = 100$ | $A_R(3) = 0$ | $A_R(3) = 0$ | $A_R(3) = 0$ |
| $A_I(3) = 50$ | $A_I(3) = 0$ | $A_I(3) = 0$ | $A_I(3) = 0$ |

any of the four positions in the sequence can have the nonzero term. The key feature of this signal is that it only activates one input to the FFT. Therefore, it shows how each input signal contributes to the output. One test approach is to apply this signal at each of the FFT inputs and ensure that the output is correct. Then, because the FFT is linear, it must work for any arbitrary input signal. The drawback to this approach is that it requires many input signals. For a 1024-point FFT, 1024 different test signals are required.

### 16.5.2 Constants

The constant signal is one where all of the complex values are the same. The key features of this input signal are that it is easy to generate and that incorrect input data reorganization does not cause errors in the output. It therefore becomes a good first test signal to verify that much of the arithmetic in an algorithm is working, independent of the input memory mapping. The biggest drawback is that the input add-subtract arithmetic common to all of the building-block FFTs has zero as the output of all of the subtractions. The $b(1)$ terms in Table 16-2 are examples of this affect. Therefore, roughly half of the algorithm's multipliers and the output arithmetic are not checked.

### 16.5.3 Single Sine Waves

The single sine wave, centered at the first nonzero output frequency of the FFT, is a signal that has exactly one cycle during the set of $N$ data values input to the FFT. In general, this test signal requires all of the multiplier constants to work to provide the correct answers. Additionally, the data reorganization memory mapping must be correct or the signal will be scrambled into another signal. This signal is best applied after the constant signal verifies most of the arithmetic. Table 16-3 shows an example of this signal for the 4-point FFT. One disadvantage of this signal is that it can also cause some intermediate points in the computations to be zero. Once that happens, subsequent computations are not checked. The $b(0)$ terms in Table 16-2 are examples of this phenomena.

### 16.5.4 Pair of Sine Waves

An input signal that is the sum of two sine waves is used to remove the problems of zeroed-out intermediate results generated by the constant and single sine-wave signals. However, since these signals are more complicated to generate and to use to decipher errors, they are best applied after the constant and single sine-wave signals have eliminated most errors. The right-hand column in Table 16-1 shows a pair of these signals for the 4-point FFT. Each entry is just the sum of the entries for the constant and single sine-wave signals. The linearity properties of the FFT ensure that this occurs all the way through the algorithm. In general, the best characteristics for these two sine waves are that they are centered at FFT output frequencies and that the frequencies are at output filter numbers that are relatively prime to each other and to the length of the FFT. The example in Table 16-1 is an exception to this approach. This is because the 4-point FFT is too small to be able to choose a pair of frequencies that meet the criteria.

## 16.6  TEST SIGNAL ERROR PATTERNS

The simplest way to illustrate the types of patterns that errors produce is with an example. Most algorithm errors produce errors with specific patterns, regardless of the input signal. However, the test signals are specifically designed to produce specific error patterns that can be easily traced to the source of the error in the algorithm. Figure 16-5 shows the 4-point FFT from Figure 16-1 with an arithmetic error in adding $a(0)$ to $a(2)$. Bold flow graph lines are the paths taken by the error as a result of the Algorithm Steps on page 402. The error is that they are subtracted rather than added. Table 16-3 shows the responses generated by each of the corresponding signals in Table 16-1 as it goes through those Algorithm Steps. Comparing Tables 16-3 and 16-2 allows the error patterns to be easily identified for each test signal.

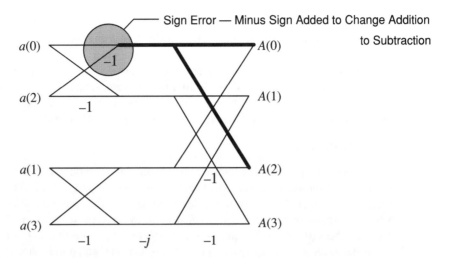

**Figure 16-5**    Four-point FFT with arithmetic error in first stage.

**Table 16-3**    Response to the Test Signals with an Error in the 4-Point FFT

| Responses to the unit pulse | Responses to the constant | Responses to sine wave 1 | Responses to sine wave 1 + constant |
|---|---|---|---|
| $b_R(0) = 100$ | $b_R(0) = 0^*$ | $b_R(0) = 200^*$ | $b_R(0) = 200$ |
| $b_I(0) = 50$ | $b_I(0) = 0^*$ | $b_I(0) = 0$ | $b_I(0) = 0^*$ |
| $b_R(1) = 100$ | $b_R(1) = 0$ | $b_R(1) = 200$ | $b_R(1) = 200$ |
| $b_I(1) = 50$ | $b_I(1) = 0$ | $b_I(1) = 0$ | $b_I(1) = 0$ |
| $b_R(2) = 0$ | $b_R(2) = 200$ | $b_R(2) = 0$ | $b_R(2) = 200$ |
| $b_I(2) = 0$ | $b_I(2) = 100$ | $b_I(2) = 0$ | $b_I(2) = 100$ |
| $b_R(3) = 0$ | $b_R(3) = 0$ | $b_R(3) = 0$ | $b_R(3) = 0$ |
| $b_I(3) = 0$ | $b_I(3) = 0$ | $b_I(3) = 200$ | $b_I(3) = 200$ |
| $A_R(0) = 100$ | $A_R(0) = 200^*$ | $A_R(0) = 200^*$ | $A_R(0) = 400$ |
| $A_I(0) = 50$ | $A_I(0) = 100^*$ | $A_I(0) = 0$ | $A_I(0) = 100$ |
| $A_R(1) = 100$ | $A_R(1) = 0$ | $A_R(1) = 400$ | $A_R(1) = 400$ |
| $A_I(1) = 50$ | $A_I(1) = 0$ | $A_I(1) = 0$ | $A_I(1) = 0^*$ |
| $A_R(2) = 100$ | $A_R(2) = -200^*$ | $A_R(2) = 200^*$ | $A_R(2) = 0$ |
| $A_I(2) = 50$ | $A_I(2) = -100^*$ | $A_I(2) = 0$ | $A_I(2) = -100^*$ |
| $A_R(3) = 100$ | $A_R(3) = 0$ | $A_R(3) = 0$ | $A_R(3) = 0$ |
| $A_I(3) = 50$ | $A_I(3) = 0$ | $A_I(3) = 0$ | $A_I(3) = 0$ |

*Indicates incorrect intermediate or output values.

## 16.6.1 Unit Pulse

For the error in Figure 16-5 and unit pulse signal in Table 16-3, there are no errors in the computations because the error was in the way $a(2)$ is used in the algorithm. Since the chosen unit pulse has $a(2) = 0$, the error had no effect on any of the outputs or intermediate results. In fact, the only version of the unit pulse that would catch this error is one with $a(2) \neq 0$. This is an illustration of the drawback of using the unit pulse test signal first. Namely, all of the possible versions of the unit pulse must be used to detect the error. For a 4-point FFT this is not a significant problem. However, for a 1024-point FFT it is. The best use of the unit pulse test signal is after the constant, single sine wave, and pair of sine waves tests have been used. If these tests do not pinpoint the error, but only localize it, then the appropriate unit pulse test signal can be used to positively identify the error.

### 16.6.2 Constants

Constant input signals exercise a significant portion of the algorithm arithmetic without the need for the input data organization to work properly. With the error shown in Figure 16-5 and the test signal responses in Table 16-5, the constant signal finds the error. The only output frequency components affected by the error (different in Tables 16-4 and 16-5) are the $A(0)$ and $A(2)$ terms. A reasonable assumption is that all of the computations associated with $A(1)$ and $A(3)$ are correct. For the flow graph in Figure 16-5, this means that the error must be associated with the top addition of one of the two input add-subtracts ($a(0) \pm a(2)$ or $a(1) \pm a(3)$).

To determine which of the two input add computations ($a(0) + a(2)$ or $a(1) + a(3)$) is incorrect, start with Table 16-5, which shows that the real parts of $A(0)$ and $A(2)$ are reduced by 200 and the imaginary parts by 100. This implies that the error occurred in such a way that it affected $A(0)$ and $A(2)$ in the same way. Again for the flow graph in Figure 16-5, the top input add ($a(0) + a(2)$) is added to $A(0)$ and $A(2)$, and the bottom input add ($a(1) + a(3)$) is added to $A(0)$ but subtracted from $A(2)$. Therefore, it must be the top input add. In Table 16-1 this is the computation that forms the complex intermediate values $b_R(0)$ and $b_I(0)$.

### 16.6.3 Single Sine Waves

There are errors that the constant test signal does not find. In particular, these errors are associated with the follow-on computations to the subtraction side of the input computations. In Table 16-1 these are the computations that use the $b_R(1)$, $b_I(1)$, $b_R(3)$, and $b_I(3)$ terms. Since $b_R(1)$, $b_I(1)$, $b_R(3)$, and $b_I(3)$ are all zero for any constant input signal, any error in computations using them will remain undetected. All of the building-block algorithms in Chapter 8 have these input add-subtract computations and therefore exhibit the same behavior for constant input signals.

The simplest test signal to remove the problems associated with the constant test signal is a sine wave that has exactly one cycle during the sequence of input samples. If the FFT is working properly, the only output that will respond to this input is $A(1)$. Again, Table 16-5 is a simple illustration of this for the 4-point FFT. This fact is true for all of the building blocks in Chapter 8 and for all combinations of building blocks used to form larger FFTs in Chapter 9. For the error in Figure 16-5, the complex $A(1)$ term still has the correct output. This implies that all of the computations used to form it must be correct. Similarly, $A(0)$ and $A(2)$ have also been modified by the same amount, which suggests that the top input add ($a(0) + a(2)$) is in error, just as for the case of the constant test signal.

Notice that the real part of $b(1)$ ($b_R(1)$) and the imaginary part of $b(3)$ ($b_I(3)$) are nonzero. In this example, the phase of the sine wave is set to zero. If the sine wave had nonzero phase, the real and imaginary parts of $b(1)$ and $b(3)$ would be nonzero. This eliminates the possibilities of error that cannot be tested by the constant signal.

### 16.6.4 Pair of Sine Waves

From the discussion of the constant and single sine-wave input signals and the data values in Tables 16-3 and 16-4, it is clear that $b(1)$ and $b(3)$ are always zero for the constant signal, regardless of the phase. Similarly, $b(0)$ and $b(2)$ are always zero for the single sine wave. Therefore, each test signal has its own class of errors it can detect. If the signals are combined, the resulting test input can be made to have nonzero outputs for all of the $b(i)$. The pair of sine waves recommended to catch errors that the others miss is two that are in the center of output filters that have relatively prime numbers and are relatively prime to the FFT length. This set of conditions removes these "always zero" conditions and picks up remaining algorithm errors. However, this signal should be used after the constant and single sine-wave tests because the patterns are more complex and the error combinations more vast than for the simpler signals. Use the simpler signals to remove most of the potential errors and then rely on this more complex waveform to ferret out the remaining problems.

## 16.7 ISOLATING ERRORS: A 16-POINT EXAMPLE

### 16.7.1 Assumptions

The 16-point radix-4 FFT, shown in flow graph form in Figure 16-6 and completely described in Section 9.7.5, is used to illustrate the error isolation approaches explained in this chapter. A single programmable DSP chip, with external data and program memory, is used as the implementation architecture because it represents the most common DSP board configuration and the majority of product applications. Further, the 4-point building-block code (blocks 1 through 4 on the left and right of Figure 16-6) will be written once and used each of the eight times it is required by the relabeling techniques in Section 9.4 to memory-map the data for each building block to different portions of data memory.

In multiprocessor applications it is prudent to test the FFT algorithms at the single processor level first to simplify the overall testing process. Additional assumptions are that the error is found after the algorithms have been developed, in this case using ones in Chapters 8 and 9, and after the 4-point building-block coding is checked.

The bold line between the multiplier error third output 4-point building block shows that the outputs of that building block are the only ones affected by the error. Therefore, any test signal that has an incorrect output will only be incorrect in the $A(6)$, $A(6)$, $A(10)$, and $A(14)$ terms. An error in one of these terms is the initial indication of an error in the algorithm. The four bold lines on the input of the third output 4-point building block show which intermediate results can possibly be in error. The goal of the test signal sequence is to isolate the error to the correct place in the algorithm. The error introduced is a sign error in the multiplier used to modify the third output of the second input 4-point building block between the building-block stages.

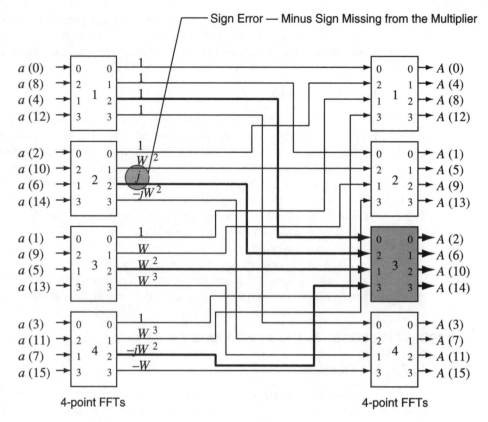

**Figure 16-6** Sixteen-point radix-4 FFT error isolation example.

### 16.7.2 Test Signal Strategy

The test signal strategy is to find the error using the least number of signals. Therefore, the constant signal is applied first, followed by the single sine wave. If needed, the pair of relatively prime sine waves is used, and the 16 unit pulses are a last resort. Even if the unit pulses are needed, hopefully the error will have been isolated far enough so that only a few of the 16 choices are required.

Since the 4-point building block is known to be correct, the error must be in the multiplier constants between the building-block stages or in the reorganization of data at the algorithm input or between the algorithm stages. Therefore, the results of applying the test signals are used to isolate the error to one of those three portions of the algorithm.

### 16.7.3 Error Isolation

Applying the Constant Test Signal.   According to Figure 16-5 and Section 16.6.2, the constant test signal does not find the error because a correct 4-point building block always has zero at the third output when the input signal is a constant. While this does not locate the error, it does eliminate certain portions of the algorithm. Namely, since all of the top outputs of the input 4-point FFTs are nonzero for the constant test signal and they are all inputs to the top output 4-point FFT, the four associated multipliers are correct and that portion of the data reorganization between stages is correct.

**Applying the Single Sine Wave at Frequency 1 Test Signal.**   The single sine wave at frequency 1 also does not provide useful information for isolating the error, because of how the input data is reorganized before entering the input FFT building blocks. From Figures 16-5 and 16-6 the input data points combined by the add-subtract computations are eight samples apart. For a sine wave that has only one cycle during the 16 samples, the samples that are eight apart are the negatives of each other, independent of the phase of the sine wave. Therefore, the add output of the 4-point FFT input computations are always zero. Since it is these two add outputs that are used to form the zero and second outputs of the 4-point FFT, the signal that feeds the incorrect multiplier value is always zero. Therefore, that multiplier value can be anything, and the 16-point FFT outputs are unaffected.

While this also does not locate the error, it also eliminates other portions of the algorithm. Specifically, all of the first and third outputs of the input 4-point FFTs are nonzero, and they feed the second and fourth output 4-point FFTs. Since these output 4-point building blocks have the correct results, it is likely that they are getting the correct inputs. Therefore, the respective multiplier constants on the input of those output 4-point building blocks should be correct and the data reorganizations at the algorithm input and between the stages must be correct. This leaves the third output from the input 4-point FFTs or their corresponding multipliers and mappings into the third output 4-point FFT.

**Applying the Pair of Relatively Primed Frequency Sine Waves.**   The choice of frequency pairs has been aided considerably by the two previous test signals. Namely, the conclusion to this point is that the error is somewhere in the path between the second input FFT outputs and the outputs of the third output 4-point FFT. Since that 4-point FFT produces output frequencies $A(2)$, $A(6)$, $A(10)$, and $A(14)$, the pair of frequencies chosen must come from that set of four if it is to isolate the error. These sine waves have the feature that the samples that are eight apart are the same. Therefore, the second output from each of the 4-point input FFTs will be nonzero, regardless of the phase of those sine waves. As a result, all of the inputs to the third output 4-point FFT will be nonzero.

To see how this test signal, with any combination of the pairs of frequencies mentioned, can isolate the error, use Figure 16-5. If the top signal to that 4-point FFT $(a(0))$ is incorrect, all of its outputs are modified by the same amount. If the next input signal $(a(2))$ is in error, the error is added to its zero and second $A(0)$ and $A(2)$ outputs and subtracted from its other $A(1)$ and $A(3)$ outputs. If the third input signal $(a(1))$ is incorrect, the error is added to the first and subtracted from the second $A(0)$ and $A(2)$ outputs and $-j$ times the error is added to the second and subtracted from the third $A(1)$ and $A(3)$ outputs. Finally, if the fourth input $(a(3))$ is incorrect, its error is added to the first and subtracted from the second $A(0)$ and $A(2)$ outputs and $-j$ times the error is subtracted from the first and added to the third $A(1)$ and $A(3)$ outputs.

Therefore, the strategy is to apply the pair of sine-wave signals and compare the outputs of the third 4-point output FFT with the correct ones. The errors must follow one of the four patterns described in the last paragraph. Once the error pattern is identified, it immediately points to which multiplier output is wrong. In this case, the second input to the third output 4-point FFT has the wrong multiplier. Thus $A(2)$ and $A(10)$ will have the same error, and $A(6)$ and $A(14)$ will have the negative of that error.

**Applying the Unit Pulse Test Signals.** In this example, the unit pulse signals are not needed because the other three test signals were sufficient to isolate the error. If this were not the case, then the results of the previous three test signals would have narrowed the error to one of a few places. The unit pulse is then used to test for those few remaining error locations sequentially until one of them had the wrong answer. However, a unit pulse signal at $a(2)$ or $a(6)$ can be used to verify the results found by using the other sequence of inputs.

## 16.8 CONCLUSIONS

This chapter details an orderly, efficient way to detect and isolate errors in FFTs , from algorithm development through product operation. Carefully chosen test signals and the sequence in which they are applied save time in error detection and isolation. Taking the time to draw a flow graph is one of the best investments for saving time when isolating errors. Examples have been used to illustrate these techniques, which are the final step in the design process of an FFT-based product.

The final chapter integrates the concepts, facts, and tools of this and all the preceding chapters, using four design examples.

# 17

# Design Examples

## 17.0 INTRODUCTION

How to make the FFT decisions in a design is not easily explained in general because each application has its own specific requirements. Therefore, four real-time design examples are developed in this chapter to illustrate the concepts, elements, and tools given throughout the book. These were chosen to cover:

- Three common uses of the DFT
- Two primary functions of the FFT
- Three applications of weighting functions
- Single and multidimensional processing
- Single and multiprocessor architectures
- Mixed-radix, convolutional, and prime factor algorithms
- Fixed-point, floating-point, and FFT-specific chips
- Single- and multiple-board implementations

The keyboard specifications from Section 15.0 are given for each example, but an actual board will not be picked or designed because the information needed to illustrate that selection process is beyond the scope of this book. The design decisions from Section 1.2 appear at the end of each example, with the choices for that example and a text that summarizes those decisions. The sequence in which these decisions get made vary from example to example.

Issues such as heat dissipation, temperature range, and vibration levels are not covered in the book or in these examples. While these are important product design decisions, they are normally related to the specific environment where the product will operate and do not affect choice of FFT length, algorithm, or architecture. Issues such as package type (ceramic versus plastic and pin-grid array versus surface mount) are also not covered because these options are available from most chip and board vendors and are unlikely to affect FFT-related decisions.

## 17.1 EXAMPLE 1: DOPPLER RADAR PROCESSOR

Processing in early Doppler radars was performed with an array of analog bandpass filters. The capacitors, resistors, and inductors used to create these filters were sensitive to temperature changes and aging, making the filters' center frequencies and bandwidths hard to control. The advent of digital integrated circuits in the early 1970s stimulated a rapid transition of Doppler radar processing from analog filtering to digital filtering, using FFT algorithms (Section 2.2) [1]. Initially, FFT-based Doppler processors could only be afforded for military applications. However, the proliferation of the DSP chips listed in Chapter 14 reduced implementation costs to the point where FFT processing is now common in both military and commercial Doppler radars.

### 17.1.1 Definition of the Product

The Doppler processing portion of a ground-based air surveillance radar, which might be used for commercial airport air traffic control or for Doppler weather radar, is designed in this example. In this class of radar applications, Doppler processing is used for three reasons. First, aircraft targets and storms are moving relative to the ground, which means their return frequency is different than the ground's. Therefore, Doppler processing can be used to separate those returns from ground returns. Second, Doppler processing determines how fast each target aircraft is moving toward the radar. This, in conjunction with angle and range measurements, can be used by the radar to track aircraft and storms.

Finally, Doppler processing is also used to improve the signal-to-noise ($S/N$) performance of the radar. Since radar system noise is random in time, its value in any target's range interval is reduced by the number of range intervals, $M$, within the interpulse period (time between radio frequency (RF) pulse transmissions). Further, within a particular range interval, the radar system's noise is also random in frequency. Since the return energy from a target is concentrated at a particular frequency, $S/N$ is improved by a factor of $N$ when the Doppler processor divides the frequency range into $N$ smaller passbands. The result is an overall $S/N$ improvement of a factor of $M * N$ by performing Doppler processing at each range interval of interest.

### 17.1.2 Specification

Table 17-1 shows the fundamental system parameters and the values they have for this example. Range resolution is the width of the transmitted pulse. Because RF energy travels at the speed of light (300,000,000 m/s), it has a round-trip time to the target and back of 150 m/$\mu$s (492 ft/$\mu$s). This means that 50-ft resolution translates into roughly 0.1-$\mu$s pulses. Azimuth resolution is defined as the 3-dB azimuth beamwidth of the radar antenna, and radial speed resolution is defined as the spacing between Doppler filters. The conversion between speed ($v$) and Doppler frequency ($f$) is

$$f = 2 * v/\lambda$$

where $\lambda$ is the wavelength of the transmitted RF energy. For an X-band radar, $\lambda \approx 0.1$ ft. Therefore, a 2-ft/s speed resolution requirement converts to a 40-Hz spacing between Doppler filters ($\Delta f = 2 * 2$ ft/s/(0.1 ft) = 40 Hz).

**Table 17-1**   Doppler Processor Technical Specifications

| System parameter | Required value |
|---|---|
| Range resolution | 50 ft |
| Antenna scan rate | 6 RPM |
| Maximum detection range | 80 nautical miles |
| Azimuth resolution | 1° |
| Radial speed resolution | 2 ft/s |
| Product volume | 100 systems |
| Time to market | 1 year |

Normally these types of radars are designed so that the return from the longest-range target reaches the receiver before the next pulse is transmitted. For an 80-nautical-mile maximum range the RF energy must travel 160 nautical miles, which is roughly 296,000 m. Since RF energy travels at 300,000,000 m/s, it takes the RF energy 0.987 ms to make the maximum round-trip excursion. Therefore, a pulse repetition interval of 1 ms (1000 transmissions per second) satisfies the maximum-range requirements. If the entire time between transmitted pulses is divided into 0.1-$\mu$s pulse widths, 10,000 pulse widths are required.

### 17.1.3 Description

Doppler radars periodically transmit pulses of RF energy and collect the radar returns and "noise" as a function of time. Given that RF energy travels at the speed of light, the time delay between pulse transmission and the reception of energy that has bounced off the target is directly related to the target's distance from the radar antenna [1].

Because a target's radial speed (motion away from or toward the radar) causes a change in the frequency of the transmitted pulse (the Doppler effect), frequency analysis of the return samples is used to aid in detecting targets and determining their radial speed. The FFT is the most widely used algorithm for determining this frequency shift.

Radar antenna scan rates and beam widths determine how many times the transmitted radar energy hits the target each time the antenna beam scans by it. The available number of return samples is rarely a power of two. However, Doppler radar processor transform lengths (number of samples at a particular range) are usually powers of two because of availability of power-of-two FFT algorithms. In these radars, the zero-padding technique discussed in Section 2.3.10 is used to obtain enough data points for a power-of-two algorithm. The alternative approach is to use one of the non-power-of-two algorithms in Chapters 8 and 9. This alternative may produce a more accurate analysis of the Doppler shift and use fewer computations and data memory. However, the high-speed FFT-specific chips in Section 14.7 only perform power-of-two algorithms. This means that non-power-of-two algorithms require either the Bluestein algorithm (Section 9.5.1) or the programmable DSP chips from Sections 14.3 and 14.5. Both reduce the throughput possible.

### 17.1.4 Design Decisions

FFT Algorithm.   Since the azimuth scan rate is 36°/s (6 RPM) and the azimuth beam is 1° wide, the radar beam hits a point target for roughly 1/36 s during each revolution. In 1/36 s the radar transmits $1000/36 = 27.7$ pulses that will bounce off of the target and return to the radar for processing. Therefore, 27- or 28-point FFT algorithms are the natural Doppler processing choice. Chapters 8 and 9 show that 30- and 32-point FFT algorithms are also good candidates, based on the computations required, and require little zero padding. Therefore, the likely FFT length is between 27 and 32.

The sampling theorem, described in Section 2.3.1, limits the frequency spectrum by the complex sampling rate, in this case 1000 Hz. Chapter 2 also states that this sampling interval is divided into $N$ equally spaced frequency intervals by the FFT (Section 2.3.2). Therefore, processing the radar returns using a 27- to 32-point FFT produces (1000 Hz)/32 $= 31.25$ Hz to $1000/27 = 37$ Hz spacing between the frequency bins. All of these satisfy the 40-Hz requirement. In fact, a 25-point FFT is the smallest that will satisfy the speed resolution requirement. This expands the choices for FFT lengths to include 25 and 26 points. Table 17-2 summarizes the factors of these candidate transform lengths.

**Table 17-2**   Transform Length Factors

| Transform lengths | Factors |
|:---:|:---:|
| 25 | 5, 5 |
| 26 | 2, 13 |
| 27 | 3, 3, 3 |
| 28 | 2, 2, 7 |
| 29 | 29 |
| 30 | 2, 3, 5 |
| 31 | 31 |
| 32 | 2, 4, 8, 16 |

Since the 27-point FFT can be computed by using either three stages of 3-point building blocks or a 3-point and a 9-point building block, the factors in Table 17-2 include all of the building blocks in Chapter 8. Additionally, the 29- and 31-point FFTs can be computed by using any of the three general algorithms for all odd numbers. The Winograd (26-, 28-, and 30-point FFTs), prime factor (26-, 28-, and 30-point FFTs), and mixed-radix (25-, 26-, 27-, 28-, 30-, and 32-point FFTs) algorithms from Chapter 9 can be used to implement the listed transform length choices.

From the Comparison Matrices in Chapter 9 (Tables 9-7 and 9-8), the most likely non-power-of-two FFT is one of the 28- or 30-point prime factor algorithms (Kolba-Parks or

SWIFT) using the Winograd building-block algorithms from Chapter 8 because they require the fewest adds and multiplies. The algorithms can be compared by using the Comparison Matrices from Chapters 8 (Table 8-1) and 9 (Tables 9-7 and 9-8). However, the 32-point FFT must also be considered because this is a high-computation-rate application which may result in the use of an FFT-specific chip from Chapter 14.

From the Comparison Matrix in Table 9-8, the 16-point radix-4 FFT algorithm takes 144 adds and 24 multiplies. The mixed-radix algorithm in Chapter 9 can be used to combine the 16-point FFT with a 2-point building block to form the 32-point FFT. This requires:

- Two 16-point FFTs (288 adds, 48 multiplies)
- Sixteen 2-point FFTs (64 adds, 0 multiplies)
- Fifteen complex multiplies (30 adds, 60 multiplies) (between the 16- and 2-point FFTs)
- Thirty-two half-complex multiplies (0 adds, 64 multiplies) (weighting function multiplies)

The total is 382 adds and 172 multiplies.

If the prime factor algorithm in Chapter 9 is used with the 7-point Winograd and 4-point building blocks from Chapter 8, the 28-point FFT uses:

- Seven 4-point FFTs (112 adds, 0 multiplies)
- Four 7-point FFTs (288 adds, 64 multiplies)
- Twenty-eight half-complex multiplies (0 adds, 56 multiplies) (weighting function multiplies)

This is a total of 400 adds and 120 multiplies.

If the prime factor algorithm in Chapter 9 is used with the 3- and 5-point Winograd and 2-point building blocks from Chapter 8, the 30-point FFT uses:

- Fifteen 2-point FFTs (60 adds, 0 multiplies)
- Six 5-point FFTs (204 adds, 60 multiplies)
- Thirty half-complex multiplies (0 adds, 60 multiplies) (weighting function multiplies)
- Ten 3-point FFTs (120 adds, 40 multiples)

This is a total of 384 adds and 160 multiplies.

Memory locations for data and constants must also be considered when choosing an algorithm. The numbers in the Comparison Matrices in Chapters 8 (Table 8-1) and 9 (Tables 9-7 and 9-8) show additional memory locations are required for the 28-, 30-, and 32-point FFTs. The 16-point FFT only has six multiplier coefficients to store. However, the 15 complex multiplications required between the 16- and 2-point FFTs require an additional 30 constant locations. One of the key advantages of the prime factor algorithm is the few multiplier constants that must be stored. The 28- and 30-point FFTs are good illustrations of that fact. Eight constants are needed for the 7-point FFT, two for the 3-point FFT, and five for the 5-point FFT. The 2- and 4-point building blocks have no multiplier constants, and no complex multiplies are required between stages. All of the algorithms must store weighting function coefficients. Assuming all these are stored, the number of memory locations for the weighting function coefficients is equal to the FFT length.

Table 17-3 summarizes the performance measures for each of the three most likely FFT algorithms. If the choice of processor is limited to the programmable processors in Chapter 14, Table 17-3 can be used to choose the 28-point prime factor algorithm because of the smaller numbers in columns 2, 3, and 4. However, the 32-point FFT can also be implemented with the FFT-specific chips in Chapter 14. Therefore, the FFT algorithm decision must be postponed until the chip and architecture choices are examined.

**Table 17-3** Doppler Radar Processor FFT Algorithm Comparison Matrix

| Algorithm | # of adds | # of multiplies | # of data locations | # of const. locations |
|---|---|---|---|---|
| 32-point mixed-radix | 382 | 172 | 64 | 68 |
| 28-point prime factor | 400 | 120 | 56 | 36 |
| 30-point prime factor | 384 | 160 | 60 | 65 |

**Weighting Functions.** In addition to FFT length requirements, constraints are placed on the radar based on ground clutter returns. Since these are not germane to this example, they are given as input dynamic range to the FFT processor of 80 dB and peak frequency filter sidelobe level of −60 dB. The 60-dB filter sidelobe requirement implies using a weighting function. Table 17-4 summarizes the performance measures of the weighting functions from the Comparison Matrix in Chapter 4 (Table 4-1) that meet the −60-dB highest sidelobe level requirement. The $\alpha = 3.0$ Dolph-Chebychev weighting function is chosen from Table 17-4 because it has the best performance in columns 5, 6, and 7.

**Table 17-4** Doppler Radar Processor Weighting Function Comparison Matrix

| Weighting function | Highest sidelobe level (dB) | Sidelobe fall-off ratio | Frequency straddle loss (dB) | Coherent integration gain | Equivalent noise bandwidth | 3-dB bandwidth |
|---|---|---|---|---|---|---|
| Three-sample Blackman-Harris (a) | −61 | −6 | 1.27 | 0.45 | 1.61 | 1.56 |
| Three-sample Blackman-Harris (b) | −67 | −6 | 1.13 | 0.42 | 1.71 | 1.66 |
| Four-sample Blackman-Harris (a) | −74 | −6 | 1.03 | 0.40 | 1.79 | 1.74 |
| Four-sample Blackman-Harris (b) | −92 | −6 | 0.83 | 0.36 | 2.00 | 1.90 |
| Kaiser-Bessel (c) $\alpha = 3.0$ | −69 | −6 | 1.02 | 0.40 | 1.80 | 1.71 |
| (d) $\alpha = 3.5$ | −82 | −6 | 0.89 | 0.37 | 1.93 | 1.83 |
| Gaussian (c) $\alpha = 3.5$ | −69 | −6 | 0.94 | 0.37 | 1.90 | 1.79 |
| Dolph-Cheb. (b) $\alpha = 3.0$ | −60 | 0 | 1.44 | 0.48 | 1.51 | 1.44 |
| (c) $\alpha = 3.5$ | −70 | 0 | 1.55 | 0.45 | 1.62 | 1.55 |
| (d) $\alpha = 4.0$ | −80 | 0 | 1.65 | 0.42 | 1.73 | 1.65 |

**Arithmetic Format.**    In the Comparison Matrix in Chapter 13 (Table 13-1), the dynamic range requirement of 80 dB (14 bits) at the input restricts the arithmetic format to floating-point, block-floating-point, or larger-than-16-bit fixed-point.

**Architectures and Chips.**    The potential architectural options are determined primarily by the number of FFTs that must be performed per second and how many FFTs can be performed by a single chip. The number of FFTs per second is determined by multiplying the FFT rate per second for a single range interval by the total number of range intervals. The single range interval processing requirement is one 25- to 32-point FFT during the 28 ms the antenna beam is on the target. Since there are 10,000 range locations within the interpulse period, the total FFT computation requirement is one 25- to 32-point FFT every 2.8 $\mu$s.

The chip Comparison Matrices in Chapter 14 (Tables 14-3 to 14-7) only provide computational performance for 1024-point transforms. For chips that perform a 1024-point FFT on-chip, the scaling formula from Chapter 14 can be used to approximate the required computation time for 32-point FFTs, namely $(1024/32) * [\log_2(1024)/\log_2(32)] = 64$ times faster. Conversely, the 2.8-$\mu$s time for the 32-point FFT can be multiplied by 64 (179.2 $\mu$s) and compared to the 1024-point complex FFT times. Table 17-5 summarizes the chips that have floating-point, block-floating-point, or 20/24-bit fixed-point arithmetic (to match the arithmetic format requirements) and the rough number of them needed to meet the FFT computation requirement. The number-of-chips estimate is based on applying the equation in this paragraph.

Based on Table 17-3, the Analog Devices 21060 is technically the best programmable fixed-point or floating-point choice because it provides the most performance per chip and is designed to be implemented in multiprocessing architectures (Section 14.5.2). Assuming that the FFT processing represents roughly half of the total signal processing, at least six of these chips will be needed in the processor architecture. To provide some cushion for future growth, assume eight ADSP-21060 chips will be used. Since the FFT processing is executed independently for each of the 10,000 range intervals, the best data organization is to distribute 1250 of the range intervals to each of the eight floating-point DSPs. To distribute the I/O load on each of these DSPs, sequential range cells should be sent to different processors. The result is that each DSP's input memory will need an area for 1250 of the 28-, 30-, or 32-point sets of complex input data; 1250 sets of data being processed; and 1250 sets of frequency results being output for subsequent processing. For the worst case of using the 32-point algorithm, this is a total of $3 * 1250 * 64 = 240,000$ thirty-two-bit data words (960,000 bytes) in each processor's local memory. Since this is less than a megabyte, there is no reason to use the 25- to 31-point algorithms to save memory space.

Table 17-5 shows that all but two of the block-floating-point FFT chips can execute the required processing load using the 32-point FFT, but none of these chips is capable of implementing the 25- to 31-point FFT choices without the Bluestein algorithm. Before going to a processor architecture block diagram, check the manufacturer's Application Notes to verify the 32-point FFT timing estimates. The Sharp FFT chip takes 3.75 $\mu$s (table on page 1A-2 of Application Notes, Reference 36 from Chapter 14) to perform a 32-point complex FFT. Similarly, the array Microsystems FFT processor takes 5.6 $\mu$s, using their formula (Table 1.4 of a66110 User's Guide, Reference 35 from Chapter 14)

**Table 17-5** Doppler Radar Processor DSP Chip Comparison Matrix

| Chip | 1K FFT time (MS) | # bits | # chips |
|---|---|---|---|
| **Fixed-Point** | | | |
| DSP56001 | 1.797 | 24 | 11 |
| DSP56002 | 0.908 | 24 | 6 |
| DSP56L002 | 1.497 | 24 | 9 |
| DSP56004 | 1.497 | 24 | 9 |
| $\mu$PD77220 | 8.5 | 24 | 48 |
| $\mu$PD77P220 | 8.5 | 24 | 48 |
| SPROC1400 | 2.4 | 24 | 14 |
| SPROC1200 | 4.8 | 24 | 28 |
| SPROC1210 | 4.8 | 24 | 28 |
| ZR38000 | 0.88 | 20 | 5 |
| **Floating-Point** | | | |
| ADSP-21020 | 0.58 | 32 | 4 |
| ADSP-21060 | 0.46 | 32 | 3 |
| DSP32C | 3.2 | 32 | 18 |
| DSP3210 | 2.4 | 32 | 14 |
| DSP3207 | 1.9 | 32 | 11 |
| i860XR | 0.74 | 32 | 5 |
| i860XP | 0.55 | 32 | 4 |
| DSP96002 | 1.04 | 32 | 6 |
| $\mu$PD77240 | 7.07 | 32 | 40 |
| $\mu$PD77230A | 11.78 | 32 | 66 |
| TMS320C30 | 1.97 | 32 | 11 |
| TMS320C31 | 1.97 | 32 | 11 |
| TMS320C40 | 1.54 | 32 | 6 |
| **Block-Floating-Pt.** | | | |
| a66110/a66210 | 0.131 | 16 | 1 |
| a66111/a66211 | 0.131 | 16 | 1 |
| LH9124/LH9320 | 0.087 | 24 | 1 |
| LH9124L/LH9320 | 0.129 | 24 | 1 |
| TMC2310 | 0.514 | 16 | 3 |
| PDSP16510/16540 | 0.096 | 16 | Cannot do |

$$\text{Time (ns)} = (M + K + 1) * (N + 24) * 25 \tag{17-1}$$

where $N = 32 = 2 * (4)^M$ and $K = 1$ because of the need of a weighting function. Therefore, two of the array Microsystems FFT processors are only marginally able to perform the 32-point FFTs at the required 2.8-$\mu$s rate. This suggests that the Sharp FFT chip is technically the best of the dedicated chip solutions and requires two chips. The reason for the discrepancy with the formula is that the 1024-point FFTs are computed in these chips using a radix-4 algorithm which takes only five passes of data through the processor. The 32-point FFT takes three passes because it needs two radix-4 and one radix-2 passes.

Based on these observations, two processor architectures are shown in Figures 17-1 and 17-2. To ensure there is plenty of processing power for the non-signal-processing portions of the radar functions, and to account for inefficiencies encountered with combining algorithms into an application, four floating-point DSP chips are used for the other radar processing in both processor architectures.

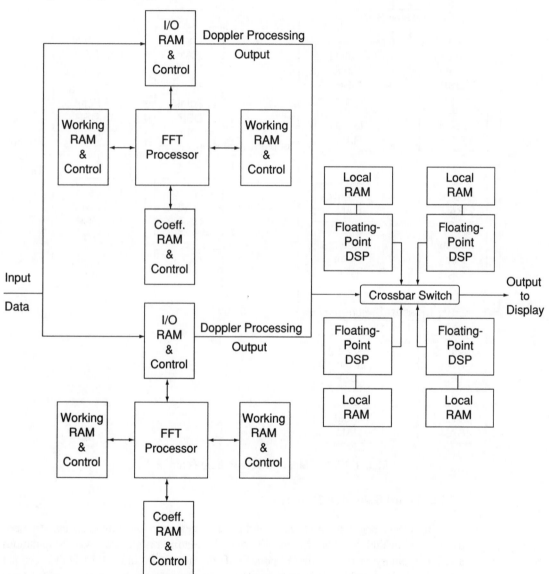

**Figure 17-1**   Radar processor architecture 1.

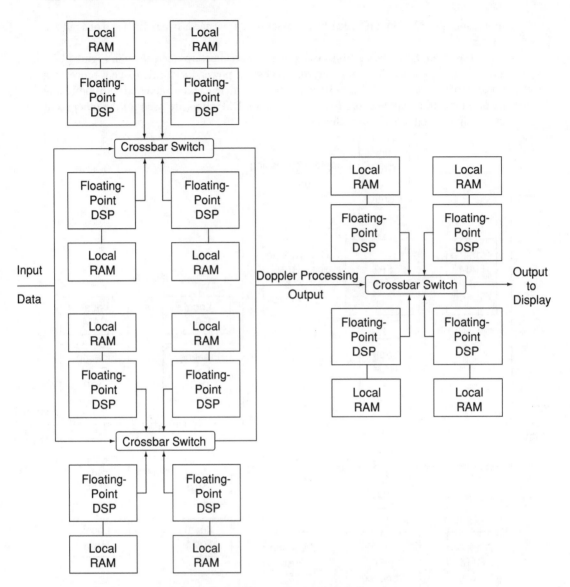

**Figure 17-2** Radar processor architecture 2.

### 17.1.5 Board Selection Process

To select a board, the FFT length and radar processor architecture decisions still need to be made. In Table 17-3 the 28- and 30-point FFT algorithms require fewer computations and less memory than does the 32-point FFT algorithm. In Table 17-5 both processor architectures are capable of meeting the processing requirements by using any of the three FFT lengths. However, 32-point FFT code exists in algorithm libraries for the Analog Devices ADSP-21060 chip. Therefore, since memory storage requirements for the three different FFT lengths all need more than 512-kbyte and less than 1-mbyte memory chips, the 32-point FFT is also the best choice for architecture 2.

Now a direct comparison can be made between the two architecture options. The only discernible difference is that the FFT-specific architecture already has the 32-point algorithm and the associated memory management built-in to the operation of the Sharp chip set. Because of the benefit of reduced development time and effort for architecture 1, it is the better choice (time-to-market requirement from Table 17-1). Table 17-6 summarizes the specifications needed to choose a COTS board that will be used twice for this multiboard design.

**Table 17-6**    Example 1, Board Selection Specifications

| Category | Specification |
|---|---|
| Processor | Sharp LH9124/LH9320 |
| Off-chip memory | 256K of 32-bit words |
| Analog I/O ports | None required |
| Instruction cycle time | 25 ns |
| Parallel and serial I/O ports (buses) | 32-bit words at 20 million per second rate |
| Host interface | None required |

### 17.1.6  Test Signals

Section 16.5 introduces four types of test signal in an order of increasing complexity. It also gives the guidelines that were followed to create the specific parameters of each signal in Table 17-7. They are reordered to match the strategy in Section 16.7.2 that lists them in an order that allows testing with the least number of signals. The pair of sine waves can be any pair of relatively prime numbers up to the length of the transform (32 points) and were arbitrarily selected.

**Table 17-7**    Example 1, Test Signals

| Signal | Parameters |
|---|---|
| Constant | Amplitude $= 1000$ for real and imaginary parts |
| Single sine wave | 1 cycle per 32 data samples |
| Pair of sine waves | 5 and 11 cycles per 32 data samples |
| Unit pulses | As needed |

### 17.1.7  Design Decisions Summary

A pair of the Sharp FFT-specific chip sets is chosen to implement a 32-point FFT. They are arranged in parallel because of the independence of the 10,000 range cells to be processed. A pipeline architecture, which has a crossbar interconnection of four Analog Devices 21060s for the remainder of the radar processing, is used for the overall processing architecture. The -60-dB sidelobe Dolph-Chebychev weighting function is chosen because it meets the sidelobe requirements and has the best performance of the applicable weighting functions in coherent gain, equivalent noise bandwidth, and 3-dB bandwidth. Table 17-8 summarizes all of the key element design decisions made for this example.

**Table 17-8** Example 1, Design Decisions

| Key element | Selection |
| --- | --- |
| Number of dimensions | 1 |
| Type of processing | Frequency analysis |
| Arithmetic format | Block-floating-point and 32-bit floating-point |
| Weighting function | Dolph-Chebychev |
| Transform length | 32-point |
| Algorithm building blocks | 2- and 16-point |
| Algorithm | Powers-of-primes mixed-radix |
| DSP chip | Sharp FFT-specific and Analog Devices 21060 |
| Architecture | Pipeline and crossbar |
| Mapping the algorithm onto the architecture | Maximum throughput |

## 17.2 EXAMPLE 2: POWER SPECTRUM ESTIMATOR

Power spectrum estimation is a technique for measuring the power in a noisy signal as a function of frequency. The image deblurring example in Section 17.4 uses power spectrum estimation as a key factor in deconvolving the real signal from the distortions of the measurement system. Other power spectrum estimation applications occur in analysis of geophysical data in oil and other mineral exploration [1], linear predictive coding models for speech synthesis and compression [1], and sonar signal processing [1].

### 17.2.1 Definition of the Product

The product is to be a plug-in board, for an IBM-compatible PC, to compute the power spectrum estimate, for sequences of noisy signals, in excess of 2000 data samples. The data is prestored on the hard disk. The user can access any portion of the data file, perform the power spectrum estimation on those samples, and display the results within 10 s of the data being downloaded from hard disk.

The user is anyone who employs a PC to analyze noisy signals for the purpose of finding patterns, which might be used to predict future values of a waveform. Two examples of the kind of signal this board can analyze are seismic data, to predict earthquakes, or sonar data, gathered to track whales.

### 17.2.2 Specification

Table 17-9 summarizes the specification of the product. Throughput is defined as the rate at which data sets can be fed to the product without the product getting behind. Latency is the time from when a data set enters the product until the analyzed version is sent back to the hard disk. The assumption is that the computational board is not used to display the results, just to compute them. The results are returned to hard disk, and a standard software package is used to display the results.

**Table 17-9**    Power Spectrum Estimator System Requirements

| System parameter | Requirement |
|---|---|
| Data set size =1 | From 32 to 8192 real data points |
| Number of bits per data point | 16 |
| Throughput rate | 1 power spectrum estimate per 5 s |
| Latency | 10 s |
| Hardware | IBM PC compatible plug-in board |
| Input source | IBM PC hard disk |
| Output | IBM PC hard disk |
| Number of data sets on board at one time | 1 |

### 17.2.3 Description

The modified periodogram method [2] of spectral estimation is based on dividing the sampled signal into subsequences of a manageable length, computing the power spectrum of those subsequences, and combining the result to estimate the power spectrum of the complete signal sequence. This strategy allows the sequence length to be controlled to fit within the memory capabilities of a computer and does not require the entire set of computations to be redone every time new samples are added to the signal. The power spectrum estimator uses the FFT in the center of its computations. Therefore, the example must include the other portions of the algorithm to obtain a realistic design. Since the modified periodogram method algorithm is not discussed in this book, it is summarized below. The details can be obtained from other sources [2].

The power spectrum of a data sequence of $L$ samples, $a(m)$ for $m = 0, \ldots, L - 1$, with the modified periodogram method, is computed from the following steps.

*Step* 1: *Sectioning the Input Data Sequence*

Section the input data sequence into P overlapping subsequences of length $N$ such that the combined subsequences span the entire data sequence. Figure 17-3 illustrates this process with an overlap of $M$ samples and $P = 5.5$.

*Step* 2: *Apply the Weighting Function and Compute the FFT of Each Section*

For each segment of length $N$, select the same weighting function ($W\,F(n)$), multiply it by the segment data samples, and compute the $N$-point FFT of the result. Specifically, compute

$$A_p(k) = \sum_{n=0}^{N-1} W\,F(n) * a[n + (p - 1)(N - M + 1)] * W_N^{k*n} \qquad (17\text{-}2)$$

where, $W_N = \cos(2\pi/N) - j * \sin(2\pi/N), k = 0, 1, \ldots, N - 1$, and $p = 1, \ldots, P$. This is a total of $P$ $N$-point FFTs. The triangular weighting function (Section 4.2.2) and an overlap of $M = N/2$ are often used for this process because of improved performance in the convergence of the variance of the power spectrum [2]. In this case, $P = (2 * L/N)$ $N$-point FFTs are required to compute the power spectrum estimate for all $P$ sets of samples.

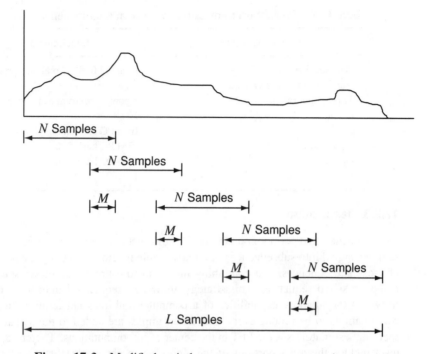

**Figure 17-3** Modified periodogram sequence segmentation example.

*Step* 3: *Compute the Periodograms*

For each of the $P$ sets of FFT coefficients, $A_p(k)$ with $k = 0, 1, \ldots, N-1$, compute the modified periodograms:

$$I_p(k) = [1/U] * |A_p(k)|^2 \qquad (17\text{-}3)$$

where $U = \sum_{n=0}^{N-1} [W F(n)]^2$ is computed ahead of time. For each set of $N$ FFT coefficients, $N$ complex multiplies are required. Since there are $P$ of these sets, this step requires $N * P$ complex multiplies. Since each complex multiply uses four real multiplies and two real adds, this is a total of $4 * N * P$ real multiplies and $2 * N * P$ real adds. For the 2:1 overlap case described in Step 2, $P = 2 * L/N$. Therefore, the number of real multiplies required is $8 * L$, and the required number of real adds is $4 * L$, independent of the FFT length.

*Step* 4: *Compute the Power Spectral Density*

Compute the power spectral density of the input data samples $a(n)$ by computing the average of the modified periodograms from Step 3:

$$PSD_P(k) = [1/P] * \sum_{p=1}^{P} I_p(k) \qquad (17\text{-}4)$$

For each of the $N$ periodogram frequency components ($k = 0, 1, \ldots, N-1$), $P-1$ adds are required, followed by one divide. This is a total of $N * (P - 1)$ real adds and one real divide. For the 2:1 overlap case described in Step 2, $P = 2 * L/N$. In this case the number of real adds required in this step is $2 * L - N$.

*Step* 5: *Update the Power Spectral Density for Each New Section of Input Data Samples*

To modify the power spectral density in Step 4 when additional data is collected, another periodogram is computed for the new data and then the average in Step 4 is recomputed. There is even a trick to simplify the computation of the new average, namely rather than computing $P - 1$ adds and a divide for each of the $N$ frequency components, compute

$$PSD_{(P+1)}(k) = [P * PSD_P(k) + I_P(k)]/(P + 1) \qquad (17\text{-}5)$$

which requires only one multiply, one add, and one divide for each of the $N$ frequency components, $k = 0, 1, \ldots, N - 1$.

### 17.2.4 Design Decisions

**FFT Algorithm.**   This is the area where most of the flexibility exists since the large data set is to be segmented into logical subsequences, overlapped by 2:1, and used to cover all of the potential data set lengths. The only requirement that will simplify the 2:1 overlap process is that the data sets to be analyzed have an even number of data points. This allows 2:1 overlap without having to zero-pad the last subsequence and makes $L = 2 * R$, where $R$ can be any number.

The other constraint on transform length is that $2 * L/N$, the number of FFTs to compute, be an integer. Combined with the requirement on $L$, this leads to a product requirement of $4 * R/N$ being an integer with $R$ any number up to 4096. If $N$ is larger than 4, it must always have factors that are in $R$. Therefore, to meet the desired system performance, $N$ must be able to be as large as prime numbers up to 4096. The only FFT algorithm in Chapter 9 that can reasonably reach these goals is the Bluestein algorithm. Figure 17-4 is a block diagram of this algorithm. The implementation of the algorithm is discussed below. Assuming it is reasonable to implement it, all of the data set length requirements can be met.

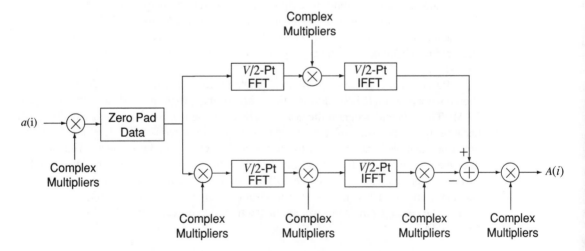

**Figure 17-4**    Bluestein FFT algorithm block diagram.

The block diagram in Figure 17-4 is for performing an $N$-point complex FFT. Since the data sets for this product are real, the Double-Length Algorithm from Section 2.4.2 can be used to more efficiently implement the complex algorithm. Therefore, the estimates made on FFT performance are based on complex data lengths that are half of the real data lengths.

To simplify the Bluestein algorithm development process, power-of-two algorithms will be used for the $V/2$-point FFTs. These algorithms are available for all of the candidate DSP chips.

**Weighting Functions.**    The theoretical development of the power spectrum estimation algorithm [2] uses the triangular weighting function from Section 4.2.2. Rather than store all of the weighting function constants in program memory, it can be easily computed.

**Arithmetic Formats.**    Nothing in the algorithm explicitly defines the arithmetic format requirement. However, since the process is looking for small patterns in a noisy signal, it makes sense to use floating-point arithmetic to minimize the algorithm-induced quantization noise, based on the Comparison Matrix in Chapter 13 (Table 13-1).

**Architecture and Chips.**    The worst-case processing load is when the required FFT is largest because the FFT computation load increases as $N * \log_2(N)$. The largest prime number less than 4096 is 4093, making 4093 the largest value of $N$. Based on $V$ being a power of two and the input data being real, $V$ only has to be 4096 points, which means the largest complex FFT to compute is 2048 points. Since the system requires four of these, it requires a total of sixteen 2048-point FFTs, as well as $4 * (4 * V + 10 * N)$ adds and $4 * (8 * V + 16 * N)$ multiplies, based on the Comparison Matrix in Table 9-7.

Table 17-10 is a list of the floating-point FFT chips from Chapter 14. For the chips that have less than 2048 locations of on-chip data RAM, the 1024-point FFT performance number already reflects going off-chip for data. Therefore, the performance numbers for these chips can be extrapolated to estimate performance for 2048-point FFTs by multiplying by a factor of $2 * 11/10 = 2.2$ (Section 14.1.1). It is easy to see that, even for the slowest 1024-point FFT time, all of the chips can execute the required computations in less than a second.

Based on the preliminary options available for chips in Table 17.7, the product should work as a single DSP chip solution with off-chip program and data memory (Figure 17-5). The data and program memory interfaces are shown for the same DSP chip pins, because the added speed of having separate buses is not required. Therefore, the combined bus approach can be used to choose a DSP chip with fewer pins. This will reduce the cost of the product. If all the devices with over 144 pins are eliminated, the list shrinks to the DSP32xx family, the $\mu$PD77240 and TMS320C3x families with 132-pin packages, and the $\mu$PD77230A with a 68-pin package, which are summarized in Table 17-11. The package pin counts were obtained from the respective chip family references in Chapter 14.

**Table 17-10**    Power Spectrum Estimator Chip Preliminary Comparison Matrix

| Floating-point chip | 1024-point complex FFT (MS) | Data I/O ports | On-chip data memory words | On-chip prog. memory words | # of address generators |
|---|---|---|---|---|---|
| **Analog Devices** | | | | | |
| ADSP-21020 | 0.58 | 0s/2p | 0 | 0 | 2 |
| ADSP-21060 | 0.46 | 8s/1p | 65,536 | 65,536 | 2 |
| **AT&T** | | | | | |
| DSP32C | 3.2 | 1s/1p | 1024/1536 | 4096/0 | 1 |
| DSP3210 | 2.4 | 1s/1p | 1024/2048 | 1024/256 | 1 |
| DSP3207 | 1.9 | 0s/1p | 1024/2048 | 1024/256 | 1 |
| **Intel** | | | | | |
| i860XR | 0.74 | 0s/1p | 1024 | 256 | 1 |
| i860XP | 0.55 | 0s/1p | 2048 | 1024 | 1 |
| **Motorola** | | | | | |
| DSP96002 | 1.04 | 0s/2p | 1024 | 1024 | 2 |
| **NEC** | | | | | |
| $\mu$PD77240 | 7.07 | 1s/1p | 1024 | 0 | 2 |
| $\mu$PD77230A | 11.78 | 1s/1p | 1024 | 1024/2048 | 2 |
| **TI** | | | | | |
| TMS320C30 | 1.97 | 2s/2p | 2048 | 4096 | 2 |
| TMS320C31 | 1.97 | 1s/2p | 2048 | 4096 | 2 |
| TMS320C40 | 1.54 | 6s/2p | 2048 | 4096 | 2 |

s = serial port;  p = parallel port.

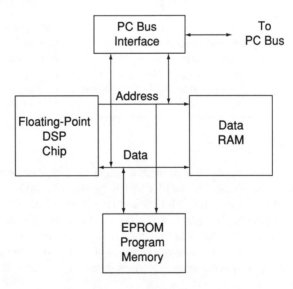

**Figure 17-5**    DSP architecture for the power spectrum estimator.

**Table 17-11** Power Spectrum Estimator Chip Comparison Matrix

| Floating-point chip | 1024-point complex FFT (MS) | Data I/O ports | On-chip data memory words | On-chip prog. memory words | # of address generators |
|---|---|---|---|---|---|
| DSP3210 | 2.4 | 1s/1p | 1024/2048 | 1024/256 | 1 |
| DSP3207 | 1.9 | 0s/1p | 1024/2048 | 1024/256 | 1 |
| μPD77240 | 7.07 | 1s/1p | 1024 | 0 | 2 |
| μPD77230A | 11.78 | 1s/1p | 1024 | 1024/2048 | 2 |
| TMS320C30 | 1.97 | 2s/2p | 2048 | 4096 | 2 |
| TMS320C31 | 1.97 | 1s/2p | 2048 | 4096 | 2 |

s = serial port; p = parallel port.

### 17.2.5 Board Selection Process

Of the six chips that meet the specifications, the best choice is the one that has the largest number of COTS boards on the market that will plug into a PC bus, because the competition of multiple boards in the market tends to reduce their cost. Multiple boards in the market also provide for second sources in case one board manufacturer goes out of business or decides to no longer make that board. There are far more TMS320C30 PC plug-in boards available than for any of the other chips in Table 17-11. Therefore, a TMS320C30-based board is the best choice to meet the specifications summarized in Table 17-12.

**Table 17-12** Example 2, Board Selection Specifications

| Category | Specification |
|---|---|
| Processor | TMS320C30 |
| Off-chip memory | 8192 of 32-bit words |
| Analog I/O ports | None required |
| Instruction cycle time | 60 ns |
| Parallel and serial I/O ports (buses) | PC bus |
| Host interface | PC compatible |

### 17.2.6 Test Signals

Testing this application presents a unique challenge because the Bluestein algorithm is used here to implement all the FFT lengths between 32 and 8192 points on real input data. The block diagram in Figure 17-4 shows that a 2048-point FFT is used as the intermediate step for all of the FFT lengths in this example. Therefore, the test signals are chosen to test the 2048-point FFT. Once it is fully tested, any remaining errors must be associated with the complex multipliers. Since they are computed based on the formulas in Section 9.5.4, they are checked by comparing the values in the application code with the values of those formulas.

Section 16.5 introduces four types of test signal in order of increasing complexity. It also gives the guidelines that were followed to create the specific parameters of each signal in Table 17-13. They are reordered to match the strategy in Section 16.7.2 that lists them in an order that allows testing with the least number of signals. The pair of sine waves can be any pair of relatively prime numbers up to the length of the transform (2048 points) and were arbitrarily selected.

**Table 17-13**    Example 2, Test Signals

| Signal | Parameters |
|---|---|
| Constant | Amplitude = 1000 for 8192 samples |
| Single sine wave | 1 cycle per 2048 data samples |
| Pair of sine waves | 7 and 13 cycles per 2048 data samples |
| Unit pulses | As needed |

### 17.2.7 Design Decision Summary

This application uses the Bluestein algorithm to meet the requirement to compute any transform length. Power-of-two FFTs are used to implement the Bluestein algorithm, to reduce the algorithm development cost. The triangular weighting function is used because the derivation of power spectrum estimation [2] reveals that as the best technical approach. A single floating-point DSP chip, which will need external program and data memory chips, provides the needed processing power and computational accuracy. The $\mu$PD77230A floating-point DSP chip would be used for a custom-designed board, and a TMS320C30-based board for an off-the-shelf design. Table 17-14 summarizes all of the key element design decisions made for this example.

**Table 17-14**    Example 2, Design Decisions

| Key element | Selection |
|---|---|
| Number of dimensions | 1 |
| Type of processing | Frequency analysis |
| Arithmetic format | 32-bit floating-point |
| Weighting function | Triangular |
| Transform length | Any up to 2048 |
| Algorithm building blocks | 2-, 4-, 18-, and 16-points |
| Algorithm | Bluestein convolutional |
| DSP chip | $\mu$PD77230A or TMS320C30 |
| Architecture | One Harvard processor & external memory |
| Mapping the algorithm onto the architecture | Maximum throughput |

## 17.3 EXAMPLE 3: SPEECH ANALYZER

Speech processing can be divided into three main categories:

1. Speech analysis for products that use speech recognition or speaker recognition
2. Speech synthesis for products that talk to the user from either stored or real-time input

3. Speech analysis followed by speech synthesis for products that compress speech to reduce storage space and/or communication bandwidth

### 17.3.1 Definition of the Product

The product is defined as the number recognition portion of a system for hands-off numerical data entry, voice car phone dialing, speaker verification for security, or fraud applications. FFT-based algorithms are not the only way to perform these tasks, but they may be more cost efficient for high-volume, low-cost products.

### 17.3.2 Specification

Table 17-15 shows the system requirements. The bottom four requirements are qualitative rather than quantitative because their quantitative values will change with the evolution of technology. The point is that, for a high-volume portable product, the lower the cost, weight, volume, and power the more likely it is to sell.

**Table 17-15**   Speech Analyzer System Requirements

| System parameter | Requirement |
|---|---|
| Real input data rate | 10 kHz |
| Number of input bits | Greater than 8 |
| Production volume | 10,000 per year |
| Product size | Small |
| Power | Low |
| Cost | Low |
| Weight | Light |
| Input | Analog from microphone |
| Output | Digital to main computer |

### 17.3.3 Description

Speech scientists have determined that the human speech generation system (lungs, vocal cords, trachea, mouth, and nose) can be modeled by the block diagram in Figure 17-6. Voiced sounds, such as vowels, can be modeled as the output of a time-varying linear filter response to a periodic impulse train. The period of the impulse train (pitch period) is determined by the dimensions of the vocal cords and trachea. Unvoiced signals, such as consonants, can be modeled as the response of the time-varying linear filter to a random number generator. The loudness (amplitude) of the resulting sound is modeled by the multiplier in front of the time-varying linear filter. The time-varying linear filter represents the way the human vocal tract and mouth modify the sources of the sound. The linear filter coefficients change slowly over time to produce different voiced and unvoiced sounds from the same signal generators. This suggests it should be possible to describe the speech samples by knowing the pitch period and the time-varying linear filter coefficients.

Figure 17-7 is a block diagram of the algorithm to be used in this example [3]. The reason it works is that the impulse train generator waveform has a periodic structure in the

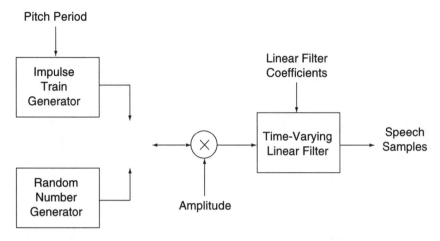

**Figure 17-6**    DSP vocal tract model.

frequency domain that repeats at roughly the pitch frequency of 50 to 100 Hz. Over the 5-kHz bandwidth of speech, this results in 50 to 100 peaks. Figure 17-8 shows what that pitch spectrum might look like. On the other hand, the frequency response of the time-varying linear filter varies smoothly and decreases with increasing frequency. The filter's response does have peaks in it, generally at three or four frequencies. These peaks are called the formants of the filter, and their locations can be used to characterize the filter's coefficients. Thus, in the frequency domain, the pitch and the linear filter have significantly different structures.

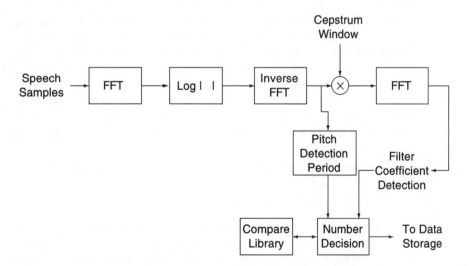

**Figure 17-7**    Number recognition algorithm block diagram.

If the composite waveform out of the log function in Figure 17-7 is linearly filtered to remove the high-frequency components, the remaining signal is the slowly varying fre-

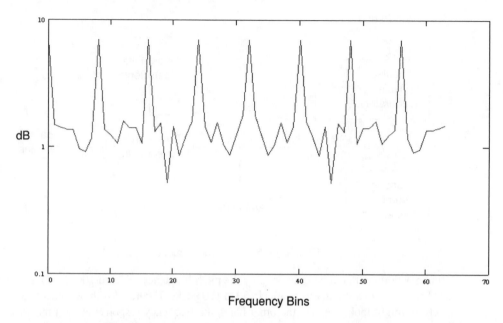

**Figure 17-8** Representative FFT of pitch unit pulse train.

quency response of the time-varying linear filter. The three blocks following the log function are the equivalent of the linear filtering in the frequency domain described in Chapter 6. The only difference is the exchanged roles of the FFT and IFFT because the waveform has started out in the frequency domain, not the time domain. Therefore, the output of the second FFT is the slowly varying frequency response of the time-varying linear filter.

Similarly, since the input to the IFFT is the sum of two waveforms, its output is the inverse transform of the sum of those two signals because the IFFT is a linear function. The slowly varying portion of the IFFT output ends up close to zero. In fact, if the slowly varying function did not fluctuate at all, all of it would be at the zero sample, because the FFT of the unit pulse at zero time is the same for all frequency components. This fact is computed from Equation 2-1. If the $n = 0$ sample is 1 and the rest of the samples are zeros (unit pulse at sample zero), then Equation 17-6 (Equation 2-1) simplifies to Equation 17-7.

$$A(k) = \sum_{n=0}^{N-1} a(n) * W_N^{k*n} \qquad \text{where } W_N = \cos(2\pi/N) + j * \sin(2\pi/N) \quad (17\text{-}6)$$

$$A(k) = a(0) \qquad\qquad\qquad\qquad\qquad\qquad\qquad\qquad\qquad\qquad (17\text{-}7)$$

At the same time, the periodic nature of the pitch unit pulse train results in a peak in the IFFT output at roughly the period of that pulse train. Therefore, the output of the IFFT can be searched to find the pitch frequency by finding the first substantial peak away from zero. This is the function of the pitch period detection block in Figure 17-7. Similarly, the filter coefficient detection function in Figure 17-7 finds the peaks in the time-varying linear filter's frequency response. These are directly related to the time-varying filter's coefficients [3]. The time-varying filter coefficients and pitch are then combined and used to search a database to determine the best match. The best match is the pattern for the number that was verbalized. The number on the database that is the best match to the computed parameters of the input data is then stored in the computer rather than as a sequence of speech samples.

### 17.3.4 Design Decisions

FFT Algorithm. The unit pulse response of the human vocal tract is known to have a response of roughly 20 to 30 ms. Therefore, it makes sense to divide the speech sample periods into somewhat larger intervals, for example 40 ms. This time period also allows multiple-pitch periods to be present in the waveform because a 50- to 100-Hz pitch frequency corresponds to a period of 10 to 20 ms. The presence of multiple-pitch periods is important because the algorithm uses the periodic nature of the pitch signal to detect it. With a 10-kHz sampling rate, the number of samples in a 40-ms period is 400. Restricting the analysis to power-of-two FFTs would immediately suggest 512-point transforms because a 256-point transform would only cover 25.6 ms, which can be too short for accurate analysis. For this design assume the transform length must be greater than 400 points but less than 512 points to try to avoid exceeding internal DSP chip memory on inexpensive earlier generations. Table 17-16 lists the transform lengths between 400 and 512 points that can be computed by using the building-block algorithms in Chapter 8 and the algorithm categories from Chapter 9, listed in the third column.

**Table 17-16**    Transform Length Factors and Algorithms

| Transform lengths | Factors | Algorithm category |
|---|---|---|
| 400 | 4, 4, 5, 5 | Mixed-radix |
| 405 | 3, 3, 3, 3, 5 | Mixed-radix |
| 420 | 2, 2, 3, 5, 7 | Prime factor |
| 432 | 2, 2, 2, 2, 3, 3, 3 | Mixed-radix |
| 441 | 3, 3, 7, 7 | Mixed-radix |
| 448 | 2, 2, 2, 2, 2, 2, 7 | Mixed-radix |
| 450 | 2, 3, 3, 5, 5 | Mixed-radix |
| 480 | 2, 2, 2, 2, 2, 3, 5 | Mixed-radix |
| 486 | 2, 3, 3, 3, 3, 3 | Mixed-radix |
| 490 | 2, 5, 7, 7 | Mixed-radix |
| 500 | 2, 2, 5, 5, 5 | Mixed-radix |
| 504 | 2, 2, 2, 3, 3, 7 | Prime factor |
| 512 | 2, 2, 2, 2, 2, 2, 2, 2, 2 | Mixed-radix |

With the exception of the 512-point transform, the Comparison Matrix in Table 9-8 shows that the prime factor algorithms require the fewest computations and smallest multiplier constant memory. The Comparison Matrix in Table 8-1 shows that the smaller FFT building blocks are the more efficient. These two facts suggest limiting the FFT lengths to $420 = 3 * 4 * 5 * 7$, $504 = 7 * 8 * 9$, and $512 = 8 * 8 * 8$. Further decisions on the FFT algorithm to choose are deferred to the architecture and chip paragraphs below because other factors will affect the best choice.

**Weighting Functions.**   Since the speech waveform is not expected to be repetitive over multiples of the 40-ms sampling time, a weighting function helps reduce the discontinuities at the edges of the sampled signal to provide better frequency domain data. The trigonometric-based weighting functions (Sections 4.3 to 4.7) are probably the best option for a low-cost application. The reason is that there are numerous look-up-table techniques for computing these functions so that memory does not have to be used to store them. However, this does require additional computational power, which implies a more powerful, more expensive, more power-consuming DSP chip. The weighting functions from the Comparison Matrix in Table 4-1 that fit these requirements, along with their performance measures, are listed in Table 17-17. Since accurate frequency domain data is a priority, the weighting function with the smallest peak sidelobes is preferable. This is the sine-to-the-fourth weighting function.

**Table 17-17**   Speech Analyzer Weighting Function Comparison Matrix

| Weighting function | Highest sidelobe level (dB) | Sidelobe fall-off ratio | Frequency straddle loss (dB) | Coherent integration gain | Equivalent noise bandwidth | 3-dB bandwidth |
|---|---|---|---|---|---|---|
| Sine lobe | −23 | −12 | 2.10 | 0.64 | 1.23 | 1.20 |
| Hanning | −32 | −18 | 1.42 | 0.50 | 1.50 | 1.44 |
| Sine cubed | −39 | −24 | 1.08 | 0.42 | 1.73 | 1.66 |
| Sine to the fourth | −47 | −30 | 0.86 | 0.38 | 1.94 | 1.86 |
| Hamming | −43 | −6 | 1.78 | 0.54 | 1.36 | 1.30 |

**Arithmetic Format.**   With only 8 bits needed at the input and peak detection being the final parameter detection process, 16-bit fixed-point numbers are likely to be sufficient. This means that the arithmetic format does not limit the chip choices because the floating- and block-floating-point arithmetic formats just provide less quantization noise based on the Comparison Matrix in Table 13-1.

**Architecture and Chips.**   The desired architecture is a single chip with all the necessary program and data memory on-chip. Since the input is voice samples, the data must go through an A/D converter somewhere. Therefore, a plus in the design is to have an A/D converter on-chip. Table 17-18 shows the FFT performance and on-chip memory capacities of DSP chips with on-chip A/D converters (Sections 14.3.1 and 14.3.5).

**Table 17-18**   Speech Analyzer DSP Chip Comparison Matrix

| Fixed-point chip | 1024-point complex FFT (MS) | Data I/O ports | On-chip data memory words | On-chip prog. memory words | # of address generators |
|---|---|---|---|---|---|
| DSP56156 | 1.53 | 2s/1p | 2k | 2k | 2 |
| DSP56166 | 1.53 | 2s/1p | 4k | 2k | 2 |
| ADSP-21msp5xx | 2.86* | 2s/1p | 1k | 2k | 2 |

* = estimate (see Section 14.4).

According to the references in Chapter 14 for each of these three devices, the immediate drawback is that their A/D converters work at 8 kHz, not the 10-kHz sampling rate assumed earlier. In the interest of taking advantage of the integrated A/D to reduce the overall cost of the product, it makes sense to reevaluate the need for sampling at 10 kHz. The higher sampling rate is actually a luxury. The telephone system has a 4-kHz bandwidth and voice is easily discernible. Based on the sampling theorem (Section 2.3.1), 8 kHz should be a sufficient rate. To keep the 40-ms sampling period means that the number of 8-kHz samples should be at least 320, rather than the 400 calculated for the 10-kHz sampling rate. This means that the $336 = (3*7*16)$- and $360 = (5*8*9)$-point prime factor algorithms, using the building blocks in Chapter 8, should be added to Table 17-16.

All of the functions in Figure 17-7 must be performed each time a new set of 40 ms of data is collected. Since all of the chips in Table 17-18 perform 1024-point FFTs in less than 3 ms, it is clear that they will have no problem completing three FFTs in the range of 336 to 512 points and all of the other computations in the allotted time of 40 ms. Therefore, the processor architecture block diagram can be as shown in Figure 17-9.

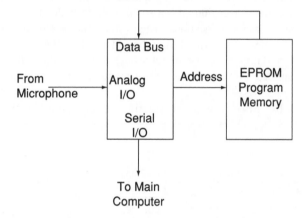

**Figure 17-9**    Speech analyzer processor architecture block diagram.

Note that the output interface to the main computer is through the serial link to reduce the number of wires and, therefore, the system cost and to improve its reliability. All of the chips in Table 17-18 have on-chip boot ROM that allows external, inexpensive EPROM to load the program to on-chip program RAM at power-up. If the product becomes a big-enough seller, the program can be put into on-chip program ROM and the external EPROM can then be eliminated.

For the product to work in real-time, it must be collecting a new data set while processing the present one. In high-speed real-time applications it would also have to output results from the previous computations while processing the present data set. However, it appears there will be enough processing time so that the answers can be output after computations and before the next set of data is available for computation. Therefore, there must be at least enough RAM for two full sets of data. Additionally, the database, as well as the pitch and formant data used to access the database, must be stored.

The key issue is the two sets of data for the FFT. Since the data is real, the Double-Length Algorithm from Section 2.4.2 can be used to efficiently utilize the FFT algorithm. This allows $N$ real data samples to be processed by an $N/2$-point FFT. Therefore, the chosen transform length will require storing from $2 * 336 = 672$ to $2 * 512 = 1024$ data words. All of the DSP chips in Table 17-16 have sufficient data memory to meet this goal, but the ADSP-21msp5xx series is marginal because of the need to store the database. Based on this, the Motorola DSP56166 is selected because it has the largest data RAM.

**FFT Algorithm Revisited.**   Now that the DSP chip has been chosen, the FFT algorithm can be chosen based on the specific characteristics of the chip. Equation 14-1, for estimating the computation time, will work for the Motorola DSP56166 because it has enough memory on-chip to execute the 1024-point complex FFT. Based on the formula, the worst-case 512-point FFT should take about $1.53 * 0.5 * 9/10 = 0.69$ ms. Therefore, three of them should take just over 2 ms out of the 40 ms available. This means that the differences in the number of adds and multiplies for the different potential FFT lengths is insignificant in deciding which length to use. Furthermore, there is plenty of time to compute the weighting function with a small look-up table and interpolation formulas. This saves program memory locations. The formulas in the Comparison Matrices in Chapter 9 (Tables 9-7 and 9-8) and with the building-block algorithm performance measures from the Comparison Matrix in Chapter 8 (Table 8-1) are used to compute the performance measures for the candidate FFT algorithms. They are summarized in Table 17-19.

**Table 17-19**   Speech Analyzer Algorithm Comparison Matrix

| Algorithm | # of adds | # of multiplies | # of data locations | # of const. locations |
|---|---|---|---|---|
| $336 = 3 * 7 * 16$ Prime factor | 7,332 | 2,596 | 672 | 14 |
| $360 = 5 * 8 * 9$ Prime factor | 8,404 | 3,412 | 720 | 13 |
| $420 = 3 * 4 * 5 * 7$ Prime factor | 9,648 | 4,064 | 840 | 12 |
| $504 = 7 * 8 * 9$ Prime factor | 12,860 | 5,756 | 1,008 | 15 |
| $512 = 8 * 8 * 8$ Mixed-radix | 11,776 | 4,352 | 1,024 | 128 |

Because the most critical issue appears to be data and program memory, not computation time, columns 4 and 5 of Table 17-19 are most important as selection criteria. In these two columns, the entry showing the most dramatic difference between the algorithms is the total number of multiplier constants required for the 512-point FFT. Therefore, the first decision is to eliminate the 512-point FFT.

Once the 512-point FFT is eliminated, the fifth column no longer is important in the decision process because all the other transform lengths are so close to each other. Columns 2, 3, and 4 of Table 17-19 show 336 and 360 as the best technical choices. The 336-point FFT is selected because it has the smallest entries in these columns.

## 17.3.5 Board Selection Process

One of the primary specifications for this product is that it be a high-volume portable product, with low cost, weight, volume, and power. A single DSP chip (DSP56166) with no external memory is the best chip choice in this application. Since weight and volume are primary specifications for the product, a custom board should be designed to take advantage of how well the DSP56166 fits the application. Table 17-20 summarizes the specifications for that board.

**Table 17-20**    Example 3, Board Selection Specifications

| Category | Specification |
|----------|---------------|
| Processor | DSP56166 |
| Off-chip memory | None required |
| Analog I/O ports | 8-kHz sample rate A/D built-in to DSP56166 |
| Instruction cycle time | 33 ns |
| Parallel and serial I/O ports (buses) | RS-232C serial port |
| Host interface | Any that are RS-232C compatible |

### 17.3.6 Test Signals

Section 16.5 introduces four types of test signal in an order of increasing complexity. It also gives the guidelines that were followed to create the specific parameters of each signal in Table 17-21. They are reordered to match the strategy in Section 16.7.2 that lists them in an order that allows testing with the least number of signals. The pair of sine waves can be any pair of relatively prime numbers up to the length of the transform (336 points) and were arbitrarily selected.

**Table 17-21**    Example 17-3, Test Signals

| Signal | Parameters |
|--------|-----------|
| Constant | Amplitude $= 1000$ |
| Single sine wave | 1 cycle per 336 data samples |
| Pair of sine waves | 17 and 41 cycles per 336 data samples |
| Unit pulses | As needed |

### 17.3.7 Design Decision Summary

The 336-point FFT algorithm is chosen because it has the smallest number of adds, multiplies, and memory locations of the choices in Table 17-19. Many of the single processors provided sufficient computational power. This allows the weighting function to be computed rather than stored. This led to choosing the sine-to-the-fourth weighting function. Any of the arithmetic formats provide the required accuracy and dynamic range. This allowed the freedom to choose a chip based on other performance measures. The DSP56166 is picked because it has a combination of an on-chip A/D converter and sufficient on-chip data memory to remove the need for external data RAM chips. Table 17-22 summarizes all of the key element design decisions made for this example.

**Table 17-22** Example 3, Design Decisions

| Key element | Selection |
|---|---|
| Number of dimensions | 1 |
| Type of processing | Frequency analysis and correlation |
| Arithmetic format | 16-bit fixed-point |
| Weighting function | Sine-to-the-fourth |
| Transform length | 336 points |
| Algorithm building blocks | 3-, 7-, and 16-points |
| Algorithm | Prime factor |
| DSP chip | DSP56166 |
| Architecture | One Harvard processor & no external mem. |
| Mapping the algorithm onto the architecture | Maximum throughput |

## 17.4 EXAMPLE 4: IMAGE DEBLURRING

The evolution of DSP technology moved image processing out of non-real-time laboratory and government-funded applications, such as enhancing images from outer space by NASA, into mainstream products. Examples include magnetic resonance imaging and ultrasound; image compression for teleconferencing, videophones, and multimedia data storage; image analysis for defect detection in countless applications; and image pattern matching for doing two-dimensional bar code reading or guiding cruise missiles to their Gulf War targets.

One of the fundamental problems with images, whether they are collected photographically, with a video camera, a CCD infrared system, or synthetic aperture radar is that the collection device may be out of proper focus or in motion during the image collection process. The result is blurred images that have reduced value. Image deblurring is the process of reducing this distortion.

Numerous image deblurring techniques have been developed and studied over the years, and each has its good and bad points. Many of these techniques use two-dimensional linear filtering techniques performed in the frequency domain because of the large number of pixels in an image. The two fundamental problems with most blurred images is that the distortion is nonlinear and noise has been added by the collection process. The nonlinear effects make unraveling the blurring process extremely complicated. The added noise makes many of the developed techniques unstable.

Since the purpose of this example is to illustrate the use of FFT algorithms to solve two-dimensional signal processing problems, the algorithms for deblurring an image are not derived, just presented and implemented. Derivations of these and other image processing algorithms can be found in image and digital signal processing texts [1].

### 17.4.1 Definition of the Product

The product is a general-purpose board that plugs into IBM PC-compatible hardware and is used for deblurring images that are downloaded to it from the PC's hard disk. The deblurred results are to be restored in the PC's hard disk before the next image is downloaded. The product is to be as inexpensive as possible so that it can be sold to law enforcement agencies for use with images stored from digital cameras, videophones, and other image input devices. Applications include license plate identification from an image taken in a moving police car and in crime labs for identification of suspects in video surveillance imagery.

### 17.4.2 Specification

Table 17-23 summarizes the specification of the product. Throughput is defined as the rate at which images can be fed to the product without the product getting behind. Latency is the time from when the image enters the product until the deblurred version exits. Notice that the throughput is three times more than the latency. This is to account for the image being loaded onto the board and for the deblurred image to be sent back to the hard disk.

**Table 17-23**    Image Deblurring Product Specification

| System parameter | Requirement |
|---|---|
| Image processing | Deblurring |
| Image size | $1024 \times 768$ pixels |
| Number of bits per pixel | 8 |
| Throughput rate | 1 per 60 s |
| Latency | 20 s |
| Hardware | IBM PC-compatible plug-in board |
| Input source | IBM PC hard disk |
| Output | IBM PC hard disk |
| Number of images on board | 1 |

### 17.4.3 Description

Figure 17-10 shows a simplified block diagram of an image recording process. The simplest example of this process is a camera, where the image formation device is the lens system and the image recording device is photographic film. If the lens system is not properly focused, the image will be blurred. The photographic film recording process is nonlinear as well as grainy. If the camera moves during the collection process, another blur is introduced because the same portion of the input image energy will be recorded in multiple locations on the film.

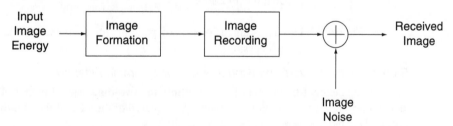

**Figure 17-10**    Image collection and recording block diagram.

The approach illustrated in this example is called power spectrum equalization [1]. More can be learned about the power spectrum of a signal in Section 17.2. Its basic definition is the FFT of the autocorrelation of the signal, where the autocorrelation of the signal is pattern matching of the signal with itself using the techniques given in Chapter 6. The computational approach is to find an estimate for the actual image that has the same power spectrum as the recorded image and can be represented by that recorded image after passing through a two-dimensional linear operator.

The algorithm for computing the deblurred $N \times M$ pixel image has the following steps:

### Step 1: *Transform the Image to the Two-Dimensional Frequency Domain*

Compute the $(2 * N \times 2 * M)$-point, two-dimensional FFT of the received image, where the outside of the array is filled with zeros as shown in Figure 17-11. Chapter 7 shows that the two-dimensional FFT of a $2 * N \times 2 * M$ array of real data can be computed as a sequence of $2 * M$ one-dimensional $2 * N$-point FFTs of real data and $2 * N$ one-dimensional $2 * M$-point FFTs of real symmetric complex data. Further, Chapter 2 shows that a $2 * N$-point FFT of real data can be computed by using an $N$-point FFT algorithm for complex data. Therefore, the computational requirement for this step is to compute $2 * N$ $M$-point FFTs and $2 * M$ $N$-point FFTs of complex data. Actually, the first dimension of FFT computations, say the row FFTs, only requires $N$ $M$-point FFTs because the other $N$ would be computing the FFT of all zeros (Figure 17-11).

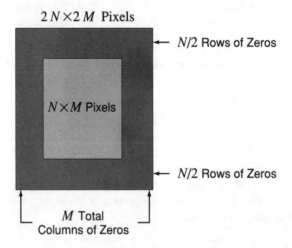

**Figure 17-11** Two-dimensional zero padding for frequency domain processing.

### Step 2: *Perform Two-Dimensional Frequency Domain Filtering*

On an element-by-element basis, multiply the two-dimensional output of Step 1 by its complex conjugate to obtain the magnitude squared of the FFT of the two-dimensional image. This requires $4 * N * M$ complex multiplies.

*Step* 3: *Apply the Two-Dimensional Inverse Filter in the Frequency Domain*

On an element-by-element basis, divide the output of Step 2 by the power spectral estimate of the inverse filter. Some DSP chips perform this process better by computing 1/(each power spectral estimate) for each element and then performing a multiplication. This requires a total of $4N * M$ divide operations.

*Step* 4: *Convert the Deblurred Image Back to the Spatial Domain*

Compute the $2 * N \times 2 * M$ IFFT of the result of Step 3. Chapter 2 shows that the IFFT has the same properties as the FFT. Therefore, this computation also requires $2 * N$ $M$-point FFTs and $2 * M$ $N$-point FFTs of complex data. Again, as in Step 1, the second dimension of IFFT computations, say the columns, only requires $M$ of the $N$-point FFTs because the output of interest is the image which is known to reside in a $N \times M$ array.

### 17.4.4 Design Decisions

FFT Algorithm.    The product needs to perform FFTs that are at least 1024 points for the rows of the image and at least 768 points for the columns of the image, using the Double-Length Algorithm from Section 2.4.2 on real data sets that are at least 2048 pixel rows and 1536 pixel columns. Therefore, efficient algorithms near 1024 and 768 points, with common factors to reduce the number of building-block algorithms, are the best choices. Since $1024 = 4 * 4 * 4 * 4 * 4$ and $768 = 4 * 4 * 4 * 4 * 3$, they are excellent candidates because only 3- and 4-point building blocks are needed. The Comparison Matrix in Table 8-1 shows that these building blocks are computationally efficient. Since Chapter 9 offers other choices near 768 and 1024, these should be examined to determine any advantages they may have. Other lengths between 768 and 1100 that use the building blocks from Chapter 8 are listed in Table 17-24, along with their factors and the algorithms from Chapter 9 that can be used to implement them.

The disadvantage of the 768- and 1024-point mixed-radix algorithms over the prime factor algorithms for 840 and 1008 points is all the between-stage multiplier constants required by the mixed-radix approach. The other mixed-radix choices in Table 17-24 have similar numbers of multiplies and require the number of multipliers between stages based on the equations in the Comparison Matrices in Tables 9-7 and 9-8. Therefore, it is realistic to limit the choice of FFT lengths to 768 and 1024 or 840 and 1008. The one disadvantage to 1008 points is that it does not meet the 1024-point criteria. However, shortening the length by this small amount will have little effect on the quality of the deblurred image.

Because the 1024-point and 768-point FFTs have only two building blocks (3 and 4 points), and these are both efficient, it is unlikely to make sense to further consider the 840- and 1008-point FFTs. Further, the 1024-point code is likely to be available for free, and the 768-point FFT can be computed by using a 256-point FFT followed by a 3-point FFT. The 256-point code is likely to be available for free also, and Chapter 8 shows the 3-point algorithm in detail. Further, combining the 256- and 3-point FFTs to form the 768-point FFT is described in Chapter 9. In fact, a more pragmatic approach is to use 1024-point FFTs in both dimensions. The theory in Chapters 6 and 7 for using two-dimensional FFTs to perform pattern matching requires that the FFT length be at least the sum of

**Table 17-24**  Transform Lengths, Factors, and Algorithms

| FFT length | Factors | FFT algorithms |
|:---:|:---:|:---:|
| 768 | 4, 4, 4, 4, 3 | Mixed-radix |
| 784 | 4, 4, 7, 7 | Mixed-radix |
| 800 | 2, 4, 4, 5, 5 | Mixed-radix |
| 810 | 2, 5, 9, 9 | Mixed-radix |
| 840 | 3, 5, 7, 8 | Prime factor |
| 864 | 2, 4, 4, 3, 3, 3 | Mixed-radix |
| 875 | 5, 5, 5, 7 | Mixed-radix |
| 882 | 2, 7, 7, 9 | Mixed-radix |
| 896 | 2, 7, 8, 8 | Mixed-radix |
| 900 | 4, 5, 5, 9 | Mixed-radix |
| 945 | 3, 5, 7, 9 | Mixed-radix |
| 960 | 3, 4, 4, 4, 5 | Mixed-radix |
| 972 | 3, 4, 9, 9 | Mixed-radix |
| 980 | 4, 5, 7, 7 | Mixed-radix |
| 1000 | 5, 5, 5, 8 | Mixed-radix |
| 1008 | 7, 9, 16 | Prime factor |
| 1024 | 4, 4, 4, 4, 4 | Mixed-radix |
| 1029 | 3, 7, 7, 7 | Mixed-radix |
| 1050 | 2, 3, 5, 5, 7 | Mixed-radix |
| 1080 | 3, 5, 8, 9 | Mixed-radix |

the lengths of the functions being correlated. The 1024-point FFT certainly meets that criterion.

**Weighting Function.**    The defined algorithm does not use weighting functions, so the Comparison Matrix in Table 4-1 does not play a role in the development of this product.

**Arithmetic Formats.**    The deblurring algorithm used here is sensitive to system noise. Therefore, it is also sensitive to quantization noise. This suggests that 32-bit floating-point arithmetic be used to minimize quantization errors.

**Architecture and Chips.**    The arithmetic format requirement immediately eliminates all but the floating-point DSP chip families in Chapter 14. These are listed in Table 17-25. The processing starts with loading the $1024 \times 768$ image onto the board, then continues with the deblurring algorithm, followed by outputting the results to the hard disk. Therefore, the board needs data memory to store all of the input pixels, but not additional memories to collect the next image while processing the present one.

Since the processing will be performed in floating-point arithmetic, the on-board data memory must hold $1024 * 768 = 786{,}432$ thirty-two-bit complex words, or $1008 * 840 = 846{,}720$ thirty-two-bit complex words, depending on the chosen FFT lengths. This amount of data memory can be cut in half by taking advantage of the symmetries in the FFT outputs as a result of the input data being real rather than complex. However, this only happens by increasing the complexity of the memory addressing scheme. The cost of developing and debugging the more complex addressing scheme is not worth the effort, except for a very high volume application.

**Table 17-25**    Floating-Point DSP Chips Comparison Matrix

| Floating-point chip | 1024-point complex FFT (MS) | Data I/O ports | On-chip data memory words | On-chip prog. memory words | # of address generators |
|---|---|---|---|---|---|
| **Analog Devices** | | | | | |
| ADSP-21020 | 0.58 | 0s/2p | 0 | 0 | 2 |
| ADSP-21060 | 0.46 | 8s/1p | 65,536 | 65,536 | 2 |
| **AT&T** | | | | | |
| DSP32C | 3.2 | 1s/1p | 1024/1536 | 4096/0 | 1 |
| DSP3210 | 2.4 | 1s/1p | 1024/2048 | 1024/256 | 1 |
| DSP3207 | 1.9 | 0s/1p | 1024/2048 | 1024/256 | 1 |
| **Intel** | | | | | |
| i860XR | 0.74 | 0s/1p | 1024 | 256 | 1 |
| i860XP | 0.55 | 0s/1p | 2048 | 1024 | 1 |
| **Motorola** | | | | | |
| DSP96002 | 1.04 | 0s/2p | 1024 | 1024 | 2 |
| **NEC** | | | | | |
| $\mu$PD77240 | 7.07 | 1s/1p | 1024 | 0 | 2 |
| $\mu$PD77230A | 11.78 | 1s/1p | 1024 | 1024/2048 | 2 |
| **TI** | | | | | |
| TMS320C30 | 1.97 | 2s/2p | 2048 | 4096 | 2 |
| TMS320C31 | 1.97 | 1s/2p | 2048 | 4096 | 2 |
| TMS320C40 | 1.54 | 6s/2p | 2048 | 4096 | 2 |

s = serial ports; p = parallel ports.

The crucial step is to estimate how many DSP chips will be required. This defines the architecture choices. The two key contributors are the FFT computations and the divides. As a conservative estimate, assume all the FFTs are 1024 points. This will help account for the fact that the double-length algorithm requires an extra stage after the FFT to compute the needed outputs. Therefore, Steps 1 and 4 in Section 17.4.1 require $6 * 1024 = 6144$ FFTs of 1024-points. If these took 1 ms each, all 6144 of them would take 6.144 s. At 2 ms per FFT, the time required for this portion of the processing is roughly 12.3 s. Using 2 ms is preferable because it allows more of the floating-point chips in Table 17-25 to be included and is still well within the 20-s throughput requirement.

To these computations must be added the $4 * N * M$ complex multiplies, which is $16 * N * M$ real multiplies, and $8 * N * M$ real adds. Assuming these are performed in series, rather than making use of the multiplier-accumulator architecture of the DSP chips to perform these functions in parallel, this is $24 * N * M = 18.87$ or 20.3 million arithmetic computations. These computations can be accomplished in less than 2 s on any of the floating-point DSP chips in Table 17-25.

To the FFTs and complex multiplies must be added the $4 * N * M = 3.15$ or 3.39 million divides, depending on the FFT lengths chosen. To perform the divides in the remaining $20 - 12.3 - 2 = 5.7$ s requires a computation rate of 0.55 or 0.59 million divides per second. This translates into 1.81 or 1.68 $\mu$s per divide. Modeling the divide function as an inverse followed by multiplication takes 35 cycles for the inverse and another for the multiply in the TI series of floating-point chips (Reference 33 from Chapter 14). At the

40-ns clock rate of the TMS320C40, the divide will take roughly 1.44 $\mu$s. The Analog Devices and Intel chip families also use software techniques to implement division. The Motorola DSP96002 floating-point chip has hardware support for division.

It appears there is a single DSP chip solution and that 2 ms is marginal for 1024-point FFT performance, if the divides are performed in software. Table 17-26 summarizes the candidate DSP chip choices from Table 17-16 that should not be marginal, based on all the computational estimates.

**Table 17-26** Image Deblurring Candidate DSP Chip Comparison Matrix

| Floating-point chip | 1024-point complex FFT (MS) | Data I/O ports | On-chip data memory words | On-chip prog. memory words | # of address generators |
|---|---|---|---|---|---|
| ADSP-21020 | 0.58 | 0s/2p | 0 | 0 | 2 |
| ADSP-21060 | 0.46 | 8s/1p | 65,536 | 65,536 | 2 |
| i860XR | 0.74 | 0s/1p | 1024 | 256 | 1 |
| i860XP | 0.55 | 0s/1p | 2048 | 1024 | 1 |
| DSP96002 | 1.04 | 0s/2p | 1024 | 1024 | 2 |
| TMS320C40 | 1.54 | 6s/2p | 2048 | 4096 | 2 |

s = serial ports; p = parallel ports.

Therefore, the product can be built with a single DSP chip with off-chip program and data memory. The off-chip data memory is required to hold the nearly 2 million 32-bit data words needed for the intermediate frequency domain computations on the image. Figure 17-12 shows the proposed processor architecture block diagram. The data and program memory interfaces are shown with separate DSP chip pins to optimize performance. Based on Table 17-26, the separate parallel memory interfaces assumption reduces the DSP chip choices to the ADSP-21020, DSP96002, and TMS320C40.

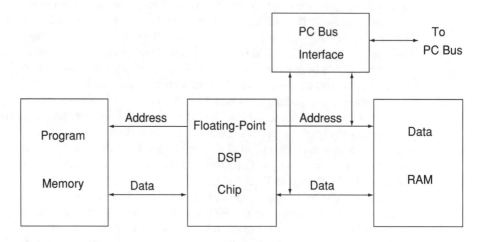

**Figure 17-12** Image deblurring processor architecture block diagram.

### 17.4.5 Board Selection Process

Of the three chips that meet the specifications, the best choice is the one that has the largest number of COTS boards on the market that will plug into a PC bus, because the competition of multiple boards in the market tends to reduce their cost. Multiple boards in the market also provide for second sources in case one board manufacturer goes out of business or decides to no longer make that board. There are far more TMS320C40 PC plug-in boards available than ones for the ADSP-21020 and DSP96002 chips. Therefore, a TMS320C40-based board should be chosen to meet the specifications summarized in Table 17-27.

**Table 17-27**    Example 4, Board Selection Specifications

| Category | Specification |
|---|---|
| Processor | TMS320C40 |
| Off-chip memory | 256K of 32-bit words |
| Analog I/O ports | None required |
| Instruction cycle time | 40 ns |
| Parallel and serial I/O ports (buses) | PC bus |
| Host interface | PC compatible |

### 17.4.6 Test Signals

Section 16.5 introduces four types of test signal in an order of increasing complexity. It also gives the guidelines that were followed to create the specific parameters of each signal in Table 17-28. They are reordered to match the strategy in Section 16.7.2 that lists them in an order that allows testing with the least number of signals. The pair of sine waves can be any pair of relatively prime numbers up to the length of the transform (1024 points) and were arbitrarily selected.

**Table 17-28**    Example 4, Test Signals

| Signal | Parameters |
|---|---|
| Constant | Amplitude $= 1000$ for the real and imaginary parts |
| Single sine wave | 1 cycle per 1024 data samples |
| Pair of sine saves | 13 and 29 cycles per 1024 data points |
| Unit pulses | As needed |

### 17.4.7 Design Decision Summary

The 1024-point FFT is used because it meets the performance requirements, the candidate DSP chips have enough computational power to compute this length in the allotted time, and code for implementing the 1024-point FFT is available in algorithm libraries from vendors. The deblurring algorithm [1] did not use a weighting function, so none is used in the example. A single TMS320C40 floating-point DSP chip, which will need external program and data memory chips to accommodate the complex algorithm and huge amount of data, is selected. Table 17-29 summarizes all of the key element design decisions made for this example.

**Table 17-29** Example 4, Design Decisions

| Key element | Selection |
|---|---|
| Number of dimensions | 2 |
| Type of processing | Convolution |
| Arithmetic format | 32-bit floating-point |
| Weighting function | None |
| Transform length | 1024 points |
| Algorithm building blocks | 2-, 4-, 8-, and 16-point |
| Algorithm | Power-of-primes mixed-radix |
| DSP chip | TMS320C40 |
| Architecture | One Harvard processor with external memory |
| Mapping the algorithm onto the architecture | Maximum throughput |

## 17.5 CONCLUSIONS

The use of FFTs in ever-increasing numbers of industrial and mainstream consumer products will be driven by the ability of design engineers to optimize code for computing this flexible class of algorithms. The examples in this chapter, which serve as an applied summary of the information in the preceding chapters, are just a taste of the astounding number of products that are possible because of constantly evolving improvements to the work begun by J. B. Fourier nearly two centuries ago.

It is our fervent hope that insights gained through the use of this book will help readers invent the FFT-based products that will transform the fields of telecommunication, medicine, seismology, oceanography, environmental protection, and consumer products well into the 21st century.

## REFERENCES

[1] A. V. Oppenheim, *Applications of Digital Signal Processing*, Prentice Hall, Englewood Cliffs, NJ, 1978.

[2] P. D. Welsh, "The Use of the Fast Fourier Transform for the Estimation of Power Spectra: A Method Based on Time Averaging over Short, Modified Periodograms," *IEEE Transactions on Audio and Acoustics*, Vol. AU-15, pp. 70–73 (1967).

[3] L. R. Rabiner and R. W. Schafer, *Digital Processing of Speech Signals*, Prentice Hall, Englewood Cliffs, NJ, 1978.

# Glossary

**Algorithm**

A series of steps to compute a set of equations.

**Architecture**

A hardware organization of adders, multipliers, control logic, and memory for implementing algorithms.

**Assembler**

Software that converts assembly language code into machine language 1's and 0's for a specific processor.

**Assembly language**

A programming language for controlling a microprocessor or DSP chip at the register level.

**Bandwidth**

The measure of the spread of frequencies that pass through a filter or are contained in a signal.

**Bit slice**

A method of dividing a number into smaller pieces so that arithmetic can be performed with less-complex chips.

**Block diagram**

A drawing to depict the electronic interconnections of hardware components.

**Block-floating-point**

A floating-point number system that uses only one exponent for an entire set of data.

**Bluestein algorithm**

An algorithm developed to compute FFTs using convolution.

**Bus**

The communication network in or between processors or other devices.

**Bus interface**

Hardware that links a processor or other device to a bus.

**Butterfly**

The fundamental building block of the 2-point FFT.

**Coefficients**

The numerical constants in an equation or filter.

**Complex arithmetic**

Arithmetic with numbers that have real and imaginary parts.

**Computational latency**

The time between the start of computations and when output of results begins.

**Computational load**

The amount of computations a processor is required to do, expressed as operations/second.

**Convolution**

A method of modifying the amplitude and/or phase of the frequency components of a signal; also known as linear filtering.

**Cooley-Tukey algorithm**

The most common power-of-two FFT.

**Correlation**

The operation of comparing or measuring the similarity of two waveforms; also known as pattern matching.

**Cross bar**

A bus architecture that allows any processor to directly connect to any other processor.

**dB**

The abbreviation for decibel, a measure of the power level of a signal relative to 1 watt.

**Debugger**

Software for removing errors from code.

**Decimation in frequency (DIF)**

A method of computing a power-of-two FFT that has the multiplier on the butterfly output.

**Decimation in time (DIT)**

A method of computing a power-of-two FFT that has the multiplier on the butterfly input.

**Discrete Fourier transform (DFT)**

A sine-wave-based set of equations to convert sampled time-domain data into frequency-domain data that has equally spaced frequencies; an array of pattern matchers where the patterns being matched are sine waves.

**Dolph-Chebyshev weighting function**

A weighting function with a spectrum characterized by uniform sidelobes.

**Doppler radar**

A radar that directly measures the radial velocity of a target.

**Dynamic range**

The ratio of largest to the smallest number that can be represented by any arithmetic format.

**Emulator**

A hardware model for a processor chip that allows access to all the functions of the chip for program development or debugging.

**Equivalent noise bandwidth**

The ratio of the input noise power to the noise power in the output of an FFT filter times the input data sampling rate.

**Fast Fourier transform (FFT)**

An algorithm for fast DFT computation.

**Filter**

An analog or digital device that reshapes the spectrum of a signal, typically to enhance desirable frequencies and attenuate undesirable frequencies.

**Fixed point**

A number system based on the numbers being represented by a fixed number of digits relative to the decimal point.

**Floating point**

A number system based on the numbers being represented by both a fixed number of digits and an exponential multiplier.

**Flowchart**

A drawing to depict the sequence for executing the steps of an algorithm or progression of information through a system.

**Fourier transform**

A sine-wave-based set of equations to convert continuous time-domain data into continuous frequency-domain data.

**Frequency analysis**

Finding the amplitude and phase of the sine waves that comprise any waveform.

**Frequency domain**

A coordinate system for representing the frequency components of a signal.

**Frequency resolution**

How close the frequency of two sine waves can be and still be separately distinguished by a measurement system.

**Frequency straddle loss**

The reduced output of a filter caused by the input signal not being at the filter's center frequency.

**Harvard architecture**

A computer architecture with separate data and program memory buses.

**High-level language**

A programming language for controlling a microprocessor or DSP chip only at the function level.

**Hybrid architecture**

A combination of features from two or more standard architectures.

**Hypercube**

A parallel processing architecture where the processors are connected in a multidimensional cube configuration.

**In-place and in-order**

A prime factor FFT algorithm that does not require reordering of input and output data or extra memory for data storage.

**Inverse DFT**

A transform that converts frequency-domain data into time-domain data.

**Kolba-Parks algorithm**

A prime factor algorithm that uses small size FFTs.

**Latency**

The time between data entering a processor and the processed results exiting.

**Linear array**

A one-dimensional connection of processors.

**Linear filter**

A linear analog or digital device that reshapes the spectrum of a signal, typically to enhance desirable frequencies and attenuate undesirable frequencies.

**Linear filtering**

The act of processing a signal through a linear filter.

**Linker**

Software that combines assembly language subroutines into a larger program.

**Mapping**

A method of distributing an algorithm or data among multiple processors.

**Massively parallel**

A multidimensional connection of hundreds or thousands of processors.

**Mixed-radix**

An FFT where the number of data points or computed frequencies is the product of at least two integers.

**Multiplier-accumulator (MAC)**

Hardware that computes sums of products.

**Narrowband filter**

A filter that attenuates all but a narrow range of frequencies.

**Nesting algorithm**

A portion of the Winograd FFT algorithm.

**Non-real-time**

Processing that is not completed as fast as the data comes in.

**Nyquist rate**

The sampling rate must be at least twice as fast as the highest-frequency component in the signal; also known as the sampling theorem.

**Overflow control**

Logic that detects when a computed answer is larger than the allowed dynamic range.

**Parallel array**

A two-dimensional or more connection of processors.

**Parseval's theorem**

The energy in the time-domain representation of a signal is the same as the energy in its frequency domain representation.

**Passband**

The range of frequencies that are not attenuated by a filter.

**Pipeline**

An architecture where data is sequentially passed from one processor to the next to execute an algorithm.

**Power-of-two**

An FFT algorithm where the number of data points or computed frequencies is 2 raised to a power.

**Power spectrum estimation**

Technique for estimating the power in the frequency components of a signal.

**Practical transform length (PTL)**

The acronym for a non-power-of-two FFT algorithm using multidimensional decomposition and complex conjugate math, developed by Win Smith.

**Prime factor**

An FFT algorithm where the factors are relatively prime and there are no twiddle factors.

**Prime number**

Any number that has no factors other than itself and 1.

**Primes-to-a-power**

An FFT algorithm where the number of data points or computed frequencies is a prime number raised to a power.

**Quantization noise**

The error signal caused by rounding-off numbers and coefficients in a digital processor.

**Rader algorithm**

A prime number FFT using circular convolution.

**Real-time operating system**

Software that helps a processor control real-time algorithms.

**Real-time operation**

Processing of data that keeps up with the input data rate rather than storing it and performing the processing later.

**Relatively prime**

Any two numbers with no common factors.

**Ring bus**

A circular bus architecture that allows data to pass from one processor to another and end up where it started.

**Sampled data**

A sequence of data values collected at regular or irregular intervals.

**Sampling theorem**

The sampling rate must be at least twice as fast as the highest-frequency component in the signal; also known as the Nyquist rate.

**Sidelobes**

Unwanted frequency components that are reduced but not removed by a filter.

**Simulator**

A software model of a processor that is used to develop and debug code prior to hardware implementation.

**Sine wave**

A continuous, smooth, periodic signal defined by the mathematical function $\sin(kt)$.

**Singleton algorithm**

Computes non-power of two FFTs using multidimensional decomposition.

**Small-point transform**

A small FFT, usually 16 or fewer points.

**Split-radix algorithm**

An FFT composed of a mixture of power-of-two small-point transforms.

**Star bus**

A bus architecture with a central processor with additional processors connected like spokes of a wheel.

**SWIFT**

The acronym for a non-power-of-two FFT algorithm using multidimensional decomposition and complex conjugate math, developed by Winthrop W. Smith.

**Throughput**

The number of times per second that a processor can compute an algorithm.

**Time domain**

A coordinate system that describes signals as a sequence of values at different points in time.

**Twiddle factor**

A standard, complex multiplication operation between small-point transforms of an FFT.

**Unit pulse**

A signal with a value of 1 for one time sample and zero for all other time samples.

**Versa module eurocard (VME)**

A standard hardware interface and software communications protocol for connecting boards onto a VME system's bus.

**Von Neumann**

An architecture with a single bus for data and program memory.

**Weighting functions**

Functions that multiply FFT input data to reduce sidelobes.

**Winograd algorithm**

An algorithm developed to compute FFTs using a minimum number of multiplications.

# Appendix

Comparison Matrices

# Index